APPLIED UNCERTAINTY ANALYSIS FOR FLOOD RISK MANAGEMENT

APPLIED UNCERTAINTY ANALYSIS FOR FLOOD RISK MANAGEMENT

Editors

Keith Beven
Lancaster University, UK

Jim Hall
University of Oxford, UK

ICP

Imperial College Press

Published by

Imperial College Press
57 Shelton Street
Covent Garden
London WC2H 9HE

Distributed by

World Scientific Publishing Co. Pte. Ltd.
5 Toh Tuck Link, Singapore 596224
USA office: 27 Warren Street, Suite 401-402, Hackensack, NJ 07601
UK office: 57 Shelton Street, Covent Garden, London WC2H 9HE

British Library Cataloguing-in-Publication Data
A catalogue record for this book is available from the British Library.

Cover images:

Upton-upon-Severn floods 23 July 07 © by RogerSmith1946, http://www.panoramio.com/photo/10609559. Mexborough probabilistic flood map, © Lancaster University and JBA Consulting. Background Google Imagery © 2010, Data SIO, NOAA, U.S. Navy, NGA, GEBCO, DigitalGlobe, GeoEye, Getmapping plc, Infoterra Ltd & Bluesky, The GeoInformation Group.

Back cover image: © Keith Beven.

APPLIED UNCERTAINTY ANALYSIS FOR FLOOD RISK MANAGEMENT

ISBN 978-1-84816-270-9

Typeset by Stallion Press
Email: enquiries@stallionpress.com

Printed in Singapore by Mainland Press Pte Ltd.

Preface

Flood risk management is a process of decision making under uncertainty. Traditional approaches to analysing floods assumed a guise of determinism though concepts such as "design floods", and dealt with uncertainties implicitly and opaquely through freeboard allowances and other factors of safety. A risk-based approach recognises that unpredictability is fundamental to decisions about how to deal with flooding. Uncertainties have to be exposed, scrutinised and incorporated in decision making through rational processes.

Such is the ideal of flood risk management. Yet when we review research and practice in the various fields that relate to flood risk (hydrology, hydraulics, geomorphology, structural reliability, human and economic vulnerability and so on) we observe patchy, inconsistent and sometimes non-existent analysis and reporting of uncertainty. Uncertainty analysis takes time and requires expertise. Whilst some principles and methods are well established, and well supported with computer-based tools, other aspects of uncertainty and decision analysis are subject to rather fundamental and sometimes bitter debate.

These are circumstances that can be disconcerting or exciting, depending on your perspective. The relative immaturity of methodology for dealing with uncertainty in many areas of flood risk management is an obstacle to the proper incorporation of uncertainties into decision making. Yet that scarcity of methodology and data for analysis of uncertainty represents an opportunity, for researchers in particular, but also for practitioners who are motivated by the need to improve flood risk management decision making.

This book seeks to respond to the challenges and opportunities presented by uncertainty analysis in flood risk management. We have sought to be as comprehensive as possible, dealing with uncertainties in both the probabilities and consequences of flooding, and addressing a range of flood risk management decisions, including planning, design and flood forecasting. Chapters deal with fluvial, urban and tidal flooding.

Whilst in editing the book we have sought to promote a coherent approach to the treatment of uncertainty, it is inevitable in a volume with a total of 45 contributing authors that there will be a diversity of perspectives, particularly in a field as contested as uncertainty analysis. We regard this diversity of perspectives as a strength of the book. This is not a manual. The reader will have to apply their judgment in assessing the applicability to the situations that they encounter of the methods and approaches that are described.

This book is an outcome of the UK Flood Risk Management Research Consortium (FRMRC), a wide ranging program of research that started in 2004 and ran, in two phases, to 2012. FRMRC involved more than 20 universities and research institutions and was jointly funded by the Engineering and Physical Research Council (EPSRC) in collaboration with the Environment Agency (EA), the Northern Ireland Rivers Agency (DARDNI), United Kingdom Water Industry Research (UKWIR), the Scottish Government (via SNIFFER), the Welsh Assembly Government (WAG) and the Office of Public Works (OPW) in the Republic of Ireland. We thank all of these organisations for the support they have provided to flood risk management research. Further details of the program and its outcomes can be found at http://www.floodrisk.org.uk/. The two editors of the book were both involved in Research Package 3 of the first phase of FRMRC, which was concerned with introducing risk and uncertainty into the flood risk management process. In phase 2, the methodologies developed were used in a variety of applications. That research is reflected in the chapters in this book, but we have also tried to reflect a wider range of international work in this area, by asking authors from different institutions to review progress in their areas of expertise. We are grateful for the time and effort of all the authors who have contributed to this extensive summary of progress in the field. Finally, we would like to thank Lynn Patterson at Newcastle University, Maya Kusajima at Kyoto University and Sue King at the University of Oxford, who did important work in assembling and formatting the manuscript.

Floods remain the most damaging natural hazard in terms of both fatalities and economic impacts. Floods will continue to occur in the future. Residual risk of flooding cannot be eliminated and the resources available for flood risk management will always be finite. So it is important that flood risk management decisions are based upon an appraisal of risks that is as far as possible accurate but is also accompanied by a well-informed appraisal of the associated uncertainties and their potential impact upon decision making. We hope that this book is a useful contribution towards that end.

Keith Beven
Jim W. Hall

Contents

SECTION I

INTRODUCTION

CHAPTER 1

Flood Risk Management: Decision Making Under Uncertainty

Jim W. Hall
Environmental Change Institute, University of Oxford, UK

1.1. Flood Risk Management

Flood risk management is a process of decision making under uncertainty. It involves the purposeful choice of flood risk management plans, strategies and measures that are intended to reduce flood risk. Hall *et al.* (2003b) define flood risk management as "the process of data and information gathering, risk assessment, appraisal of options, and making, implementing and reviewing decisions to reduce, control, accept or redistribute risks of flooding". Schanze (2006) defines it as "the holistic and continuous societal analysis, assessment and reduction of flood risk". These definitions touch upon several salient aspects of flood risk management:

- A reliance upon rational analysis of risks;
- A process that leads to acts intended to reduce flood risk;
- An acceptance that there is a variety of ways in which flood risk might be reduced;
- A recognition that the decisions in flood risk management include societal choices about the acceptability of risk and the desirability of different options;
- A sense that the process is continuous, with decisions being periodically reviewed and modified in order to achieve an acceptable level of risk in light of changing circumstances and preferences.

Whilst neither of the definitions cited above explicitly mention uncertainty, it is clear that the choices involved in flood risk management involve comparing different, and often rather complex options, in the context of environmental, technical and human processes that are at best partially understood, and according to mutable societal values and preferences. Therefore, not only is flood risk management a problem of decision making under uncertainty — it is a hard problem of decision making under uncertainty! Indeed Hall *et al.* (2003b) argued that the complexity of the process of modern flood risk management is one of the main motives for replacing traditional informal approaches to dealing with uncertainty with more rigorous, quantified methods. Of course this does not remove the need for judgement, especially when decisions are value-laden and contested, but it does help to eliminate the most egregious inconsistencies in the ways in which uncertainty is handled.

1.2. The Transition to Flood Risk Management

Before proceeding to examine the problem of decision making under uncertainty in more detail, it is worth providing some recent historic context in an attempt to explain why and how flood risk management has come to be the dominant paradigm in public policy and engineering practice dealing with floods.

It has long been recognised that "risk" is a central consideration in providing appropriate flood protection. In the UK, the Waverley Report (Waverley Committee, 1954) following the devastating East Coast floods of 1953 recommended that flood defence standards should reflect the land use of the protected area, noting that urban areas could expect higher levels of protection than sparsely populated rural areas. The notion of risk-based optimisation of the costs and benefits of flood defence was laid out in van Dantzig's (1956) seminal analysis, which also followed soon after the devastating 1953 floods, but on the other side of the North Sea. However, the practical process of flood defence design, whilst containing probabilistic content, was not fundamentally risk based, proceeding roughly as follows:

(1) Establishing the appropriate standard for the defence (e.g. the "100-year return period" water level), based on land use of the area protected, consistency and tradition.
(2) Estimating the design load, such as the water level with the specified return period.
(3) Designing (i.e. determining the primary physical characteristics such as crest level) to withstand that load.

(4) Incorporating safety factors, such as a freeboard allowance, based on individual circumstances.

Meanwhile, as flood warning systems were progressively introduced and refined in the decades since the 1950s, the decision-making process was also essentially deterministic, based on comparing water level forecasts (without uncertainty) with levels that would trigger the dissemination of a warning.

Over the last two decades the limitations of such an approach in delivering efficient and sustainable flood risk management have become clear. Because informal methods for decision making and handling of uncertainty have evolved in different ways in the various domains of flood risk management (flood warning, flood defence design, land use planning, urban drainage, etc.), they inhibit the integrated systems-based approach that is now promoted (Sayers *et al.*, 2002).

The systems approach is motivated by the recognition that there is no single universally effective response to flood risk. Instead, portfolios of flood risk management measures, be they "hard" structural measures such as construction of dikes, or "soft" instruments such as land use planning and flood warning systems, are assembled in order to reduce risk in an efficient and sustainable way. The makeup of flood risk management portfolios is matched to the functioning and needs of particular localities and will be adapted as more knowledge is acquired and as systems change. Implementing this approach involves the collective action of a range of different government authorities and stakeholders from outside government. This places an increasing emphasis upon effective communication and mechanisms to reach consensus. In this portfolio-based approach, risk estimates provide a common currency for comparing and choosing between alternatives that might contribute to flood risk reduction (Dawson *et al.*, 2008). The criteria for assessment of flood risk management options are seldom solely economic, but involve considerations of public safety, equity and the environment.

The principles of flood risk calculation have become well established (CUR/TAW, 1990; Goldman, 1997; USACE, 1996; Vrijling, 1993) and are not repeated here. However, it is worth reviewing how the risk-based approach addresses some of the main challenges of analysing flooding in systems (Sayers *et al.*, 2002):

(1) **Loading is naturally variable:** The loads such as rainfall and marine waves and surges on flood defence systems are not forecastable beyond a few days into the future. For design purposes, loads have to be described

in statistical terms. Extreme loads that may never have been observed in practice form the basis for design and risk assessment. Extrapolating loads to these extremes is uncertain, particularly when based on limited historical data and in a climate that may be changing.

(2) **Load and response combinations are important:** The severity of flooding is usually a consequence of a *combination* of conditions. So, for example, overtopping or breach of a sea defence is usually a consequence of a combination of high waves and surge water levels, rather than either of these two effects in isolation. In complex river network systems, the timing of rainfall and runoff at different locations in the catchment determines the severity of the flood peak. The severity of any resultant flooding will typically be governed by the number of defences breached or overtopped, as well as the vulnerability of the assets and preparedness of the people within the floodplain. Therefore, analysis of loads and system response is based on an understanding of the probability of combinations of random loading conditions and the system responses. Improved understanding of system behaviour has illustrated the importance of increasingly large combinations of variables.

(3) **Spatial interactions are important:** River and coastal systems show a great deal of spatial inter-activity. It is well recognised that construction of flood defences upstream may increase the water levels downstream in a severe flood event. Similarly, construction of coastal structures to trap sediment and improve the resistance of coasts to erosion and breaching in one area may deplete beaches down-drift (Dickson *et al.*, 2007). These interactions can be represented in system models, but engineering understanding of the relevant processes, particularly sedimentary processes over long timescales, is limited. Even where we have a detailed understanding of the physical processes, there may be fundamental limits to our ability to predict behaviour due to the chaotic nature of some of the relevant processes and loading.

(4) **Complex and uncertain responses must be accommodated:** Models of catchment processes are known to be highly uncertain due to the complexity of the processes involved and the scarcity of measurements at appropriate scales (Beven, 2006). The response of river, coast and man-made defences to loading is highly uncertain. The direct and indirect impacts of flooding depend upon unpredictable human behaviours for which relevant measurements are scarce (Egorova *et al.*, 2008).

(5) **Flooding systems are dynamic over a range of timescales:** Potential for long-term change in flooding systems, due to climate and socio-economic changes, adds further uncertainty as one looks to the future. Change may impact upon the loads on the system, the response to loads or the potential impacts of flooding. It may be due to natural environmental processes, for example, long-term geomorphological processes, dynamics of ecosystems, or intentional and unintentional human interventions in the flooding system. Social and economic change will have a profound influence on the potential impacts of flooding and the way they are valued.

Today, the term "flood risk" is used in a number of ways. A range of meanings derived from either common language or the technical terminology of risk analysis are in use (Sayers *et al.*, 2002). These different meanings often reflect the needs of particular decision-makers — there is no unique specific definition for flood risk and any attempt to develop one would inevitably satisfy only a proportion of risk managers. Indeed, this very adaptability of the concept of risk is one of its strengths.

In all of these instances, however, risk is thought of as a combination of the chance of a particular event, with the impact that the event would cause if it occurred. Risk, therefore, has two components — the chance (or *probability*) of an event occurring and the impact (or *consequence*) associated with that event. Intuitively it may be assumed that risks with the same numerical value have equal "significance" but this is often not the case. In some cases the significance of a risk can be assessed by multiplying the probability by the consequences. In other cases it is important to understand the nature of the risk, distinguishing between rare, catastrophic events and more frequent, less severe events. For example, risk methods adopted to support the targeting and management of flood warning represent risk in terms of probability and consequence, but low probability/high consequence events are treated very differently to high probability/low consequence events. Other factors include how society or individuals perceive a risk (a perception that is influenced by many factors including, for example, the availability and affordability of insurance), and uncertainty in the assessment.

The benefit of a risk-based approach, and perhaps what above all distinguishes it from other approaches to design or decision making, is that it deals with *outcomes*. Thus in the context of flooding it enables intervention options to be compared on the basis of the impact that they are expected to have on the frequency and severity of flooding in a specified area. A risk-based approach therefore enables informed choices to be made based on comparison of the expected outcomes and costs of alternative courses of action. This is distinct from, for example, a standards-based approach that focuses on the severity of the load that a particular flood defence is expected to withstand.

Whilst the theory of risk-based flood management decision making has been well established for many years, the transition in practice to an explicitly risk-based approach to flood management has been stimulated by severe floods, for example on the Oder (1997), Yangtze (1998), Elbe (2002), in New Orleans (2005), on the Danube (2006) and in England (2007). The severity of these events has underlined the relentless upward trend in vulnerability to flooding worldwide (Munich Re Group, 2007), as well as the recognition of potential impacts of climate change on flood frequency. In the aftermath of the severe Rhine River flooding of 1993 and 1995, the Dutch government adopted a flood control policy of "more room for rivers" with an emphasis on establishing new storage and conveyance space. In the UK the Foresight Future Flooding project (Evans *et al.*, 2004) stimulated the Government's "Making Space for Water" policy (Defra, 2005). The European Directive on the assessment and management of flood risk entered into force on 26 November 2007 and is leading to the development of flood risk maps and risk management plans across the whole of the European Union. In the USA there has been corresponding progressive evolution of floodplain management in the USA (Galloway, 2005; Interagency Floodplain Management Review Committee, 1994; Kahan, 2006). In summary, integrated flood risk management is characterised by:

(1) *A broad definition to the flooding system and scope of flooding impacts.* Arbitrary sub-division of the flooding system, for example due to geographical boundaries or administrative divisions, is avoided. Temporal and spatial interactions in system performance are accounted for.
(2) *Continuous management of flood system performance.* Consideration of one or a few "design events" is replaced by consideration of a whole range of system behaviours and development of appropriate management responses. There is a commitment to ongoing monitoring and control of the system at time intervals appropriate to the system dynamics.
(3) *Tiered analysis and iterative decision making.* Flood risk management cascades from high-level policy decisions, based on outline analysis, to

detailed designs and projects, which require more detailed analysis. High-level policy and plans provide the framework and common understanding within which more detailed actions are implemented.

(4) *Consideration of the widest possible set of management actions that may have some impact on flood risk.* This includes measures to reduce the probability of flooding and measures to reduce flood impact (vulnerability).

(5) *Development of integrated strategies that combine a range of flood risk management actions and implement them in a programmed way.* Management strategies are developed following consideration of both effectiveness, in terms of risk reduction, and cost. They will involve co-ordinating the activities of more than one organisation and multiple stakeholders.

(6) *Evolving within current policy framework.* Integrated flood risk management will remain an abstract concept unless it is placed within the current policy and administrative context. This involves making best use of existing policy instruments and actively identifying opportunities to influence policy change. It may involve reacting opportunistically to policy, administrative or regulatory reviews and changes that are initiated for non-flood-related reasons.

Compelling as modern integrated flood risk management certainly is, it brings with it considerable complexity. The risk-based approach involves analysing the likely impacts of flooding under a very wide range of conditions. As the systems under consideration expand in scope and timescale, so too does the number of potentially uncertain variables. There are many potential components to a portfolio of hard and soft flood risk management measures and they can be implemented in many different sequences through time, so the decision space is potentially huge. Communicating risks and building the consensus necessary to engage effectively with stakeholders in flood risk management requires special aptitude for communication, facilitation and mediation.

1.3. Flood Risk Management Decisions

Analysis of uncertainty should start by identifying the decisions that an uncertainty analysis is supposed to inform. Table 1.1 summarises the range of flood risk management actions which flood risk analysis might seek to inform. It summarises attributes of the information that is required to inform choice. So, for example, national policy analysis requires only

Table 1.1. Scope of flood risk management decisions (Hall *et al.*, 2003b).

Decision	Precision of information required	Requirement for dependable information	Spatial scope of decision	Tolerable lead-time to obtain information	Timescale over which decision will apply	Technical aptitude of decision makers
National policy	Approximate	Must reflect year-on-year changes in performance	National	Months	From annual budgets to policies intended to apply over decades	Politicians advised by civil servants
Catchment and shoreline management planning	Approximate	Must be able to distinguish broad strategic options	Regional, catchment	Months to years	Sets regional policies intended to apply over decades. Roughly 5-yearly review.	Technical officers, but a range of non-technical stakeholders
Development control	Detailed	Consistency is expected	Local and regional development plans	Months	Decades. Decisions very difficult to reverse	Planners
Project appraisal and design	Very detailed	Costly decisions that are difficult to reverse	Local, though impacts may be wider	Months to years	Decades	Engineering designers

(*Continued*)

Table 1.1. (*Continued*)

Decision	Precision of information required	Requirement for dependable information	Spatial scope of decision	Tolerable lead-time to obtain information	Timescale over which decision will apply	Technical aptitude of decision makers
Maintenance	Detailed	Need to set maintenance priorities	Local. Regional prioritisation.	Weeks	Months to years	Maintenance engineers and operatives
Operation	Very detailed	Can have a major impact of flood severity	Local	Hours	Hours	Flood defence engineers and operatives
Flood warning	Very detailed	Missed warnings can be disastrous. False alarms undesirable	Regional	Hours	Hours	Flood warning specialists
Risk communication	Detailed	Inaccurate information will undermine trust	Local to national	Hours (evacuation) to years (property purchase)	Days to years	General public

approximate analysis of risks, though at sufficient resolution to rank alternative policies. One of the principles of risk-based decision making is that the amount of data collection and analysis should be proportionate to the importance of the decision (DETR *et al.*, 2000). For flood warning decisions, timeliness is of paramount importance. In selecting appropriate analysis methods, the aptitude of decisions makers to make appropriate use of the information provided is also a key consideration. The outputs of analysis need to be customised to the needs and aptitudes of decision makers.

In Table 1.1 there is an approximate ordering of decisions on the basis of the spatial scale at which they operate. National policy decisions and prioritisation of expenditure require broad-scale analysis of flood risks and costs. This leads to a requirement for national-scale risk assessment methodologies, that need to be based upon datasets that can realistically be assembled on a national scale (Hall *et al.*, 2003a). Topographical, land use and occupancy data are typically available at quite high resolutions on a national basis.

The logical scale for strategic planning is at the scale of river basins and self-contained (from a sedimentary point of view) stretches of coast. At this scale, there is need and opportunity to examine flood risk management options in a location-specific way and to explore spatial combinations and sequences of intervention. Decisions to be informed include land use planning, flood defence strategy planning, prioritisation of maintenance and planning of flood warning. The datasets available at river basin scale are more manageable than at a national scale and permit the possibility of more sophisticated treatment of the statistics of boundary conditions, the process of runoff and flow and the behaviour of flood defence systems.

At a local scale, the primary decisions to be informed are associated with scheme appraisal and optimisation. This therefore requires a capacity to resolve in appropriate detail the components that are to be addressed in design and optimisation.

Implicit in this hierarchy of risk analysis methods is recognition that different levels of analysis will carry different degrees of associated uncertainty. Similarly, different decisions have varying degrees of tolerance of uncertainty. Policy analysis requires evidence to provide a ranking of policy options, whilst engineering optimisation yields design variables that are to be constructed to within a given tolerance. We now address more explicitly how uncertainty is accommodated in flood risk management decisions.

1.4. Uncertainty in Flood Risk Management Decisions

Uncertainty has always been inherent in flood defence engineering. Traditionally it was treated implicitly through conservative design equations or through rules of thumb, for example through the introduction of freeboard allowances. The introduction of risk-based approaches (CUR/TAW, 1990; Meadowcroft *et al.,* 1997; USACE, 1996) enabled more rational treatment of natural variability in loads and responses. It also paved the way for more explicit treatment of uncertainty in the evidence that is used to support risk-based decision making. Explicit uncertainty analysis provides a means of analysing the robustness of flood risk management decisions as well as the basis for targeting investment in data collection and analysis activities that make the greatest possible contribution to reducing uncertainty.

Increasingly governments are requiring a careful consideration of uncertainty in major planning and investment decisions. For example, USWRC (1983) (quoted in Al-Futaisi and Stedinger, 1999) state that:

> "Planners shall identify areas of risk and uncertainty in their analysis and describe them clearly, so that decisions can be made with knowledge of the degree of reliability of the estimated benefits and costs and of the effectiveness of alternative plans."

The UK Department of the Environment, Food and Rural Affairs (Defra) guidance on flood and coastal defence repeatedly calls for proper consideration of uncertainty in appraisal decisions. Guidance document FCDPAG1 (Defra, 2001) on good decision-making states: "Good decisions are most likely to result from considering all economic, environmental and technical issues for a full range of options, together with a proper consideration of risk and uncertainty." As Pate Cornell (1996) states, in the context of quantified risk analysis: "Decision makers may need and/or ask for a full display of the magnitudes and the sources of uncertainties before making an informed judgment."

However, the practice of uncertainty analysis and use of the results of such analysis in decision making is not widespread, for several reasons (Pappenberger and Beven, 2006). Uncertainty analysis takes time, so adds to the cost of risk analysis, options appraisal and design studies. The additional requirements for analysis and computation are rapidly being (more than) compensated for by the availability of enhanced computer processing power. However, computer processing power is only part of the solution, which also requires a step change in the approach to managing data and integrating the software for uncertainty calculations (Harvey *et al.*, 2008). The data necessary for quantified uncertainty analysis are not always available, so new data collection campaigns (perhaps including time-consuming expert elicitation exercises) may need to be commissioned. Project funders need to be convinced of the merits of uncertainty analysis before they invest in the time and data collection it requires.

It is not always clear how uncertainty analysis will contribute to improved decision making. Much of the academic literature on hydrological uncertainties (Liu and Gupta, 2007) has tended to focus upon forecasting problems. Providing uncertainty bounds on a flood forecast may be intriguing, but to be meaningful this needs to be set within the context of a well defined decision problem (Frieser *et al.*, 2005; Todini, 2008). In the following section we review the principles of decision making under uncertainty, in order to provide the context for the range of methods that are subsequently addressed in this volume.

1.5. The Principles of Decision Making Under Uncertainty

In order to situate uncertainty analysis within the decision-making process, we briefly review conventional decision theory. Conventionally there is a set of decision options or "acts" $\{d_1, \ldots, d_n\}$, and a set of future states of nature $\{\theta_1, \ldots, \theta_m\}$, defined on some space Ω, that may materialise after the choice. Depending on which state of nature in fact materialises, act d_i will yield one of m possible outcomes $y_{i,1}, \ldots, y_{i,m}$ (e.g. "no flood" or a flood of a given severity). The problem of valuing outcomes $y_{i,1}, \ldots, y_{i,m}$ is a fundamental one, to which we will return, but for the time being suppose that the net value (including both costs and benefits) associated with a given decision outcome $y_{i,j}$ can be written as a scalar function $v(y_{i,j})$, in which case, the following scenarios were first identified by Knight (1921):

(i) *Decision making under certainty*: The state of nature after the decision is known, i.e. $m = 1$. The decision maker chooses the option with the highest value $v(y_{i,1})$.

(ii) *Decision making under risk*: Only the probabilities $p(\theta_j) : j = 1, \ldots, m$: $\sum_{j=1}^{m} p(\theta_j) = 1$ of occurrence of set of states of nature $\{\theta_1, \ldots, \theta_m\}$ are known. Provided the decision maker accepts a set of consistency and continuity axioms (Savage, 1954) and is neutral in their attitude to risk then he or she should choose the option that maximises:

$$\sum_{j=1}^{m} v(y_{ij})p(\theta_j). \tag{1.1}$$

(iii) *Decision making under uncertainty*: There is no information about the probabilities of states of nature $\theta_1, \ldots, \theta_m$. Under these circumstances there are various decision strategies that are in some sense rational; for example, maximin utility, minimax regret, Hurwicz α, and that based on Laplace's principle of insufficient reason (French, 1988).

The decision maker's attitude to risk may be incorporated via a utility function $u(y_{i,j})$. This addresses the situation (ii) above, where known payoffs are replaced by gambles, in which case it is well known that some individuals are risk averse, whilst others are risk seeking. When an individual is risk neutral then their utility function $u(y_{i,j})$ is precisely equal to their value function $v(y_{i,1})$. Risk neutrality is often advocated for government decisions (Ball and Floyd, 1998; USWRC, 1983), though public safety decisions illustrate aversion to low probability/high consequence events (Pasman and Vrijling, 2003). The extension from risk neutrality to other utility functions is in principle straightforward (French, 1988), though in practice it requires elicitation of the decision maker's utilities.

Knight's formalisation of the decision problem implicitly distinguishes between two types of uncertainty. Case (ii), which Knight referred to as "decision making under risk", requires a probability distribution over the future states of nature $\theta_1, \ldots, \theta_m$ whilst Case (iii), "decision making under uncertainty", acknowledges that this probability distribution may not be known. Empirical evidence, from Ellsberg and subsequent studies, indicates an aversion to situations in which probabilities are not well known ("ambiguity aversion"). The theory of imprecise probabilities (Walley, 1991) provides a coherent treatment of the situation in which probabilities are not precisely known.

In the context of flood risk management the acts $d_1, \ldots, d_n$ are flood risk management options. They may be portfolios of options, i.e. different combinations of some set of basic elements, and they may differ from one another in the sequence, though time, in which they are implemented. The decision problem may involve a continuous design variable, such as the crest level of a dike, so the decision problem may be continuous rather than discrete.

The future states $\theta_1, \ldots, \theta_m$ are conventionally thought of as dealing at least with the unpredictable loads in nature to which flooding systems are subject, e.g. fluvial flows, water levels and wave heights. These will seldom be discrete but will typically extend over a continuous multi-dimensional space and the state of the flooding system will be described by a vector of k continuous variables $\mathbf{x} := (x_1, \ldots, x_k)$: $\mathbf{x} \in (\mathbb{R}^+)^k$. In the case of decision-making under risk the discrete probability distribution $p(\theta_j) : j = 1, \ldots, m$: $\sum_{j=1}^{m} p(\theta_j) = 1$ is replaced by a continuous distribution $f(\mathbf{x}) : \int_0^\infty f(\mathbf{x})dx = 1$ and the summation in Equation (1.1) is replaced by an integral:

$$\int_0^\infty v(y_i(\mathbf{x}))f(\mathbf{x})d\mathbf{x} \tag{1.2}$$

where we now have to explicitly acknowledge that y_i is a function of $\mathbf{x}$. It is not uncommon to require a combination of discrete and continuous variables in order to describe the state of a flooding system comprising multiple components (see for example Dawson and Hall, 2006). Here, for clarity, we will now deal with continuous variables only.

Thus far we have not been specific about the nature of the value function $v(y_{i,1})$. Valuation of flood risks, as well as the costs associated with flood risk management options, naturally starts with consideration of the economic losses due to flooding and the economic costs of implementing flood risk management options. However, it is also clear that flood risk management decision making is a multi-attribute problem, which incorporates considerations of safety, equity and the environment as well as the economic dimension (Egorova *et al.*, 2008; Finkel, 1990). Uncertainties associated with valuation enter the decision problem either if an economic valuation approach is adopted for dealing with these non-market risks and costs, or if an explicit multi-criteria approach is adopted. In the former, these originate in the prices assigned to non-market goods and services, whilst in the latter the uncertainties are associated with the value functions used to transform (uncertain) predicted outcomes to aggregate utilities. For the sake of clarity, we do not extend here the presentation of the decision problem to the multi-attribute context, though the approach for so-doing is well established (Keeney and Raiffa, 1993). It should, however, be clear though that economics provides only one perspective on flood risk management decisions, which inevitably raise a host of valuation problems. One such problem is the valuation of time, as typically the quantities in $y_i(\mathrm{x})$ will extend through time, so it is necessary to establish a method of aggregating a stream of annual payments or losses, $y_{i,t}(\mathrm{x})$: $j = 0, \ldots, T$ where t denotes the year in which the cost or risk is incurred and T is the time horizon. Customarily this is done by discounting to a "present value", though it is well known that discounting implies rather strong normative assumptions that do not necessarily apply in general (Adams, 1995; Cothern, 1996; Stern and Fineberg, 1996).

With the caveats in place, let us proceed with a simplified version of the flood risk management decision problem. Without loss of generality the expression in Equation (1.2) may be separated into terms that represent Present Value costs and Present Value risks. In any given year, t, the risk $r_{i,t}$ is given by

$$r_{i,t} = \int_0^\infty D(\mathbf{x}_t) f(\mathbf{x}_t) d\mathbf{x}_t \tag{1.3}$$

where $D(\mathbf{x}_t)$ is a damage function and we have introduced the subscript t to signify that in general we expect $\mathbf{x}$ to change with time. The simplicity of Equation (1.3) belies the potential complexity of the underlying calculations in practice, which have been extensively explored elsewhere (Beard, 1997; Dawson and Hall, 2006; Stedinger, 1997). In order to estimate flood risks it is necessary to be able to:

(1) Estimate probability distributions, $f(\mathrm{x}_t)$, for the sources of flooding, i.e. loading variables including extreme rainfall, water levels, marine surge tides and waves.
(2) Relate given values of loading variables to probabilities of flooding at locations where flooding may cause damage. This may involve hydrological and hydraulic modelling as well as analysis of the reliability of flood defence structures and pumping stations, and the operation of reservoirs. In urban areas flood risk analysis will involve analysis of the effects on the sewer network and pumped systems as a potentially major modifier of flooding behaviour, as well as analysing overland flows.
(3) Calculate the damage that is caused by floods of a given severity.

Steps (2) and (3) are together contained in $D(\mathbf{x}_t)$. These three steps typically involve sequences of models and analysis processes. For systems with significant time-dependency at the sub-annual scale (for example hydrological response of catchments), accurate simulation of flooding will involve additional explicit treatment of the temporal dimension.

The Present Value risk $PV(r_i)$ is:

$$PV(r_i) = \sum_{t=0}^{T} \frac{r_{i,t}}{(1+q)^t} \tag{1.4}$$

where q is the discount rate. The Present Value cost $PV(c_i)$ is defined similarly.

In the case of purely economic decision making, commonplace decision criteria are Net Present Value (NPV) and Benefit Cost Ratio (BCR), in which case it is necessary to introduce the notion of a "base case" that involves no investment. The Present Value risk corresponding to the base case is r_0 and expected to be unacceptably high, which is why investment in risk reduction is being contemplated. The NVP is $NPV_i = PV(r_0) - PV(r_i) - PV(c_i)$ whilst the BCR is $BCR_i = (PV(r_0) - PV(r_i))/PV(c_i)$. If the preference ordering between risk reduction options is established on the basis of NPV then if $NPV_i > NPV_j > NPV_l$, the preference ordering is denoted $i \succ j \succ 1$, and similarly for BCR.

The familiar theoretical framework reviewed above depends upon a number of assumptions. Typically there are uncertainties in:

(1) the system characterisation in terms of k variables in $\mathbf{x}_t$;
(2) the specification of the joint probability density function $f(\mathbf{x}_t)$, which describes the variation in $\mathbf{x}_t$;
(3) the function $D(\mathbf{x}_t)$ which embodies all of the modelling to relate given values of variables $\mathbf{x}_t$ to flood damage, as well as the problem of damage valuation (including non-market goods and services);
(4) the cost c_i; and
(5) the choice of discount rate q.

There may also be uncertainties due to numerical approximations in the integration in Equation (1.4).

If the estimates of r_0, r_i, or c_i are uncertain then the preference ordering between options could be switched. *Uncertainty is of relevance to decision makers because of its potential influence on preference orderings.* In the case of continuously variable options (e.g. the crest height of a flood defence) any variation in the risk or cost estimates will alter the choice of design variable.

The joint probability density function $f(\mathbf{x}_t)$ in Equations (1.2) and (1.3) already represents random variability. In calculating risks we acknowledge that quantities of relevance to decision making vary. In flood risk analysis $f(\mathbf{x}_t)$ is, as a minimum, used to represent random variation in loading variables such as water levels and wave heights that vary naturally through time. It is usually extended to include natural variability in space of variables such as soil properties (National Research Council, 2000). Natural variability (which is sometimes also referred to as "inherent uncertainty" or "aleatory uncertainty") is thought of as a feature of nature and cannot be reduced (Vrijling and van Gelder, 2006).

Epistemic uncertainties (knowledge uncertainties) require more careful consideration. Epistemic uncertainties include model uncertainties and statistical uncertainties due to observation error and small samples of random phenomena. Model uncertainties mean that the function $D(\mathbf{x}_t)$ is uncertain. Statistical uncertainties mean that the function $f(\mathbf{x}_t)$ may not be an accurate description of the variation in $\mathbf{x}_t$ (National Research Council, 1999). This may be due to limitations in the number of statistical samples, ambiguity in the choice of potential statistical models, or inappropriate statistical assumptions, such as statistical stationarity through time. We have acknowledged that many of the quantities of interest in a flood risk calculation will change in a systematic way over extended timescales, and this extends to the statistical properties of $f(\mathbf{x}_t)$, for example due to non-stationary climate. The nature of that change will also be uncertain due to epistemic uncertainties.

The contributions in this book present a range of ways of dealing with these uncertainties. In understanding them however, it is important to recognise the relationship between uncertainty analysis and decision-making, in the sense that has been presented above. In order to promote more comprehensive incorporation of uncertainty in flood risk management decision-making processes, Hall and Solomatine (2008) presented a framework for the process of incorporating uncertainty analysis in decision-making. Whilst the methods, for example for estimation or propagation of probabilities may differ, the logical structure is intended to be generically applicable, though it may need to be adapted to the characteristics of a specific situation. The approach is as far as possible quantified, by using probability distributions where these can be credibly generated and using intervals or sets of probability distributions where probability distributions cannot be justified. Uncertainties are propagated through to key decision outputs (e.g. metrics of net benefit in terms of risk reduction) and results are presented as distributions and maps. As well as estimating the amount of uncertainty associated with key decision variables, the framework supports the decision-making process by identifying the most influential sources of uncertainty, and the implications of uncertainty for the preference ordering between options. Sensitivity analysis is used to understand the contribution that different factors make to total uncertainty. The effect of uncertainty on choices is analysed using robustness analysis.

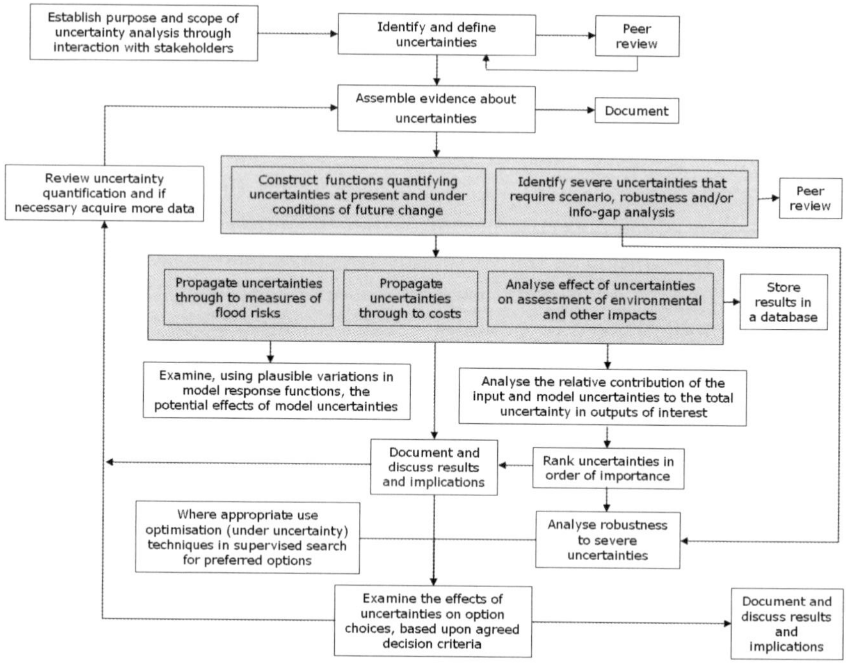

Figure 1.1. Uncertainty analysis framework (Hall and Solomatine, 2008).

1.6. Prospects for Uncertainty Analysis in Flood Risk Management Decisions

The methods of uncertainty analysis are becoming progressively embedded in flood risk management decision making, but the process of doing so is only partially complete. Some of the reasons for incomplete take-up of uncertainty analysis are discussed above and in Pappenberger and Beven (2006). The aim of this book is to promote further the uptake of uncertainty analysis methods. If this is successful, what might be the characteristics of improved flood risk management decision-making processes in future? Sluijs *et al.* (2003) suggest that decision making should be structured so that it facilitates awareness, identification, and incorporation of uncertainty. That said, Sluijs and colleagues acknowledge that uncertainty analysis does not necessarily reduce uncertainties. They argue that it provides the means to assess the potential consequences of uncertainty and avoid pitfalls associated with ignoring or ignorance of uncertainties. Moreover, they go on to argue that uncertainty analysis should facilitate the design of effective strategies for communicating uncertainty. In support of these aims they provide definitions, guidelines and tools for uncertainty analysis. The need for a similar set of guidelines and procedures in the context of flood risk management was argued for by Faulkner *et al.* (2007).

In order to achieve these goals there will need to be a more widespread recognition of the importance of uncertainty analysis in flood risk management. Analysts should be expected to provide a full representation of uncertainty associated with the evidence upon which decisions will be based. To enable this, the data necessary to analyse uncertainties will need to be made more widely available, in a format that can be conveniently assimilated into uncertainty analysis. Bayesian analysis provides a rational approach to valuing information, which could and should be used more widely to inform data acquisition strategies. The software systems that are used to support flood risk analysis, for example in hydrodynamic simulations, need to be restructured so that uncertainty analysis can be applied more routinely and transparently (Harvey *et al.*, 2008). The results of these analyses will be propagated directly through to decisions, so that the implications of uncertainty for decision making are explicit.

The attributes of good practice in uncertainty analysis are now recognisable in an increasing number of flood risk management decisions. There is much work that needs to be done in terms of promoting good practice. The aim of the remaining chapters of this book is to contribute to that effort.

References

Adams, J. (1995). *Risk*, UCL Press, London.

Al-Futaisi, A. and Stedinger, J.R. (1999). Hydrologic and economic uncertainties in flood risk project design, *J. Water Res. Pl.*, **125**, 314–324.

Ball, D.J. and Floyd, P.J. (1998). *Societal Risk, Health & Safety Executive*, Risk Assessment Policy Unit, London.

Beard, L.R. (1997). Estimated flood frequency and average annual damages, *J. Water Res. Pl.*, **123**, 84–88.

Beven, K. (2006). A manifesto for the equifinality thesis, *J. Hydrol.*, **320**, 18–36.

Cothern, C.R. (1996). *Handbook for Environmental Risk Decision Making: Values, Perceptions and Ethics*, CRC, Boca Raton, Florida.

CUR/TAW (1990). *Probabilistic Design of Flood Defences, Centre for Civil Engineering Research and Codes (CUR), Technical Advisory Committee on Water Defences (TAW)*, Gouda, the Netherlands.

Dantzig, D.V. (1956). Economic decision problems for flood prevention, *Econometrica*, **24**, 276–287.

Dawson, R.J. and Hall, J.W. (2006). Adaptive importance sampling for risk analysis of complex infrastructure systems, *Proc. Royal Soc. A*, **462**, 3343–3362.

Dawson, R.J., Speight, L., Hall, J.W. *et al.* (2008). Attribution of flood risk in urban areas, *Hydroinformatics*, **10**, 275–288.

Defra (2001). *Flood and Coastal Defence Project Appraisal Guidance: Overview (including general guidance)*, FCDPAG1.

Defra (2005). *Making Space for Water: Taking Forward a New Government Strategy for Flood and Coastal Erosion Risk Management in England: First Government Response to the Autumn 2004*, Making Space for Water Consultation Exercise, London.

DETR, Environment Agency and Institute of Environmental Health (2000). *Guidelines for Environmental Risk Assessment and Management*, the Stationery Office, London.

Dickson, M.E., Walkden, M.J. and Hall, J.W. (2007). Modelling the impacts of climatic change on an eroding coast over the 21st century, *Climatic Change*, **81**, 141–166.

Egorova, R., Noortwijk, J.M.V. and Holterman, S.R. (2008). Uncertainty in flood damage estimation, *J. River Basin Manage.*, **6**, 139–148.

Evans, E.P. *et al.* (2004). *Foresight Flood and Coastal Defence Project: Scientific Summary: Volume I, Future Risks and their Drivers*, Office of Science and Technology, London.

Faulkner, H., Parker, D., Green, C. *et al.* (2007). Developing a translational discourse to communicate uncertainty in flood risk between science and the practitioner, *AMBIO: J. Hum. Environ.*, **36**, 692–704.

Finkel, A.M. (1990). *Confronting Uncertainty in Risk Management: A Guide for Decision-makers*, Center for Risk Management, Resources for the Future, Washington, D.C.

French, S. (1988). *Decision Theory: An Introduction to the Mathematics of Rationality,* Ellis Horwood, Chichester.

Frieser, B.I., Vrijling, J.K. and Jonkman, S.N. (2005). *Probabilistic Evacuation Decision Model for River Floods in the Netherlands*, 9th International Symposium on Stochastic Hydraulics, IAHR, Nijmegen, the Netherlands.

Galloway, G.E. (2005). Corps of engineers responses to the changing national approach to floodplain management since the 1993 Midwest Flood, *J. Contemp. Water Res. Educ.*, **130,** 5–12.

Goldman, D. (1997). Estimating expected annual damage for levee retrofits, *J. Water Res. Pl. Manage.*, **123**, 89–94.

Hall, J.W., Dawson, R.J. Sayers, P.B. *et al.* (2003a). A methodology for national-scale flood risk assessment, *Proc. ICE Water Marit. Eng.*, **156**, 235–247.

Hall, J.W., Meadowcroft, I.C., Sayers, P.B. *et al.* (2003b). Integrated flood risk management in England and Wales, *Nat. Hazards Rev.*, **4**, 126–135.

Hall, J.W. and Solomatine, D. (2008). A framework for uncertainty analysis in flood risk management decisions, *Int. J. Basin Manage.*, **6**, 85–98.

Harvey, D.P., Peppé, R. and Hall, J.W. (2008). Reframe: a framework supporting flood risk analysis, *Int. J. Basin Manage.*, **6**, 163–174.

Interagency Floodplain Management Review Committee (1994). *Sharing the Challenge: Floodplain Management into the 21st Century*, report to the Administration Floodplain Management Task Force, US Government Printing Office, Washington, DC.

Kahan, J.P., Wu, M., Hajiamiri, S., *et al.* (2006). *From Flood Control to Integrated Water Resource Management: Lessons for the Gulf Coast from Flooding in Other Places in the Last Sixty Years*, RAND Corporation, Santa Monica.

Keeney, R.L. and Raiffa, H. (1993). *Decisions with Multiple Objectives-Preferences and Value Tradeoffs*, Cambridge University Press, Cambridge & New York.

Knight, F.H. (1921). *Risk, Uncertainty, and Profit*, Hart, Schaffner & Marx, Boston, MA.

Liu, Y. and Gupta, H. (2007). Uncertainty in hydrologic modeling: toward an integrated data assimilation framework, *Water Resour. Res.*, **43**, doi:10.1029/2006WR005756.

Meadowcroft, I.C., Hall, J.W. and Ramsbottom, D.M. (1997). *Application of Risk Methods in Flood and Coastal Engineering: Scoping Study*, Report SR483, HR Wallingford.

Munich Re Group (2007). *Natural Catastrophes 2006: Analyses, Assessments, Positions*, Munich.

National Research Council (1999). *Improving American River Flood Frequency*, National Academy Press, Washington, DC.

National Research Council (2000). *Risk Analysis and Uncertainty in Flood Damage Reduction Studies*, National Academy Press, Washington, DC.

Pappenberger, F. and Beven, K. (2006). Ignorance is bliss: 7 reasons not to use uncertainty analysis, *Water Resour. Res.*, **42**, doi:10.1029/2005WR004820.

Pasman, H.J. and Vrijling, J.K. (2003). Social risk assessment of large technical systems, *Hum. Factor. Ergon. Man.*, **13**, 305–316.

Pate-Cornell, M.E. (1996). Uncertainties in risk analysis, six levels of treatment, *Reliab. Eng. Syst. Safe.*, **54**, 95–111.

Savage, L.J. (1954). *The Foundations of Statistics*, Wiley, New York.

Sayers, P.B., Hall, J.W. and Meadowcroft, I.C. (2002). Towards risk-based flood hazard management in the UK, *Civil Eng.*, **150**, 36–42.

Schanze, J. (2006). "Flood Risk Management — A Basic Framework" in: Schanze, J., Zeman, E. and Marsalek, J. (eds), *Flood Risk Management — Hazards, Vulnerability and Mitigation Measures*, Springer, Berlin, pp. 1–20.

Sluijs, J.P.V.D, Risbey, J.S., Kloprogge, P. *et al.* (2003). *RIVM/MNP Guidance for Uncertainty Assessment and Communication*, Netherlands Environmental Assessment Agency (RIVM/MNP).

Stedinger, J.R. (1997). Expected probability and annual damage estimators, *J. Water Res. Pl. Manage.*, **123**, 125–135.

Stern, P.C. and Fineberg, H.V. (1996). *Understanding Risk: Informing Decisions in a Democratic Society*, National Academy Press, Washington, DC.

Todini, E. (2008). Predictive uncertainty in flood forecasting models, *Int. J. River Basin Manage.*, **6**, 123–137.

USACE (1996). *Risk-Based Analysis of Flood Damage Reduction Studies*, US Army Corps of Engineers, Washington, DC.

USWRC (1983). *Economic and Environmental Principles and Guidelines for Water Related Land Resources Implementation Studies*, US Water Resources Council, Washington, DC.

Vrijling, J.K. (1993). "Development in Probabilistic Design of Flood Defences in The Netherlands" in: Yen, B.C. and. Tung, Y.-K. (eds), *Reliability and uncertainty analysis in hydraulic design*, ASCE, New York, pp. 133–178.

Vrijling, J.K. and van Gelder, P.H.A.J.M. (2006). *Implications of Uncertainties on Flood Defence Policy*, Stochastic Hydraulics '05, Proc. 9th Int. Symp. on Stochastic Hydraulics, IAHR, Nijmegen, the Netherlands.

Walley, P. (1991). *Statistical Reasoning with Imprecise Probabilities*, Chapman and Hall, London.

Waverley Committee (1954). *Report of the Departmental Committee on Coastal Flooding*, Cmnd. 9165, HMSO, London.

CHAPTER 2

Use of Models in Flood Risk Management

Keith Beven
Lancaster Environment Centre, Lancaster University, UK

2.1. A Brief History of Models in Flood Risk Management

The history of models for flood risk assessment and management is long. It begins with conceptual models of flood runoff generation and routing in the pre-computer age, moving onto current computer power, which allows for the implementation of detailed distributed models of both hydrological and hydraulic processes built on concepts from "physical-based" equations with coupling to sediment transport and to pollutant transport among other processes (Beven, 2012). In the UK, one of the very earliest books on hydrological analysis was *Flood Estimation and Control* by B.D. Richards (1944) which included a method for developing predictions of the flood hydrograph from a catchment area, based on a time–area histogram concept for a catchment scale transfer function. In the US, this approach had already been used by Ross (1921), Zoch (1934) and Clark (1945), while dynamic flood routing had been instigated as a result of work on the Muskingum catchment in Ohio by McCarthy (1938; see Cunge, 1969). Further impetus was given to this method of predicting catchment responses by the Horton (1933) concepts of runoff generation and the generalisation of the time–area histogram to the *unitgraph* or unit hydrograph by Sherman (1932). In fact, these developments had been anticipated in France by Imbeaux who presented a form of the catchment transfer function and reservoir routing equations on the Durance in 1892.[a]

[a]Thanks to Charles Obled for rediscovering this.

In this pre-computer era, the possibility of making local predictions of flood runoff generation and inundation was limited by the number of calculations that could be performed by people (the computers of Lewis Fry Richardson's first experiments in weather forecasting in 1910 (Lynch, 2006)). Almost as soon as digital computers were more widely available, however, more detailed hydrological and hydraulic models were developed. Stoker (1957) produced the first computer solution of the Saint Venant equations and by the 1970s there were already books available discussing how best to implement the approximate numerical algorithms required (e.g. Abbott, 1979). In hydrology, a lumped catchment model (the Stanford Watershed model) was programmed by Norman Crawford under the supervision of Ray Linsley (Linsley and Crawford, 1960), while Freeze and Harlan (1969) set out the set of coupled equations needed to define the first distributed hydrological model. This was later implemented using finite difference methods by Freeze (1972) and finite element methods by Beven (1977).

There are now many distributed flood routing models available in one and two dimensions; even 3D computational fluid dynamics codes have been used for some specific local problems. There are also many distributed hydrological models based on the Freeze and Harlan (1969) blueprint, but implemented in different ways or with different simplifications. There are a number of attractions to moving towards such higher dimensional models. They are based on equations that have (more or less) physical justification; they allow a more detailed spatial representation of the processes, parameters and predicted variables; they allow the coupling of different types of predictions through the prediction of velocity fields; they allow the use of local measurements of parameters, calibration and validation variables; and they have a spatial element so that they can link more easily to GIS databases and remote sensing data.

However, there are also disadvantages. Higher dimensional models take more computer time (this is important if we are interested in estimating the uncertainty associated with the models); they require the definition of initial and boundary conditions in space and time at a level of detail for which information is not available even in research studies; they require the definition of a very large number of parameter values (usually multiple parameters for every spatial element in the spatial discretisation); and they may still be subject to numerical problems, both numerical diffusion and, in some cases, numerical stability problems. Finally, there is no guarantee that the "physically-based" equations on which many of these models are based are an adequate representation of the complex flow processes we find in reality. This is particularly true of subsurface runoff generation but there are also limitations in the representation of surface flow and transport processes (see, for example, Beven, 1989, 2001, 2002, 2006).

All of these disadvantages can be interpreted as sources of uncertainty in the modelling process. These may be summarised as uncertainties due to:

- conceptual model error and numerical algorithm error (which combined make up a model structural error),
- parameter estimation error,
- commensurability errors (when the values of parameters and variables in a model refer to different quantities than observable variables of the same name),
- initial condition errors, and
- boundary condition errors of all types.

A final source of uncertainty is that associated with the measurements of any observations that are used to calibrate the model, or verify the predictions. We should allow for such errors in calibration and validation rather than blithely assuming that all the observations available are perfect measurements — they are not. Such observations can also be subject to commensurability and interpolation errors due to scale differences between the model variables and what is actually observable.

Successful application of a model is, in great part, a process of successfully controlling these sources of uncertainty through careful selection of model structures and parameter values and careful treatment of the data that can be used to estimate initial and boundary conditions. It is unlikely, however, in any real applications that these uncertainties can be eliminated completely. Indeed, when a distributed model might require the specification of many more parameter values and boundary condition values than there are actually observations available for calibration and validation, it is often the case that additional assumptions are required to avoid a poorly posed calibration problem (e.g. by assuming that channel and floodplain roughness does not vary in space, or hydraulic conductivities are constant for a given soil type). These assumptions introduce constraints that mean that, even if the model structure is a perfect representation of the processes, the model predictions might not match observed responses locally.

2.2. How to Use Flood Risk Management Models

The issues of uncertainty in the use of models in flood risk management have been extensively discussed in the literature and recur throughout this volume. There are, however, a number of generic points that can be abstracted from this discussion pertinent to the use of models in this context.

(1) Since models are known to be approximate representations of the system, which are calibrated using uncertain observations, and are driven using uncertain input data, uncertainty is generic in the modelling process (e.g. Pappenberger and Beven, 2006). Model studies in flood risk management should therefore be associated with an analysis of uncertainty that is proportionate to the importance of the decision being made.
(2) The assessment of uncertainties is a topic of continued debate (e.g. Beven, 2009; Hall, 2003). This is, in part, because not all uncertainties in flood risk management can be treated statistically. There are important sources of uncertainty arising from a lack of knowledge (also called *epistemic uncertainties*) as well a random statistical variability (also called *aleatory uncertainties*).
(3) It is difficult to characterise epistemic uncertainties, precisely because they arise from a lack of knowledge (about process representations, effective parameter values, input data or calibration data such as flood discharge estimates). Thus, it is difficult to be entirely objective about the assessment of uncertainties; some subjective decisions will always be required.
(4) Sources of uncertainty will interact in both model calibration using observations from past events and prediction of future events (e.g. Figure 2.1). Such interactions may be very difficult to specify *a priori* without making some strong assumptions.
(5) However, in making predictions, some assumptions about the nature of the uncertainties affecting the predictions need to be made. This might range from an assessment based on past experience, to a full Monte Carlo sampling exercise; this is dependent on the application.
(6) There are risk-based decision making frameworks available (as outlined, for example, in Chapter 1) that can make use of uncertain model predictions. There is, however, a communication problem for users understanding the nature of uncertain model predictions *and* the assumptions on which they are based (there is a need for a

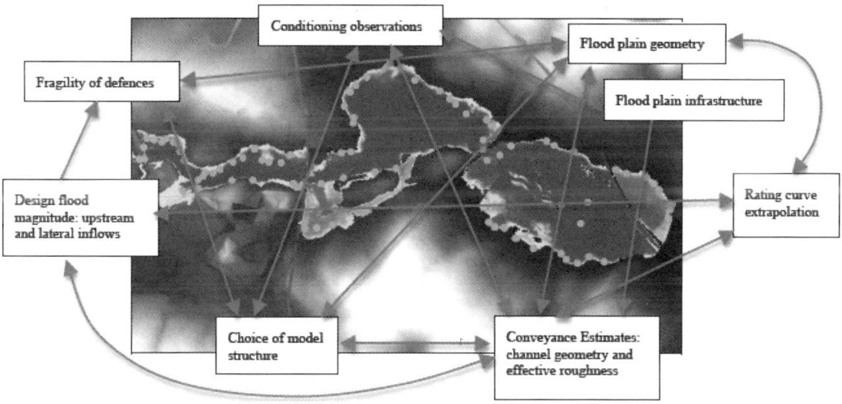

Figure 2.1. Interactions amongst sources of uncertain in flood risk mapping.

translationary discourse about estimated uncertainties expressed in Faulkner *et al.*, 2007).

Table 2.1 illustrates some of these issues in terms of the potential sources of aleatory and epistemic uncertainties in the problem of flood risk mapping. This is an example where a lack of knowledge can be important in making predictions and some uncertainty in mapping the flood risk might be important in planning and other decisions. This will not always be the case; some sources of uncertainty may not affect model simulations significantly and, where the floodplain is bounded, may not have a significant effect on the extent of inundation (though it might still have an effect on the depths of inundation and consequent damages). An example of the prediction of uncertain flood risk maps, conditioning on historical data, is provided by Leedal *et al.* (2010) using the GLUE methodology (see Chapter 6 in this volume for the case of Carlisle in England, where data on inundation were available from surveys after a major flood in January 2005).

The Carlisle event provides an example of the interaction between different sources of uncertainty. The initial estimate of the peak discharge of the January 2005 event was of the order of $900\,\mathrm{m^3 s^{-1}}$. The simplified 2D hydraulic model used to predict the flood inundation (LISFLOOD) was calibrated using this value as the input discharge and produced a reasonable fit to the maximum inundation data surveyed after the event from wrack and water level marks (see, Horritt *et al.*, 2010 and Neal *et al.*, 2009 for calibration of the 2D SFV model to the same data set). The peak estimate was a significant extrapolation of the available rating curve,

Table 2.1. Sources of uncertainty with random (aleatory) and knowledge (epistemic) errors in the case of flood risk mapping (after Beven and Alcock, 2011).

Source of uncertainty	Aleatory errors	Epistemic nature
Design flood magnitude	What is the range of sampling variability around underlying distribution of flood magnitudes?	Are floods generated by different types of events? What frequency distribution should be used for each type of event? Are frequencies stationary? Will frequencies be stationary in the future?
Conveyance estimates	What is the random sampling variability around estimates of conveyance at different flood levels?	Is channel geometry stationary over time? Do conveyance estimates properly represent changes in momentum losses and scour at high discharges? Are there seasonal changes in vegetation in the channel and on the floodplain? Is floodplain infrastructure (walls, hedges, culverts etc.) taken into account?
Rating curve interpolation and extrapolation	What is the standard error of estimating the magnitude of discharge from measured levels?	Is channel geometry stationary over time? What is the estimation error in extrapolating the rating curve beyond the range of measured discharges? Does extrapolation properly represent changes in momentum losses and scour at high discharges?
Floodplain topography	What is the standard error of survey errors for floodplain topography?	Are there epistemic uncertainties in correction algorithms in preparing a digital terrain map?
Model structure		How far do results depend on choice of model structure, dimensions, discretisation, and numerical approximations?

(*Continued*)

Table 2.1. (*Continued*)

Source of uncertainty	Aleatory errors	Epistemic nature
Floodplain infrastructure	What is the random error in specifying positions of elements, including elevations of flood defences?	How should storage characteristics of buildings, tall vegetation, walls and hedges be treated in geometry? Are there missing features in the terrain map (e.g. walls, culverts, etc.)?
Observations used in model calibration/ conditioning	What is the standard error of estimating a flood level given post-event survey of wrack marks, or gauging station observations?	Is there some potential for the misinterpretation of wrack marks surveyed after past events? Are there any systematic survey errors?
Future catchment change	—	What process representations for effects of land management should be used? What future scenarios of future change should be used? Are some scenarios more likely than others?
Future climate change	What is the variability in outcomes due to random weather generator realisations?	What process representations in weather generators should be used? What potential scenarios of future change should be used? Are some scenarios more likely?
Fragility of defences	What are the probabilities of failure under different boundary conditions?	What are the expectations about failure modes and parameters?
Consequences/ vulnerability	What is the standard error of estimation for losses in different loss classes?	What knowledge about uncertainty in loss classes and vulnerability is available?

however, so the Environment Agency also commissioned a review of the rating and discharge estimates. This included some hydraulic modelling of the gauging site. The peak estimate was revised to more than $1500\,\mathrm{m^3s^{-1}}$. Refitting the LISFLOOD model also gave a good fit to the survey data, albeit with somewhat different effective roughness coefficients. It has been shown before how calibration of hydraulic models in this way can lead to effective parameter values that might be dependent on the magnitude of the event simulated (e.g. Romanowicz and Beven, 2003). While this might be in part a result of changing friction losses at different depths of inundation, it might also be the result of the interaction of many other sources of uncertainty in applying models of this type.

2.3. Guidelines for Good Practice in Using Models

The many sources identified in Table 2.1 are an indication of the complexity of the problem of communicating the assumptions made in an analysis (and, hence, what the significance of the resulting uncertain outputs to a given user is). This problem is intrinsic to the use of models in flood risk management and, as part of the UK Flood Risk Management Research Consortium, a number of initiatives have been taken to facilitate this type of communication, including workshops and conference sessions, the development of software systems (e.g. Reframe, see Harvey *et al.*, 2008) and the development of guidelines for good practice in the area of flood risk mapping (Beven *et al.*, 2011).

Such guidelines for good practice can serve as a repository for experience in dealing with different types of uncertainty in different types of application. There are many existing guidelines or standards used for assessing flood risk and resulting planning decisions in different countries. The EU Floods Directive is itself a framework for setting standards in assessing flood risk. Few such standards to date have, however, taken any account of the different sources of uncertainty in assessing the predictions of models for different flood risk management purposes.

One way of trying to achieve this is to define the guidelines for good practice as a set of decisions to be agreed between the modeller and the user. The decisions will cover uncertainties in data and modelling, together with choices for the presentation and visualisation of the results. The response to those decisions can be agreed and recorded as part of the audit trail for a particular application. Such a decision structure provides a framework for communication and allows for the evolution of practice over time, while also making the assumptions of any analysis subject to explicit definition prior to the analysis, and therefore open to later evaluation and review. Figure 2.2 provides a summary of the decision trees involved in an analysis of uncertainty in flood risk maps. It follows the source–pathway–receptor structure commonly employed in this type of application. Feed back arrows in the figure indicate where decisions might be revisited when new data become available.

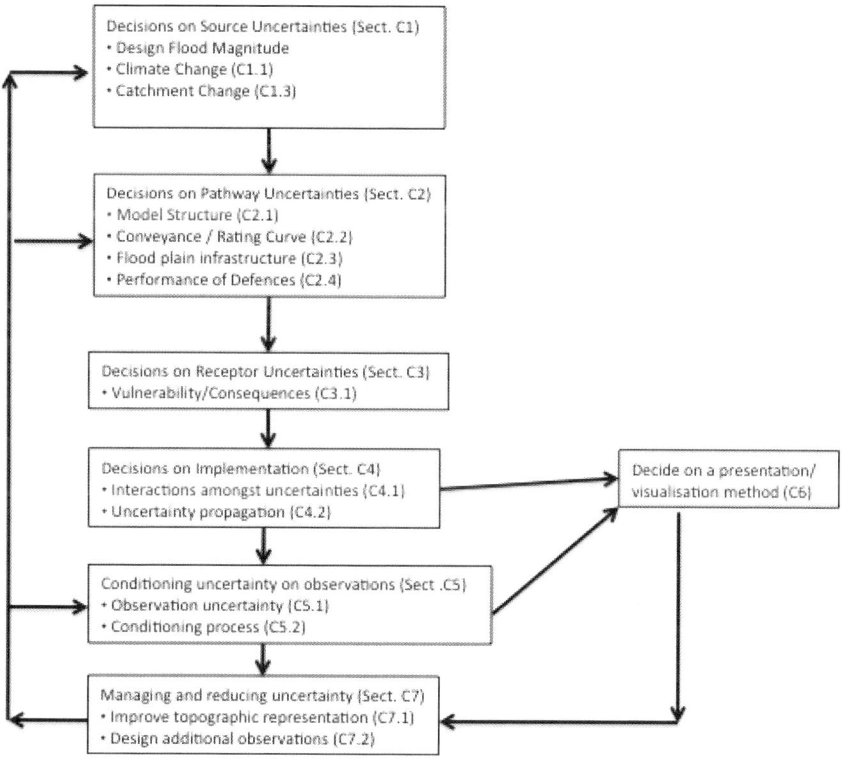

Figure 2.2. High-level decision tree for assumptions in an analysis of uncertainty in assessing risk of flood inundation (after Beven *et al.*, 2011).

2.4. Conclusions

The use of models in flood risk management involves multiple sources of uncertainty that involve knowledge as well as statistical errors. Different sources of uncertainty will also interact in complex ways, both in model calibration and prediction. Thus, characterising the impact of different sources of uncertainty is difficult and will inevitably involve some subjective judgments. It is therefore important to convey the assumptions made about uncertainty in any analysis to users of the outputs of model predictions. One way of doing so is to provide guidelines for good practice in different areas of flood risk management that provide a framework for decisions about different sources of uncertainty within a source–pathway–receptor structure. Ideally, users and practitioners would have the opportunity to agree the decisions at an early stage in a study, even if the decision is to neglect all sources of uncertainty, perhaps because to do more would not be proportionate to the nature of the problem under study. In many cases, an analysis of uncertainty will be justified, particularly in the case of decisions about valuable infrastructure on the floodplain. The recording of the decisions made provides an audit trail for the application, with the possibility for later evaluation and review of the assumptions made.

References

Abbott, M.B. (1979). *Computational Hydraulics — Elements of the Theory of Free Surface Flows*, Pitman, London.

Beven, K.J. (1977). Hillslope hydrographs by the finite element method, *Earth Surf. Process.*, **2**, 13–28.

Beven, K.J. (1989). Changing ideas in hydrology: the case of physically based models. *J. Hydrol.*, **105**, 157–172.

Beven, K.J. (2001). How far can we go in distributed hydrological modelling?, *Hydrol. Earth Syst. Sc.*, **5**, 1–12.

Beven, K.J. (2002). Towards an alternative blueprint for a physically-based digitally simulated hydrologic response modelling system, *Hydrol. Process.*, **16**, 189–206.

Beven, K.J. (2006). The Holy Grail of scientific hydrology: $Q_t = H(\underline{\underline{S}}\underline{R})A$ as closure, *Hydrol. Earth Syst. Sci.*, **10**, 609–618.

Beven, K.J. (2012). *Rainfall-Runoff Modelling — The Primer 2nd Edition*, Wiley-Blackwell, Chichester.

Beven, K.J. and Alcock, R.E. (2011). Modelling everything everywhere: a new approach to decision-making for water management under uncertainty, *Freshwater Biol.*, **57**, 124–132.

Beven, K.J., Leedal, D.T. and Alcock, R. (2010). Uncertainty and good practice in hydrological prediction, *Vatten*, **66**, 159–163.

Beven, K.J., Leedal, D.T. and McCarthy, S. (2011). *Guidelines for Assessing Uncertainty in Flood Risk Mapping*, Flood Risk Management Research Consortium User Report. Available at www.floodrisk.net (Accessed on 10/05/12).

Clark, C.O. (1945). Storage and the unit hydrograph, *T. Am. Soc. Civ. Eng.*, **110**, 1416–1446.

Cunge J.A. (1969). On the subject of a flood propagation computation method (Muskingum method), *J. Hydrol.*, **7**, 205–230.

Faulkner, H., Parker, D., Green, C. *et al.* (2007). Developing a translational discourse to communicate uncertainty in flood risk between science and the practitioner, *Ambio*, **16**, 692–703.

Freeze, R.A. (1972). Role of subsurface flow in generating surface runoff: 2. upstream source areas, *Water Resour. Res.*, **8**, 1272–1283.

Freeze, R.A. and Harlan, R.L. (1969). Blueprint for a physically-based digitally simulated hydrologic response model, *J. Hydrol.*, **9**, 237–258.

Hall, J.W. (2003). Handling uncertainty in the hydroinformatic process, *J. Hydroinform.*, **5**, 215–232.

Harvey, D.P., Peppé, R. and Hall, J.W. (2008). Reframe: a framework supporting flood risk analysis, *J. River Basin Manage.*, **6**, 163–174.

Horritt, M.S., Bates, P.D, Fewtrell, T.J. *et al.* (2010). Modelling the hydraulics of the Carlisle 2005 event, Proceedings of the Institution of Civil Engineers, *Water Manage.*, **163**, 273–281.

Horton, R.E. (1933). The role of infiltration in the hydrological cycle, *Eos. T. Am. Geophys. Un.*, **14**, 446–460.

Imbeaux, E. (1892). La Durance. Ch. VI Recherche de lois udometrique des crus. Essai de prevision pluviométrique, *Annales des Ponts et Chaussées. Mémoires et Documents Relatifs àl'art des constructions,* **1**, 166–179.

Leedal, D.T., Neal, J., Beven, K., *et al.* (2010). Visualization approaches for communicating real-time flood forecasting level and inundation information, *J. Flood Risk Manage.*, **3**, 140–150.

Linsley, R.L. and Crawford, N.H. (1960). Computation of a synthetic streamflow record on a digital computer, *Int. Assoc. Sci. Hydrol. Publ.*, **51**, 526–538.

Lynch, P. (2006). *The Emergence of Numerical Weather Prediction: Richardson's dream*, Cambridge University Press, Cambridge.

McCarthy, G.T. (1938). *The Unit Hydrograph and Flood Routing*, paper presented at the conference of the North Atlantic Division of the US Army Corps of Engineers, New London.

Neal, J.C., Bates, P.D., Fewtrell, T.J. *et al.* (2009). Distributed whole city water level measurements from the Carlisle 2005 urban flood event and comparison with hydraulic model simulations, *J. Hydrol.*, **368**, 42–55.

Pappenberger, F. and Beven, K.J. (2006). Ignorance is bliss: 7 reasons not to use uncertainty analysis, *Water Resour. Res.*, **42,** W05302, doi:10.1029/2005WR004820.

Richards, B.D. (1944). *Flood Estimation and Control*, Chapman and Hall, London.

Romanowicz, R. and Beven, K. (2003). Estimation of flood inundation probabilities as conditioned on event inundation maps, *Water Resour. Res.*, **39**, 1073–1085.

Ross, C.N. (1921). The calculation of flood discharge by the use of time contour plan isochrones, *T. Inst. Eng. Aus.*, **2**, 85–92.

Sherman, L.K. (1932). Streamflow from rainfall by unit-graph method, *Eng. News-Rec.,* **108**, 501–505.

Stoker, J. (1957). *Water Waves*, Interscience, New York.

Zoch, R.T. (1934). On the relation between rainfall and streamflow, *Mon. Weather Rev.,* **62**, 315–322.

SECTION II

THEORETICAL PERSPECTIVES

CHAPTER 3

A Framework for Uncertainty Analysis

Keith Beven
Lancaster Environment Centre, Lancaster University, UK

3.1. Uncertainty about Uncertainty Estimation in Flood Risk Management

There are two primary reasons for using models in flood risk management. The first is to show that we understand how the processes affecting flood generation and impacts work. Uncertainty is an issue in this usage of models, which can be considered as hypotheses of how the system works, since if it is ignored we might accept poor models/hypotheses falsely (Type I error) or reject good models/hypotheses falsely (Type II error). We would like to avoid both types of error (see Beven, 2010, 2012). The second usage is to help in practical decision making for real applications in flood risk management. Uncertainty in this case is important, in that if it is ignored then a poor decision might be taken, but the decision context may determine how much effort needs to be put into the assessment of uncertainties. This volume primarily concentrates on the second question, but clearly this is not independent of the first in that we would like to think that the models we use to make predictions to inform decision making are actually good representations of the system that is being predicted.

There is, however, a basic problem in estimating the uncertainty associated with the predictions of models for real-world problems: there is no general agreement on what methods should be used to estimate that uncertainty. This is because there are too many sources of uncertainty in the modelling process that cannot easily be disaggregated when we can only evaluate model predictions against a limited number of observations. Thus, it is necessary to make assumptions about how to represent uncertainty, and there are sufficient degrees of freedom in doing so that different methods based on different types of assumptions (including purely qualitative evaluations) cannot easily be verified or rejected. They will, however, produce different results. It is therefore necessary, in applying any method of uncertainty analysis, to be explicit about the assumptions that are being made so that they will be clear to any user of the results.

Some statisticians have argued that the *only* way to represent uncertainty is in terms of formal probability theory (e.g. O'Hagan and Oakley, 2004). Some practitioners in hydrology and hydraulics have argued that formal statistical methods are advantageous in presenting results to users (Hall *et al.*, 2007; Todini and Mantovan, 2007). As we will see below, however, in real applications it can be difficult to justify or verify the strong assumptions necessary to implement formal probabilistic methods (Andréassian *et al.*, 2007; Beven, 2006). A particular issue arises in that probabilistic estimates of uncertainty are usually made on the basis of an assumption that the model is correct and that all errors considered can be considered to be probabilistic (aleatory) in nature. Uncertainty estimates are then conditional on this assumption. In flood risk management this is a difficult assumption to justify in that model structural error often makes it difficult to make an unequivocal choice of one model over another. Model structural error is a form of non-probabilistic (epistemic) error that is not easily represented in terms of probability distributions. Knight (1921) made this distinction between probabilistic uncertainty and what he called "true" (epistemic) uncertainty in terms of risks that insurers would be prepared to offer insurance for, and those that they would not. It is much more difficult to make assumptions about epistemic uncertainties, particularly in the types of nonlinear models that arise in flood risk management. While this has not stopped the insurance industry making judgements about the risks they face in setting premiums against flood damages based on probabilistic models, the current generation of models are undoubtedly subject to non-probabilistic errors that might require the use of alternative methods. This has resulted in alternative sets of assumptions leading to alternative uncertainty estimation methodologies that have been developed and used, including the Generalised Likelihood Uncertainty Estimation (GLUE) methodology; fuzzy set (possibilistic) methods; info-gap methods; and the more qualitative NUSAP method. Over the course of this volume, all of these methods will be used for different types of applications.

There is also, however, a certain commonality to all of these approaches that allows us to talk of a simple framework for uncertainty analysis. Different methods may be described in terms of a number of common decisions as follows:

(1) Decide on what model (or models) will be considered in the analysis.
(2) Decide on what prior knowledge is available about uncertainty in parameters and input variables and run a forward uncertain analysis based on this prior information.
(3) Decide on how the model results will be compared with observables (taking proper account of uncertainty in the observations) and form an updated posterior estimate of uncertainty in models, parameter values and other variables.

Each decision implies making specific assumptions. This is what leads to the different methodologies. All the wide range of applications and methods used in this volume fit within this simple framework, even though they might make quite different assumptions in these decisions. It is therefore very important that the assumptions are made explicit at each stage so that they will be transparent to the decision maker and can be openly debated and changed as necessary.

Step (3) cannot, of course, be used if there are no observables (though it might still be possible to carry out some model evaluation process, for example by invoking the opinions of experts about whether a model is making adequate predictions or not). This is often the case in flood risk management. Data from past events is often very limited if only because floods, by definition, rarely occur. Data on particular features of floods, such as embankment failures, damming and failure of bridges, effects of debris flows and land slips etc. are also limited, and such factors are often ignored (even in an uncertainty analysis) in applications of models to specific locations. We might expect, however, that over time more data will become available and it will be possible to use the third step in this framework as part of a learning process about how different places might be represented within a flood risk management strategy (see the discussion of "models of everywhere" in Beven, 2007).

Within this framework we can distinguish three different types of applications. There are those for which no observables are available so that only a forward uncertainty analysis based on prior assumptions about sources of uncertainty is possible. Secondly, there are applications for which only historical data are available, allowing some conditioning of the predictions and their associated uncertainties (simulation). Finally, there are applications for which observations that can be used for conditioning are available in real time (forecasting/data assimilation). These different techniques allow different types of uncertainty methodologies to be used.

3.2. Sources of Uncertainty

Flood risk management problems involve both aleatory and epistemic uncertainties (and, in many real applications, epistemic uncertainties that are treated as aleatory uncertainties for the sake of convenience). How these different sources of uncertainty should be taken into account in flood risk assessment depends on the choices and assumptions made in the three stages of the simplified framework for uncertainty presented above. However, it is important in any problem to have an appreciation of the potential sources of uncertainty.

Decisions 1 (the choice of model or models) controls what other sources of uncertainty need to be considered but this choice itself raises the issue of model structural uncertainty. This will not be random; the very nature of the model calculations will introduce non-stationary bias and structure into the residuals. This can easily be demonstrated in hypothetical cases where the inputs are known perfectly by using one feasible model structure to reproduce the results of a different but similarly feasible model structure. There is also the possibility in all real applications, however, that all the available model structures will suffer from epistemic uncertainties in their representation of the system. As a result of the lack of full understanding of how the system works, there is the possibility of making what have been called Type III errors. Type III errors are those where there is neglect of the processes important in controlling the response (e.g. neglect of floodplain infrastructure in predicting flood depths because of a lack of available information). Even models that have an accepted theoretical basis can be subject to this type of epistemic uncertainty (for example, the use of the simple 1D advection–dispersion equation to represent solute transport in rivers that has a good theoretical basis but usually produces the wrong type of response, see Beven, 2002).

In Decisions 2 (prior knowledge about parameters and variables required to make a forward uncertainty analysis) this issue of model structural uncertainty is compounded by commensurability errors. These occur when effective values of variables or parameters in a model have a different meaning to variables or parameters with the same name that can be measured as a result of time or space scales, or measurement technique limitations. There are many examples in flood risk modelling. We should expect, for example, that the effective roughness required by a hydraulic model applied at a reach scale (1D) or element scale (2D or 3D) will vary with scale. We should expect that the effective hydraulic conductivity of a runoff generation model would also vary with scale.

Commensurability issues also arise in the definition of the boundary and initial conditions required to set up a model run. Simple measurement and interpolation errors will also affect estimates of boundary conditions. Discharge is often required as a boundary condition for hydraulic models. This is often estimated by interpolation from measurements of stage by use of a rating curve which may vary over time and be uncertain in form, especially when extrapolated to flood stages beyond the range of actual measurements of discharge (which may themselves be subject to error). Rainfall inputs to flood models might also be subject to measurement, interpolation and commensurability errors, whether they are derived from point rain gauge or radar measurements.

In Decisions 3 (comparison of predictions with observables) we can again distinguish measurement and commensurability errors that will affect any assessment of model predictive accuracy. These are often neglected. Observed values are taken as true; predicted values are taken as directly comparable; and an assessment is based on the resulting residuals.

It is instructive to take a more complete view of the sources of error and uncertainty (e.g. Beven, 2000, 2005). At any point in space or time (x, t) we expect that

$$\begin{aligned} &O(x,t) + \varepsilon_O(x,t) + \varepsilon_C(\Delta x, \Delta t, x, t) \\ &\quad = M(\theta, \varepsilon_\theta, I, \varepsilon_I, x, t) + \varepsilon_M(\theta, \varepsilon_\theta, I, \varepsilon_I, x, t) + \varepsilon_r, \end{aligned} \tag{3.1}$$

where O is a vector of observations, M is a vector of modelled values, θ is a vector of model parameter values, I is a vector of inputs and boundary conditions, ε_O is an observation error, $\varepsilon_C(\Delta x, \Delta t, x, t)$ is a commensurability error dependent on the time and space scales $(\Delta x, \Delta t)$ of the observed variables relative to the model variables, ε_θ is a matrix of parameter errors, ε_I is a matrix of input and boundary condition errors, $\varepsilon_M(\theta, \varepsilon_\theta, I, \varepsilon_I, x, t)$ is a model error and ε_r is a random error.

What can actually be calculated, on the other hand, is the residual:

$$\varepsilon(x,t) = O(x,t) - M(\theta, \varepsilon_\theta, I, \varepsilon_I, x, t). \tag{3.2}$$

Comparing (3.1) with (3.2) shows that the residual $\varepsilon(x,t)$ can be expressed as

$$\varepsilon(x,t) = \varepsilon_M(\theta, \varepsilon_\theta, I, \varepsilon_I, x, t) - \varepsilon_C(\Delta x, \Delta t, x, t) - \varepsilon_O(x,t) + \varepsilon_r. \tag{3.3}$$

The first three terms on the right-hand side of (3.3) are unlikely to be aleatory in nature but will involve various epistemic errors. Therein lies the fundamental problem of uncertainty estimation. It is evident that the various terms on the right-hand side cannot be disaggregated from the single observable residual $\varepsilon(x,t)$ unless some very strong assumptions are made. The usual assumption in statistical inference is that all the terms on the right-hand side can be treated as a lumped "measurement error" and represented by an appropriate statistical model (e.g. additive, zero mean, Gaussian, autocorrelated random errors).

A probabilistic treatment of the residuals $\varepsilon(x,t)$ might often be a useful approximation, but it should be remembered that it is an approximation to a fuller treatment of the different sources of error and uncertainty. While writing the model error $\varepsilon_M(\theta,\ \varepsilon_\theta,\ I,\ \varepsilon_I,\ x,\ t)$ as an additive term in (3.1) is formally correct at any (x,t), that does not mean that the function it represents is in any way simple. The model estimate $M()$ depends on parameter errors, and input and boundary condition errors, in complex and nonlinear ways. This would be the case even if, for some hypothetical case, the model could be assumed to be formally correct. In real applications, however, this is not the case. Then the model deficiencies will interact nonlinearly with errors in parameters and input and boundary conditions to induce nonstationarity and heteroscedasticity in this term.

It is the difficulty of separating the effects of these different sources of error that allows scope for different representations of uncertainty in different methodologies. It is a problem that is analogous to the choice of a particular model structure in making predictions: since no model can be shown to be formally correct in real applications, there is scope for different model structures to provide predictions. Similarly, since no uncertainty methodology can be shown to be formally correct in real applications, there is scope for different methods to provide estimates of predictive uncertainty. The user of such estimates then has to be careful to be clear about the assumptions that have been made and what information the estimates convey.

3.3. Statistical Representations of Uncertainty

In this section we will briefly introduce some of the available statistical uncertainty estimation methodologies and their assumptions. More detail can be found in Beven (2009).

3.3.1. *Frequentist statistics*

Frequentist statistical approaches to uncertainty treat all uncertainties as aleatory. They aim to estimate the probability of a potential outcome on the basis of available observations of that outcome. Frequentist theory is based on estimating what the asymptotic probability distribution of outcomes would be given a limited sample of observations by maximising a likelihood function. The larger the number of observations, the more secure the estimates of the probabilities will be (at least where the process can be assumed stationary) and the smaller the uncertainties in the estimates will be.

A typical use of frequentist statistics is the estimation of flood frequencies given a series of peak discharges. A likelihood function can be specified based on assumptions about the deviations of the observed values from a theoretical distribution function (the model), under the assumption that the model being fitted is correct. Maximising the likelihood allows the values of the parameters of the chosen model distribution to be estimated. The shape of the likelihood surface around the maximum likelihood parameter set allows the uncertainties in the parameter values to be estimated. Increasing the number of flood peaks will improve the estimates of the parameters and reduce the uncertainties in the predicted frequencies. Since we are often interested in predicting flood frequencies for return periods longer than the length of the data series, however, the uncertainties in the predictions may remain significant and should be passed on to decision makers.

This is also an interesting simple example of model structural and observational uncertainty issues. There is a decision to be made about whether the analysis will be based on only annual maximum peaks or peaks over a discharge threshold. For both cases there is a choice of theoretical distribution to be made (for the annual maximum case, possible distributions that have been used in the past include the log normal, log Pearson Type II, Wakeby, Generalised Extreme Value, and Generalised Logistic Distributions). Different distributions will all produce different predicted frequencies, especially for higher return period floods, but it may be difficult to decide whether one distribution is a better predictor than another (sample moment plots are sometimes used to make this decision, but often without consideration of uncertainty in the sample moment estimates, which can be significant for small samples and higher moments). There is also the issue of observational error in the peak discharges used in the analysis. As noted earlier, rating curves might be rather uncertain at flood stages, though this is often treated implicitly in frequency analysis, i.e. it is assumed to be included in the total residual errors. Similar issues arise when frequentist statistics are used in regression analyses in flood risk management (including, for example, in the fitting of rating curves).

The critical assumptions in this form of analysis are, therefore, the choice of model to be fitted, and the assumptions about the model residuals that define the likelihood function (especially that the residuals are stationary and that all other sources of observation error can be included implicitly in the total residual, as in (3.2)). The analysis can be objective in the sense that, if the assumptions about the residuals are correct, then the probabilities of the predicted outcomes will be correctly estimated. Common assumptions about the residuals are that they are unbiased (have zero mean), have a Gaussian distribution with constant variance, and are independent.

Where the only uncertainties considered are lumped into the total residual, these assumptions can easily be checked for a given choice of model. More complex assumptions about the structure of the residuals can be incorporated into the analysis where such an analysis indicates that this is appropriate (e.g. changing variance or heteroscedasticity, and correlation between residuals). However, it can be dangerous to assume that the theoretical objectivity will actually produce the correct results in real applications. Draper (1995) provides a sobering example of a regression analysis of the available observations on O-ring seal failures on the Challenger Space Shuttle. Different assumptions about the form of the regression model resulted in multiple representations of the data, all of which had been fitted objectively. When extrapolated to the conditions of the Challenger launch, however, they produced quite different predictions with non-overlapping uncertainty estimates. Reliance on a single objective prediction in this case resulted in a catastrophic failure.

Further information on classical approaches for statistical inference in model calibration is provided by Chandler in Chapter 4 of this volume.

3.3.2. *Bayesian statistics*

The origins of Bayesian statistics lie in a paper found amongst the papers of the Rev. Thomas Bayes (c.1701–1761) after his death. Presented at the Royal Society of London by his friend Richard Price in 1763, the "Essay Towards Solving a Problem in the Doctrine of Chances" contained the first expression of what is now called Bayes theorem. A more general discrete form was developed, apparently independently, by Pierre-Simon Laplace (1749–1827) and published in France in 1774.

We can define Bayes theorem in a form that given a set of feasible models, M, and observations O, then the probability of M given O is given by

$$L_p(M|O) = L_o(M)L(O|M)/C, \tag{3.4}$$

where $L_o(M)$ is some prior probability defined for the range of feasible models (here model is used to indicate a combination of model structure and parameter set), $L(O|M)$ is the likelihood of simulating the evidence given the hypotheses, and C is a scaling constant to ensure that the cumulative of the posterior probability density $L_p(M|O)$ is unity.

Bayes theorem represents a form of statistical learning process. When applied to models it provides a rigorous basis for the expression of degrees of confirmation of different model predictions, expressed as probabilities, as long as the different components of (3.4) can be defined adequately.

This learning process starts with the definition of prior distributions for the factors that will be considered uncertain. There is a common perception of Bayes statistics that the choice of prior distributions introduces subjectivity into the analysis. However, both Bayes and Laplace originally applied their methods using priors that were non-informative, giving equal chances to all possible outcomes until some evidence became available. Berger (2006), points out that Bayes equation underpinned the practice of statistics for some 200 years, without it being considered necessary to be too specific about the prior distributions and suggests that the development of frequentist statistical theory in the 20th century from R. A. Fisher onwards was, in part, a response to dissatisfaction with the constant prior assumption.

The choice of subjective prior distributions is a relatively recent innovation, but is now strongly argued for by many Bayesian statisticians (see Goldstein, 2006, and the additional comments on the Berger and Goldstein papers by Draper, 2006, and O'Hagan, 2006). In principle such priors can be defined by the modeller on the basis of expert elicitation, past experience or subjective judgment. The prior should, however, only be really important if there is only a very limited amount of observational data available. This is because as more evidence about the likelihood of a model is gained by repeated application of (3.4), then the prior will come to be dominated by the likelihood $L(O|M)$.

Thus, the predicted uncertainties will depend on the assumptions made about both the prior likelihood $L_o(M)$ and the likelihood function $L(O|M)$. The latter, as in the case of frequentist statistics, will be dependent on assumptions about the nature of the model residuals. Again, however, such assumptions can (and should be) checked and changed as necessary. If the assumptions can be shown to be valid, then the Bayesian approach provides objective estimates of uncertainties, albeit conditional on the choice of model structure and error structure. Model structural error might still be an issue but O'Hagan and Oakley (2004) suggest that this form of probabilistic analysis is the only way of properly estimating uncertainties. However, this depends on the aleatory model of the residual errors being able to account for the epistemic uncertainties discussed earlier in the chapter. That this may not always be the case means that other methods might also be useful.

Further details on formal Bayesian methods for model calibration are provided by Rougier in Chapter 5 of this volume.

3.3.3. *Imprecise probabilities*

Probabilities can, of course, themselves be uncertain. As noted above, where we are trying to estimate the distribution of a variable, both the correct function form of that distribution, and the parameter values of the function, might be uncertain. This is a form of uncertainty about uncertainties within a purely statistical framework. The concept can be generalised in the theory (or rather competing theories) of imprecise probabilities (e.g. Walley, 1991; see also Hall, 2003, 2006). An extreme form of imprecise probability function is when occurrences of a variable are assumed to be known only within some range or interval of values.

3.4. Other Interpretations and Representations of Uncertainty

3.4.1. *Fuzzy, interval and set-based approaches*

A different way of representing variability, that need not be limited to the aleatory uncertainties of probabilities, is in terms of fuzzy sets of values. Instead of a definitive answer as to whether a value belongs to a certain set of values that would result from defining that set of values as a crisp set, the value is given a membership value (with a range from 0 to 1) to reflect the degree of membership of the set in a given context. It is easy to think of many environmental examples where a fuzzy definition might be useful (see Beven, 2009; Hall, 2003). Membership function values are not equivalent to probabilities. To make this distinction clear, the outcomes from a fuzzy analysis are often called relative *possibilities*.

Fuzzy variables were introduced by Zadeh (1965). They can be used to define uncertainties in inputs, in parameter values and in evaluating model outputs against observations. They can be combined in more flexible ways than probabilities, including fuzzy union and intersection operators. There have been a number of attempts to provide a general theory of uncertainty that spans probabilistic, possibilistic and other representations of uncertainty (see, for example, Klir, 2006; Zadeh, 2005).

3.4.2. *Equifinality and the GLUE methodology*

One of the issues that arises in uncertainty analysis in environmental modelling is that, given the uncertainty in model inputs and observational data, it may be difficult to distinguish between different model representations of the system of interest. This is often interpreted as a difficulty in finding the best model of a system and is referred to as non-uniqueness, ambiguity or non-identifiability. Thinking about models as hypotheses of how the system is working, however, suggests that there need not be a best model. There may rather be many models that give equally acceptable predictions; it is just that we cannot easily differentiate between them. This is the equifinality thesis (Beven, 1993, 2006; Beven and Freer, 2001; von Bertalanffy, 1968) that underlies the GLUE methodology, first outlined by Beven and Binley (1992) as an extension of the Generalised Sensivity Analysis of Hornberger and Spear (1981). GLUE is based on making a large number of Monte Carlo realisations of model structures and parameter values; evaluating the predictions of those models against some observed data in terms of a formal or informal likelihood measure; rejecting those models that do not give acceptable predictions (i.e. they are given a likelihood of zero); and using the remaining behavioural models to form cumulative likelihood weighted distribution functions as an expression of uncertainty in the predicted variables. GLUE is generalised in that it can include statistical identification methods as special cases, but in this case the error structure assumed is treated as an additional non-process component of the model structure (Romanowicz *et al.*, 1994, 1996). It can also use fuzzy and other non-statistical performance measures in weighting the predictions of the set of behavioural models (e.g Beven, 2006; Beven and Freer, 2001). Multiple model structures are easily incorporated into the methodology, as long as they can be evaluated in the same ways. Multiple evaluation measures are also easily incorporated by combining the likelihoods for each evaluation into a final likelihood weight. Bayes equation can be used for these combinations, but also other operators such as fuzzy union or intersections and averages of different types.

More details of the GLUE methodology are given in Chapter 6. GLUE has been criticised for the number of subjective assumptions that are required in using it and it is true that the resulting uncertainty estimates on predicted variables are conditional estimates: conditional on the choice of models, choice of prior parameter distributions, choice of likelihood measures, choice of the limits of acceptability for those models retained as behavioural, and choice of how to combine different likelihood measures. It is also apparent, however, that if we make over-strong assumptions within a formal statistical framework, the results can be incorrect even for near ideal hypothetical cases (e.g. Beven *et al.*, 2007, 2009). GLUE was always intended for the non-ideal conditions of real applications, where the flexibility it allows might be advantageous. However, it is always recommended that the assumptions made in a particular application are stated quite explicitly so that any users can assess them critically.

There have now been many GLUE applications in the area of flood risk management, including the prediction of flood runoff, flood frequency analysis using continuous simulation, and flood inundation predictions using hydraulic analysis. These applications have thrown up some interesting issues. One is that there can be many different model/parameter set combinations, spread widely through the model space, that give equally good predictions. This reinforces the suggestion above that strong statistical assumptions might lead to over-conditioning of effective parameter distributions. It also suggests that finding efficient guided sampling methods might be difficult in real applications where the likelihood response surfaces may be very complex (see also Duan *et al.*, 1992). Another issue is that it is often necessary to use quite relaxed rejection criteria to have any behavioural models at all. As noted earlier, this might not be just a matter of using models that do not properly represent the way the system works, but could also be a result of input errors. In some cases, however, it is clear that using different parameter sets in a model is necessary to get good predictions either for different hydrological conditions or different places in a flood inundation prediction (e.g. Choi and Beven, 2007; Pappenberger *et al.*, 2006b), suggesting that some improvements to model structure might be sought.

A major limitation on the use of GLUE is the computational requirements in that, particularly for models with more than just a few parameters, a very large number of simulations might be required (e.g. the 2,000,000,000 simulations of Iorgulescu *et al.*, 2005, of which 216 were retained as behavioural). This is becoming less of a problem with the use of parallel computers based on cheap PC processors. It does seem that computing power may well increase faster than we can reduce other sources of uncertainty, so that the methodology should become applicable to a wider range of model applications.

Further details of the GLUE methodology for model calibration with uncertainty are provided in Chapter 6 of this volume.

3.4.3. *Info-gap theory*

Info-gap theory (Ben-Haim, 2006) is another non-statistical approach to uncertainty estimation. It was explicitly conceived as an attempt to deal with the potential impact on decisions of epistemic uncertainties (Knight's *true uncertainties*) that cannot easily be handled using aleatory statistical concepts. It does this by focussing on the predicted outcomes of models, relative to some nominal prediction, regardless of the parameter values, inputs and boundary conditions that control those outcomes.

This process can be defined generally in the following way:

$$U(\alpha, \tilde{M}(\Theta)) = \{M(\Theta) : |M(\Theta) - \tilde{M}(\Theta)| \leq G(\alpha)\} \quad \alpha \geq 0. \tag{3.5}$$

$U(\alpha, \tilde{M}(\Theta))$ is the set of all possible model predictions of a variable of interest whose deviation from the nominal prediction, $M(\Theta)$, is nowhere greater than some function, $G(\alpha)$, of a scaling variable α. The sources of uncertainty might in this approach come from different model structures, different parameter sets or different input conditions, but unlike Bayesian theory or the GLUE methodology, it is not necessary to try to put any form of likelihood weight on the predictions. Here, we wish to assess only the relative deviation of the outcome away from some nominal case for any value of α.

Given that deviation, however, two functions may be specified, representing *robustness* and *opportuneness* (using the nomenclature of Ben-Haim, 2006). The robustness function expresses the greatest level of uncertainty (in terms of the values of α) for which the performance requirements of a decision maker must be satisfied. The definition of what constitutes satisfactory performance, of course, will depend on the particular characteristics of a project. The opportuneness function on the other hand, is the least level of uncertainty at which success is assured. For any decision we would ideally like to maximise the value of α for which the decision is robust, while minimising the value of α at which success is assured. These requirements are often in opposition to each other. For example, an "optimal" solution that would maximise performance (which effectively has $\alpha = 0$) should not be expected to be robust in the face of significant uncertainties. Ben-Haim shows that appropriate robustness and opportuneness functions can provide, in cases of real uncertainty, a rational and consistent framework for assessing the relative merits of different decisions. Hall and Hine (2010) provide an extensive example of application of info-gap theory to uncertainty analysis in regional frequency analysis and flood management decisions.

3.4.4. *NUSAP*

There has been another approach to epistemic uncertainty developed that is more qualitative in nature. This is the NUSAP methodology that has been implemented as a way of presenting uncertainty in environmental predictions to decision makers in RIVM in the Netherlands (see, for example, Janssen *et al.*, 2005; van der Sluijs *et al.*, 2005).

NUSAP is an acronym of five different ways of assessing uncertainty (Numeral, Unit, Spread, Assessment and Pedigree). It is framed so as to complement quantification of uncertainty assessments (Numeral, Unit and Spread) with expert judgments of reliability (Assessment) and the multi-criteria evaluation of the historical background to the methods and data used (Pedigree). In the Netherlands, the Assessment and Pedigree stages have been formalised into a guidance manual for Uncertainty Assessment and Communication using an index methodology. The guidance is intended for use in framing the uncertainty context of a new project, in structuring methods of uncertainty estimation and choosing quantification methods during a project, and in reviewing and communicating results after a project.

3.4.5. *Uncertain scenarios*

A final method of assessing uncertainties that cannot easily be treated by probabilistic means is the use of model scenarios. The most high profile example of this is the use of different scenarios of future emissions in predicting possible climate change within the Intergovernmental Panel for Climate Change (IPCC) assessments. In this case, the number of possible runs that can be made is strongly limited by computer run times of the current generation of fully coupled ocean–atmosphere circulation models. Thus, the assessment of uncertainties has, to date, been based on deterministic runs of different emission scenarios with different General Circulation Model (GCM) implementations (IPCC, 2007). There are plans to run ensembles of simulations using one or more coarser grid-scale GCMs in future, but computer run times will still mean that the number of runs that can be made will be in the hundreds, a number that is small relative to the number of degrees of freedom in the specification of the model parameterizations and emission scenarios.

Essentially, scenario modelling is a form of forward uncertainty analysis, but without the possibility of assigning a probability to each model run. We can, of course, as in any form of forward uncertainty analysis suggest that some scenarios are more probable than others, but this would necessarily be speculative until some data became available to allow evaluation of the relative likelihood of the scenarios. It has been argued from a Bayesian point of view that it should be possible to make assessments of the probability of different scenarios on the basis of expert opinion (e.g. Rougier, 2007, in the context of climate change scenarios). This would then allow uncertainty evaluation to remain within a probabilistic framework. In many cases of scenario modelling, however, there are very real epistemic uncertainties that must make such expert derived probabilities questionable.

This then means that, in using uncertain scenario predictions in decision making, interpretation is going to be an important issue. In this respect, there is an interesting overlap with the info-gap methodology here, in that the evaluation of outcomes within info-gap can also be viewed as a form of evaluation of feasible scenarios. Through the robustness and opportuneness functions, info-gap theory does then provide a framework for structuring decision making in the face of complex uncertainties.

3.5. Sources of Information to Reduce Uncertainty

3.5.1. *Prior information*

It is evident that the better the prior information about model structures, parameter values and boundary conditions that can be specified for an application, the smaller will be uncertainty in the predictions. Forward uncertainty analysis of course depends totally on prior information. Prediction uncertainties are then purely a consequence of the uncertainties assumed about the inputs needed, including the choice of which model or models to use. The better the definition of prior distribution of uncertain quantities, the narrower the prediction uncertainties will usually be. The difficulty here is that it may then make it difficult to be secure about the definition of uncertain boundary conditions and model parameters. The effective values required to give good predictions may vary from application to application and from model structure to model structure which may then make it difficult to give good prior estimates on the basis of past experience at other sites or textbook values.

Expert opinion is then important in defining prior uncertainties and there is an extensive literature on eliciting expert opinions in the context of the Bayesian approaches to model uncertainty estimation (see for example Kurowicka and Cooke, 2006; O'Hagan *et al.*, 2006). Within the Bayesian approach (or any other learning strategy), however, the prior estimates should become less and less important as more site specific data are added, for both model drivers and for model output evaluation (Beven, 2007).

3.5.2. *Data and its deficiencies*

It is important to remember that the estimation of uncertainty, for any of the applications presented in this book, should not be considered the final goal of any study. It is, instead, a starting point in two ways. The first is as an input to a decision-making process. The second is as a baseline for the reduction of uncertainty. Learning how to constrain uncertainties is a critical issue for future improvements in flood risk management. It is an issue that involves both data and improvements to model structures. It will be difficult to draw strong conclusions about the adequacy of any model structures as hypotheses about how well the system is working cannot be made without first assessing the potential for more and better data to constrain prediction uncertainties.

Observational data for a particular application site plays two roles in inducing or constraining prediction uncertainties. The first is in terms of defining the initial and boundary conditions for a given model run; the second is in the evaluation of model outputs as part of the learning process to find appropriate models for that site. In both cases, the limitations of observational techniques become an important consideration. The first role applies equally to forward uncertainty analysis where there are no data available to evaluate the model outputs. It is common to assess the effects of parameter uncertainty in such a forward analysis, less common to also look at boundary condition uncertainty. Yet, it might be that the latter are more important in some cases (for example, in predicting flood inundation, both the estimate of upstream discharge and the representation of the flood plain topography might dominate the effects of variations in roughness in accurately predicting inundation (see Pappenberger *et al.*, 2006a)).

The data available for use in both setting boundary conditions and model evaluation inevitably have errors associated with them. These errors are commonly neglected, but might be important. It is a modelling truism that even if we had the "perfect" model structure (if such a thing existed), it would *not* provide good predictions if the boundary conditions used to drive it are subject to significant error. The difficulty then is assigning a source to the modelling errors, even for the simplest case where we would like to separate out model structure error from boundary condition errors. It turns out that even this is not simple without making strong assumptions about the nature of the errors, assumptions that are not easily checked.

3.5.3. *Data assimilation in forecasting*

The constraint of uncertainty takes on a particular importance in data assimilation for forecasting for real-time flood warnings and flood mitigation control. When data (precipitation, water levels at gauging sites, status of control gates etc.) can be made available online, then forecasts can be compared with what is actually happening. Over-predictions or under-predictions can be detected. There are a number of different schemes for real-time data assimilation to correct the forecasts, including the Kalman filter, ensemble Kalman filter and variational methods (see Section IV in this book).

One very important consideration in forecasting is the need for an adequate lead time for decision makers. There are many papers in the literature that present results for data assimilation for one-step ahead forecasts. Whether this is useful or not will depend on the scale of the system. If the forecasting time step is one hour, then for a community that could be subject to flash flooding in a small catchment even a one hour ahead warning could help save lives and damages. But on larger catchments, where at least 24 hours warning might be needed in deciding whether to deploy demountable flood defences, then a one hour ahead forecast would not be very useful, however accurate it is. So, in forecasting, we are often in the situation of using data that are available now to constrain the uncertainty in forecasts with an appropriate lead time (6, 12, 24 hours or longer). The feasible lead time will depend on the natural response time scale of the system. If forecasts at longer lead times are needed for decision making, then it will normally only be possible to achieve this by also having a forecast of the *inputs* to the system.

3.6. The Analysis Sequence

The three sets of decisions outlined above in the framework for uncertainty analysis can also be viewed as an analysis sequence. We wish to test whether we have the right modelling tools for the job (Decisions 1); evaluate the prior uncertainties in model predictions using those tools (Decisions 2); and constrain the uncertainty using whatever relevant data are available (Decisions 3). In the past, these stages have not been thought of so much as decisions, since a decision implies that there are alternatives, and methodologies tend to be presented without alternatives. In the Flood Studies Report (NERC, 1975), for example, the methodology for flood frequency estimation at ungauged sites in regions of the UK was based on the index flood method and generalised extreme value distributions. This methodology was reviewed 20 years later and in the Flood Estimation Handbook (Institute of Hydrology, 1999), it was decided that the generalised logistic distribution should be the basis for the analysis and that the estimates should be based on pooling groups of catchments rather than regions. In both cases, the methodologies were based on decisions by the researchers, resulting in a methodology for widespread use (see Chapter 8).

Each set of decisions of this type involves assumptions. What will become apparent in this book is that it is difficult to provide methodologies for which the assumptions are always valid. That suggests an onus on researchers to make the assumptions explicit in a way that allows them to be checked and compared with alternatives if necessary. The lesson is to use uncertainty estimation methodologies with care, at least as yet, until we learn more about the value of different approaches and different data types for different applications. This is, however, no excuse for not doing any uncertainty analysis at all (see Juston *et al.*, 2012; Pappenberger and Beven, 2006). Being realistic about uncertainties for model predictions that will be used for decision making might indeed change the decision made.

References

Andréassian, V., Lerat, J., Loumagne, C. *et al.* (2007). What is really undermining hydrologic science today?, *Hydrol. Process.*, **21**, 2819–2822.

Ben-Haim, Y. (2006). *Info-Gap Decision Theory 2nd Edition*, Academic Press, Amsterdam.

Berger, J. (2006). The case for objective Bayesian analysis, *Bayesian Anal.*, **1**, 385–402.

Beven, K.J. (1993). Prophecy, reality and uncertainty in distributed hydrological modelling, *Adv. Water Resour.*, **16,** 41–51.

Beven, K.J. (2002). Towards a coherent philosophy for environmental modelling, *Proc. Roy. Soc. Lond. A,* **458**, 2465–2484.

Beven, K.J. (2005). On the concept of model structural error, *Water Sci. Technol.*, **52**, 165–175.

Beven, K.J. (2006). A manifesto for the equifinality thesis, *J. Hydrology*, **320**, 18–36.

Beven, K.J. (2007). Working towards integrated environmental models of everywhere: uncertainty, data, and modelling as a learning process, *Hydrol. Earth Syst. Sc.,* **11**, 460–467.

Beven, K.J. (2009). *Environmental Modelling: An Uncertain Future?*, Routledge, London.

Beven, K.J. (2010). Preferential flows and travel time distributions: defining adequate hypothesis tests for hydrological process models, *Hydrol. Process.*, **24**, 1537–1547.

Beven, K.J. (2012). Causal models as multiple working hypotheses about environmental processes, *Comptes Rendus Geoscience, Académie de Sciences*, Paris, doi: 10.1016/j.crte.2012.01.005.

Beven, K.J. and Binley, A.M. (1992). The future of distributed models: model calibration and uncertainty prediction, *Hydrol. Process.*, **6**, 279–298.

Beven, K.J and Freer, J. (2001). Equifinality, data assimilation, and uncertainty estimation in mechanistic modelling of complex environmental systems, *J. Hydrol.*, **249**, 11–29.

Beven, K.J., Smith, P.J. and Freer, J.E. (2007). Comment on "Hydrological forecasting uncertainty assessment: incoherence of the GLUE methodology" by Mantovan, P. and Todini, E., *J. Hydrol.*, **338**, 315–318.

Choi, H.T. and Beven, K.J. (2007). Multi-period and multi-criteria model conditioning to reduce prediction uncertainty in distributed rainfall-runoff modelling within GLUE framework, *J. Hydrol.*, **332**, 316–336.

Draper, D. (1995). Assessment and propagation of model uncertainty, *J. Roy. Stat. Soc.*, **B37**, 45–98.

Draper, D. (2006). Comment on "Coherence and calibration: comments on subjectivity and 'objectivity'" by Berger and Goldstein, *Bayesian Anal.*, **1**, 423–428.

Goldstein, M. (2006). Subjective Bayesian analysis: principles and practice, *Bayesian Anal.*, **1**, 403–420.

Hall, J.W. (2003). Handling uncertainty in the hydroinformatic process, *J. Hydroinformatics*, **5**, 215–232.

Hall, J.W. (2006). Uncertainty-based sensitivity indices for imprecise probabilities, *Reliab. Eng. Syst. Safe.*, **91**, 1443–1451.

Hall, J.W, O'Connell, P.E. and Ewen, J. (2007). On not undermining the science: coherence, validation and expertise, *Hydrol. Process.*, **21**, 985–988. Discussion of invited commentary by Keith Beven (2006), *Hydrol. Process.*, **20**, 3141–3146.

Hine, D and Hall, J.W. (2010). Information-gap analysis of flood model uncertainties and regional frequency analysis, *Water Resour. Res.*, **46**, W01514.

Hornberger, G.M. and Spear, R.C. (1981). An approach to the preliminary analysis of environmental systems, *J. Environ. Manage.*, **12**, 7–18.

Institute of Hydrology (1999). *Flood Estimation Handbook*, Wallingford, UK.

Inter-governmental Panel on Climate Change (IPCC) (2007). Climate Change 2007: The Physical Science Basis. Summary for Policy Makers, WMO, Geneva.

Iorgulescu, I., Beven, K.J. and Musy, A. (2005). Data-based modelling of runoff and chemical tracer concentrations in the Haute-Mentue (Switzerland) Research Catchment, *Hydrol. Process.*, **19**, 2257–2574.

Janssen, P.H.M., Petersen, A.C., van der Sluijs, J.P. *et al.* (2005). A guidance for assessing and communicating uncertainties, *Water Sci. & Tech.*, **52**, 125–131.

Juston, J.M., Kauffeldt, A., Montano, B.Q. *et al.* (2012). Smiling in the rain: seven reasons to be positive about uncertainty in hydrological modelling, *Hydrol. Process.*, DOI: 10.1002/hyp.9625.

Klir, G. (2006). *Uncertainty and Information*, Wiley, Chichester.

Kurowicka, D. and Cooke, R. (2006). *Uncertainty Analysis with High Dimensional Dependence Modelling*, Wiley, Chichester.

Natural Environment Research Council (1975). *Flood Studies Report*, Wallingford, UK.

O'Hagan, A. (2006). Science, subjectivity and software (comment on articles by Berger & Goldstein), *Bayesian Anal.*, **1,** 445–450.

O'Hagan, A., Buck, C.E., Daneshkhah, A. *et al.* (2006). *Uncertain Judgements: Eliciting Expert's Probabilities,* Wiley, Chichester.

O'Hagan, A. and Oakley, A.E., (2004) Probability is perfect but we can't elicit it perfectly, *Reliab. Eng. Syst. Safe.*, **85**, 239–248.

Pappenberger, F. and Beven, K.J. (2006). Ignorance is bliss: or 7 reasons not to use uncertainty analysis, *Water Resour. Res.,* **42**, W05302, doi:10.1029/2005WR004820.

Pappenberger, F., Beven, K.J., Frodsham, K. *et al.* (2006b). Grasping the unavoidable subjectivity in calibration of flood inundation models: a vulnerability weighted approach. *J. Hydrol.*, **333**, 275–287.

Pappenberger, F., Matgen, P., Beven, K.J. *et al.* (2006a). Influence of uncertain boundary conditions and model structure on flood inundation predictions, *Adv. Water Resour.*, **29**, 1430–1449, doi:10.1016/j.advwatres.2005.11.012.

Romanowicz, R., Beven, K.J. and Tawn, J. (1994). "Evaluation of Predictive Uncertainty in Non-Linear Hydrological Models Using a Bayesian Approach" in: Barnett, V. and Turkman, K.F. (eds), *Statistics for the Environment II. Water Related Issues*, Wiley, Chichester, pp. 297–317.

Romanowicz, R., Beven, K.J. and Tawn, J. (1996). "Bayesian Calibration of Flood Inundation Models" in: Anderson, M.G., Walling, D.E. and Bates, P.D. (eds), *Floodplain Processes*, Wiley, Chester, 333–360.

Rougier, J. (2007). Probabilistic inference for future climate using an ensemble of climate model evaluations, *Climate Change*, **81**, 247–264.

Todini, E. and Mantovan, P. (2007). Comment on "On Undermining the Science?" by Keith Beven, *Hydrol. Process.*, **21**, 1633–1638.

van der Sluijs J.P., Craye, M., Funtowicz, S. *et al.* (2005). Experiences with the NUSAP system for multidimensional uncertainty assessment, *Water Sci. Technol.*, **52**, 133–144.

von Bertalanffy, L. (1968). *General Systems Theory*, Braziller, New York.

Zadeh, L. (2005). Towards a generalised theory of uncertainty (GTU) — an outline, *Inform. Sciences*, **172**, 1–40.

CHAPTER 4

Classical Approaches for Statistical Inference in Model Calibration with Uncertainty

R.E. Chandler
Department of Statistical Science, University College London, UK

4.1. Introduction

Many of the methods and tools described in this book rely at some stage upon the use of data to select and calibrate a model. "Statistical inference" is the collective name given to the area of statistical investigation concerned with the processes of model selection and parameter estimation, and with quantifying the uncertainty in parameter estimates.

4.2. The Method of Maximum Likelihood

In modern statistical practice, the method of maximum likelihood is something of a gold standard for inference. It can be used whenever the available data, denoted by, say, a vector $\mathbf{y}$, can be considered as drawn from some joint probability distribution and a model can be used to specify the form of this distribution. In this case, $\mathbf{y}$ can be regarded as the realised value of a vector $\mathbf{Y}$ of random variables with joint density $f(\bullet;\boldsymbol{\theta})$ say, where $\boldsymbol{\theta}$ is an unknown parameter vector; the function $f(\bullet;\bullet)$ may itself depend on observed values of other variables (note that we refer to $f(\bullet;\bullet)$ as a joint density for convenience even though, strictly speaking, this is not always correct: for example, if $\mathbf{Y}$ consists of discrete random variables, then "probability mass function" should be substituted for "density").

In this setting, the function $L(\boldsymbol{\theta};\mathbf{y}) = f(\mathbf{y};\boldsymbol{\theta})$ is called the likelihood function for $\boldsymbol{\theta}$ and $\ell(\boldsymbol{\theta};\mathbf{y}) = \log L(\boldsymbol{\theta};\mathbf{y}) = \log f(\mathbf{y};\boldsymbol{\theta})$ is the log-likelihood.

Notice the distinction between L and f: $L(\boldsymbol{\theta}; \mathbf{y})$ is a function of the parameter $\boldsymbol{\theta}$ with the data $\mathbf{y}$ held fixed, whereas $f(\mathbf{y}; \boldsymbol{\theta})$ is a function of $\mathbf{y}$ with $\boldsymbol{\theta}$ held fixed. Informally, comparing the values of $L(\boldsymbol{\theta}; \mathbf{y})$ at different values of $\boldsymbol{\theta}$ provides a means of assessing the relative compatibility of those values with the observations. In some sense, the most plausible value of $\boldsymbol{\theta}$ is that yielding the highest value of the likelihood, or equivalently (since $\ell(\boldsymbol{\theta}; \mathbf{y})$ and $L(\boldsymbol{\theta}; \mathbf{y})$ are in one-to-one correspondence) of the log-likelihood. This value of $\boldsymbol{\theta}$ (denoted by $\hat{\boldsymbol{\theta}}$, say) is called the maximum likelihood estimator (MLE). In well-behaved problems, $\hat{\boldsymbol{\theta}}$ satisfies the equation

$$\mathbf{U}(\boldsymbol{\theta}; \mathbf{y}) = \mathbf{0},$$

where $\mathbf{U}(\boldsymbol{\theta}; \mathbf{y}) = \partial\ell/\partial\boldsymbol{\theta}$ is the gradient vector of the log-likelihood, also known as the score function.

In general, if the observations can be regarded as independent at some level then their joint density can be written as a product of contributions from each observation: $f(\mathbf{y}; \boldsymbol{\theta}) = \prod_i f_i(y_i; \boldsymbol{\theta})$ say. In this case the log-likelihood is a sum of terms: $\ell(\boldsymbol{\theta}; \mathbf{y}) = \sum_i \log f_i(y_i; \boldsymbol{\theta})$.

For a wide class of models where the log-likelihood can be written in this form, the MLE and an estimate of the Fisher information (see below) can be computed efficiently using standard software. In fact, many commonly used estimation techniques can be regarded as maximum likelihood procedures for particular classes of model — for example, least squares provides the maximum likelihood estimates of regression coefficients in linear models when the errors are independent and normally distributed with constant variance. Where efficient model-specific algorithms are not available, general-purpose numerical methods must be used.

Many of the arguments justifying the use of maximum likelihood estimation are based on its behaviour for large samples, when the data are generated from the distribution with density $f(\mathbf{y}; \boldsymbol{\theta}_0)$ so that it is meaningful to refer to $\boldsymbol{\theta}_0$ as the true value of $\boldsymbol{\theta}$. In this case, the aim of any estimation procedure is usually to deliver a value as close to $\boldsymbol{\theta}_0$ as possible. In a wide class of problems it can be shown that for large samples the distribution of the MLE is approximately multivariate normal:

$$\hat{\boldsymbol{\theta}} \sim MVN(\boldsymbol{\theta}_0, \boldsymbol{\Sigma})$$

say, where $\boldsymbol{\Sigma}^{-1} = -\boldsymbol{E}\lfloor \partial^2\ell/\partial\boldsymbol{\theta}\partial\boldsymbol{\theta}^T|_{\boldsymbol{\theta}=\boldsymbol{\theta}_0} \rfloor$ is the expected second derivative matrix of the negative log-likelihood, called the Fisher information. In large samples the MLE is therefore an unbiased estimator (i.e. its expectation is equal to the target value $\boldsymbol{\theta}_0$). Moreover, the matrix $\boldsymbol{\Sigma}$ is equal to the Cramér–Rao lower bound for the covariance matrix of an unbiased estimator (Scholz, 2006): this means that the variance of any linear combination of parameter estimates is the smallest possible (Lloyd, 1984, pp. 111–112). Put simply: in large samples the use of maximum likelihood estimation often yields the smallest possible parameter uncertainty given the data available.

The asymptotic distribution of the MLE can be used to construct approximate confidence intervals and hypothesis tests for components of $\boldsymbol{\theta}$. For example, a 95% interval for the ith component θ_i is $\hat{\boldsymbol{\theta}}_i \pm 1.96\sigma_{ii}$, where σ_{ii} is the square root of the ith diagonal element of $\boldsymbol{\Sigma}$. Tests and confidence regions for subsets of parameters can also be constructed (see Cox, 2006, Chapter 6). Notice, however, that confidence intervals obtained in this way will always be symmetric about the MLE. If the likelihood function is markedly asymmetric then such intervals may be inaccurate. In this case it may be preferable to construct tests and confidence regions using the log-likelihood function itself. Specifically, suppose that the parameter vector is partitioned as $\boldsymbol{\theta} = (\boldsymbol{\psi}'\,\boldsymbol{\lambda}')'$, with target value $\boldsymbol{\theta}_0 = (\boldsymbol{\psi}_0'\boldsymbol{\lambda}_0')'$. Write $\ell(\boldsymbol{\theta};\mathbf{y}) = \ell(\boldsymbol{\psi},\boldsymbol{\lambda};\mathbf{y})$ for the log-likelihood, and consider maximising this with respect to $\boldsymbol{\lambda}$ alone for some fixed value of $\boldsymbol{\psi}$. In general, the resulting value of $\boldsymbol{\lambda}$ will depend on $\boldsymbol{\psi}$: call it $\hat{\boldsymbol{\lambda}}(\boldsymbol{\psi})$. The value of the corresponding maximised log-likelihood, $\ell(\boldsymbol{\psi},\hat{\boldsymbol{\lambda}}(\boldsymbol{\psi});\mathbf{y})$, is also a function of $\boldsymbol{\psi}$; this is called the profile log-likelihood for $\boldsymbol{\psi}$. Let $\hat{\boldsymbol{\psi}}$ be the overall MLE for $\boldsymbol{\psi}$; then the overall MLE for $\boldsymbol{\lambda}$ is $\hat{\boldsymbol{\lambda}}(\hat{\boldsymbol{\psi}})$. By definition, $\ell(\hat{\boldsymbol{\psi}},\hat{\boldsymbol{\lambda}}(\hat{\boldsymbol{\psi}});\mathbf{y})$ cannot be less than the maximised log-likelihood at any other value of $\boldsymbol{\psi}$. Therefore the quantity

$$\Lambda(\boldsymbol{\psi}) = 2\lfloor \ell(\hat{\boldsymbol{\psi}},\hat{\boldsymbol{\lambda}}(\hat{\boldsymbol{\psi}});\mathbf{y}) - \ell(\boldsymbol{\psi},\hat{\boldsymbol{\lambda}}(\boldsymbol{\psi});\mathbf{y})\rfloor$$

is always positive-valued, although we would expect $\boldsymbol{\Lambda}(\boldsymbol{\psi}_0)$ to be "small" in general, if $\hat{\boldsymbol{\psi}}$ is close to $\boldsymbol{\psi}_0$. This suggests that when $\boldsymbol{\psi}$ is unknown, a confidence region could be determined as the set of values for which $\boldsymbol{\Lambda}(\boldsymbol{\psi})$ is less than some threshold — or equivalently, as the set of values for which the profile likelihood exceeds a corresponding threshold. The quantity $\boldsymbol{\Lambda}(\psi)$ is sometimes referred to as a likelihood ratio statistic. It can be shown (see Cox, 2006, Section 6.4) that in large samples, $\boldsymbol{\Lambda}(\boldsymbol{\psi}_0)$ has approximately a chi-squared distribution with q degrees of freedom, where q is the dimension of $\boldsymbol{\psi}$. Therefore, a confidence region for $\boldsymbol{\psi}_0$ can be determined as the set $\{\boldsymbol{\psi} : \boldsymbol{\Lambda}(\boldsymbol{\psi}) < c\}$, where c is the appropriate percentile of the χ_q^2 distribution.

Equivalently, the null hypothesis $H_0 : \boldsymbol{\psi} = \boldsymbol{\psi}_0$ can be accepted if $\boldsymbol{\Lambda}(\boldsymbol{\psi}) < c$ and rejected otherwise: this procedure is called a likelihood ratio test.

A common use of likelihood ratio tests is to compare nested models: two models are said to be nested when one can be regarded as a simplified version of the other. For example, testing for the presence of a linear trend in a time series is equivalent to testing whether the slope of any trend is zero: this could be done by finding the MLEs for models with and without a trend and comparing the resulting log-likelihoods. Likelihood ratio tests cannot be used, however, to compare non-nested models. Fortunately, a variety of likelihood-related "information criteria" are available to assist with model selection in this case. Most of these criteria take the form

$$\mathbf{IC} = -2\ell(\hat{\boldsymbol{\theta}}; \mathbf{y}) + Kp,$$

where p is the dimension of $\boldsymbol{\theta}$ and K is a "penalty" designed to discourage the use of overly complex models. At some level, criteria of this form are motivated by the desire to choose the simplest model that fits the data well. If the information criteria for two models are IC_1 and IC_2 respectively, then the first model will be preferred if $\mathrm{IC}_1 < \mathrm{IC}_2$ and vice versa. Putting $K = 2$ yields Akaike's information criterion, usually denoted by "AIC": the motivation behind this is to select the model that is expected to yield the best out-of-sample predictions in some sense, taking into account parameter uncertainty (see Davison, 2003, Section 4.7). Another common choice of $K$is $\log(n)$, where n is the sample size: this leads to the Bayes information criterion (BIC), which can be justified — at least when the observations $\mathbf{y}$ are independent — using asymptotic arguments since, if the data are really generated from one of the candidate models under consideration, then this model will yield the lowest BIC with probability tending to 1 as the sample size increases.

Although maximum likelihood estimation has desirable optimality properties as outlined above, it is not a universal panacea. Some potential difficulties are as follows:

- It requires the specification of a plausible model for the joint density of the data: with complex data structures, this may not be possible. Moreover, not all models are explicitly probabilistic in nature: maximum likelihood cannot help, for example, with the calibration of non-statistical models.

- The optimality arguments are asymptotic and may not be applicable in situations where sample sizes are small. This is especially true when models contain large numbers of "nuisance parameters" — these are parameters that are necessary to complete the model specification but that are not of direct interest. In the context of testing hypotheses about $\boldsymbol{\psi}$ in the discussion above, the vector $\boldsymbol{\lambda}$ represents the nuisance parameters. Indeed, there are situations in which the number of nuisance parameters is proportional to the sample size, and in this case the method can fail completely (in the sense that the estimate $\hat{\boldsymbol{\theta}}$ does not approach $\boldsymbol{\theta}_0$ as the sample size increases).
- Likelihood-based procedures can be sensitive to outliers and data errors (although the problem here is arguably not with the use of maximum likelihood *per se*, but rather with the use of a model that, by failing to represent errors explicitly, does not represent the data structure correctly).
- No account is taken of information other than the available data. In some situations, the investigator may have prior knowledge or beliefs about the values of components of $\boldsymbol{\theta}$, and may wish to incorporate this supplemental information into the analysis. This type of situation is most easily handled in a Bayesian framework (see Chapter 5).

For further discussion of the problems that can occur with maximum likelihood estimation, and some ways of overcoming them, see Cox (2006, Chapter 7).

4.3. Estimating Functions

As noted above, the specification of a likelihood function is not always feasible, either because the model structure is too complex to derive the joint density or because the model is not specified in probabilistic terms. There is, nonetheless, a general theory of estimation that is applicable whenever a model is calibrated by solving an equation of the form

$$\boldsymbol{g}(\boldsymbol{\theta}; \mathbf{y}) = \mathbf{0}$$

for some vector-valued function $\mathbf{g}(\bullet;\bullet)$. Such an equation is called an estimating equation, and the function $\mathbf{g}(\bullet;\bullet)$ is an estimating function. To study the statistical properties of estimating equations it is necessary, as before, to consider the data $\mathbf{y}$ as realised values of random variables $\mathbf{Y}$, even if the model being fitted is not explicitly probabilistic. This should not be conceptually difficult: few models provide a perfect fit to observations, and any discrepancy between model and data (whether due to structural error in the model, measurement error in the data or some other source) can be described probabilistically if one is prepared to accept, for example, that two sets of observations, gathered under conditions that are identical in terms of the model inputs, will in general be different. In this case it is meaningful to speak of the distribution of $\mathbf{Y}$, even if we have no way of specifying the form of this distribution. In some sense then, if the model at hand provides some information about $\mathbf{y}$ then we can think of it as partially specifying the distribution for $\mathbf{Y}$, regardless of how the model was constructed.

As in the likelihood case, if the model being fitted is correct in the sense that when $\boldsymbol{\theta} = \boldsymbol{\theta}_0$ it partially or fully describes the process generating the data, then $\boldsymbol{\theta}_0$ can be thought of as the "true" value of the parameter vector. Any estimating equation such that $\boldsymbol{E}[\mathbf{g}(\boldsymbol{\theta}_0; \mathbf{Y})] = \mathbf{0}$ in this situation is called an unbiased estimating equation; the expectation here is with respect to the distribution of $\mathbf{Y}$. It can be shown that an unbiased estimating equation yields an asymptotically unbiased estimator: if $\hat{\boldsymbol{\theta}}$ now denotes the root of the equation, then the expected value of $\hat{\boldsymbol{\theta}}$ is approximately $\boldsymbol{\theta}_0$ in large samples (in fact, the same holds if the estimating equation is asymptotically unbiased after appropriate normalisation). A key requirement of any estimating equation is that it yields consistent estimators, in the sense that the value $\hat{\boldsymbol{\theta}}$ can be made arbitrarily close to $\boldsymbol{\theta}_0$ with probability close to 1, simply by gathering enough data that are not too strongly interdependent. Asymptotic unbiasedness is a necessary (but not sufficient) requirement for consistency: in general, the additional conditions required are technical but fairly unrestrictive. It can be shown that under mild conditions (relating to, for example, the differentiability of the estimating function with respect to $\boldsymbol{\theta}$) the estimator has approximately a multivariate normal distribution, for large samples and in regular problems. The mean vector of this distribution is $\boldsymbol{\theta}_0$ and its covariance matrix is $\mathbf{V} = \mathbf{H}^{-1}\mathbf{J}\mathbf{H}^{-1}$, where $\mathbf{H} = \partial\mathbf{g}/\partial\boldsymbol{\theta}|_{\boldsymbol{\theta}=\boldsymbol{\theta}_0}$ and $\mathbf{J}$ is the covariance matrix of $\mathbf{g}(\boldsymbol{\theta}_0; \mathbf{Y})$ (Davison, 2003, Section 7.2). The matrix $\mathbf{V}^{-1}$ is sometimes called the Godambe information. For use in applications, $\mathbf{H}$ can be estimated as $\hat{\mathbf{H}} = \partial\mathbf{g}/\partial\boldsymbol{\theta}|_{\boldsymbol{\theta}=\hat{\boldsymbol{\theta}}}$. Furthermore, the estimating function $\mathbf{g}(\bullet; \bullet)$ can often be written as a sum of a large number of uncorrelated contributions: $\mathbf{g}(\boldsymbol{\theta}; \mathbf{y}) = \sum_i \mathbf{g}_i(\boldsymbol{\theta}; \mathbf{y}_i)$ say, where the $\{\mathbf{y}_i\}$ represent different subsets of the data; and in such cases, $\mathbf{J}$ can be estimated as $\sum_i \mathbf{g}_i(\hat{\boldsymbol{\theta}}; \mathbf{y}_i)[\mathbf{g}_i(\hat{\boldsymbol{\theta}}; \mathbf{y}_i)]'$.

The framework regarded here is extremely general, and it is worth considering how estimating equations might be derived in some specific situations. Note first that in the context of likelihood-based inference, the score equation $\mathbf{U}(\boldsymbol{\theta};\mathbf{y}) = \mathbf{0}$ is itself an estimating equation; it can be shown that this is unbiased and that in this particular case the matrices $\mathbf{H}$ and $\mathbf{J}$ are equal, so that the general results given above reduce to those in the previous section, and the Godambe and Fisher information matrices coincide. More generally, one could consider estimating parameters using maximum likelihood for a model that is known to be incorrect but nonetheless useful. An example arises in the analysis of non-Gaussian time series data from multiple spatial locations: few tractable models are available for the joint distribution of such data, so maximum likelihood fitting can be unfeasible. An alternative is to use a likelihood function computed as though sites are independent even though, in fact, this is unlikely to be the case. Such an approach allows standard software to be used to fit models. However, if this is done then the calculated standard errors (which will be based on the Fisher information for a wrong model) require adjustment to obtain a correct assessment of parameter uncertainty. More details can be found in Chandler *et al.* (2007, Section 5.4).

An alternative way to derive an estimating equation, which makes no explicit reference at all to any joint distribution, is by minimising an objective function that represents some measure of discrepancy between data and model. This measure might represent the cumulative discrepancy between individual observations and the corresponding values output by the model, or the discrepancy between overall properties of the data and model outputs. Some widely-used examples of the former type of measure are reviewed by Smith *et al.* (2008), who refer to them as "informal likelihoods"; the latter is essentially the generalised method of moments introduced by Hansen (1982). If the discrepancy measure is denoted by $Q(\boldsymbol{\theta};\mathbf{y})$ then the corresponding estimating equation is $\mathbf{g}(\boldsymbol{\theta};\mathbf{y}) = \partial Q/\partial\boldsymbol{\theta} = \mathbf{0}$. Study of this estimating equation can help to understand the properties of a given estimation procedure. For example, Smith *et al.* (2008) discuss a measure known as the "index of agreement" and conclude on heuristic grounds that it has some undesirable properties; in fact, the theory outlined above can be used to show that minimising this quantity leads to a biased estimating equation in general, and hence that the procedure delivers inconsistent estimators. The theory can also be used to compare the asymptotic covariance matrices obtained using different discrepancy measures, and hence to choose the measure leading to the smallest parameter uncertainty.

If an estimating equation is derived by minimising an objective function $Q(\boldsymbol{\theta}; \mathbf{y})$ then, as with maximum likelihood, an alternative way to assess parameter uncertainty is via the objective function itself, taking as "plausible" any value of $\boldsymbol{\theta}$ for which $Q(\boldsymbol{\theta}; \mathbf{y})$ is below some threshold (see, for example, Wheater *et al.* 2005). This can be regarded as a means of defining a set of equifinal parameter values in the context of the Generalised Likelihood Uncertainty Estimation (GLUE) framework discussed in Chapter 6. Furthermore, the identifiability of each parameter can be investigated by plotting a profile objective function, obtained by holding the parameter fixed at each of a range of values, and optimising over the remaining parameters.

The discussion of estimating functions above has necessarily been brief; for a more comprehensive survey, including applications to the calibration of stochastic rainfall models in hydrology, see Jesus and Chandler (2011).

References

Chandler, R.E., Isham, V.S., Bellone, E. *et al.* (2007). "Space-Time Modelling of Rainfall for Continuous Simulation" in: Finkenstadt, B., Held, L., Isham, V.S. (eds), *Statistical Methods for Spatial-Temporal Systems*, CRC Press, Boca Raton.

Cox, D.R. (2006). *Principles of Statistical Inference*, Cambridge University Press, Cambridge.

Davison, A.C. (2003). *Statistical Models*, Cambridge University Press, Cambridge.

Hansen, L.R. (1982). Large sample properties of generalized method of moments estimators, *Econometrica*, **50**, 1029–1054.

Jesus, J. and Chandler, R.E. (2011). Estimating functions and the generalized method of moments, *Interface Focus*, **1**, 871–885.

Lloyd, E. (1984). *Statistics*, Volume VI, Part A of *Handbook of Applicable Mathematics* (chief editor: Ledermann, W.), John Wiley and Sons, Chichester.

Scholz, F.W. (2006). "Maximum likelihood estimation" in: Kotz, S., Balakrishnan, N., Read, C. *et al.* (eds), *Encyclopedia of Statistical Sciences 2nd Edition*, Wiley, New Jersey, pp. 4629–4639.

Smith, P., Beven, K.J. and Tawn, J.A. (2008). Informal likelihood measures in model assessment: theoretic development and investigation, *Adv. Water Resour.*, **31**, 1087–1100.

Wheater, H.S., Chandler, R.E., Onof, C.J. *et al.* (2005). Spatial-temporal rainfall modelling for flood risk estimation, *Stoch. Env. Res. Risk A.*, **19**, 403–416.

CHAPTER 5

Formal Bayes Methods for Model Calibration with Uncertainty

Jonathan Rougier
Department of Mathematics, University of Bristol, UK

5.1. Introduction

This chapter describes the Bayesian approach to assessing uncertainty, and how it can be implemented to calibrate model parameters using observations, taking account of the imperfections of the model, and measurement errors. Section 5.2 outlines the justification for the Bayesian approach, Section 5.3 outlines the Bayesian approach to model calibration, and Sections 5.4 and 5.5 discuss simple and more advanced strategies for performing the inferential calculations. There is a brief summary in Section 5.6.

5.2. Bayesian Methods

Bayesian methods provide a formal way of accounting for uncertainty, through the use of probability, and the probability calculus. Uncertainty, treated generally, is a property of the mind; it pertains to an individual, and to the knowledge that individual possesses. Many people baulk at the uncompromisingly subjective or "personalistic" nature of uncertainty. A superficial understanding of science would suggest that this subjectivity is out of place, but in fact it lies at the very heart of what makes a scientist an expert in his or her field: the capacity to make informed judgements in the presence of uncertainty (Ziman, 2000, provides a naturalistic assessment of "real" science, where the only equation in the entire book is Bayes theorem). Different hydrologists will produce different models of the same catchment, which is to say that the process of designing and constructing a model is subjective. The Bayesian approach extends this subjectivity to descriptions of uncertainty, e.g. uncertainty about the relationship between the model output and the behaviour of the actual catchment. But while model-building is something hydrologists do a lot of, thinking about uncertainty is less familiar, and seems less structured. And yet it is a vital part of any model-based analysis — we cannot make inferences about a catchment without accounting for the limitations of the model. The probabilistic approach is therefore a way of making explicit what must be happening implicitly. In requiring us to quantify our uncertainties as probability distributions, it puts these judgements into a form where they may be debated, and amended (Goldstein, 2006). Some authors have stressed that the Bayesian approach is a way to make scientific inference more objective (see, for example, Good, 1983).

The fact that these judgements are subjective, and the case for making them transparent in scientific inference, are unassailable. What we have yet to establish here is why we should do this within a probabilistic framework. The pragmatic answer is that the probabilistic approach has proved to be extremely powerful and, in conjunction with modern computational methods (particularly Monte Carlo methods), is unsurpassed in complex inferences such as data assimilation, spatial–temporal modelling, and scientific model calibration and model-based prediction (see also the many scientific applications in Liu, 2001). As these fields have developed, a consensus has emerged, and the result is that the overt subjectivity has been somewhat reduced, in the same way that a consensus on how to treat a certain aspect of a hydrologic model reduces the differences across models.

The pragmatic answer focuses on the efficacy of the probability calculus. Perhaps that is the only justification that is required. Before the advent of modern computational methods, though, the first answer would have been that there is foundational support for the probability calculus as a model for the way we reason. The probability calculus is based on three simple axioms. We suppose the existence of a set Ω, and a measure $\Pr(\cdot)$ defined on subsets of Ω. The axioms assert that $\Pr(\cdot)$ satisfies the following properties:

(1) $\Pr(A) \geq 0$;
(2) $\Pr(\Omega) = 1$;
(3) $\Pr(A \cup B) = \Pr(A) + \Pr(B)$ if $A \cap B = \emptyset$;

where $A, B \subseteq \Omega$ (see, e.g. Dawid, 1994). Why these axioms and not others? There are a number of interpretations, i.e. suggested relations between these axioms and the real world (see, e.g. Gillies, 1994). In the Bayesian interpretation, $\Pr(A)$ is an operationally-defined assessment of an individual's uncertainty about A, a view that was formalised by Bruno de Finetti in the 1930s (de Finetti, 1937, 1972). Book-length studies of this approach can be found in Savage (1972), Lad (1996), and Jeffrey (2004); de Finetti (1974) and Goldstein and Wooff (2007) adopt a more general approach in which expectation rather than probability is taken as primitive.

Imagine that we are interested in Y, the amount of rain in mm on the Met Office roof on Christmas Day 1966 (conventionally, capitals are used for uncertain quantities, and small letters for possible values). We might set $\Omega = \{Y = 0, Y = 1, \ldots, Y = 100\}$. Any subset of Ω is termed a proposition, and interpreted as the union of its components. Thus if $A = \{Y = 0, Y = 1, Y = 2\}$ then $\Pr(A) = \Pr(Y = 0 \text{ or } Y = 1 \text{ or } Y = 2)$. But how does one assess $\Pr(A)$? One operationalisation of the Bayesian approach is to think of $\Pr(A)$ as the value of v that minimises, for me (or whoever's probability is being assessed), the loss function $(v - I_A)^2$, where I_A is the indicator function of the proposition A, i.e. $I_A = 1$ if A is true, and 0 otherwise. If I was sitting at my desk with the meteorological records for the Met Office roof in front of me, I would know whether or not A was true. In this case $v = 1$ would minimise my loss if it was, and $v = 0$ if it was not. In general, however, I would likely settle on a value for v somewhere between 0 and 1: where exactly would be a quantification of how probable I thought that A was. This operationalisation in terms of a loss function is less familiar than the one in terms of betting, but perhaps more palatable to people who don't bet, or who are suspicious of betting; the betting approach is described in Jeffrey (2004), and the relationship between the two approaches is explained in Goldstein and Wooff (2007, Section 2.2).

The operational definition of probability is combined with a simple rationality principle: I would never choose a probability (or, more generally, collection of probabilities) which resulted in a loss that could be unambiguously reduced no matter what the outcome. Probabilities obeying this principle are termed coherent. It is easy to show that coherence implies the three axioms given above. For example, if I chose a value for $\Pr(A)$ that was strictly less than zero, then a value $\Pr(A) = 0$ would result in a loss that was smaller, no matter whether A turned out to be true or false; hence $\Pr(A) \geq 0$ is implied by coherence.

In order for these axioms to lead to a useful calculus, we need rules for describing how knowing the truth of one proposition would change our probabilities for others. In other interpretations of the probability axioms this is defined to be the conditional probability $\Pr(A \mid B) = \Pr(A \cap B)/\Pr(B)$, provided that $\Pr(B) > 0$. In the Bayesian approach, however, the conditional probability $\Pr(A \mid B)$ is operationally defined as the value of v which minimises, for me, the loss function

$$I_B(v - I_A)^2, \tag{5.1}$$

a definition which subsumes $\Pr(A)$, which is equal to $\Pr(A \mid \Omega)$ since $I_\Omega = 1$ with certainty. This definition of conditional probability captures the notion of describing the probability of A "supposing B to be true", in the sense that if B were not true, then $I_B = 0$ and there would be no penalty. It can then be proved that the relation

$$Pr(A \cap B) = Pr(A \mid B)Pr(B) \tag{5.2}$$

follows as a consequence of the coherence of the collection of probabilities for $A \cap B$, $A \mid B$, and B (see, e.g. de Finetti, 1972, Chapter 2). The result,

$$Pr(A \mid B) = \frac{Pr(B \mid A)\Pr(A)}{Pr(B)} \quad \text{providing that } \Pr(B) > 0, \tag{5.3}$$

which is an immediate consequence of Equation (5.2), is referred to as Bayes theorem precisely because it is a theorem: it is a consequence of the operational definition of $\Pr(A \mid B)$ and the principle of coherence.

The Bayesian approach does not assert that this is how people actually assess probabilities: it is a model for reasoning, and has the same advantages and disadvantages as models used elsewhere in science. This viewpoint is expounded in detail by Howson and Urbach (2006). For simple propositions we can usually assess $\Pr(A)$ directly, without recourse to thinking about loss functions or betting: most people seem to understand probability without having to operationalise it. For more complicated propositions, however, the probability calculus helps us to break probability assessments down into more manageable parts.

Most of us make our everyday probabilistic assessments directly. For example, when we assess Pr(rain today) we take account, informally, of the event "rain yesterday": we do not do the conditional probability calculation. In scientific applications, though, the conditional calculation has two advantages. The first is transparency. The second is slightly more convoluted.

In many applications it is possible to construct a "forward model" of the system, often deterministic, that can be used to propose candidate values for Y, the collection of quantities we want to assess. For example, suppose that Y represents the behaviour of a catchment. We could try to specify judgements about Y directly, but getting the spatial–temporal structure even approximately right is a huge challenge, and we would be unlikely to attach much confidence to the outcome. If we have a model which converts precipitation forcing into catchment behaviour then we can use the model to constrain our judgements about Y to those that are much more plausible than we could achieve on our own. Here we treat the forcing as known, for simplicity; likewise the initial conditions in the catchment.

From now on I will refer to forward models as "simulators", to avoid the heavy overloading that the word "model" has.

The catch with simulators is that they are imperfect. One manifestation of this is that they are only specified up to a vector of parameters which cannot be operationally defined in terms of the underlying system. Learning about these parameters is conventionally called calibration if formal methods are used, and *tuning* if more informal methods are used. Any representative value of the parameters is denoted θ, and the "best" value, the one that is most appropriate for the system, is denoted θ^* (a capital θ would adhere to the convention mentioned above, but it is visually intrusive). A simulator evaluation is then written $f(\theta)$, where $f(\theta^*)$ is the most informative evaluation.

Therefore before the simulator can be used to make inferences about Y, it must be calibrated, and this is where conditioning comes in. Observations of the system (which are related to Y), are used to update our judgements about the best input θ^*, which is represented in terms of the conditional distribution $\Pr(\theta^* = \theta \,|\, Z = z^{\text{obs}})$, where Z is the representation of the measurement process, and z^{obs} is the result (this is a function of θ alone, z^{obs} being specified). Probabilistic calibration uses conditional probability and Bayes theorem to "reverse" the simulator, and push the information in z^{obs} back to θ^*. The *posterior* distribution of θ^* is

$$\Pr(\theta^* = \theta | Z = z^{\text{obs}}) = \frac{\Pr(Z = z^{\text{obs}} \,|\, \theta^* = \theta)\Pr(\theta^* = \theta)}{\Pr(Z = z^{\text{obs}})}. \tag{5.4}$$

In the numerator, the probability $\Pr\,(Z = z^{\text{obs}} \,|\, \theta^* = \theta)$ is based largely on the simulator. Of course the simulator does not get us all the way from θ^* to Z, because we still have to account for the effect of its inaccuracies, and of measurement errors. But it is reasonable to expect the simulator, if it is carefully constructed, to get us most of the way there. Therefore, while there is nothing to stop us trying to assess θ^* directly, taking account informally that $Z = z^{\text{obs}}$, most people would judge that the method based on conditioning θ^* on the observations z^{obs} using Bayes theorem would be more accurate. If a single value of the "best" parameters was required (e.g. to be plugged in to a set of simulations under different forcing conditions), then the mean would be the usual choice. For a defensible treatment of predictive uncertainty, though, the uncertainty about θ^* ought to be included, and the rest of this chapter is concerned with estimating more general properties of the posterior, such as its mean vector and its variance matrix.

The next section describes a statistical framework for linking together the simulator parameters, simulator evaluations, the actual system values, and the observations, all of which is hidden inside $\Pr(Z = z \mid \theta^* = \theta)$.

5.3. Simulator Calibration in a Simple Statistical Framework

As above, let θ^* denote the (unknown) "best" values for the simulator parameters, $f(\theta)$ the simulator evaluation at parameter value θ, Y the system values, e.g. river height at various locations and various times; Z the measurements; and z^{obs} the actual measured values. While calibration is learning about θ^* using $Z = z^{\text{obs}}$, calibrated prediction is learning about θ^* and Y using $Z = z^{\text{obs}}$. This chapter focuses on calibration.

The main purpose of calibration is to assess a point value and a measure of uncertainty for the "best" values for the simulator parameters. Craig *et al.* (1997, 2001) and Goldstein and Rougier (2006) discuss the statistical approach to calibration and calibrated prediction, particularly for large problems; Kennedy and O'Hagan (2001) provide a more conventional but less scalable approach. Note that the assertion that there exists a "best" value for the simulator parameters is not clear-cut; this is discussed within the context of a more general statistical framework in Goldstein and Rougier (2004, 2009).

One feature of much of the statistical literature on computer experiments is the replacement in the inference of the simulator with an emulator, which is a statistical representation of the simulator built from an ensemble of runs at a collection of different values for θ. Emulators are useful in situations where the simulator is slow to run, but here we will assume that the simulator is quick to run, and can be evaluated every time we need to evaluate the function $\Pr(Z = z^{\text{obs}} \mid \theta^* = \theta)$. O'Hagan (2006) provides an introduction to emulators, and recent work on multivariate emulation is discussed in Rougier (2008) and Rougier *et al.* (2009).

Referring back to Equation (5.4), we need to specify the prior distribution $\Pr(\theta^* = \theta)$ for each θ, and the statistical model $\Pr(Z = z \mid \theta^* = \theta)$, for each combination of θ and z. The prior distribution quantifies our judgements about the simulator parameters before observing Z (or, more realistically, neglecting the information that $Z = z^{\text{obs}}$). It must respect the physical limitations of the simulator parameters, but it should also reflect information collected from previous observations on the catchment, or similar catchments, where such information exists. Sometimes a fairly vague specifications for $\Pr(\theta^* = \theta)$ will suffice, in situations where there is lots of information about θ^* in Z. The "classical" situation where this occurs is where the components of Z are independent conditional on θ^*. However, this is emphatically not the case with simulators, for reasons to be explained below. Therefore the choice of $\Pr(\theta^* = \theta)$ is likely to have some impact on the posterior distribution, and it is worth investing some effort in this choice or, if that is not possible or if it proves too hard, performing a sensitivity analysis by re-doing the inference for a range of choices.

We also have to specify the statistical model $\Pr(Z = z \mid \theta^* = \theta)$. This is a function of both z and θ, but for inferential purposes it is only ever evaluated at $z = z^{\text{obs}}$. The function

$$L(\theta) := \Pr(Z = z^{\text{obs}} \mid \theta^* = \theta) \tag{5.5}$$

is known as the *likelihood function*, where ":=" denotes "defined as". The likelihood function is a function of just θ, and it appears as though $\Pr(Z = z \mid \theta^* = \theta)$ is not required, except where $z = z^{\text{obs}}$. But the validity of Bayes theorem depends on the likelihood function being one particular value from a well-defined conditional distribution, and so we have to specify the whole distribution $\Pr(\text{Z} = z \mid \theta^* = \theta)$, even if we only have to compute $\Pr(Z = z^{\text{obs}} \mid \theta^* = \theta)$.

Other methods for uncertainty assessment, such as the GLUE approach (see Chapter 6), specify a "likelihood-like" function: a function of θ for which smaller values indicate a poorer fit to the data. The principle is explained in Smith *et al.* (2008). Inferences based on these "informal likelihood" measures cannot be formally interpreted as probabilities. They might, however, be informally interpreted as probabilities: within the subjective framework there is nothing to stop an individual from adopting whatever method he or she sees fit to assess his or her probabilities. The issue, though, is whether the resulting assessments are transparent enough to be authoritative. There is, perhaps, a lack of authority in a probabilistic assessment that cannot be demonstrated to be consistent with the probability calculus (however, further discussion on this is contained in the final paragraph of this section).

We could specify $\Pr(Z = z \mid \theta^* = \theta)$ directly, but in practice it is easier to induce this distribution by specifying two other quantities. We will assume, for simplicity, that the components of $f(\theta)$, Y, and Z correspond one-to-one, for every θ. The difference $Y - f(\theta)$ denotes the difference between the system values and the simulator output when evaluated at θ. This is uncertain because Y is uncertain. The difference

$$\varepsilon := Y - f(\theta^*) \tag{5.6}$$

denotes the simulator discrepancy. This is the difference between the system value and the simulator output when evaluated at the best choice of parameter values, θ^*. This is uncertain because both θ^* and Y are uncertain. Next, we need a statistical model for the measurement error, to take us from Y to Z:

$$e := Z - Y. \tag{5.7}$$

Putting these together, we have a statistical model for the distribution of Z conditional on θ^*, since

$$Z \equiv Y + e \equiv f(\theta^*) + \varepsilon + e, \tag{5.8}$$

where "$\equiv$" denotes "equivalent by definition".

In the simplest case where θ^*, ε and e are treated as probabilistically independent, a treatment that is almost always used in practice, our choices for the marginal distributions of ε and e induce the conditional distribution $\Pr(Z = z \mid \theta^* = \theta)$. For example, suppose that we decide that both ε and e are multivariate Gaussian (this might require a transformation of $f(\theta)$, Y, and Z), each with mean zero, and with variance matrices Σ^ε and Σ^e. Exploiting the fact that the sum of two independent Gaussian distributions is Gaussian, we find that

$$L(\theta) = \varphi(z^{\mathrm{obs}}; f(\theta), \Sigma^\varepsilon + \Sigma^\varepsilon), \tag{5.9}$$

where $\phi(\cdot)$ is the Gaussian Probability Density Function (PDF) with specified mean and variance.

We will adopt these choices from now on, so that our task simplifies to: (i) choosing a prior distribution for θ^* and choosing the variance matrices Σ^ε, and Σ^e; and (ii) calculating $\Pr(\theta^* = \theta \mid Z = z^{\text{obs}})$ on the basis of these choices. Strategies for doing the calculation are discussed in Sections 5.3 and 5.4.

Now we can clarify why simulators do not give rise to observations that are conditionally independent given θ^*. When simulators are inaccurate, their errors are almost always systematic across the output components. For example, if the simulator predicts a value that is too high at time t, then we would usually judge that this error will persist into time $t+1$, if the unit of time is not too large. This persistence of errors is represented by dependence among the components of the discrepancy ε. This dependence means that the observations are not conditionally independent given θ^*. The only situation in which we can ignore the discrepancy (in the sense that it has little effect on the inference) is when it is dominated by the measurement error, so that $\Sigma^\varepsilon + \Sigma^e \approx \Sigma^e$. It is fairly standard to treat the measurement errors as independent, and in this case the observations would be conditionally independent given θ^*. But if the observation errors are large, then the data are not very informative about θ^*, and so the choice of prior $\Pr(\theta^* = \theta)$ will be important.

Specifying the discrepancy variance Σ^ε is the hardest task in calibrating a simulator. Often it is ignored, i.e. implicitly set to zero. In this case each observation is treated as more informative than it actually is, and the result can be incompatible posterior distributions based on different subsets of the observations. Another example of a poor implicit choice is to minimise the sum of squared differences between z^{obs} and $f(\theta)$. This is equivalent to finding the mode of the posterior distribution in the special case where both Σ^ε and Σ^e are proportional to the identity matrix, which treats each component's "discrepancy plus measurement error" as independent and identically distributed. The problem with this choice is that it ignores the persistence of simulator errors, and so over-weights collections of observations for which these are correlated, e.g. those that are close in space or time. A crude way around this is to thin the observations, arranging it so that they are sufficiently well-separated enough that the persistence is negligible. This is an effective strategy if the observations are plentiful, and it reduces the specification of Σ^ε to a diagonal matrix: perhaps even simply $\sigma_\varepsilon^2 I$ for some scalar σ_ε and identity matrix I, if all the observations are the same type. In general, however, the diagonal components of Σ^ε will have to be set according to how good the simulator is judged to be, and the off-diagonal components according to how persistent the simulator errors are judged to be.

One particular difficulty when specifying the discrepancy variance for hydrologic simulators is the time-series nature of the outputs. The simulator transforms an irregular forcing into a time-series that reflects catchment properties. One aspect of the simulator's limitations is getting the delays slightly wrong. A minor error in the delay can result in a large additive discrepancy if the forcing changes rapidly, even though the eye would consider the fit between simulator output and observations to be good. This can be encoded into Σ^ε in the form of a contribution with a large localised correlation, but this is something of a fudge. A better strategy is to transform the simulator output and observations, effectively retaining information that is insensitive to minor errors in the delay. A recent development in statistics may be useful here, known as Approximate Bayesian Computation (ABC). This concerns stochastic simulators, and the idea is to construct likelihoods based on distances between summary statistics. This has much in common with the Informal Likelihood approach of Smith *et al.* (2008) mentioned above. Toni *et al.* (2009) is a useful reference, although the field is developing rapidly.

5.4. Simple Sampling Strategies

The posterior distribution in Equation (5.4) is very unlikely to have a closed-form solution (which would only happen if the simulator was linear and the prior $\Pr(\theta^* = \theta)$ was Gaussian). Therefore either we estimate the constant of integration, $\Pr(Z = z^{\text{obs}})$, or we use a random sampling scheme that does not require this value to be computed explicitly. For simplicity, we will assume from now on that θ^* is absolutely continuous with prior PDF $\pi_{\theta^*}(\theta)$ and posterior PDF $_{\pi\theta^* \mid Z}(\theta)$, for which Bayes theorem states

$$\pi_{\theta^* \mid Z}(\theta) = c^{-1} L(\theta) \pi_{\theta^*}(\theta) \quad \text{where } c := \int_\Omega L(\theta) \pi_{\theta^*}(\theta) \mathrm{d}\theta, \tag{5.10}$$

where $\Omega \subseteq \mathbf{R}^p$ is the parameter space, and c is the *normalising constant* (also known as the *marginal likelihood*), which we previously denoted as $\Pr(Z = z^{\text{obs}})$.

If p (the dimension of Ω) is low, say less than five, the former approach may be the best option. In this case, a deterministic numerical scheme can be used to approximate c (see, e.g. Davis and Rabinowitz, 1984; Kythe and Schäferkotter, 2004). Once this value has been computed, the posterior distribution can be summarised in terms of means, standard deviations and correlations, using further integrations. If a single point-estimate of θ^* is required, the posterior mean is usually a good choice.

If p is much larger than about five, though, this approach becomes unwieldy, because so many points are required in the integration grid. The alternative strategy is to randomly sample from the posterior directly, which can then be summarised in terms of the properties of the sample. These properties include quantiles, since the empirical distribution function of any parameter can be computed directly from the sample, and this can then be inverted. Sampling does not of itself fix the problem of a high-dimensional parameter space. In particular, n function evaluations in a random sampling scheme are likely to do a worse job than n points in an integration scheme, since in the latter these points will be chosen to span Ω. But the overriding advantage of sampling is its flexibility: we can keep going until the summaries of the posterior are accurate enough for our purposes, and we can adapt our approach as we go along. Numerical integration requires us to operate on pre-specified grids, and if we find out that an n-point grid does not deliver the required accuracy, then it is hard to reuse these points in a more accurate calculation on a new, denser grid (although Romberg integration is one possibility; see Kythe and Schäferkotter, 2004, Section 2.7).

The subject of Monte Carlo sampling is huge and still developing; Robert and Casella (2004) is a standard reference. Some of the most recent progress will undoubtedly be highly relevant to simulator calibration, such as particle Markov chain Monte Carlo methods (Andrieu *et al.*, 2010). But here we outline one of the simplest approaches, importance sampling, since it is intuitive and corresponds quite closely to much current practice. The mantra for the most basic form of importance sampling is sample from the prior, weight by the likelihood. The following steps are repeated for $i = 1, \ldots, n$:

(1) Sample $\theta^{(i)}$ from $\pi_{\theta^*}(\theta)$;
(2) Evaluate the simulator to compute $f(\theta^{(i)})$;
(3) Now compute the weight $w_i := L(\theta^{(i)})$, e.g. using Equation (5.9).

The weights describe the quality of the fit between each $f(\theta^{(i)})$ and z^{obs}, taking account both of the simulator discrepancy and the observation error. Upweighting candidates for θ^* that give a good fit to the observations is very intuitive, but this general principle gives no guidance regarding the form of the weighting function. The Bayesian formalism indicates that in this simple approach (a more sophisticated approach is described in Section 5.4) the correct choice for the weighting function is the likelihood function.

After n samples, the mean of the weights is an estimate of c:

$$c \equiv E(L(\theta)) \approx n^{-1}(w_1 + \cdots + w_n), \tag{5.11}$$

where the expectation is with respect to $\pi_{\theta^*}(\theta)$. To estimate the posterior expectation of some specified function $h(\theta)$ we compute

$$\begin{aligned} E(h(\theta)) &= c^{-1} \int_{\Omega} h(\theta) L(\theta) \pi_{\theta^*}(\theta) \mathrm{d}\theta \\ &= c^{-1} E(h(\theta) L(\theta)) \\ &\approx \frac{w_1 h(\theta^{(1)} + \cdots + w_n h(\theta^{(n)}))}{w_1 + \cdots + w_n}, \end{aligned} \tag{5.12}$$

where n^{-1} cancels top and bottom. Thus to estimate the mean vector μ we choose $h(\theta) = \theta$, and to estimate the variance of θ_i^* or the covariance between θ_i^* and θ_j^* we choose $h(\theta) = (\theta_i - \mu_i)(\theta_j - \mu_j)$, where for the variance $j = i$.

We can also estimate quantiles, by inverting the distribution function. The cumulative probability $\Pr\,(\theta^* \leq \theta' \mid Z = z^{\mathrm{obs}})$ can be estimated for any θ' by setting $h(\theta) = I_{\theta \leq \theta'}$, remembering that $I_{\theta \leq \theta'}$ is the indicator function. For simplicity, suppose that θ^* is a scalar (i.e. $p = 1$). Then the estimated posterior distribution function of θ^* has steps of

$$\frac{w_{(1)} + \cdots + w_{(i)}}{w_1 + \cdots + w_n} \tag{5.13}$$

at each $o_{(i)}$, where $o_{(1)}, \ldots, o_{(n)}$ are the ordered values of $\theta^{(1)}, \ldots, \theta^{(n)}$, and $w_{(1)}, \ldots, w_{(n)}$ are the correspondingly-ordered weights. This distribution function can be inverted to give marginal posterior quantiles; i.e. we identify that value $o^{(i)}$ for which $\Pr\,(X \leq o^{(i)} \mid Z = z^{\mathrm{obs}})$ is approximately equal to our target probability. An intuitive measure of uncertainty about θ^* is the 95% symmetric credible interval, which is defined by the 2.5th and 97.5th percentiles. It is a mistake often made in practice, but this should not be referred to as a 95% confidence interval, which is quite a different thing (see, e.g. a standard statistics textbook such as DeGroot and Schervish, 2002, Section 7.5).

5.5. Comparison with Current Practice

Current practice for calibration is diverse, but one strategy that is used frequently is to sample $\theta^{(i)}$ from the PDF $\pi_{\theta^*}(\theta)$ and then select only those samples that match the observations sufficiently well. This practice requires us to provide a metric for what we mean by "sufficiently well", and to specify a threshold at which samples are de-selected. For the metric, any number of different choices are possible, although the Nash–Sutcliffe measure seems to be the most popular in hydrology. For the threshold, it would be wise to make a choice that depends on the metric and the observations, in the sense that the samples that are selected ought to "look right" to an expert. The prescription that one ought to set this type of threshold in advance (rather than in conjunction with the observations) is part of the Neyman–Pearson orthodoxy for statistical testing, but can lead to difficulties if used uncritically; as Good (1956) notes, it may be "good advice to those whose judgement you do not trust". These issues are resolved in the Importance Sampling approach, which tells us that the metric is the likelihood, and that we should re-weight rather than select.

Selection cannot be consistent with importance sampling unless the likelihood function is zero for some set of parameter values, and constant on the complement of this set in Ω. It is hard to imagine such a likelihood function emerging from any reasonable choice for the conditional probability of Z given θ^*; hence, this approach cannot be said to generate a sample from the posterior PDF $\pi_{\theta^* \mid Z}(\theta)$. Selection *does* arise in rejection sampling, but only if performed on a stochastic basis. Rejection sampling can be used with a likelihood function such as (5.9), which can easily be majorised (see, e.g. Rougier, 2005), but importance sampling is more efficient for estimating means and variances.

5.6. More Advanced Strategies

Sampling from the prior distribution and weighting by the likelihood is very intuitive. It works well in situations where the observational data are not highly informative, so that the posterior PDF $\pi_{\theta^* \mid Z}(\theta)$ is not that different from the prior, $\pi_{\theta^*}(\theta)$. This is because the sampled values $\{\theta^{(1)}, \ldots, \theta^{(n)}\}$ do a good job of spanning the θ-values that predominate in the posterior. Typically the observational data will not be highly informative when the measurement errors are large; when the discrepancy is large (i.e. the simulator is judged to be poor); or when the simulator output is fairly constant in θ.

What about the other situation, though, when the observations are expected to be highly informative? In this case, simple importance sampling "wastes" simulator evaluations by putting many of the $\theta^{(i)}$ into regions of the parameter space that have near-zero likelihood, and thus near-zero posterior probability. In this case, it would be more efficient to find a way to sample the $\theta^{(i)}$ so that they were likely to occur in regions of high posterior probability. Importance sampling allows us to do this, and to correct for the fact that we are not sampling from $\pi_{\theta^*}(\theta)$.

Suppose we think that some specified PDF $\pi'_{\theta^*}(\theta)$ is likely to be a better approximation to the posterior PDF than is $\pi_{\theta^*}(\theta)$, and that $\pi'_{\theta^*}(\theta)$ is easy to sample from and to compute; $\pi'_{\theta^*}(\theta)$ is known as the *proposal* distribution. The PDFs $\pi_{\theta_*}(\theta)$ and $\pi'_{\theta^*}(\theta)$ must satisfy certain technical conditions: $\pi'_{\theta^*}(\theta) > 0$ wherever $\pi_{\theta^*}(\theta) > 0$, and the ratio $L(\theta)\pi_{\theta^*}(\theta)/\pi'_{\theta^*}(\theta)$ must be strictly bounded above (these are discussed further below). Where these conditions hold, the sampling strategy is:

(1) Sample $\theta^{(i)}$ from $\pi'_{\theta^*}(\theta)$;
(2) Evaluate the simulator to compute $f(\theta^{(i)})$;
(3) Now compute $w_i := L(\theta^{(i)})\pi_{\theta^*}(\theta^{(i)})/\pi'_{\theta^*}(\theta^{(i)})$.

Then we proceed as before. Note that this generalises the strategy of the previous section, where the proposal distribution was taken to be $\pi_{\theta^*}(\theta)$. We only have to compute the likelihood and the two PDFs up to multiplicative constants, since the product of these will cancel out when we take the ratio of the weights.

How do we choose a good proposal distribution? One simple approach is to approximate the posterior distribution using numerical methods. Asymptotic theory suggests that as the amount of information in z^{obs} becomes large, so the prior becomes less and less important in determining the posterior, and the likelihood function tends to a Gaussian PDF (Schervish, 1995, Section 7.4.2). Now this is unlikely to be true in the case of calibrating a hydrologic simulator: there is unlikely to be sufficient information in z^{obs}, particularly if we are realistic about the size of the simulator discrepancy. But the attraction of importance sampling is that the proposal distribution only needs to be approximately like the posterior. In fact, pragmatically, the proposal only needs to be a better approximation to the posterior than the prior is.

One simple approach is to take the proposal distribution to be a multivariate Gaussian distribution. The mean vector is the maximum likelihood value,

$$\hat{\theta} = \sup_{\theta \in \Omega} \ln L(\theta), \tag{5.14}$$

where "ln" denotes the natural logarithm, and the variance matrix is the negative of the inverse of the Hessian matrix of ln $L(\theta)$ evaluated at $\theta = \widehat{\theta}$:

$$\widehat{\Sigma} := -\left[\frac{\partial^2}{\partial\theta_i\partial\theta_j} \ln L(\theta) \mid_{\theta=\widehat{\theta}}\right]^{-1}. \tag{5.15}$$

Both $\widehat{\theta}$ and $\widehat{\Sigma}$ can be assessed in a single numerical maximisation of the log-likelihood function: this maximisation does not have to be overly precise. It is interesting to make the link back to more deterministic methods of simulator calibration, in which finding $\widehat{\theta}$, the best-fitting simulator parameter, is seen as the goal. In this respect the Bayesian approach is clearly a generalisation: one that allows us also to assess the uncertainty in our choice of simulator parameters.

The Gaussian proposal is very tractable, and seems a safe choice because $\pi'_{\theta^*}(\theta) > 0$ for all $x \in \mathbf{R}^p$, so that the condition $\pi'_{\theta^*}(\theta) > 0$ wherever $\pi_{\theta^*}(\theta) > 0$ is automatically met. But there is a risk that the posterior distribution might have thicker tails than the proposal distribution, so that the condition that $L(\theta)\pi_{\theta^*}(\theta)/\pi'_{\theta^*}(\theta)$ is strictly bounded above might not be met, which would result in one or two weights dominating. A simple and fairly robust expedient is to thicken the tails of the proposal distribution by switching from a multivariate Gaussian to a multivariate Student-t distribution with a small number of degrees of freedom (Geweke, 1989).

Why stop at just one choice of proposal distribution? This procedure can be iterated, if we have a measure of how well our proposal distribution is matching the posterior distribution; this is known as Adaptive Importance Sampling (AIS) (Oh and Berger, 1992). One simple measure is the Effective Sample Size (ESS),

$$ESS := \left\{\sum_{i=1}^{n} (\tilde{w}_i)^2\right\}^{-1}, \tag{5.16}$$

where $\tilde{w}_i$ is the normalised weight, $\tilde{w}_i \propto w_i, \sum_i \tilde{w}_i = 1$. The ESS ranges from 1, when a single w_i dominates the weights, to n, when all weights are equal. It can be shown that the efficiency of the proposal distribution is roughly proportional to the ESS (Liu, 2001, Section 2.5.3). Once we have a reasonably-sized sample from our initial proposal distribution, we can re-estimate the posterior mean and variance of θ^*, and we can use these estimates to select a more appropriate proposal distribution, typically by updating the mean vector and scale matrix of the multivariate Student-t distribution.

We can think of this approach as a pilot sample followed by the main sample, or we can iterate a few times, until the ESS has increased and stabilised at — one hopes — a value fairly close to n. We can use only the final sample to estimate properties of the posterior distribution or, if this is not sufficiently large, we can pool estimates from all the samples, weighting by the estimated standard error. We might also use the ESS to tune our choice of the degrees of freedom. Adaptive methods can sometimes be unstable, and so, if resources allow, a duplicate analysis would increase confidence in the result.

If after several iterations the ESS remains small, this suggests that our proposal distribution is a poor match to the posterior; i.e. the posterior is not unimodal and roughly bell-shaped. In this case a large n will be required, in order to raise the ESS to a reasonable value, or alternatively a more sophisticated proposal can be used, such as a mixture of multivariate Student-t distributions (Oh and Berger, 1993). Another possibility is to transform one or more of the components of θ^*. For example, if a component is strictly positive, then using a Gaussian marginal distribution for the logarithm might be better than, say, a gamma distribution for the original value. Likewise, if a component is a proportion, then a Gaussian distribution for the logit might be better than a beta distribution for the original value. Transforming θ in this way, so that $\Omega = \mathbf{R}^p$ and $\pi_{\theta^*}(\theta)$ is relatively symmetric, is a good principle in general. Another one is to arrange, as far as possible, that the transformed values will be roughly uncorrelated in the posterior. Usually, this requires more information about the simulator than we possess, but we might be able to infer such a transformation from previous studies, if they have taken care to present multivariate uncertainty estimates for the simulator parameters.

The more advanced methods in this section are really concerned with making the most efficient use of a fixed budget of simulator evaluations. The more efficient methods are a little more complicated to implement, and — taking a defensive view of the possibility of implementation errors — are only justified if the budget is a real constraint. However, such methods can make a huge difference to the accuracy of the resulting approximations. The most robust and useful recommendation is to proceed in stages: spend some of the budget on a pilot sample from the prior distribution, and evaluate a diagnostic like the ESS. If this is a reasonable fraction of n, then it is quicker and safer to spend the rest of the budget on more points sampled from the prior; otherwise, extra efficiency can be purchased with extra coding.

5.7. Summary

The Bayesian approach provides a framework within which we may assess our uncertainties. It is important to appreciate that there is no unambiguous "Bayesian answer" to a problem, and that the answer that we derive will be one that is imbued throughout by our judgements. This is obviously the case for the process of building the simulator. But it is also the case both for the formal process of describing and quantifying our beliefs about our simulator, the underlying system, and the observations; and for the calculations we implement to approximate features of our inferences regarding the simulator parameters.

To emphasise a point made at the start, once we have decided to assess our uncertainty probabilistically, we choose to adopt an approach such as estimating the posterior distribution $\pi_{\theta^* \mid Z}(\theta)$ because we think that it helps us to make better judgements about the "best" value of the simulator parameters, and it also helps us to convince other people to adopt these judgements, since our reasoning is transparent. In other words, calculating $\pi_{\theta^* \mid Z}(\theta)$ does not automatically lead us to the "right" answer regarding θ^*, but, rather, to a better and more convincing answer than we might have got through other methods.

From this point of view it is not necessary to be able to represent, definitively, our uncertainty about the relationship between the simulator, the system, and the observations in the form of a probability distribution. This is far too demanding! Many of the uncertainties are effectively ontological: there is no operational definition for the "best" value of the parameters of a simulator. In this situation some authors have proposed a more general calculus — imprecise probabilities, for example — but it is simpler (especially for large problems) to perform a sensitivity analysis within a probabilistic approach. This is perfectly acceptable, if it is remembered that probability is a model for reasoning, and not reasoning itself.

References

Andrieu, C., Doucet, A., Holenstein, R. (2010). Particle Markov chain Monte Carlo methods, with discussion, *J. R. Stat. Soc.*, Series B, **72**, 269–342.

Craig, P., Goldstein, M., Rougier, J. *et al.* (2001). Bayesian forecasting for complex systems using computer simulators, *J. Am. Stat. Assoc.*, **96**, 717–729.

Craig, P., Goldstein, M., Seheult, A. *et al.* (1997). "Pressure Matching for Hydrocarbon Reservoirs: A Case Study in the Use of Bayes Linear Strategies for Large Computer Experiments (With Discussion)" in: Gatsonis, C., Hodges, J., Kass, R. *et al.* (eds), *Case Studies in Bayesian Statistics III*, New York, Springer-Verlag, pp. 37–87.

Davis, P., Rabinowitz, P. (1984). *Methods of Numerical Integration 2nd Edition*, Academic Press Inc, Orlando, Florida,

Dawid, A. (1994). "Foundations of Probability" in: Grattan-Guinness, I. (ed.), *Companion Encyclopedia of the History and Philosophy of the Mathemetical Sciences*, Vol. 2, London, Routledge, pp. 1399–1406.

de Finetti, B. (1937). La pr'evision, ses lois logiques, ses sources subjectives, *Annals de L'Institute Henri Poincar'e* **7**, 1–68.

de Finetti, B. (1964). "Foresight, its logical laws, its subjective sources" in: Kyburg, H. and Smokler, H. (eds), *Studies in Subjective Probability*, Vol. 1, Wiley, New York, pp. 93–158, English translation by Kyburg, H.

de Finetti, B. (1972). *Probability, Induction and Statistics*, Wiley, London.

de Finetti, B. (1974). *Theory of Probability*, Wiley, London.

DeGroot, M. H., Schervish, M. (2002). *Probability and Statistics 3rd Edition*, Addison-Wesley Publishing Co, Reading, MA.

Geweke, J. (1989). Bayesian inference in econometric models using Monte Carlo integration, *Econometrica*, **57**, 1317–1339.

Gillies, D. (1994). "Philosophies of Probability" in: Grattan-Guinness, I. (ed.), *Companion Encyclopedia of the History and Philosophy of the Mathematical Sciences*, Vol. 2, London, Routledge, pp. 1407–1414.

Goldstein, M. (2006). Subjective Bayesian analysis: principles and practice, *Bayesian Anal.* **1**, 403–420.

Goldstein, M., Rougier, J. (2004). Probabilistic formulations for transferring inferences from mathematical models to physical systems, *SIAM J. Sci. Comput.*, **26**, 467–487.

Goldstein, M., Rougier, J. (2006). Bayes linear calibrated prediction for complex systems, *J. Am. Stat. Assoc.*, **101**, 1132–1143.

Goldstein, M., Rougier, J. (2009). Reified Bayesian modelling and inference for physical systems, *J. Stat. Plan. Infer.*, **139**, 1221–1239.

Goldstein, M., Woofi, D. (2007). *Bayes Linear Statistics: Theory & Methods*, John Wiley & Sons, Chichester.

Good, I. (1956). Which comes first, probability or statistics? *J. Inst. Actuaries*, **82**, 249–255 (reprinted in Good, I. (1983), pp. 59–62).

Good, I. (1983). *Good Thinking: The Foundations of Probability and its Applications*, University of Minnesota Press, Minneapolis.

Howson, C., Urbach, P. (2006). *Scientific Reasoning: The Bayesian Approach 3rd Edition*, Open Court Publishing Co, Chicago.

Jeffrey, R. (2004). *Subjective Probability: The Real Thing*, Cambridge University Press, Cambridge.

Kennedy, M., O'Hagan, A. (2001). Bayesian calibration of computer models, *J. R. Stat. Soc.*, Series B, **63**, 425–450.
Kythe, P. K., Schäferkotter, M. (2004). *Handbook of Computational Methods for Integration*, Chapman & Hall/CRC, London.
Lad, F. (1996). *Operational Subjective Statistical Methods*, John Wiley & Sons, New York.
Liu, J. (2001). *Monte Carlo Strategies in Scientific Computing*, Springer, New York.
Oh, M.-S., Berger, J. (1992). Adaptive importance sampling in Monte Carlo integration, *J. Stat. Comput. Sim.*, **41**, 143–168.
Oh, M.-S., Berger, J. (1993). Integration of multimodal functions by Monte Carlo importance sampling, *J. Am Stat. Assoc.*, **88**, 450–456.
O'Hagan, A. (2006). Bayesian analysis of computer code outputs: a tutorial, *Reliab. Eng. Syst. Safe.*, **91**, 1290–1300.
Robert, C., Casella, G. (2004). *Monte Carlo Statistical Methods 2nd Edition*, Springer, New York.
Rougier, J. (2005). Probabilistic leak detection in pipelines using the mass imbalance approach, *J. Hydraul. Res.*, **43**, 556–566.
Rougier, J. (2008). Efficient emulators for multivariate deterministic functions, *J. Comput. Graph. Stat.*, **17**, 827–843.
Rougier, J., Guillas, S., Maute, A. *et al.* (2009). Expert knowledge and multivariate emulation: the thermosphere-ionosphere electrodynamics general circulation model (TIE-GCM), *Technometrics*, **51**, 414–424.
Savage, L. (1972). *The Foundations of Statistics Revised Edition*, Dover, New York.
Schervish, M. (1995). *Theory of Statistics*, Springer, New York. Corrected second printing, 1997.
Smith, P., Beven, K., Tawn, J. (2008). Informal likelihood measures in model assessment: theoretic development and investigation, *Adv. Water Resour.*, **31**, 1087–1100.
Toni, T., Welch, D., Strelkowa, N. *et al.* (2009). Approximate Bayesian computation scheme for parameter inference and model selection in dynamical systems, *J. R. Soc. Interface*, **6**, 187–202.
Ziman, J. (2000). *Real Science: What it is, and What it Means*, Cambridge University Press, Cambridge.

CHAPTER 6

The GLUE Methodology for Model Calibration with Uncertainty

Keith Beven
Lancaster Environment Centre, Lancaster University, UK

6.1. The Origins of GLUE

The Generalised Likelihood Uncertainty Estimation (GLUE) methodology was first introduced by Beven and Binley (1992) as a way of trying to assess the uncertainty in model predictions when it is difficult to formulate an appropriate statistical error model. It was effectively an extension of the Generalised Sensitivity Analysis (GSA) approach of Hornberger and Spear (1981) which uses many model runs with Monte Carlo realisations of sets of parameters and criteria to differentiate between models that are considered acceptable (*behavioural models*) and those that are not (*non-behavioural models*). Where the posterior parameter distributions of the behavioural and non-behavioural sets are strongly differentiated, it suggests that the results are sensitive to a particular parameter. Where they are not, it suggests insensitivity. The results, of course, depend on choices of the prior ranges of the parameters and the criteria used to separate the two sets of models.

If a model is considered behavioural then we expect that the predictions of that model will have some utility in practice. But the multiple behavioural models, even after conditioning on some calibration data in this way, might well give widely different predictions. Beven (1993, 2006a, 2009) later called this the *equifinality problem*, chosen to indicate that there may be many ways of reproducing the calibration data acceptably in some sense. That does not imply that all the models give equally good results, nor does it imply that they should be expected to give such similar results in prediction. The extension incorporated into GLUE therefore was to associate each behavioural model with a likelihood weight determined during calibration and use that weight in forming distributions of predicted variables. Then, as more data become available for model evaluation, the likelihood weights can be updated (for example using Bayes equation, see Chapter 5).

GLUE has been criticised for its use of informal likelihood measures rather than the formal measures of statistical estimation (e.g. Mantovan and Todini, 2006; Montanari, 2005; Stedinger *et al.*, 2008). Some statisticians believe that formal probability is the only way of estimating uncertainties (e.g. O'Hagan and Oakley, 2004). Two points are worth making here. The first is that there is no reason why a formal statistical error model cannot be used within the GLUE approach. It is treated simply as an additional model component with its own parameters. Two examples of this are Romanowicz *et al.* (1994, 1996) in applications to rainfall-runoff modelling and flood inundation modelling. The second issue is whether the simple error assumptions that are normally made in defining formal likelihood measures can be justified given the complexity and apparent non-stationarity of real modelling errors in real applications. The use of (incorrect) formal likelihood measures will then tend to lead to over-conditioning and bias in identified parameter values. Beven *et al.*, (2007, 2008) have argued that in these circumstances the choice of a formal measure might then be incoherent in real applications since we *know* that it might lead to incorrect inferences (see also Beven, 2006a).

Of course, this is not the first time that the strong formal requirements of statistical inference have been criticised. The use of fuzzy set methods, Dempster–Shafer inference, and Laplacian rather than Gaussian error formalisms have been proposed elsewhere (see for example, Klir, 2006; Tarantola, 2005, 2006). Such methods are finding increasing use in areas relevant to flood risk management. GLUE is generalised in that it can make use of these alternative methodologies in both forming and combining likelihood measures (e.g Beven and Freer, 2001; Beven, 2009).

6.2. Model Calibration as Hypothesis Testing in GLUE

There are two fundamental functions for models in hydrology and other areas of environmental science: one is as hypotheses in formalising understanding of how a system works; another is to make predictions that will be of use in decision making (Beven, 2002; 2006b). The two functions are linked in that the testing of predictions against data can confirm (at least conditionally) the formalisation of understanding and suggest whether the predictions will be useful in practical applications. The process of testing is therefore an important part of the modelling process. It has not, to date, however, been formalised in a rigorous way. Instead, models tend to be calibrated (often optimised) to fit a set of available data; sometimes tested against another set of available data ("validation") and the results evaluated qualitatively in terms of being fit for purpose. There has been little testing of multiple models as different hypotheses about how the system works; there has been little rejection of models as inadequate (or at least such results are rarely reported). There are good reasons for this. Ideally we want to find model structures that are consistent with the available observations for the right reasons (Beven, 2001, 2006c, 2012; Kirchner, 2006). However, we know that there is the possibility of making Type I (accepting a false model) and Type II (rejecting a good model) errors because of the uncertainties associated with model inputs and the observations with which the model will be compared. We also know that there is the possibility of making Type III errors (of finding models that appear to work well in calibration, but that do not have the correct process representation or are missing critical processes when new conditions outside the range of calibration instigate a new type of response).

Beven (2006a) proposes a rejectionist strategy to hypothesis testing that extends the GLUE approach of differentiating between models that are accepted as behavioural and those that are rejected as non-behavioural. It is suggested that prior to making any model runs, limits of acceptability are defined for each observation to be used in model evaluation. The foundation for the limits of acceptability are the observation error itself, but Beven notes that other sources of error, especially input error, might affect acceptability and that some "effective observation error" might be required to set limits that avoid making Type II errors. If limits of acceptability can be defined in this way, then a non-statistical hypothesis testing framework follows quite naturally in a way that is consistent with the equifinality thesis that underlies GLUE. Essentially, a large number of feasible models are run and the models that satisfy the limits of acceptability are retained for use in prediction; those that do not are rejected. This can be applied to any model structures (and multiple parameter sets within a model structure) that can be evaluated in the same way.

6.3. Setting Prior Limits of Acceptability

Model evaluation as hypothesis testing is dependent on setting the right sort of limits of acceptability. This is not an issue in statistical identification, where a formal likelihood function follows directly from the assumptions about the structure of the errors. In most applications this means that no models are actually rejected, they are just given very low likelihoods relative to those that maximise the likelihood function. In this case, model runs that might have rather similar error variance can be given orders of magnitude different likelihoods. This is the over-conditioning that gives rise to problems in the formal approach when those error assumptions are not correct.

The extended GLUE methodology is based on a model evaluation based on limits of acceptability defined *prior* to the model being run (Beven, 2006a). The limits of acceptability should take account of observational error, incommensurability errors due to transformation or heterogeneity effects for the original measured variable, and the effects of input error. This is because the decision over whether a model is considered acceptable or not will depend on the effects of input error as processed by the model. Work has already been carried out on allowing for measurement errors (e.g. Blazkova and Beven, 2009; Freer *et al.*, 2004; Liu *et al.*, 2009) but input error is a greater problem. Within GLUE, all models that provide predictions consistently within that range of acceptability can be retained for use in prediction; those that do not are rejected. Evaluation on multiple criteria is easily handled in this framework, as is allowance for non-stationarities and temporal or spatial correlations in prediction errors. Different types of model diagnostics, including soft information and tracer data, can also be used in such evaluations. These evaluations may result in all the models being rejected (if no explicit compensating additive error component is included), but examination of where and how the models fail may lead to suggestions for improvements to either the model itself or to a reconsideration of the input and/or observational data and effective observation errors being used. We expect such diagnostics might also have a role in revealing the potential for compensatory error components and model improvements to developers and users (see Figure 6.1 in an application of dynamic Topmodel to the Brue catchment). This appears to be a useful framework within which to develop novel model diagnostic techniques.

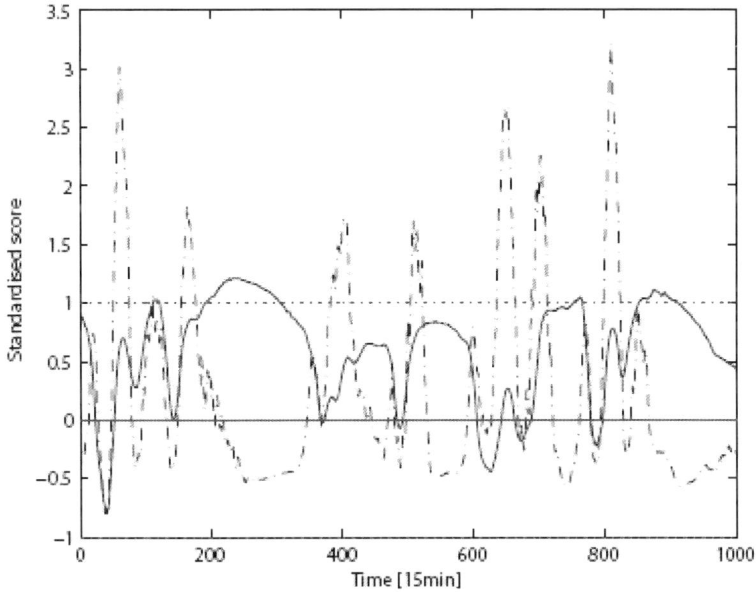

Figure 6.1. Plots of standardised scores relative to limits of acceptability (range −1 to 1) for 2 different parameterisations of dynamic TOPMODEL.

Ideally, we would hope acceptable models would be satisfactory for all of the specified limits, but this might not be possible. In an analogous way to the tails of distributions in statistics, there might be some outlier observations that might be difficult for any model to satisfy, even after allowing for input errors. Experience suggests, however, that the deviations of model predictions outside the limits of acceptability will normally be structured and, in particularly, have non-stationary bias (e.g. a tendency to over-predict on rising limbs of hydrographs and under-predict on falling limbs or vice versa). This has led to work on error diagnostics that attempts to analyse these deviations and go beyond the manifesto ideas to take more explicit account of the non-stationarities in bias and range of modelling errors in prediction. An algorithm for doing this has already been developed within the extended GLUE approach and is the subject of two papers (Blazkova and Beven, 2009; Liu *et al.*, 2009). The results will feed back into the dialogue on potential model structures with experimentalists.

6.4. The Issue of Input Error

We have already noted that the testing of models as hypotheses depends crucially on input error. It is a modelling aphorism that even the "true" model (if it existed) would not necessarily give acceptable predictions if driven with inadequate inputs. Thus defining the limits of acceptability for hypothesis testing, must take some account of input error so as to avoid Type II errors. Unfortunately, in hydrological models, input errors are not purely random or aleatory but involve non-stationary bias effects from event to event and within events. It is consequently difficult to propose a suitable model structure to represent input errors. Interactions with the model structure are then implicitly included in the range of outputs.

There is one obvious approach to take to the input error problem, which is to run multiple possible input realisations though every model realisation. For all but the very simplest models, however, the number of degrees of freedom that this strategy involves would make the computational cost prohibitive. An alternative, computationally feasible, strategy can be formulated as follows:

1. Define one or more representative models.
2. Run different plausible input scenario realisations through those models to evaluate the range and distribution of errors to be expected, purely on the basis of input error (a similar strategy has been used before by Pappenberger *et al.*, 2005, in the European Flood Forecasting System project).
3. Combine these errors with the observation and commensurability errors for the variables used in model evaluation, weighted by their distribution to define limits of acceptability.

This assessment of the effects of input error can then be done prior to running and evaluating the performance of multiple model realisations using a full set of Monte Carlo realisations. There remains the question of how to define plausible input errors, for rainfalls, antecedent conditions, channel geometry, representation of floodplain geometry and infrastructure, etc., in flood risk assessment.

6.5. Prediction Uncertainty Using Likelihood Weights

Within the extended GLUE methodology, the limits of acceptability provide a means of defining the set of behavioural models. There is then the possibility of assigning different weights to the predictions of each of those models in forming a distribution function of the model outputs (note that this can include a statistical model of the errors if the user wishes and is happy that the assumptions are satisfied). An obvious form of likelihood measure is the type of triangular or trapezoidal fuzzy measures suggested in Beven (2006a) in which a weight is defined for each observation dependent on the departure of the prediction for a particular model from that observation. The weights from each observation can then be combined in different ways (multiplicative, weighted addition, fuzzy union or fuzzy intersection, etc.) with any prior weight associated with that model to give a posterior weight after evaluation, over all the available observations. Similar weights can be defined for evaluations using the types of soft information described above.

Given the posterior likelihood values, the predictions from each realisation can then be weighted by the associated likelihood value to calculate prediction quantiles as:

$$P(\hat{Z}_t < z_t) = \sum_{i=1}^{i=N_B} \{L[M(\underline{\Theta}_i, \underline{I})] \mid \hat{Z}_t < z_t\}, \tag{6.1}$$

where $\hat{Z}_t$ is the value of variable Z at time t simulated by model i. Within this framework, accuracy in estimating such prediction quantiles will depend on having an adequate sample of models to represent the behavioural part of the model space. This, however, is a purely computational problem that will depend on the complexity of the response surface within the parameter space, the dimensionality of the parameter space, and the number of model structures to be considered. Iorgulescu *et al.* (2005, 2007), using a similar procedure, identified 216 behavioural models from two billion simulations in a 17 parameter space. These were not all in the same area of the space, and showed rather different predictions of the behaviour of the system when considered in terms of effective mixing volumes.

6.6. Dealing with Multiple Models

The GLUE concepts of equifinality and likelihood weighting can easily be extended to consider multiple model structures in a way (somewhat) analogous to Bayesian model averaging, providing that each of the model structures can be evaluated in the same way (i.e. that a consistent likelihood measure can be defined for all models). In GLUE however, unlike Bayesian model averaging, there is no attempt to try to optimise a combination of different models in fitting the data since it is expected that such a combination might be over-conditioned in calibration and prove to give non-optimal predictions. Instead, each realisation of parameter sets within each model structure is run and evaluated. It will prove to be either behavioural (within the defined limits of acceptability) or will be rejected as non-behavioural. Likelihood weighted Cumulative Density Functions (CDFs) of the predicted variables can be formulated in the same way as before, but may now include results from multiple model structures as well as different parameter sets. This may give some indication about whether one model structure (as a hypothesis of system response) performs better than another; it may suggest that there is little to choose between different model structures. Note that the model structures need not have the same parameters — the models are sampled in their own parameter spaces; it is only the predicted variables that need to be the same.

6.7. The Meaning of GLUE Uncertainty Limits

There are thus a wide variety of ways in which to approach the evaluation and conditioning of models within the GLUE framework. The choice of a measure will be generally a subjective choice, but argued and reasonable for the model purpose. The resulting prediction quantiles will therefore also be dependent on this choice. Thus any prediction bounds produced using the GLUE methodology are conditional on the choices made about the range and distribution of parameter values considered, the model structure or structures considered, the likelihood measure or measures used in defining belief in the behavioural models, and the way of combining the different measures. These are all subjective choices but must be made explicit (and can be debated or justified if necessary).

The prediction bounds are then taken from the quantiles of the cumulative likelihood weight distribution of predictions over all the behavioural models as in (6.1). They may be considered as empirical probabilities (or possibilities) of the set of behavioural model predictions. They have the disadvantage that unless a formal error model is used (where the assumptions are justified) they will not provide formal estimates of the probability of estimating any particular observation conditional on the set of models; they have the advantage that the equifinality of models as hypotheses, non-stationarities in the residual errors, and model failures are more clearly revealed. They reflect what the model can say about the response of the system, after conditioning on past data. It is then up to the user to decide whether the representation is adequate or not.

Experience suggests that for ideal cases where we can be sure that the model is a good representation of the system (unfortunately, in the case of environmental systems, this is normally only the case in hypothetical computer experiments) the GLUE methodology with its implicit treatment of residuals, can provide good bracketing of observations. The effects of strong input error and model structural error, however, may mean that it is simply not possible for the range of responses in the model space to span the observations consistently. This is, nevertheless, valuable information. Since the residual errors in such cases are not necessarily random or stationary, it may not be appropriate to represent them as a random error model.

6.8. Conclusions

This final point about the possible epistemic (rather than aleatory) nature of the errors is the reason why the more general approach to uncertainty estimation that is provided by the GLUE methodology might be valuable in real applications. The classical and Bayesian statistical approaches to model calibration make the assumption that the errors can be treated as aleatory (perhaps after some transformation or allowance for model discrepancy) and that, therefore, every residual will be informative in model calibration. As noted earlier, some hydrologists have taken a hard line that this is the only way to approach model calibration in an objective way (Mantovan and Todini, 2006; Stedinger *et al.*, 2008). But their arguments are based on hypothetical examples, when every residual is informative by prior assumption. It does not follow that this is also the case in real examples when there are many explanations for why particular residuals might be disinformative (Beven, 2006a; Beven *et al.*, 2009). Thus the question of the real information content of a set of residuals in the face of epistemic input, commensurability and model structural errors remains open. The GLUE methodology as currently used does not provide more than an interim methodology to allow for the effects of such errors. Improved methods for evaluating information content and uncertainty estimation for real applications with complex errors are required in the future.

References

Beven, K.J. (1993). Prophecy, reality and uncertainty in distributed hydrological modelling, *Adv. Water Resour.*, **16**, 41–51.

Beven, K.J. (2001). How far can we go in distributed hydrological modelling?, *Hydrol. Earth Syst. Sc.*, **5**, 1–12.

Beven, K.J. (2002). Towards a coherent philosophy for environmental modelling, *Proc. Roy. Soc. Lond.*, **A458**, 2465–2484.

Beven, K.J. (2006a). A manifesto for the equifinality thesis, *J. Hydrol.*, **320**, 18–36.

Beven, K.J. (2006b). The Holy Grail of scientific hydrology: $Q_t = (S, R, \Delta t)A$ as closure, *Hydrol. Earth Syst. Sci.*, **10**, 609–618.

Beven, K.J. (2006c). On undermining the science?, *Hydrol. Process. (HPToday)*, **20**, 3141–3146.

Beven, K.J. (2009). *Environmental Modelling: An Uncertain Future?*, Routledge, London.

Beven, K.J. (2012). *Rainfall-Runoff Modelling — The Primer 2nd Edition,* Wiley-Blackwell, Chichester.

Blazkova, S. and Beven, K.J. (2009). A limits of acceptability approach to model evaluation and uncertainty estimation in flood frequency estimation by continuous simulation: Skalka catchment, Czech Republic, *Water Resour. Res.*, **45**, W00B16, doi: 10.1029/2007WR006726.

Beven, K.J. and Binley, A.M. (1992). The future of distributed models: model calibration and uncertainty prediction, *Hydrol. Process.*, **6**, 279–298.

Beven, K.J. and Freer, J. (2001). Equifinality, data assimilation, and uncertainty estimation in mechanistic modelling of complex environmental systems, *J. Hydrol.*, **249**, 11–29.

Beven, K.J., Smith, P.J. and Freer, J.E. (2007). Comment on "Hydrological forecasting uncertainty assessment: incoherence of the GLUE methodology" by Mantovan, P. and Todini E., *J. Hydrol.*, **338**, 315–318, doi: 10.1016/j.jhydrol.2007.02.023.

Beven, K.J., Smith, P.J. and Freer, J.E. (2008). So just why would a modeller choose to be incoherent?, *J. Hydrol.*, **354**, 15–32.

Freer, J.E., McMillan, H., McDonnell, J.J. *et al.* (2004). Constraining dynamic TOPMODEL responses for imprecise water table information using fuzzy rule based performance measures, *J. Hydrol.*, **291**, 254–277.

Hornberger, G.M. and Spear, R.C. (1981). An approach to the preliminary analysis of environmental systems, *J. Environ. Manage.*, **12**, 7–18.

Iorgulescu, I., Beven, K.J. and Musy, A. (2005). Data-based modelling of runoff and chemical tracer concentrations in the Haute-Mentue (Switzerland) research catchment, *Hydrol. Process.,* **19**, 2257–2574.

Iorgulescu, I., Beven, K.J. and Musy, A. (2007). Flow, mixing, and displacement in using a data-based hydrochemical model to predict conservative tracer data, *Water Resour. Res.*, **43**, W03401, doi: 10.1029/2005WR004019.

Kirchner, J.W. (2006). Getting the right answers for the right reasons: linking measurements, analyses and models to advance the science of hydrology, *Water Resour. Res.*, **42**, W03S04, doi: 10.1029/2005WR004362.

Klir, G. (2006). *Uncertainty and Information*, Wiley, Chichester.

Liu, Y., Freer, J.E., Beven, K.J. *et al.* (2009). Towards a limits of acceptability approach to the calibration of hydrological models: extending observation error, *J. Hydrol.*, **367**, 93–103, doi: 10.1016/j.jhydrol.2009.01.016.

Mantovan, P. and Todini, E. (2006). Hydrological forecasting uncertainty assessment: incoherence of the GLUE methodology, *J. Hydrol.*, **330**, 368–381.

Montanari, A. (2005). Large sample behaviors of the generalized likelihood uncertainty estimation (GLUE) in assessing the uncertainty of rainfall-runoff simulations, *Water Resour. Res.*, **41**, W08406, doi: 10.1029/2004WR003826.

O'Hagan, A. and Oakley, A.E. (2004). Probability is perfect but we can't elicit it perfectly, *Reliab. Eng. Syst. Safe.*, **85**, 239–248.

Romanowicz, R., Beven, K.J. and Tawn, J. (1994). "Evaluation of Predictive Uncertainty in Non-Linear Hydrological Models using a Bayesian Approach", in: Barnett, V. and Turkman, K.F. (eds), *Statistics for the Environment II. Water Related Issues*, Wiley, Chichester, pp. 297–317.

Romanowicz, R., Beven, K.J. and Tawn, J. (1996). "Bayesian Calibration of Flood Inundation Models", in: Anderson, M.G., Walling, D.E. and Bates P.D. (eds), *Floodplain Processes*, Wiley, Chichester, pp. 333–360.

Stedinger, J.R., Vogel, R.M., Lee, S.U. *et. al.* (2008). Appraisal of the generalized likelihood uncertainty estimation (GLUE) method, *Water Resour. Res.* **44**, 55–63.

Tarantola, A. (2005). *Inverse Problem Theory and Model Parameter Estimation*, Society for Industrial and Applied Mathematics, Philadelphia, PA.

Tarantola, A. (2006). Popper, Bayes and the inverse problem, *Nat. Phys.*, **2**, 492–494.

SECTION III

UNCERTAINTIES IN FLOOD MODELLING AND RISK ANALYSIS

CHAPTER 7

Uncertainty in Rainfall Inputs

R.E. Chandler, V.S. Isham and P.J. Northrop
Department of Statistical Science, University College London, UK

H.S. Wheater[†] and C.J. Onof
Department of Civil and Environmental Engineering, Imperial College London, UK
[†] *Global Institute for Water Security, University of Saskatchewan, Canada*

N.A. Leith
Risk Management Solutions Ltd., UK

7.1. Introduction

At some level, precipitation is the primary driver of almost all flood events, with the exception of coastal floods. Flooding is usually a direct response to rainfall experienced over a period ranging from minutes (in the case of urban runoff) to months (for some groundwater flooding), although the role of precipitation may also be indirect as in the case of floods caused by snowmelt.

One approach to flood risk assessment is to consider the statistics of observed flows, typically peak discharge. However, this assumes implicitly that neither the climate nor the underlying catchment characteristics will change over the period of interest; this may be unrealistic. In addition, more complete information for a particular location is often required, for example, to estimate flood inundation volumes, or for the design of flood storage and attenuation systems. Thus for many applications, not least for the study of the effects of land use and climate change, flood risk management requires the ability to characterise the rainfall inputs to a system and to understand the system response to these inputs. This chapter deals with the first of these issues; aspects of the second are dealt with elsewhere in this volume.

Our focus is on the assessment and management of flood risk, and the associated design of engineering solutions. The intended lifetime of such solutions is typically on the order of decades, where their design requires consideration of long-term future scenarios of precipitation and system response. The frequency of events of interest will vary, from relatively frequent (perhaps 1 in 10 years) for agricultural land, to perhaps 1 in 100 years for the protection of towns from fluvial flooding, and 1 in 10,000 years or more for dams where lives are potentially at risk from a breach. The methods described in this chapter are not intended for use in applications such as real-time flood forecasting, where the timescales of interest typically range from hours to days. Rainfall inputs for such short-term applications are usually derived from observed precipitation, based on rain gauge networks and/or rainfall radar, in conjunction with numerical weather prediction or radar rainfall propagation. Similarly, snowmelt is not considered here.

A common perception is that floods are a direct result of intense rainfall over a short period of time, as in the Boscastle flood in the UK in August 2004 (Met Office, 2005). However, as noted above, the relevant duration of rainfall can vary from minutes to months. For example, the widespread groundwater flooding in the UK in winter 2000–2001 was associated with exceptional rainfall totals for the period September 2000 to May 2001, with two-month totals exceeding a return period of 200 years (Finch *et al.*, 2004). Rainfall properties must thus be considered in the context of the response of the underlying hydrological system, which depends on catchment properties (e.g. soils, land use, geology) and spatial scale. Precipitation is also important in establishing catchment states, i.e. the antecedent conditions prior to a specific storm event. Heavy rainfall on a dry catchment may generate only limited runoff compared with less intense rainfall on a wet catchment. For example, the extreme severity of the UK floods of summer 2007 (Pitt, 2008) arose due to intense rainfall on catchments that were already saturated from the wettest summer in England and Wales since records began.

A key feature of the rainfall process is its variability in space and time. This variability depends on storm type, and ranges from highly localised intense rainfall from convective thunderstorms, to a more uniform spatial and temporal distribution in widespread frontal systems. The importance of rainfall variability for flood risk depends on the underlying catchment response in a complex manner (Segond *et al.*, 2007). In general, as catchment response time increases, the hydrological response increasingly dampens the effects of rainfall spatial and temporal variability. Hence small, fast responding urban catchments are the most sensitive, requiring temporal resolution on a timescale of minutes, and large rural catchments are the least sensitive, particularly where the underlying geology is permeable. However, as catchment scale increases, there is increasing likelihood that the spatial extent of a precipitation event is less than the catchment scale, in which case location effects become important. Segond *et al.* conclude that in general for rural catchments in the UK, hourly to daily resolution of rainfall is required, and suggest guidelines for spatial monitoring density of 16 rain gauges for 1000 km^2 and 4–7 for 80–280 km^2. However, these criteria would not be adequate for convective thunderstorm events. For urban systems, sub-hourly data are needed, with a spatial resolution of a few kilometres.

Historically, most approaches to quantifying flood risk have involved assessing the response of hydrological systems to "design storms" representing, in idealised form, the structure of rainfall events that are judged to be extreme in some sense — for example, such that the resulting storm total is expected to be exceeded once every 100 years on average. A design storm is typically characterised by a duration, an associated depth, and a temporal profile (NERC, 1975) This approach dates back at least as far as the 1940s (Chow, 1964) and subsequently became the cornerstone of national practice in the UK and elsewhere (e.g. Institute of Hydrology, 1999, NERC, 1975). However, the approach has several important limitations. Firstly, flood response generally depends on the temporal distribution of precipitation, and the use of a fixed temporal profile may fail to capture important aspects of system performance; this led Onof *et al.* (1996) to propose the use of stochastic generation of storm rainfall time series for flood design. Secondly, many factors other than rainfall contribute to determining whether or not a particular storm will give rise to flooding. These include antecedent catchment state in particular, as discussed above. The relationship between flood probability and rainfall probability is therefore complex, and in practice is typically represented in event-based analyses by simple empirical relationships between the frequency of storm depth (for a specified duration) and flood peak. However, these relationships may be inappropriate for other flood characteristics (such as flood inundation volumes and temporal properties), and would certainly be expected to change in an altered climate. To understand flood risk fully, it is therefore necessary to consider the temporal sequence of rainfall inputs rather than considering individual events in isolation. With the advent of modern computing power, it is now feasible to address this by running long time series of rainfall and other weather data through hydrological models; this is referred to as "continuous simulation", and can be regarded as accounting for uncertainty in antecedent conditions by representing them explicitly. Further discussion of event-based and continuous simulation modelling can be found in Chapter 9.

As noted above, future precipitation scenarios are required to inform the long-term management of flood risk. At some level, the dominant source of uncertainty in such scenarios is the natural variability in the rainfall process itself, which is much higher than for most other weather variables. For example, in a 50-year time series of daily rainfall from Heathrow in the UK, an analysis of variance reveals that just 0.3% of the overall variation can be attributed to seasonality as represented by the month of the year, rising to 5% when the data are aggregated to a monthly timescale. This is typical in our experience, and not only for temperate climates. For comparison, we have obtained corresponding figures of around 3–6% and 20–30% respectively when analysing wind speed time series at individual locations in northern Europe; and figures of around 70% and 90% for temperatures. Although seasonality is only one of the factors that might be expected to affect rainfall, these figures illustrate that the systematic structure in rainfall sequences is very weak compared with other variables. To account for this, it is necessary to think in terms of probability distributions when describing rainfall time series.

For the purpose of flood risk management, another source of uncertainty relates to future changes in the climate system (this is discussed in more detail in Chapter 14). From the previous paragraph, it is tempting to conclude that this is largely irrelevant since the natural variability of rainfall will dominate any systematic climate change signal. This reasoning is incorrect, however, since even small shifts in the mean of a distribution can have surprisingly large effects on aspects relevant to flood risk assessment, such as the probability of exceeding some high threshold. Furthermore, it is possible that rainfall variability will also alter in the future, and this also affects risk. For an illustration of how risk can be sensitive to changes in rainfall regime that are relatively small compared with the associated uncertainty, see Cameron *et al.* (2000a).

Elsewhere in this volume, substantial attention is paid to the treatment of model parameter uncertainty. As far as rainfall is concerned, this is tied to the availability and quality of data: for example, long time series of daily data are often available (although care needs to be taken to ensure their homogeneity and quality — see Yang *et al.* 2006, for example) so that daily rainfall model parameters can be estimated reasonably precisely. Sub-daily records tend to be less extensive however, and some classes of sub-daily models are difficult to calibrate; hence parameter uncertainty is more of an issue here. At very fine space and timescales, rainfall measurement becomes difficult using standard techniques: our experience is that in the UK, the useful temporal resolution of data from a 0.2 mm tipping-bucket rain gauge is around 15 minutes (this is based on analyses of times between bucket tips and the resulting discretisation of the rainfall time series at finer timescales). If data are required at finer resolution then it becomes necessary to rely on indirect methods, such as drop-counting rain gauges or radar, in which rainfall is estimated based on measurements of other quantities; this estimation introduces further uncertainty. Moreover, some measurement techniques are unsuitable for use in the kinds of applications considered in this chapter because continuing technological developments make it impossible to assemble data sets that can be regarded as homogeneous over time. In such situations, one could try to calibrate the rainfall estimates against some kind of "ground truth", but this introduces further problems because the statistical properties of the recorded sequences can be sensitive to the calibration procedure. An example of this, in relation to radar–rain gauge calibration, can be found in Chandler *et al.* (2007).

When calibrating hydrological models of a system, missing rainfall data can be an additional source of uncertainty. The most obvious example is when a few observations, or a block of observations, are missing from an otherwise complete record. Other common situations can also be regarded as "missing data" problems, for example, when time series from several locations have different lengths (in this case, the shorter series can be regarded as long series with a lot of missing data) or, in extreme cases, where models require rainfall inputs at specified locations (e.g. the nodes of a regular spatial grid) but no observations are available at these locations. If the absence of data values is problematic, they are commonly "filled in". Where gridded data are required, for example, it is common to interpolate observations from a network of rain gauges onto the grid. Interpolation should, however, be used with care in the kinds of applications considered here: it is a form of smoothing, and reduces variability. Additionally, it introduces inhomogeneities since the reduction of variability depends on the distance from a grid node to the nearest gauge. In Section 7.2.3 below, we present an alternative that can be used to deal with any of the missing data scenarios described here.

The remainder of the chapter discusses event-based modelling and continuous rainfall simulation in more detail. Sections 7.2.2 to 7.2.4 focus primarily on the use of probabilistic models to represent rainfall variability, possibly incorporating projections of future climate change; parameter uncertainty is discussed where it is an issue. Section 7.2.5 gives a brief discussion of how uncertainty in rainfall inputs is propagated through into flow estimates; and Section 7.2.6 highlights some current unresolved issues.

7.2. Event-based Modelling

To motivate the discussion in this section, consider an idealised problem relating to the construction of flood defences that will fail under conditions that arise on average every T years. Assuming that failures occur independently from year to year (so that the probability of failure in any one year is $1/T$), the probability that the defences fail within L years of their construction is $1 - (1 - 1/T)^L = p$, say. Rearranging this, we find that to achieve a failure probability of p within a design lifetime of L years, it is necessary to design for conditions that arise on average every $T = [1-(1-p)^{1/L}]^{-1}$ years. For small p, this is approximately equal to L/p. If, for example, we require a failure probability of at most 0.1 within a design lifetime of 20 years, then we must design for events that occur roughly every 200 years on average. Although this example is idealised, it shows that in design problems there may be a real need to consider event magnitudes that have not been observed during the relatively short period for which relevant data are available. It is therefore necessary to extrapolate well beyond the range of the available data into the far tails of the distribution of possible values, and to quantify the often substantial uncertainty in this extrapolation. In this section we describe the most commonly used modern techniques for achieving this.

7.2.1. *Extreme value theory*

Suppose that we are interested in rainfall at a particular timescale, perhaps hourly, six-hourly or daily totals. The general aim is to make inferences about future large rainfall totals, based on historical data. One approach, reviewed in Section 7.2.3, achieves this by extrapolating from a model fitted to all available data. Unless the particular model used can be justified by some physical theory, however, extrapolation from central properties to unobserved tail properties can be questionable. An alternative approach uses only data judged to be extreme, for example the largest values in each calendar year. Indeed, in some applications only extreme data are available. Extreme value theory provides an asymptotic justification for particular families of models for extreme data, analogous to the way that the central limit theorem suggests the normal distribution as an approximate model for averages of large numbers of random variables. Coles (2001) gives an accessible introduction to modern statistical methods for extremes.

To explain the theoretical basis for extreme value models we consider the simplest situation: a sequence Y_1, $Y_2, \ldots$ of Independent, Identically Distributed (iid) continuous random variables representing, say, daily rainfall totals (the iid assumption will be relaxed later). Define $M_n = \max(Y_1, \ldots, Y_n)$, the largest daily total over a block of n days. On studying the possible behaviour of M_n as n tends to infinity (Fisher and Tippett, 1928) it emerges that, if a non-degenerate limiting distribution for M_n exists, then it must be a member of the Generalised Extreme Value (GEV) family, with distribution function

$$G(x) = P(M_n \leq x) = \begin{cases} \exp\left\{-\left[1+\xi\left(\dfrac{x-\mu}{\sigma}\right)\right]^{-1/\xi}\right\} & \text{when } \xi \neq 0 \\ \exp\left\{-\exp\left[-\left(\dfrac{x-\mu}{\sigma}\right)\right]\right\} & \text{if } \xi = 0, \end{cases}$$

where $1 + \xi(x - \mu)/\sigma > 0$ and $\sigma > 0$. This suggests the GEV(μ, σ, ξ) distribution as an approximate model for maxima of long sequences of iid random variables. For example, we could set $n = 365$ to model annual maxima. The parameter ξ determines the shape of the distribution, while μ and σ are location and scale parameters respectively. In terms of extrapolation, the value of ξ is crucial: if $\xi < 0$ the distribution of M_n is bounded above at $\mu - \sigma/\xi$, while increasing positive values of ξ correspond to an increasingly heavy upper tail.

Using block maxima is potentially wasteful of information about large rainfall values. One possible improvement is to use the $r(>1)$ largest observations within a block. Alternatively, if the entire sequence Y_1, $Y_2, \ldots, Y_m$ is available it will often be more informative to model the number and magnitude of exceedances of some large threshold u. Suppose that the limiting distribution of M_n is GEV(μ, σ, ξ). Then, for sufficiently large u, given that there is an exceedance $Y > u$, the magnitude of the exceedance $Z = Y - u$ has approximately a generalised Pareto, $\mathrm{GP}(\sigma_u, \xi)$ distribution, where the scale parameter $\sigma_u = \sigma + \xi(u - \mu)$ (Pickands, 1975). Under the assumed independence of $Y_1, Y_2, \ldots,$ the number of exceedances has a binomial distribution with parameters m and $\mathrm{P}(Y > u)$. Graphical methods can be used to inform the choice of u (Davison and Smith, 1990). The extreme value models outlined above can be viewed as special cases of a point process model for observations over the threshold u (see Smith, 1989). The natural parameterisation of the latter model, via the GEV parameters μ, σ and ξ, is invariant to changes in u, a property which is advantageous if a time-varying threshold is necessary.

7.2.2. *Accounting for dependence*

In practice $Y_1, Y_2, \ldots$ are often unlikely to be independent. Suppose instead that they form a stationary sequence of dependent random variables. Provided that a weak condition limiting long-range dependence at extreme levels is satisfied, the GEV family can still be used to model block maxima $\{M_n\}$. Local dependence means that threshold exceedances occur in clusters: the degree of such dependence can be measured using the extremal index θ, which is defined such that the approximate mean cluster size is $1/\theta$. Peaks-over-threshold modelling identifies clusters (e.g. Ferro and Segers, 2003) and models the peak excesses for each cluster using a GP distribution. Alternatively, a common GP distribution can be fitted to all exceedances, assuming independence, and measures of uncertainty inflated post-analysis (Fawcett and Walshaw, 2007). Note that $n\theta$ can be viewed as the effective number of independent observations. Another approach (Smith *et al.*, 1997) is to model within-cluster extremal dependence using a Markov chain. This illustrates a general point: if extreme rainfall observations are dependent, due to their spatial and/or temporal proximity, then this must be taken into account when quantifying uncertainty in inferences. This may be achieved by ignoring the dependence initially and then making an adjustment using asymptotic theory (Chandler and Bate, 2007; Smith, 1990) or simulation-based methods such as the bootstrap (Eastoe and Tawn, 2008); or by modelling the dependence explicitly (Casson and Coles, 1999).

7.2.3. *Quantities of interest*

In the stationary case it is common to make inferences about the level x_p exceeded with a given small probability p in any one year, which is approximately the level exceeded on average once every $1/p$ years. This is often called the $1/p$ return level and is a function of the unknown parameters of the extreme value model. Perhaps surprisingly, in general it is more likely than not that x_p will be exceeded in a given period of $1/p$ years. To see this in the case when different years are independent, note first that by definition, $G(x_p) = 1 - p$, where $G(\bullet)$ is the distribution function of annual maxima. The probability that x_p is exceeded at some point during any n-year period is therefore $1 - [G(x_p)]^n = 1 - (1-p)^n$. For $n = 1/p$ this is always greater than 0.5: for example, the probability of exceeding the hundred-year return level $x_{0.01}$ during a given hundred-year period is $1 - 0.99^{100} = 0.63$.

Typically samples of extreme data are relatively small, so uncertainty about the model parameters may be large. Uncertainty about the value of ξ in particular is often a substantial component of the uncertainty in estimated return levels. It is not uncommon in hydrological practice to fit Gumbel distributions (i.e. GEV distributions with $\xi = 0$), either in the first instance or after determining that a confidence interval for ξ includes zero so that the Gumbel is consistent with the available data. However, this can lead to serious underestimation of the uncertainty in return levels, as illustrated in Figure 7.1. In the left panel, GEV and Gumbel distributions have been fitted to annual maxima of hourly rainfall totals recorded by a rain gauge at Heathrow airport in southern England, for the period 1961–1990; both have been used to calculate estimated return

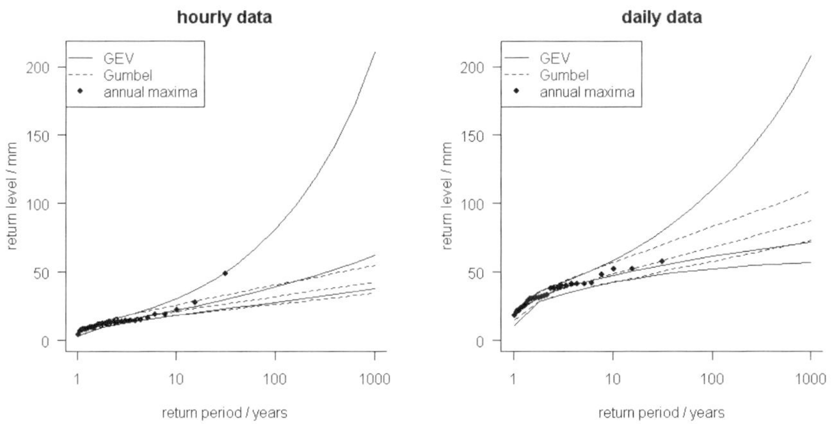

Figure 7.1. Return level plots for annual rainfall maxima at Heathrow airport, 1961–1990. Solid lines show return level estimates and profile likelihood-based 95% confidence intervals obtained from GEV distributions fitted to the data; dashed lines show the results obtained using Gumbel distributions.

level curves with approximate 95% confidence intervals. The maximum likelihood estimate of ξ is 0.15 with a 95% confidence interval (−0.02, 0.41), which includes zero. However, the Gumbel fit vastly underestimates the uncertainty at large return levels. The right-hand panel repeats this comparison for daily rainfall totals. Now, the maximum likelihood estimate of ξ is −0.11 with a 95% confidence interval (−0.39, 0.28). Since the estimate of ξ is negative, the Gumbel fit yields higher point estimates at large return levels than does the more flexible GEV. However, since the data are also consistent with large positive values of ξ, the upper 95% confidence limits for large return levels are much greater under the GEV than under the Gumbel. In the absence of any subject-matter knowledge suggesting that ξ should be zero, the wider GEV limits yield a more realistic uncertainty assessment.

7.2.4. *Non-stationarity*

It is often inappropriate to assume that $Y_1, Y_2, \ldots$ are identically distributed. It may, however, be reasonable to assume that they form a non-stationary process, indexed by covariates. In applications where changes in climate are of concern, time may be a key covariate along with climatological variables anticipated to be associated with the extremes of interest. Spatial location also often affects extremal properties. Unfortunately, there is no general extreme value theory for non-stationary processes. Therefore, the standard approach is to use regression modelling (Davison and Smith, 1990), in which the parameters of the extreme value model depend on the values of the covariates (note that when forecasting extremes in this way, uncertainty in the forecast values of the covariates should be incorporated). It is common to assume that ξ is constant. This is partly a pragmatic choice stemming from having insufficient data to estimate covariate effects on shape reliably, but it may also be that ξ is thought to reflect an intrinsic property of the variable of interest. For example, Koutsoyiannis (2004) analysed data from a large number of rainfall stations and suggested that the value $\xi = 0.15$ is consistent with daily rainfall records worldwide.

7.2.5. *Statistical inference*

Many methods have been used to make inferences using extreme value models. Maximum likelihood estimation (see Chapter 4) has several advantages. Provided that $\xi > -1/2$, standard theory demonstrates its asymptotic optimality (Smith, 1985), although, since extreme data are rare, small sample performance may also be important (see Coles and Dixon, 1999; Hosking *et al.*, 1985). Importantly, maximum likelihood can easily be used to make inferences about regression models for extremes, without the need for new theory. Likelihood ratio tests provide an objective means to choose between nested models. Once a model has been chosen and checked for systematic and isolated departures using residual plots, uncertainty in quantities of interest can be quantified using confidence intervals. Intervals can be based either on the asymptotic normality of the Maximum Likelihood Estimator (MLE), or the asymptotic chi-squared distribution of the profile log-likelihood function for a particular parameter. It is important to use the latter method to produce a confidence interval for the estimate of a return level x_p. For small p the sampling distribution of the MLE of x_p is typically highly skewed, but the former method is not able to reflect this because it is constrained to produce an interval that is symmetric about the estimate.

7.2.6. *Interpretation in a changing climate*

In the non-stationary case, return levels depend on covariate values. Care is required, therefore, to define quantities that are relevant to future extremes. Suppose that a GEV regression model fitted to annual maxima up to the present day suggests an upward time trend in, say, the location parameter μ. What are the implications of this trend for future extreme values? Consider the largest value $M_h = \max(X_1, \ldots, X_h)$ to occur in a planning horizon of h years, where the $\{X_j\}$ are independent GEV random variables with common scale σ and shape ξ and X_j has location $\mu_j = \mu_0 + \beta j$ with $\beta > 0$. One approach is to use simulation to study the distribution of M_h, ideally incorporating parameter uncertainty using the normal approximation to the distribution of the MLE. If the trend is large then the largest value will occur close to the end of the planning horizon with high probability. If the trend is small then Cox *et al.* (2002) show that, in the Gumbel ($\xi = 0$) case, prediction of M_h based on the projected parameter values at the midpoint, $h/2$, of the planning horizon only slightly underestimates the risk of large rainfalls.

7.2.7. *Modern developments*

Current UK national procedures for extreme value analysis were developed in NERC (1975), and updated in the Flood Estimation Handbook (Institute of Hydrology, 1999). Annual maximum rainfall data are used and, to reduce the uncertainty in estimated return levels at any given location, data are pooled from networks of rain gauges at sites within circular regions centred on the site of interest. At each site, annual maxima are standardised by dividing by an estimate of the Median Annual Maximum Rainfall (RMED) at the site. A separate modelling procedure maps RMED across the UK to enable frequency estimation at ungauged sites. A graphical fitting method is used, based on a graph like those in Figure 7.1. The method incorporates various other refinements which are not given here due to a lack of space. A difficulty is that uncertainty in estimated return levels is difficult to quantify, partly as a result of the fitting method (although this can be overcome in principle using bootstrap techniques) but mainly because of inter-site correlations in the pooled sample of data used for fitting. Such difficulties can be overcome using modern developments in extreme value modelling that make better use of limited data and provide realistic measures of uncertainty. Cooley *et al.* (2007), Casson and Coles (1999) and Smith (1999b) consider parametric regression modelling of trends in extremes in space and/or time. Ramesh and Davison (2002) use local likelihood techniques to carry out semi-parametric regression smoothing of sample extremes, which is particularly useful for exploratory analyses. Eastoe and Tawn (2008) propose a two-stage approach: a data-based transformation removes trends in the location and scale of the entire dataset and then threshold exceedances of the transformed data are modelled. The general idea is that apparent non-stationarity in extremes can be wholly or partly attributable to non-stationarity in the main body of the distribution. Coles and Tawn (1996b) make inferences about rainfall aggregated over spatial regions based on pointwise rainfall data. Bayesian methodology (see Chapter 5) can be used to incorporate prior information, either from experts (Coles and Tawn, 1996a), or from other data sources (Coles and Powell, 1996). Many recent extreme value analyses use "non-informative" priors, chosen with the intention of allowing the observed data to dominate the resulting inferences. Examples include the hierarchical random effects models of Fawcett and Walshaw (2006) and Cooley *et al.* (2007) and the mixture model of Walshaw (2006). The Bayesian framework also provides a natural way to handle prediction of future extreme events: in principle it is straightforward to calculate the predictive distribution of, for example, the largest value to be observed in the next 100 years, given the observed data. Subject to the choice of prior, this allows for uncertainty in the model parameters; see Smith (1999a) for a comparison of frequentist and

Bayesian approaches to prediction of extremes. In some applications it is important to makes inferences about the simultaneous occurrence of extreme values in two or more processes, for example, rainfall extremes at two different locations or aggregated over a catchment. Recent work in this area (multivariate extremes) includes Heffernan and Tawn (2004) and Eastoe (2008).

7.2.8. *Other considerations for flood risk management*

The preceding discussion has focussed on the statistical properties of single site rainfall, such as that observed by an individual rain gauge. The variable of interest is typically the depth of rainfall for a specified duration, where a duration or plausible set of durations is selected for a given design application according to the catchment response time. Extreme value analysis is typically undertaken using national or regional data for a range of durations, and the results generalised as a set of intensity-duration-frequency relationships that can be used in design (e.g. Institute of Hydrology, 1999, NERC, 1975).

However, in practice, single site rainfall is almost never the variable of interest — it is the rainfall that falls over the area of concern that is important. The estimation of areal rainfall is a problem for design. Typically an areal average rainfall is used, and this is selected based on empirical analysis of regional or national data in which a frequency analysis of areally-averaged rainfall (for a given spatial scale) is compared with a frequency analysis of single-site rainfall. For a given frequency it is then possible to compare the areal average with the value of single site rainfall. The ratio is commonly known as the Areal Reduction Factor (ARF). In the UK and elsewhere, ARF values are available for national application (e.g. Chow, 1964, NERC, 1975) and can be used to scale single site design rainfall to generate an areal average. However, a simple areal average clearly masks any spatial variability in the rainfall. Hence current design practice recommends the use of areal averages for small catchments ($<1000\,\text{km}^2$). There is no clear guidance, however, for application to larger areas.

7.3. Continuous Simulation of Daily Rainfall

The event-based methodologies of the previous section have the advantage that the distributions fitted to the extreme events have a firm theoretical justification, which provides some protection against the consequences of extrapolation based on an incorrectly specified model. However, as described in the introduction to this chapter, the response of any hydrological system to a particular rainfall event will depend on antecedent conditions. This motivates the development of continuous simulation methodologies. We start with methods for modelling and simulating daily rainfall sequences; sub-daily rainfall is dealt with later.

The fundamental idea of continuous simulation is to generate long time series of rainfall and other relevant input variables (for example, evapotranspiration), and to run these series through hydrological models to build up a picture of system response. However, the variability of daily rainfall is such that any two series, and the associated responses, could be very different. It is therefore helpful to simulate large numbers of rainfall time series for the same time period and, for each of these series, to calculate the values of any quantities of interest. The end result is a distribution of simulated values of each quantity, which represents uncertainty due to rainfall variability. Examples are provided below.

For flood risk management, a potential disadvantage of continuous simulation is that the properties of models fitted to a complete rainfall time series may be dominated by the "average" behaviour of the series and, hence, may fail to adequately represent the rare events that are primarily of interest. Indeed, historically, this was probably one of the main obstacles to the use of continuous simulation in this kind of application (e.g. NERC, 1975, Section 2.9), and some authors continue to raise concerns over the potential for stochastic models to generate unrealistically high rainfall intensities due to the use of distributions that, in principle, allow arbitrarily large values to be simulated (e.g. Cameron *et al.*, 1999; Blazkova and Beven, 2004). However, advances in modelling techniques mean that it is now possible to contemplate the generation of rainfall time series with convincing extremal properties. In the present context, the ultimate test of any continuous simulation procedure is in terms of the duration-frequency properties of flow scenarios derived from simulated rainfall sequences (see Section 7.2.5 below). However, any shortcomings here could be due to deficiencies in representing either the rainfall or the hydrological response of the system. It is therefore useful to check the two components individually. We usually assess the ability of simulation models to reproduce a wide range of properties of rainfall aggregated to different timescales (e.g. daily, seasonal and annual), as well as comparing simulation-based return level estimates with those obtained from conventional extreme value analyses of the data. Again, examples are given below; Yang *et al.* (2005) contains more details.

7.3.1. *Review of daily rainfall modelling techniques*

Daily rainfall models should have the capability to reproduce the most obvious features of daily rainfall sequences. These are the presence of zeroes corresponding to dry days; the highly skewed distribution of non-zero rainfall amounts; the presence of temporal dependence and the potential presence of seasonality (although, as illustrated earlier, seasonality can be relatively weak by comparison with day-to-day variability). Multi-site data also typically exhibit spatial dependence and systematic regional variability due, for example, to topographic controls.

Probably the most widespread family of daily rainfall models in use today can be traced back to the work of Gabriel and Neumann (1962). The fundamental idea is to use a Markov chain to represent the sequence of wet and dry days at a site, and to regard non-zero amounts as drawn independently from a skewed probability distribution such as the gamma or lognormal. Temporal dependence is therefore accounted for by (but limited to) the Markov structure of the rainfall occurrence sequence; seasonality is usually dealt with by estimating separate sets of model parameters for each month of the year, although other techniques have also been suggested, for example by Woolhiser and Pegram (1979). The performance of these simple models is now well understood: they have a tendency to underestimate the frequency of long wet and dry spells (which could have implications for the representation of antecedent conditions when assessing flood risk), and also to underestimate the variability of rainfall aggregated over long timescales — this high variability is sometimes referred to as "overdispersion" (Katz and Parlange, 1998). Various attempts have been made to overcome these deficiencies — for example, Wilks (1998) uses a mixture of two exponential distributions to model wet-day rainfall amounts, which alleviates the problem of overdispersion. For more details and references, see Wilks and Wilby (1999).

A more radical solution to the deficiencies of the simple Markov-based models is to embed them within a wider, more flexible class. This was first attempted by Coe and Stern (1982) and subsequently by Grunwald and Jones (2000), who showed that the models could be regarded as special cases of Generalised Linear Models (GLMs). In their approach, the probability of rainfall at a site on any given day is related to the values of various covariates using logistic regression, and wet-day rainfall amounts are modelled using distributions with means that are also allowed to be covariate-dependent. Specifically, the rainfall occurrence probability is modelled as

$$\log \frac{p_i}{1 - p_i} = \mathbf{x}_i^{\mathrm{T}} \boldsymbol{\beta},$$

where p_i is the probability of rain on day i, $\mathbf{x}_i$ is a corresponding vector of covariates (which might include functions of previous days' rainfall to account for temporal dependence, as well as sine and cosine functions representing the seasonal cycle) and $\boldsymbol{\beta}$ is a coefficient vector. If the rainfall amount on day i is non-zero, it is modelled using a gamma distribution with mean μ_i, where

$$\log \mu_i = \mathbf{z}_i^{\mathrm{T}} \boldsymbol{\xi}.$$

Here, $\mathbf{z}_i$ is another vector of covariates and ξ another coefficient vector. The gamma distributions are all assumed to have a common dispersion (or shape) parameter. This has been found to be a reasonable approximation in many cases, although there is scope for relaxing this assumption using the methods described in Smyth (1989), for example.

The papers cited above focussed on models for rainfall at a single site; subsequently, Chandler and Wheater (2002) showed how to use GLMs to accommodate spatial–temporal structure, and Yang *et al.* (2005) demonstrated their use to simulate multi-site daily rainfall sequences. A particular feature of their work was the use of interaction terms to deal with situations in which some covariates modulate the effects of others: for example, Chandler and Wheater (2002) used interactions to represent the seasonally varying effect of the North Atlantic Oscillation (NAO) on rainfall in the west of Ireland. The results here and in several other studies suggest that GLMs are able to overcome most of the problems associated with simpler approaches to daily rainfall generation and, in addition, offer good reproduction of extremes which is important in applications relating to flood risk management.

A potential drawback with GLMs is that the right-hand sides of the defining equations above are restricted to be linear functions of the covariates. This assumption may be relaxed: in the class of Generalised Additive Models (GAMs), the right-hand sides are written as sums of smooth functions of the covariates without imposing any specific parametric form on these functions. Bowman and Azzalini (1997) give an accessible introduction to the ideas involved. The additional flexibility offered by GAMs can be regarded either as a means of representing the covariate effects more realistically, or as a means of checking the adequacy of the linear forms adopted in GLMs. GAMs have been applied in the context of daily rainfall modelling by Hyndman and Grundwald (2000), Beckmann and Buishand (2002) and Underwood (2008). The (admittedly limited) experience of GAMs in the rainfall literature to date suggests that the main opportunities for improvement over GLMs are in representing the effects of secular trends and previous days' rainfalls. A potential disadvantage, however, is that very large quantities of data are required to estimate complex interaction structures accurately in a non-parametric framework.

Another non-parametric approach to daily rainfall modelling involves resampling the observed record (e.g. Buishand and Brandsma, 2001). Here, for each day of simulation the values of a set of covariates are compared with those from a historical record to determine which days can be regarded as similar to that being simulated; one of these similar days is then chosen at random, and the observed rainfalls at all sites of interest are taken as the simulated values. The approach makes minimal assumptions regarding both the nature of the covariate effects and the most appropriate form of distribution, and is perhaps the most straightforward way to generate multi-site data where reasonably complete records are available. However, it has some drawbacks. For flood risk assessment, the most obvious one is that it will never generate daily rainfalls higher than those observed so that, unless very long records are available for resampling, the performance could be poor with respect to properties such as extremes. A further problem is that as more covariates are considered it becomes increasingly difficult to find similar days in the historical record (this is often called the "curse of dimensionality"), so that the approach may be unsuitable when the covariate-rainfall relationships are complex.

Apart from resampling, rainfall occurrence and amounts are modelled separately in all of the modelling approaches described above. This often makes sense physically because the mechanisms governing occurrence and amounts — and in particular the effects of covariates upon the two components — are different. For example, rainfall occurrence in temperate climates tends to be higher in the winter than the summer, whereas the reverse is true for non-zero amounts (this is due to the predominance of convective rainfall events in the summer compared with large-scale synoptic systems in the winter). However, the two-stage approach has been seen as unsatisfactory by some authors, and some progress has been made in the development of models that treat occurrence and amounts simultaneously. The most common class of such models (see, for example, Bárdossy and Plate, 1992) treats the rainfall at time t, Y_t say, as a function of some underlying random variable X_t which is normally distributed, for example via a relationship of the form

$$Y_t = \begin{cases} X_t^{\beta} & \text{if } X_t > 0 \\ 0 & \text{otherwise} \end{cases}$$

for some transformation parameter β. Covariate effects can be incorporated into such models. However, the introduction of Gaussian random variables (X_t) is arguably a rather artificial device. More recent work has sought to avoid this by using innovative distributions such as those in the Tweedie family (Dunn, 2004), which model the rainfall Y_t directly without the need for transformation or truncation.

Models for multi-site data are generally less well developed than those for rainfall at a single site. Alternatives to the multi-site GLMs of Chandler and Wheater (2002) and Yang *et al.* (2005), and the resampling approach of Buishand and Brandsma (2001), include the models of Hughes *et al.* (1999), Stehlík and Bárdossy (2002) and Wilks (1998). Hughes *et al.* define a Hidden Markov Model (HMM) which postulates the existence of a small number of unobserved weather states, each of which has a characteristic spatial pattern of precipitation that is described probabilistically; the daily sequence of weather states itself follows a Markov process with transition probabilities that can be related to the values of covariates. Stehlík and Bárdossy work with a transformed Gaussian random field: the correlation structure of this field ensures that inter-site dependence is preserved in the simulated rainfall sequences. Wilks' scheme is similar to this, although here the transformation to Gaussianity is determined by the assumption that the non-zero rainfall amounts follow mixed exponential distributions.

As discussed in Section 7.2.1, missing data can often cause problems in applications and simple solutions such as interpolation in time or space can be unsatisfactory. For purposes such as the calibration of hydrological models, it may be of interest to determine the sensitivity of model parameters to uncertainty in rainfall inputs. Where multi-site data are available, a method for achieving this is as follows. First, determine the distribution of the missing values conditional upon the observed values at all sites, then simulate from this conditional distribution many times to obtain multiple reconstructions of the complete rainfall series, and then carry out analyses on each of the reconstructions to quantify the uncertainty due to missing data. This process is called multiple imputation. The idea can also be used to generate multiple realisations of rainfall over a regular grid, which are consistent with observed data at an irregular network of sites. The main impediment to its widespread use is the lack of multi-site models for which the required conditional distributions can be calculated easily. Some progress can be made with models based on transformed and truncated Gaussian distributions (e.g. Sansó and Guenni, 2000) but, to our knowledge, the only models that have been specifically designed to facilitate imputation where necessary are the multi-site GLMs of Yang *et al.* (2005) and the multi-site model of Wilks (2009).

To illustrate uncertainty due to natural variability and missing data, Figure 7.2 shows some specimen simulation and imputation results for one of the sites in the case study considered by Yang *et al.* (2005). These results are based on GLMs fitted to data from a network of 34 gauges in a $50 \times 40\,\text{km}^2$ region of southern England, and subsequently used to simulate 100 time series of daily rainfall data at ten locations corresponding to a subset of these gauges, over the period 1961–1999. The simulations were conditioned on values of the NAO for this period and upon the available rain gauge observations for December 1960. In addition, ten sets of imputations were carried out on the 1961–1999 observations themselves, replacing any missing values by simulated values conditional on the available data as described above. Simulations and imputations were both generated using the GLIMCLIM software package (Chandler, 2002). For each set of imputed and simulated data, various summary statistics were calculated, separately for each month of the year (see the caption to Figure 7.2 for details). The solid lines in the figure show the ranges of values from the imputations, indicating the uncertainty due to missing data. Most of the time this uncertainty is fairly small, but there are some interesting features such as the wide range of the imputation envelope for the maximum daily rainfall in October. The width of this envelope suggests that the October maxima correspond to imputed values, and hence that data from this site are missing during a period when other sites recorded particularly high rainfall. Furthermore, a comparison with the imputed maxima for all other months of the year suggests that at this site, the highest daily rainfall during the period may have occurred during October and not been recorded. Failure to account for this in the flood risk assessment exercise could have potentially serious consequences.

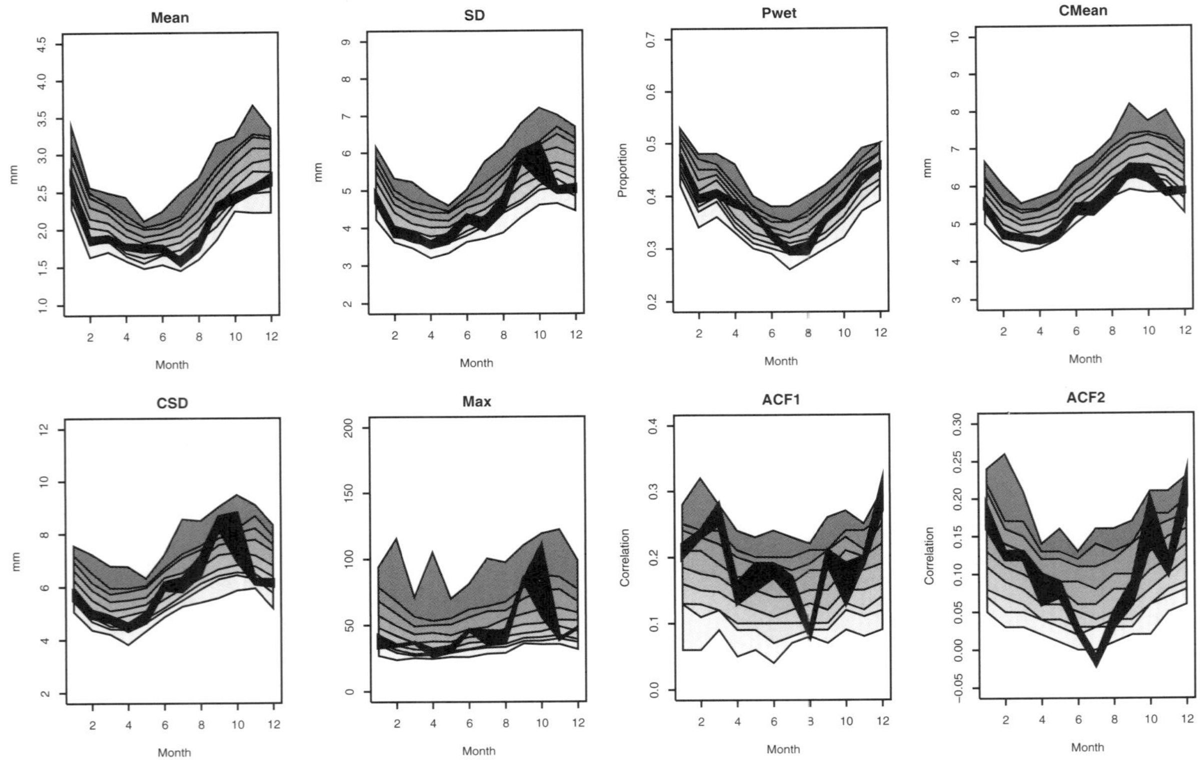

Figure 7.2. Distributions of monthly rainfall summary statistics for a site in southern England for the period 1961–1999, obtained using the GLM of Yang *et al.* (2005). Row-wise from top left: mean, standard deviation, proportion of wet days, conditional mean (i.e. mean on wet days only), conditional standard deviation, maximum, and autocorrelation at lags 1 and 2. Thick lines show the envelope obtained from ten sets of imputations of missing data; shading shows the ranges obtained from 100 simulations of the same period along with 5th, 25th, 50th, 75th, and 95th percentiles.

The grey bands in Figure 7.2 show percentiles of the distributions obtained by simulation: these represent uncertainty in the underlying climatology of the site due to natural variability. This uncertainty is substantial. For example, the 5th and 95th percentiles of the simulated distribution of mean January rainfall are around 2.4 mm and 3.1 mm respectively: this suggests that without any large-scale changes in climate, the mean January rainfall at this site over a 39-year period could conceivably deviate from its historical value of around 2.6 mm per day by as much as 20%. Of course, such conclusions rely on the adequacy of the simulation models upon which they are based, but the diagnostics reported in Yang *et al.* (2005) suggest that in this particular instance the models' reproduction of rainfall variability is fairly realistic.

7.3.2. *Parameter uncertainty*

With the exception of resampling, all of the simulation methodologies reviewed above are based on models that involve unknown parameters (and, in the case of GAMs, unknown functions). As noted previously, long daily rainfall records and relatively parsimonious models ensure that parameter uncertainty is often small compared with the uncertainty due to natural variability, and this perhaps explains why little if any work has been done to quantify its effect. However, modern methods of model fitting provide uncertainty estimates as a by-product of the fitting process so that the effect of parameter uncertainty could be investigated if required. This is perhaps most straightforward when models are fitted using Bayesian methods (see Chapter 5): in this case, parameter uncertainty can be incorporated into multiple rainfall simulations by using a different parameter set, sampled from the posterior distribution, for each simulation. Of the models described above, HMMs and their variants are usually fitted using Bayesian methods, which are also particularly convenient for the analysis of complex models based on transformed Gaussian distributions.

In modern statistical practice, the main alternative to Bayesian model fitting is the method of maximum likelihood which, providing the model being fitted is not too complex, has desirable optimality properties. As discussed in Chapter 4, the distribution of the MLE is approximately multivariate normal in large samples. Moreover, if the sample size is extremely large (as is often the case when analysing daily rainfall records) then there is a close connection between this distribution and the posterior distribution obtained using a Bayesian analysis, provided that the prior distribution in the latter case is not too informative (e.g. Davison, 2003, p. 579): the posterior is approximately multivariate normal, centred on the MLE and with covariance matrix equal to that of the MLE (see Chapter 4 for the definition of this covariance matrix). If models are fitted using maximum likelihood there is some justification for incorporating parameter uncertainty into multiple rainfall simulations by using a different parameter set, sampled from this multivariate normal distribution, for each simulation. Such a procedure can also be regarded as a computationally efficient approximation to a parametric bootstrap (for an introduction to bootstrap techniques, see Davison and Hinkley, 1997).

As described in Chapter 5, Bayesian methods also require the specification of a likelihood function. For most of the models described above, likelihoods can be constructed. However, this is extremely difficult for some classes of model such as multi-site GLMs, because of a lack of plausible models for joint distributions with highly non-Gaussian marginal (i.e. single-site) structure. In this case, a feasible alternative to maximum likelihood is to ignore the joint distributional structure and fit models to multi-site data as though sites are independent. This amounts to maximising the likelihood of an incorrectly specified probability model for the data (because it ignores inter-site correlations). Unfortunately, in this case the usual expression for the covariance matrix of the parameter estimates is no longer valid and an adjustment must be made. Details can be found in Chapter 4.

7.3.3. *Incorporating climate change projections*

Amid general consensus that the Earth's climate is changing, any prudent approach to flood risk management must take account of the impacts of future changes in climate and land use. Most projections of change are based on deterministic models of the physical and chemical processes involved, conditioned on future scenarios of socio-economic development and associated greenhouse gas emissions. Atmosphere-ocean General Circulation Models (GCMs) provide climate simulations on a coarse grid (with typical resolution currently around $250 \times 250\,\mathrm{km}^2$ at UK latitudes), while Regional Climate Models (RCMs) operate on a finer grid resolution (typically around $50 \times 50\,\mathrm{km}^2$) over smaller areas, for example, Europe. RCM simulations are usually used to add detail to a GCM simulation over an area of interest. However, despite continuing improvements in both GCMs and RCMs, questions remain regarding their ability to represent precipitation adequately at the fine space and time scales required (e.g., Beniston *et al.*, 2007; Blenkinsop and Fowler, 2007; Maurer, 2007). Moreover, to quantify the uncertainty due to natural variability of rainfall, it is useful to be able to generate a large number of possible rainfall sequences, and this cannot be easily achieved using climate models due to computational costs.

Against this background, for the purposes of flood risk management it is desirable to avoid direct use of climate model precipitation outputs where possible. In current practice, probably the most widespread approach to deriving future precipitation is to scale historical series using "change factors" obtained from climate model outputs (Prudhomme *et al.*, 2002). If, for example, a climate model suggests that mean winter rainfall in some future period will be 20% higher than in the past, then a future scenario is obtained simply by scaling the historical observations by a factor of 1.2. Although conceptually simple, this technique is difficult to justify and is completely unsuitable for applications that are sensitive to the temporal structure of rainfall inputs at sub-seasonal timescales (for example, it does not allow for changes in the durations of wet and dry spells, which could affect soil properties and hence, ultimately, flood risk). We therefore do not consider it further.

A more satisfactory way of using climate model output is to identify atmospheric variables that are reasonably represented in climate models and to use these as covariates in statistical models, of the type described in the previous subsection, for precipitation at the fine space–timescales that are appropriate for flood estimation. Future precipitation sequences can then be generated using statistical model simulations conditioned on the appropriate covariate information from climate models. To compensate for possible climate model bias, it is conventional to standardise all covariates with respect to the mean and standard deviation over some historical period in any such exercise. The approach is referred to as "statistical downscaling" and there is a substantial literature on it: see Fowler *et al.* (2007) and Maraun *et al.* (2010) for excellent application-focussed reviews.

Statistical downscaling relies on three fundamental assumptions (e.g. Charles *et al.*, 2004; Wilby *et al.*, 1998; Wilby and Wigley, 2000): first, that there is a genuine relationship between the chosen indices of large-scale atmospheric structure and local-scale precipitation; second that the relevant aspects of large-scale structure, and its change in response to greenhouse gas forcing, are realistically represented by the climate models at the scales used for analysis; and finally that the observed relationship between local precipitation and large-scale structure remains valid under altered climatic conditions. Unfortunately, the second assumption cannot be verified in advance since, even if a climate model is assessed as performing credibly when simulating the present climate, its future performance is not guaranteed (IPCC, 2001, p. 473). A pragmatic response is to determine, on the basis of our understanding of the models, the variables and scales at which they might reflect reality (Smith, 2002) and to focus on these variables and scales while acknowledging the underlying difficulty.

The third assumption above implies that statistical downscaling should be used with caution if climate model simulations contain large-scale atmospheric conditions that have not been observed previously, because it would involve extrapolating beyond the range of empirical experience (although it can be argued that the climate models themselves should be used with care in such situations, because they use parameterisations based on historical climate). In general, confidence in this assumption will be increased if the relationship between local scale precipitation and large-scale atmospheric structure can be justified physically. For the purposes of generating realistic precipitation sequences, this usually requires the use of variables representing some measure of atmospheric moisture in addition to indices of atmospheric circulation such as pressure and temperature fields. For example, Charles *et al.* (1999) compared the performance of statistical downscaling models with and without atmospheric moisture variables and found that, although the models fitted equally well during the period used for model calibration, their downscaled projections for a subsequent period were rather different: the projections from the model without moisture information were not realistic.

In many situations, the choice of model itself represents a significant source of uncertainty (see Chapter 17). In the context of statistical downscaling however, evidence is emerging that similar results can often be obtained from several different methods, providing they are applied intelligently and compared on the basis of aspects that they are designed to reproduce. For example, Timbal *et al.* (2008) compared the performance of a HMM with a method based on meteorological analogues. They initially found that for their study area in south-west Western Australia, the methods gave different results. However, further examination revealed that this was due to the use of different atmospheric covariates: both methods produced similar results when the same covariates were used. Other comparison studies include Haylock *et al.* (2006) and Schmidli *et al.* (2007), although these focused mainly on techniques that are relatively simple compared with the leading methodologies that are currently available.

To illustrate the use of statistical downscaling in a simple setting, we consider the use of GLMs to generate simulated precipitation sequences for a single location: Heathrow airport in southern England. Other authors have also used GLMs for downscaling, notably Fealy and Sweeney (2007) and Furrer and Katz (2007). The latter authors also show how GLMs may be used to generate simultaneous daily sequences of precipitation and temperature conditioned on the climate model outputs. Historical data for our example include hourly rainfall (these were also used in the extremes example above) and monthly daily time series of atmospheric variables from the NCEP reanalysis dataset (Kalnay *et al.*, 1996). The rainfall record starts in 1949, although data are missing for January to August of 1988 and February 1989. The recording resolution of the gauge was 0.1 mm until 1987 and 0.2 mm afterwards. The NCEP data were taken from the archive supplied with the "Statistical Downscaling Model" (SDSM) software package (Wilby and Dawson, 2004). The atmospheric covariates considered were monthly values of temperature, sea level pressure and relative humidity, averaged over a spatial area roughly the size of a GCM grid square centred on the gauge location. The choice of monthly, rather than daily, covariate values was made primarily for convenience because GCM years typically do not have the same number of days as calendar years. This choice is justified because GLM-based downscaling results for the UK have been found to be relatively insensitive to the use of monthly, rather than daily, atmospheric covariates (Frost *et al.*, 2006): covariate values at finer temporal resolution yield slightly larger simulated precipitation extremes, but the differences are small in practical terms (e.g. at most 5–6 mm for a simulated daily 100-year return level). The covariates were standardised, separately for each month of the year, so as to have zero mean and unit variance over the period 1961–1990. The rainfall data were aggregated to a daily timescale and, as described above, separate GLMs for rainfall occurrence and amounts were fitted to data from the 1961–1990 period. The choice of covariates in the models was based on the likelihood ratio statistic (see Chapter 4) and on residual analyses (see Chandler and Wheater 2002, for example); Leith (2008) gives full details. As well as covariates representing seasonality and temporal persistence, the final occurrence model included temperature, sea-level pressure and relative humidity as covariates, along with some interactions (for example, temperature anomalies are positively associated with rainfall occurrence in winter and negatively associated in summer) and a "post-1987" indicator, included to account for a spurious decrease in recorded occurrence after 1987 that is associated with the change to a coarser recording resolution (Yang *et al.* 2006 give a general discussion of this issue). Relative humidity was not included in the final model for rainfall amounts but temperature and sea-level pressure were; there were also interactions between sea level pressure and seasonal covariates. The interactions between the atmospheric covariates and the seasonal components imply that seasonality will alter in a changed climate.

A variety of checks were carried out to ensure that the fitted GLMs were able to generate realistic rainfall sequences. These included an out-of-sample validation exercise in which simulations of the 1991–2000 period (not used in model calibration) were compared with observed rainfall over this period. The results, given in Leith (2008) suggest that the performance is good overall, except for a slight tendency to overestimate rainfall occurrence in summer and underestimate in winter. Having verified this, the fitted models were used to simulate 100 sequences of daily rainfall for the period 1961–1990, taking the atmospheric covariate values from the NCEP dataset. In the first panel of Figure 7.3, winter (DJF) rainfall totals have been computed for each year of each simulation, excluding the incomplete winter of 1961 (data for December 1960 were not simulated); this panel shows the evolution of the simulated distribution of winter rainfall, along with the observed time series of winter rainfall totals for comparison. The overall median (P50) of the simulated winter rainfalls is also marked, along with their 10th (P10) and 90th (P90) percentiles. The observations span the full range of the simulated distributions, showing that the simulations are able to produce a realistic range of variability in the seasonal totals. The year-to-year variability in the simulated distributions is also noteworthy: this reflects changes in the atmospheric drivers and shows that although they account for a very small proportion of the variability in daily rainfall (see the introduction to this chapter), their effects can be substantial when aggregated to larger timescales. Further simulations of the 1961–1990 period (not shown) have been carried out using covariate values taken from the GCMs. The overall properties of these simulations were almost identical to those shown here.

The second panel of Figure 7.3 shows the distributions of winter rainfall obtained from simulations of the 2071–2099 period conditioned on atmospheric covariates derived from the HadCM3 climate model under the SRES A2 emissions scenario (IPCC, 2001). The distributions are shifted up slightly relative to those for the historical period: the increase in the P10 and P90 is around 20 mm and that in P50 is around 30 mm (see Table 7.1).

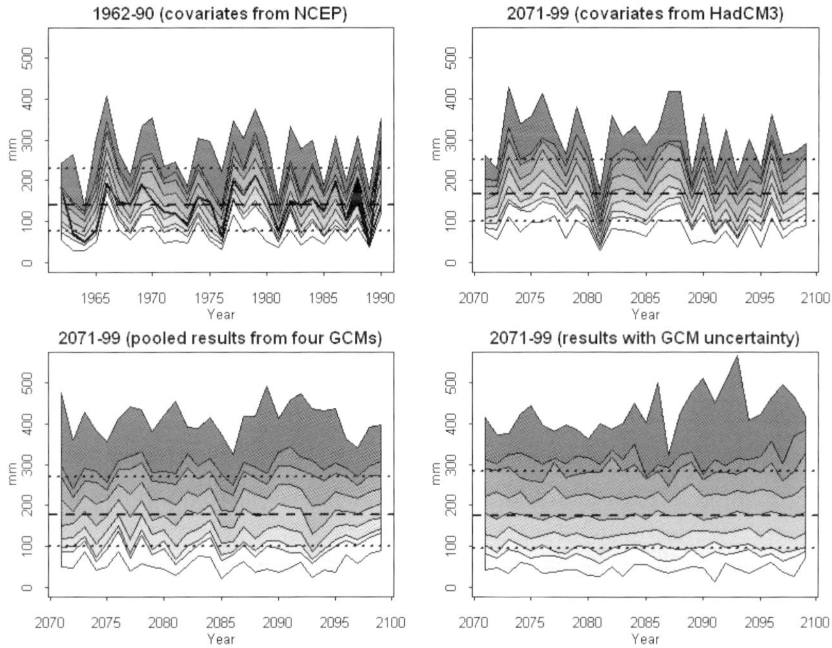

Figure 7.3. Distributions of winter (DJF) rainfall totals at Heathrow airport, each derived from daily rainfall time series simulated using a GLM and conditioned on large-scale atmospheric predictors. See text for details, and Figure 7.1 for legend. Horizontal lines represent the overall median, 10th and 90th percentiles of the distributions in each panel.

Table 7.1. Percentiles of simulated winter rainfall distributions at Heathrow airport for different time periods and using different sources of atmospheric covariate data.

Source of atmospheric covariates	P10 (mm)	P50 (mm)	P90 (mm)
NCEP, 1961–90	77.6	140.2	231.0
HadCM3, 2070–99	101.0	168.0	251.2
Four GCMs, 2070–99; overall distribution obtained by pooling results	101.9	178.0	270.7
Hierarchical model accounting for GCM uncertainty, 2070–99	87.5	160.0	262.7

In percentage terms, these represent increases of around 30%, 20% and 9% for P10, P50 and P90 respectively. The indication of wetter winters agrees with other projections for the south east of England (e.g. Hulme *et al.*, 2002) and, qualitatively, with the changes in precipitation simulated directly by the GCM (not shown). These results suggest an overall increase in the risk of winter flooding associated with extended periods of high rainfall. However, it is important to understand that not every winter will be wetter in the future than in the past: this is clear from the substantial overlap between the two sets of simulated distributions.

The second panel of Figure 7.3 once again shows substantial variability from year to year, although the precise timings of peaks and troughs should not be interpreted literally since the GCM data are a single realisation of possible climate over this period: another run of the same GCM, with slightly different initial conditions, would yield different results although the overall properties would be similar. Thus, even within a single GCM, the use of a single realisation will underestimate the variability in rainfall sequences in any given year. Moreover, if different climate models are used, the differences are accentuated. This is shown in the third panel of Figure 7.3, obtained by pooling simulations from four different GCMs (HadCM3, along with model CGCM2 from the Canadian Centre for Climate Modelling and Analysis, model CSIROMK2 from the Commonwealth Scientific and Industrial Research Organisation and model ECHAM4 from the Max Planck Institute). Although the overall value of P10 for this pooled distribution is similar to that obtained using HadCM3 alone, the values of P50 and P90 are substantially greater (Table 7.1). This shows that the choice of GCM may itself be a significant source of uncertainty when deriving projections of future precipitation for flood risk assessment. The need to account for GCM uncertainty in such situations is increasingly being recognised and is discussed in more detail elsewhere in this volume (see Chapter 17). To address the problem, it is common to weight the GCMs according to some measure of their performance (e.g. Giorgi and Mearns, 2002; Wilby and Harris, 2006). Although this provides a simple and easily interpretable means of incorporating climate model uncertainty, in general it will underestimate the true uncertainty because the results are constrained to lie between the limits set by the available data: the number of climate models used for any particular study is often small, raising the possibility that another model will yield more extreme projections. Moreover, the choice of weights is often heuristic and difficult to justify formally. To some extent, this can be overcome via the use of a hierarchical model which makes explicit the statistical assumptions underlying the chosen weights (e.g. Tebaldi *et al.* 2005; Tebaldi and Sansó, 2008). More generally, hierarchical models provide a means of conceptualising explicitly the way in which climate models work and of representing GCM uncertainty within a coherent, logically consistent and interpretable framework. Leith and Chandler (2010) illustrated this using the atmospheric covariate data from the Heathrow example considered here: they used Bayesian methods to fit a hierarchical model characterising a notional population from which the four available GCMs were considered to have been sampled, and used this model to generate multiple realisations of the atmospheric covariates corresponding to unobserved GCMs in this notional population. The final panel in Figure 7.3 shows the 2071–2099 winter rainfall distributions obtained from GLM simulations conditioned on the output of this hierarchical model. The most obvious differences between this and the previous panel are that the distributions are more similar from year to year (which is clearly a more realistic reflection of uncertainty in projections at this time horizon), that the largest simulated values are higher than previously and that there is an increasing trend in the upper percentiles of the distribution throughout the simulation period. This increasing trend does not appear in the lower percentiles, or in the simulations driven by HadCM3 alone (indeed, the HadCM3 simulations in the second panel of Figure 7.3 appear to show a decreasing trend in the upper percentiles). The 90th percentile of the overall winter rainfall distribution is somewhat higher again than for the pooled simulations (see Table 7.1), although the values of P10 and P50 are similar. The increase in P90 suggests that according to the hierarchical model, the four GCMs used in the current analysis may not bound the range of uncertainty fully at the upper end of the distribution: clearly this could be important from the perspective of risk assessment.

The results shown in Table 7.1 give some indication of the potential complexity of thinking about uncertainty in flood risk — and, indeed, in any other study of climate change impacts. Simple summaries such as changes in mean rainfall clearly are not sufficient to characterise the projected changes and associated uncertainties. One consequence of this is that it is difficult to communicate information about uncertainty to stakeholders in simple terms. Visual representations of changing distributions, such as those in Figure 7.3, can be particularly helpful in this respect.

7.4. Continuous simulation of sub-daily rainfall

The daily rainfall models reviewed above are essentially descriptive in nature, in the sense that they describe the structure of the rainfall process and its dependence on other variables without explicit reference to the underlying mechanisms. At the sub-daily timescale, the structure of rainfall time series is more complex and a purely descriptive approach is less feasible; this has led to the development of models that attempt to represent the underlying mechanisms in a more or less simplified form. These are reviewed in this section, concentrating mainly on the single-site case because extensions to multi-site and spatial–temporal data are currently less well developed.

7.4.1. *Review of sub-daily rainfall modelling techniques*

In the literature, there are essentially two broad approaches to modelling rainfall stochastically at sub-daily scales. The first involves reproducing the intermittency and intensity properties at coarse scales through the clustering of structures at finer scales (Rodriguez-Iturbe *et al.*, 1987). The other proposes a scale-independent representation of the rainfall process which can be simulated by a cascade process, analogous to that used to represent the distribution of energy in Kolmogorov's theory of turbulence (Lovejoy and Schertzer, 1995).

The essence of the clustering approach is to regard a rainfall time series as consisting of a succession of storms, each made up of a collection of "rain cells". Each cell lasts for a random duration during which it deposits rainfall at a constant random intensity; the total rainfall during any time interval is thus a sum of contributions from all cells active during that interval. The times of storm origins are usually modelled as a Poisson process. Within each storm, cell arrival times can then be determined either via a second Poisson process (Bartlett–Lewis model) or as independent and identically distributed displacements from the storm origin (Neyman–Scott model). In the most common variant of the Bartlett–Lewis model, the lifetime of each storm (i.e. the period during which it can generate cells) is taken to be an exponentially distributed random variable. Likewise, the most common form of the Neyman–Scott model has an exponential cell displacement distribution. It has been shown that in these forms, the two model types have equivalent second-order properties (Cowpertwait, 1998), and empirical studies have generally not found significant differences between them (Onof *et al.*, 2000). For both types of model, cell duration and intensity are usually assumed to be independent and typically modelled with exponential distributions, although heavier-tailed distributions such as the Pareto, may be used for the latter (Onof *et al.*, 2000). Extensions have also been proposed to represent rainfall at several sites — see for example

Cox and Isham (1994); Cowpertwait (1995). Note finally that the clustering can also be generated by using doubly-stochastic Poisson processes. This idea has been applied directly to the modelling of the tip times from a tipping-bucket rain gauge by Onof *et al.* (2002).

Several variants on the basic Bartlett–Lewis or Neyman–Scott models have been developed and tested in the literature. Amongst the most important are: the use of different cell types (Cowpertwait, 1994) to represent the variety of rainfall that can be found even within a specific month; the introduction of dependence between cell duration and intensity (Evin and Favre, 2008; Kakou, 1998) for greater realism; the randomisation of the temporal structure of the storm (Islam *et al.* 1990; Onof and Hege 2009; Onof and Wheater 1993; Rodriguez-Iturbe *et al.* 1988) which considerably improves the reproduction of the proportion of dry intervals; the separation between "normal" and "extreme" rainfall, whereby the latter is dealt with using a different intensity distribution (Cameron *et al.*, 2000b); the introduction of a jitter to render the rectangular pulse shape more realistic (Onof and Wheater, 1994; Rodriguez-Iturbe *et al.*, 1987); and the replacement of the rectangular pulse by a series of instantaneous pulses occurring at points generated by a Poisson process (Cowpertwait *et al.*, 2007) so as to reproduce sub-hourly rainfall properties. For application to flood risk assessment, the choice of model variant will be influenced by the need for convincing extremal performance. Our experience is that, although variants such as the Random Parameter Bartlett–Lewis Model (RPBLM) of Rodriguez-Iturbe *et al.* (1988) generally perform reasonably with respect to properties such as 100-year events, in very long simulations they can occasionally generate unrealistically large rainfalls associated with the use of what are effectively heavy-tailed cell duration distributions. Elsewhere, Cameron *et al.* (1999) imposed large upper bounds on the distributions of cell intensities to prevent the generation of unrealistically large amounts. Conversely, Cameron *et al.* (2001) found that the RPBLM tended to underestimate extremes at fine timescales. To fully resolve these issues, further model development and testing are needed.

Likelihood-based calibration for this class of models is difficult, because in general it is not possible to obtain a useful expression for the likelihood function (Onof *et al.*, 2000; Rodriguez-Iturbe *et al.*, 1987). For simple models, some progress was made by Northrop (2006) who developed a marginal likelihood for a subset of the parameters; and an approximate likelihood was formulated in the Fourier domain by Chandler (1997). However, current practice in fitting these models is largely dictated by the availability of analytical expressions, in terms of the model parameters, for the first and second-order properties of the rainfall intensity, as well as the proportion of dry periods at any scale of interest. Therefore, the models are almost always fitted using a Generalised Method of Moments (GMM) in which parameters are chosen to achieve as close as possible a match, according to a weighted least-squares criterion, between the observed and fitted values of selected properties (Wheater *et al.*, 2005). Specifically, let $\boldsymbol{\theta} = (\theta_1, \ldots, \theta_p)'$ denote the parameter vector for the model, let $\mathbf{T} = (\mathbf{T}_1, \ldots, \mathbf{T}_k)'$ be a vector of summary statistics computed from data, and denote by $\mathbf{T}(\boldsymbol{\theta}) = (\tau_1(\boldsymbol{\theta}), \ldots, \tau_k(\boldsymbol{\theta}))'$ the expected value of $\mathbf{T}$ under the model. Then $\boldsymbol{\theta}$ is estimated by minimising an objective function of the form

$$S(\boldsymbol{\theta}; \mathbf{T}) = \sum_{i=1}^{k} w_i (\mathbf{T}_i - \tau_i(\boldsymbol{\theta}))^2$$

for some collection of weights $\{w_1, \ldots, w_k\}$. Usually the minimisation must be done numerically, and care can be required to avoid local optima. However, if this can be done then the theory of estimating functions (see Chapter 4) can be used to obtain standard errors and confidence regions for the resulting parameter estimates. For some of the more complex models, such techniques reveal that model parameters can be poorly identified using the range of properties for which analytical expressions are available: indeed, confidence intervals for some parameters can be effectively unbounded (see Figure 3 of Wheater *et al.* 2005, for example). One solution to this is to add a second, simulation-based phase to the calibration process that seeks to identify, among parameter sets that perform equivalently in terms of the objective function $\mathbf{S}(\boldsymbol{\theta}; \mathbf{T})$, those that are best able to reproduce observed rainfall extremes (Wheater *et al.*, 2005). A less computationally intensive alternative, which achieves a similar result, is to include the skewness of the rainfall intensities as one of the components of $\mathbf{T}(\boldsymbol{\theta})$ (Cowpertwait, 1998); however, tractable analytical expressions for the skewness are currently available only for relatively few models. A final possibility, which to our knowledge has not yet been attempted, is to combine the objective function with prior knowledge of the likely ranges of the various parameters in the spirit of a Bayesian analysis (see Chapter 5); the techniques proposed by Smith *et al.* (2008) could be helpful in this respect. It would be straightforward in principle to specify plausible prior distributions for many of the parameters of a Poisson cluster model, because these parameters typically represent physically meaningful quantities (such as storm arrival rate and mean number of cells per storm) for which "typical" values are known (Onof *et al.*, 2000).

A potential disadvantage of the GMM is that the results can depend on the properties selected for inclusion in the objective function, as well as on the associated weights $\{w_1, \ldots, w_k\}$. Various authors have sought to address this empirically by comparing the results obtained from different objective functions (e.g. Burton *et al.* 2008, Wheater *et al.* 2005). More constructively, however, it is possible to use the theory of estimating functions to identify an optimal choice of weights. Using an extension of the arguments given in Section 3.6 of Hall (2005) for example, it can be shown that the optimal objective function based on fitting properties $\mathbf{T}$ is the quadratic form $(\mathbf{T} - \mathbf{T}(\boldsymbol{\theta}))'\mathbf{V}^{-1}(\mathbf{T} - \mathbf{T}(\boldsymbol{\theta}))$, where $\mathbf{V}$ is the covariance matrix of $\mathbf{T}$. This is optimal in the sense that the large-sample variance of any linear combination of parameter estimates is smaller than that obtained by minimising any other objective function of the form $(\mathbf{T} - \mathbf{t}(\boldsymbol{\theta}))'\mathbf{W}(\mathbf{T} - \mathbf{t}(\boldsymbol{\theta}))$: the function $\mathbf{S}(\boldsymbol{\theta}; \mathbf{T})$ is of this form with $\mathbf{W}$ diagonal. This has been investigated by Jesus and Chandler (2011), who found that some care was required in the implementation of the theory but that problems of parameter non-identifiability were effectively eliminated.

At sub-daily timescales, the main alternative to Poisson cluster models is the representation of rainfall time series as a multi-scaling process. Here, a series is characterised in terms of the way in which its properties vary with the temporal scale of observation. An example is given below. One parameter may be sufficient to describe how these properties vary with the scale of description. Some authors claim that for fine timescales, the data point to such a monofractal representation of precipitation (Paulson, 2004). More generally, however, a function of timescale is required to describe the variation in rainfall properties. This can, for instance, be the specification of the fractal co-dimensions of the occurrence of rainfall depths exceeding a range of thresholds (i.e. singularities — see Bernadara *et al.*, 2007). Such multifractal representations have generally been used in preference to monofractals where larger ranges of scales are involved (Tessier *et al.*, 1993). The results in our example below suggest that temporal scaling relationships can be extremely stable through time, although as far as we are aware this important question has not been studied elsewhere in detail.

The standard method for generating a time series with multifractal properties is to use a random cascade process. The stages in the cascade correspond to progressively finer temporal resolutions: to disaggregate a rainfall time series $(Y_t^{(h)})$ at a resolution of h time units to a resolution of h/b time units, each value is multiplied by b independent identically distributed random variables so that an initial value $Y_t^{(h)}$ generates b new values $Y_t^{(h)}X_{1t}^{(h)}, \ldots, Y_t^{(h)}X_{bt}^{(h)}$ say. Generally, $b = 2$. The distribution of the $(X_{jt}^{(h)})$ (which is the same for all t and h and has mean 1) defines the cascade generator. This process is iterated so that rainfall depths can be downscaled indefinitely from a given timescale by specifying the distribution of the generator. To use this for rainfall simulation, all that is needed is to initiate the cascade process with a total amount of rainfall for the simulation period. Note, however, that this coarse-scale total will not be preserved through disaggregation due to the stochastic nature of the algorithm; the requirement $E(X_{jt}^{(h)}) = 1$ merely ensures that the coarse-scale totals are preserved "on average".

Distributions whose type is preserved through multiplication are often used for cascade generation, in particular the continuous lognormal distribution (Gupta and Waymire, 1993) and the discrete log-Poisson distribution (Deidda *et al.*, 1999). It has been argued that the best choice is a log-Lévy distribution (Schertzer and Lovejoy, 1993) on the grounds that the generalised central limit theorem (Gnedenko and Kolmogorov, 1968) shows that convolutions of distributions (even of infinite variance) converge to a Lévy-stable distribution as the number of variables grows. This result is applicable when the number of cascade steps between two given scales is increased indefinitely so that the discrete cascade now becomes a continuous one (Schertzer and Lovejoy, 1997). The generation of these continuous multifractal cascades is achieved using transformations in Fourier space.

Aside from the above canonical multifractal cascade models, two important variants have been developed and tested. The first involves altering this cascade framework by relaxing the assumption that the random variables used to generate the cascade are identically distributed. By using variables whose distribution changes with the scale, in such a way that the variance becomes smaller at finer scales, it is possible to improve the reproduction of certain features of the rainfall signal (e.g. intermittency) which appear to be scale-dependent. These are bounded random cascade models (Menabde *et al.*, 1997). The second primarily involves relaxing the assumption that the cascade generating random variables are independent. The purpose is to develop a cascade tool that is able to disaggregate a rainfall total to finer scale amounts that sum to the given total (Olsson,

1998). This is useful for applications which take this disaggregated rainfall as input: it ensures that the total rainfall input does not depend upon the scale. For some such microcanonical cascades, the generating random variables are also not identically distributed, although choosing a beta or uniform distributions will ensure they are. Spatial–temporal models based upon multifractals have also been developed, either by combining spatial cascades with some representation of advection (e.g. Seed, 2003), or by using a multi-scaling representation of rainfall both in time and space (e.g. Over and Gupta, 1996; Marsan *et al.*, 1996).

Fitting random cascade models is fraught with difficulties, for a variety of reasons. Firstly, the moment scaling function requires the evaluation of higher order moments of the rainfall process and, in the case of heavy-tailed distributions, care is required since higher-order moments can be infinite (Bernadara *et al.*, 2007). Secondly, the fractal co-dimensions of large singularities are difficult to estimate because they correspond to extreme values of the process. As a result, estimation is often carried out using the moment scaling function, which is obtained by plotting the moments of the rainfall, $E\lfloor(Y_t^{(h)})^q\rfloor$ as a function of scale h on a log–log plot and considering the slopes of the resulting lines as a function of the moment order q. Random cascade models are often calibrated by minimising the difference between the observed and theoretical moment scaling functions, but this leads to a biased estimator with high sampling variability whose properties are otherwise not well known (Gaume *et al.*, 2007). A final difficulty is that the construction of a random cascade model involves an infinite range of scales, whereas in practice a cascade can be observed only over a finite range; this increases the errors in the parameter estimates (Bernadara, 2007).

Despite these calibration issues, multifractal cascades are parsimonious models with a proven track record in hydrological applications, in particular as disaggregators of observed data to finer timescales (Onof *et al.*, 2005). While Poisson-cluster models have been shown to reproduce the scaling properties of rainfall (Olsson and Burlando, 2002; Onof *et al*, 1996), they require more parameters than the scaling models. On the other hand, these parameters provide a more intuitive understanding of the structure of the rainfall process in terms of cells and storms. This makes them suitable for examining the impact of climate change upon local rainfall properties (Leith, 2006), although our experience is that the extremal performance can be sensitive to model choice. It seems that both approaches therefore have their strengths, and it may be that there are ranges of scales over which the one type ought to be preferred to the other as Marani (2003) shows by looking at how the variance changes with the scale of observation.

7.4.2. *Incorporation of climate change projections*

The incorporation of climate change projections is less well developed for sub-daily than for daily rainfall simulations. A simple approach is to start by generating daily series using any of the methods reviewed above, and then to disaggregate it to the required timescale. In principle, multifractal methods are ideally suited to this (Onof and Arnbjerg-Nielsen, 2009), at least if the issues surrounding parameter estimation and uncertainty can be addressed satisfactorily. Poisson cluster models have also been used for this purpose; see, for example, Koutsoyiannis and Onof (2001). Any such approach implies, either directly or indirectly, some kind of relationship between daily and sub-daily rainfall structure; an implicit assumption is that this relationship will not change in the future.

As far as we are aware, most current work on direct sub-daily rainfall generation under climate change scenarios is based on perturbing the parameters of Poisson cluster models. The first work in this direction seems to be that of Kilsby *et al.* (1998), who used generalised linear models to predict both the mean daily rainfall and probability of a dry day for any site in England and Wales for each month of the year, based on measures of atmospheric circulation. They suggested that by extracting the corresponding atmospheric predictors from climate model simulations, estimates of future daily rainfall statistics could be derived; Poisson cluster models could then be fitted to historical rainfall data and, for future simulations, the storm arrival rates and mean cell intensities could be perturbed to match the projected daily statistics (with the remaining parameters held at their historical values).

An alternative approach, which can be used if a sufficiently realistic daily rainfall generator is available, is to calculate the statistical properties of simulated daily series and then to exploit relationships between rainfall statistics at different timescales to infer the corresponding sub-daily properties. For example, Marani and Zanetti (2007) derived relationships between rainfall variances at different timescales, based on assumed forms of autocorrelation function (e.g. exponential or power law decay) of the underlying continuous-time process. Figure 7.4 illustrates another possibility. It shows the variation of two summary statistics (standard deviation and proportion of wet intervals) for time series of rainfall amounts aggregated to

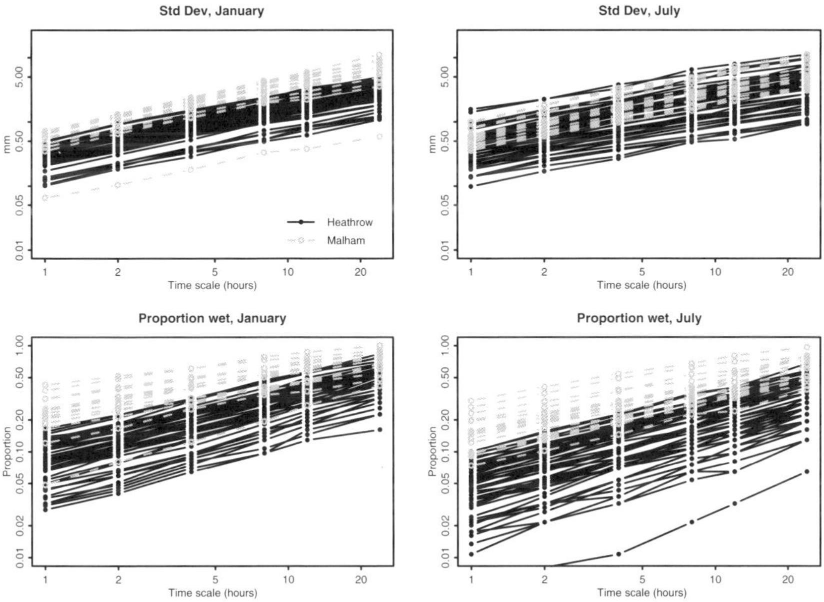

Figure 7.4. Variation of rainfall summary statistics with timescale for the months of January (left) and July (right), at two contrasting sites in the UK. All axis scales are logarithmic. Black lines are for data from Heathrow airport, 1949–2001; grey lines are for data from Malham Tarn, 1985–2006. Heathrow airport is at an altitude of 25 m in southern England and experiences an annual rainfall of just over 600 mm; Malham Tarn is at an altitude of 391 m in northern England and experiences an annual rainfall of around 1400 mm. Statistics are plotted separately for each year, for 1-, 2-, 4-, 6-, 12- and 24-hourly rainfall accumulations.

different temporal resolutions from 1 to 24 hours, at two contrasting sites in the UK. With logarithmic scales on both axes, all of the individual curves here — which are derived from different months and years — are roughly linear. The truly remarkable feature, however, is that the slopes of the lines for each property are almost identical; we have found similar results at a variety of other sites in the UK. From such a plot, knowledge of the 24-hour value of any statistic will clearly enable the corresponding sub-daily values to be calculated with high precision. Although the physical reason for this is unclear, the similarity of the slopes for different sites, seasons and years suggests that they are insensitive to changes in atmospheric conditions and hence that they will remain valid in a moderately altered climate.

Unfortunately, a deeper investigation (Leith, 2006) of the relationships apparent from Figure 7.4 reveals that the scaling relationships for the standard deviation cannot be parallel lines. For example, if the slopes were all identical then the lag 1 autocorrelation at any timescale would be the same for all months, sites and years: in the UK however, autocorrelation is weaker in the summer than in the winter, due to the predominance of convective and frontal/synoptic systems respectively. Similarly, if the relationships were exactly linear then the lag 1 autocorrelation would be the same at all timescales. Some small adjustments are therefore required; these do not compromise the overall message of highly stable scaling relationships, but they do enable the autocorrelation structure of sub-daily rainfall sequences to be reconstructed realistically. One possibility is to allow for a quadratic dependence on timescale and to allow the log properties to depend on atmospheric covariates in addition to timescale. This was done by Leith (2006), who then used the fitted relationships in conjunction with GLM simulations, such as those illustrated in Figure 7.3, to derive expected values of future daily and sub-daily summary statistics at different UK locations under the SRES A2 emissions scenario. These summary statistics were used to calibrate a Poisson cluster model, which was used to simulate future sub-daily rainfall sequences. A comparison with the model parameters obtained from historical simulations revealed some interesting changes: for example, estimates of future storm arrival rates were lower in the summer and higher in the winter, whereas estimates of mean cell intensity were increased year-round (but particularly in the summer). Systematic changes were also observed in all other model parameters: this suggests that the procedure proposed by Kilsby *et al.* (1998), in which only two parameters are allowed to change in a future climate, may be over-simplified. Changes in the other parameters were, however, site-specific. For example, at Manchester airport the results suggested that future summer storms will have fewer, longer-lasting cells on average than in the past. By contrast, at Heathrow airport there was little change in the number of cells per storm, but some indication of reduced mean cell durations in the autumn.

It is worth considering the uncertainty in parameter estimates obtained via the method just outlined. Notice first that, conditional on the parameters of the daily rainfall generator and the climate model used to drive the simulations, the properties of daily rainfall sequences can be estimated arbitrarily precisely, simply by running a large enough number of daily rainfall simulations. Uncertainty in the corresponding sub-daily properties is small, due to the strength of the scaling relationships illustrated in Figure 7.4. It follows that uncertainty in the Poisson cluster model parameters is also small. The conclusion is that the dominant sources of uncertainty here are the daily rainfall generator and the choice of climate model. As noted previously, daily rainfall model parameters can usually be identified fairly precisely and there is little difference in performance between the best available models. Ultimately therefore, the main source of uncertainty in the sub-daily model parameters is likely to be the choice of climate model used to drive the daily simulations. Having said this, in qualitative terms at least the projected changes in model parameters obtained by Leith (2006) were remarkably similar for four different climate models. Moreover, experience of this type of approach is currently limited and it is not clear whether the uncertainty in parameter estimates contributes significantly to the overall uncertainty in aspects of sub-daily rainfall simulations that are relevant for flood risk assessment — it is conceivable that the variability in sub-daily rainfall simulations dominates any parameter uncertainty.

The techniques just described are mainly for single-site rainfall. Few methods are currently available for the simulation of multi-site sub-daily rainfall incorporating scenarios of climate change. Fowler *et al.* (2005) describe one possibility in which parameters of a space–time Neyman–Scott model are obtained separately for different weather states; sequences of these weather states are extracted from future climate model simulations, and the appropriate Neyman–Scott model parameters are then used to generate rainfall simulations (this is equally applicable to single-site problems, of course). The climate change signal in this approach is reflected in changes in the individual weather state frequencies. An alternative approach was proposed by Segond *et al.* (2006), who suggested disaggregating a multisite daily rainfall simulator by using a single-site method to disaggregate the daily totals at one site (the "master gauge") and then, for each day, scaling the totals at the other sites by the corresponding temporal profile. Although the assumption of a uniform sub-daily temporal profile at all sites may appear unrealistic except at very local scales, the approach has been tested at catchment scales up to 2000 km^2 with reasonable results.

7.5. Propagation of Rainfall Uncertainties into Flow Estimates

While it is perhaps self-evident that uncertainties in rainfall inputs will lead to uncertainties in the simulation of river flows, a general understanding of the relationships between rainfall uncertainties and flow response is not yet available. One complication is that most studies of the effects of rainfall uncertainty have been based on rainfall-runoff modelling, in which it is difficult to separate effects of rainfall-runoff model error. In fact, most research into uncertainty in rainfall-runoff modelling has effectively aggregated all sources of uncertainty into that associated with the rainfall-runoff model parameters (Wagener *et al.*, 2004). Notable exceptions include the work of Cameron *et al.* (1999, 2000a,c) and Blazkova and Beven (2002), who used the GLUE methodology to identify near-optimal parameter combinations for a rainfall generator and rainfall-runoff model and to characterise the uncertainty in the combined system. However, their primary goal was to quantify, rather than explain, the uncertainty: in addition to this, it is of interest to determine what features of rainfall inputs are most important when it comes to characterising (and potentially reducing) the uncertainty in system response.

Issues of temporal resolution are relatively well understood at a basic level. In the use of simple rainfall-runoff models, it is recognised that the required temporal resolution of the rainfall depends on the response time of the catchment (e.g. NERC, 1975). There has also been a recognition that realistic sequences of precipitation are needed for a wide range of flood design applications, which provides some of the reasoning behind a move in design practice from design storms to the use of stochastic simulations of rainfall time series (e.g. Onof *et al.*, 1996). It is not always appreciated that physics-based hydrological models also have requirements for temporal resolution. For example if "physically-based" models are run on a daily time step, rainfall intensities will be lower, often dramatically so, than instantaneous rainfall intensities, and soil hydraulic parameters will need to be adjusted accordingly in order to generate the correct physical process response. This is a particular problem for arid areas, where convective rainfall can generate infiltration-excess overland flow (e.g. Al-Qurashi *et al.*, 2008).

Moving beyond those basic principles, however, a major complication for any consideration of the temporal distribution of rainfall is that for most hydrological applications, some level of spatial aggregation of rainfall is required. This could be a catchment-scale average, as in the input to a spatially-lumped rainfall-runoff model, a sub-catchment average for a semi-distributed hydrological model, or a pixel average for a fully-distributed model. Clearly the temporal properties of single-site rainfall will be different from a spatial average and hence a more general problem of the characterisation of spatial–temporal rainfall fields must be confronted.

As briefly discussed above, the significance of spatial variability of rainfall for flow estimation is a complex issue. A recent summary of the literature and associated issues is presented by Segond *et al.* (2007). Effects of rainfall spatial variability will depend on rainfall properties, catchment type and spatial scale and are also likely to be influenced by antecedent conditions. Intuitively, the significance of spatial variability of precipitation for streamflow might be expected to increase with increasing catchment scale, but as catchment scale increases, the damping effects of catchment response become greater. Thus the effects on streamflow depend on the extent to which rainfall variability is damped by the catchment, which in turn is dependent on catchment characteristics and the associated runoff processes.

Problems are at their most severe where convective rainfall predominates. This is a characteristic of many arid areas, for which intense rainfall is likely to be associated with rapid runoff from overland flow. The history of rainfall gauging at the Walnut Gulch experimental catchment in Arizona (Goodrich *et al.*, 2008) is instructive. In a 149 km^2 catchment, rain gauge numbers were progressively increased to a maximum of 95, in an attempt to characterise the structure of the predominant convective rainfall. Michaud and Sorooshian (1994) investigated flood runoff from convective thunderstorm rainfall at Walnut Gulch, which showed that high spatial resolution of rainfall data (2 km) is essential to simulate flood peaks (coarser 4 km resolution data led to underestimation of flood peaks by 50–60%). Elsewhere, Al-Qureshi *et al.* (2008) have shown for an arid catchment in Oman, that uncertainty in spatial rainfall is the dominant influence on flow simulation. More generally, Wheater (2008) argues that progress in rainfall-runoff modelling for arid areas is dependent on developments in rainfall simulation; available data from conventional sources are generally inadequate to characterise spatial rainfall adequately to support a reasonable accuracy of model calibration and validation. In more humid climates, convective rainfall remains problematic, for similar reasons of lack of observability. Although radar networks are now reaching the resolution where they can be used to observe convective rainfall at an appropriate resolution, appropriate analysis of fine scale structure has not been undertaken to support design procedures which recognise spatial rainfall structure.

In humid areas, the effects of rainfall variability are expected to be most marked for urban areas, where runoff is rapid, and the attenuation of rainfall variability due to runoff processes is minimal. Numerous studies of urban catchment response show sensitivity to rainfall variability in space and time, including storm movement (see e.g., Ngirane-Katashaya and Wheater, 1985; Singh, 1997). In contrast, Naden (1992) found for the 10,000 km^2 Thames basin that the effect of rainfall spatial variation on channel network response could be marked, but that the slow response of chalk catchments damped out effects of rainfall variability.

Segond *et al.* (2007) investigated the effects of rainfall spatial variability and rainfall uncertainty for the 1400 km^2 River Lee catchment, UK, using a network of 17 rain gauges and 2 km resolution radar data. The sub-catchments have a mixed geology, including (responsive) clay and (damped response) chalk catchments, and urban areas. Results for rural catchments showed that runoff was sensitive to the type of rainfall event, rainfall representation, catchment scale and catchment type, and least sensitive to rainfall spatial variability at the whole catchment scale. Urbanised sub-catchments showed greater sensitivity to spatial rainfall, but this effect was not seen at whole catchment scale. Overall the results confirmed the importance of appropriate representation of rainfall spatial structure, although this was strongly dependent on rainfall type. Effects were most marked for smaller catchments and urbanised response.

We can conclude that as semi-distributed modelling of catchments is increasingly undertaken to support flood management, and continuous simulation is needed to represent effects of antecedent conditions, appropriate modelling to represent the spatial distribution of rainfall will be required. The greater importance of this for small catchments is noteworthy, since such catchments have generally been considered in design practice as those where spatial averaging of rainfall can most appropriately be used. The strong sensitivity of urban areas to rainfall spatial structure points to the need for research into the fine-scale spatial–temporal properties of rainfall, which have largely been neglected.

7.5.1. *Current issues*

The review in this chapter has focused on the representation of natural variability in rainfall sequences, as this often dominates other sources of uncertainty here. We have also seen that the choice of climate model can represent a significant source of uncertainty in future rainfall projections; this is discussed more fully in Chapter 22. Since there are often substantial disagreements between different climate models, particularly at the local and regional scales that are usually of interest for flood risk assessment, an important question is to determine how much useful information is actually present in climate model simulations for any given situation. One way to achieve this is via a Bayesian approach, in which the posterior and prior distributions of quantities of interest can be compared to determine the information content of the climate model data (see Section 6 of Leith and Chandler 2010, for example).

There is an emerging consensus that, for daily rainfall at least, a variety of different modelling approaches are currently available that have comparable and impressive performance with respect to many of the key features of rainfall sequences, including extremes. Therefore the uncertainty due to the choice of rainfall model is often relatively minor (subject, of course, to the use of a model that is able to capture the key properties of interest). Nonetheless, in all cases there is scope for further development to improve performance with respect to specific features that may be of interest in particular applications; and the extremal performance of sub-daily rainfall models needs to be studied more thoroughly than has hitherto been attempted. There are also features, such as the spatial dependence of extremes and the persistence of abnormally high rainfall, which may be relevant in flood risk assessment but are not routinely examined when assessing the quality of rainfall simulations. One feature that is often poorly represented at present is extreme summer rainfall in temperate climates. For example, Yang *et al.* (2005) found that their GLM simulations of daily rainfall in south-east England tended to underestimate extremes during the summer and to overestimate during the winter, although the simulated annual maxima agreed closely with those obtained from a conventional extreme value analysis. This underestimation of summer extremes, which in the UK are associated with intense convective rainfall events, is a feature of many other simulation techniques as well; as far as GLMs are concerned, the problem appears to be related to the assumption of a common shape parameter, which can be relaxed as indicated above.

As far as we are aware, little work has been done to assess the impact of parameter uncertainty in stochastic rainfall simulations. As argued above, one may reasonably expect that this will be relatively minor where long sequences of data are available; however, it would be interesting to carry out a study to confirm this. This is particularly true in situations where several models are used in tandem to produce a rainfall simulation (for example, the use of GLMs in conjunction with scaling relationships and Poisson cluster models to generate sub-daily simulations under scenarios of climate change).

Finally, rainfall data themselves represent an important source of uncertainty in rainfall inputs. For example, we have argued above that the common practice of interpolating rain gauge observations onto a regular grid is problematic and that, as with other missing data problems, a multiple imputation approach is preferable. However, there are few models currently available that have been designed specifically to facilitate this; indeed, the only ones that we are aware of are those of Yang *et al.* (2005) and Wilks (2009). More generally, where rain gauge data are not available or are not suitable, it is generally necessary to use indirect measurements such as those obtained from radar. Our experience is that the uncertainty in such indirect measurements is often extremely large and therefore that it must be accounted for in any analysis. It is imperative, therefore, that providers of indirectly measured rainfall data supply credible, quantitative assessments of uncertainty with their products.

References

Al-Qurashi, A., McIntyre, N., Wheater, H. *et al.* (2008). Application of the Kineros2 rainfall-runoff model to an arid catchment in Oman, *J. Hydrol.*, **355**, 91–105.

Bárdossy, A. and Plate, E.J. (1992). Space–time model for daily rainfall using atmospheric circulation patterns, *Wat. Resour. Res.*, **28**, 1247–1259.

Beckmann, B.-R. and Buishand, T.A. (2002). Statistical downscaling relationships for precipitation in the Netherlands and North Germany, *Int. J. Climatol.*, **22**, 15–32.

Beniston, M., Stephenson, D., Christensen, O. *et al.* (2007). Future extreme events in European climate: an exploration of regional climate model projections, *Climatic Change,* **81**, 71–95.

Bernadara, P., Lang, M., Sauquet, E. *et al.* (2007). *Multifractal Analysis in Hydrology*, Quae Editions, Versailles.

Blazkova, S. and Beven, K.J. (2004). Flood frequency estimation by continuous simulation of subcatchment rainfalls and discharges with the aim of improving dam safety assessment in a large basin in the Czech Republic, *J. Hydrol.*, **292**, 153–172.

Blenkinsop, S. and Fowler, H.J. (2007). Changes in drought frequency, severity and duration for the British Isles projected by the PRUDENCE regional climate models, *J. Hydrol.*, **342**, 50–71.

Bowman, A.W. and Azzalini, A. (1997). *Applied Smoothing Techniques for Data Analysis — The Kernel Approach with S-Plus Illustrations*, Oxford University Press, Oxford.

Buishand, T.A. and Brandsma, T. (2001). Multisite simulation of daily precipitation and temperature in the Rhine basin by nearest-neighbour resampling, *Water Resour. Res.*, **37**, 2761–2776.

Burton, A., Kilsby, C.G., Fowler, H.J. *et al.* (2008). RainSim: a spatial–temporal stochastic rainfall modelling system, *Environ. Modell. Softw.*, **23**, 1356–1369.

Cameron, D.S., Beven, K.J. and Naden, P. (2000a). Flood frequency estimation under climate change (with uncertainty), *Hydrol. Earth Syst. Sc.*, **4**, 393–405.

Cameron, D.S., Beven, K.J., Tawn, J. *et al.* (1999). Flood frequency estimation by continuous simulation for a gauged upland catchment (with uncertainty), *J. Hydrol.*, **219**, 169–187.

Cameron, D.S., Beven, K.J. and Tawn, J. (2000b). An evaluation of three stochastic rainfall models, *J. Hydrol.*, **228**, 130–149.

Cameron, D.S., Beven, K.J., Tawn, J. *et al.* (2000c). Flood frequency estimation by continuous simulation (with likelihood based uncertainty estimation), *Hydrol. Earth Syst. Sci.*, **4**, 23–34.

Cameron, D., Beven, K. and Tawn, J. (2001). Modelling extreme rainfalls using a modifed random pulse Bartlett-Lewis stochastic rainfall model (with uncertainty), *Adv. Wat. Resour.*, **24**, 203–211.

Casson, E.A. and Coles, S.G. (1999). Spatial regression models for extremes, *Extremes*, **1**, 449–468.

Chandler, R.E. (1997). A spectral method for estimating parameters in rainfall models, *Bernoulli*, **3**, 301–322.

Chandler, R.E. (2002). GLIMCLIM: *Generalised linear modelling for daily climate time series (software and user guide)*, Research Report No. 227, Department of Statistical Science, University College London (http://www.ucl.ac.uk/Stats/research/abs02.html#227).

Chandler, R.E. (2005). On the use of generalized linear models for interpreting climate variability, *Environmetrics*, **16**, 699–715.

Chandler, R.E. and Bate, S. (2007). Inference for clustered data using the independence loglikelihood, *Biometrika*, **94**, 167–183.

Chandler, R.E., Isham, V.S., Bellone, E. *et al.* (2007). "Space–time modelling of rainfall for continuous simulation" in: Finkenstadt, B., Held, L., Isham, V.S. (eds), *Statistical Methods for Spatial–Temporal Systems*, Boca Raton, CRC Press.

Chandler, R.E. and Wheater, H.S. (2002). Analysis of rainfall variability using generalized linear models: a case study from the west of Ireland, *Water Resour. Res.*, **38**, 1192.

Charles, S.P., Bates, B.C, Smith, I. *et al.* (2004). Statistical downscaling of daily precipitation from observed and modelled atmospheric fields, *Hydrol. Process.*, **18**, 1373–1394.

Charles, S.P., Bates, B.C., Whetton, P.H. *et al.* (1999). Validation of downscaling models for changed climate conditions: case study of southwestern Australia, *Climate Research*, **12**, 1–14.

Chow, V.T. (ed.) (1964). *Handbook of Applied Hydrology*, McGraw-Hill, New York.

Coe, R. and Stern, R. (1982). Fitting models to daily rainfall, *J. Appl. Meteorol.*, **21**, 1024–1031.

Coles, S.G. (2001). *An Introduction to Statistical Modelling of Extreme Values*, Springer, London.

Coles, S.G. and Dixon, M.J. (1999). Likelihood-based inference for extreme value models, *Extremes*, **2**, 5–23.

Coles, S.G. and Powell, E.A. (1996). Bayesian methods in extreme value modelling: a review and new developments, *Int. Statist. Rev.*, **64**, 119–136.

Coles, S.G. and Tawn, J.A. (1996a). A Bayesian analysis of extreme rainfall data, *Appl. Statist.*, **45**, 463–478.

Coles, S.G. and Tawn, J.A. (1996b). Modelling extremes of the areal rainfall process, *J. R. Statist. Soc. B*, **58**, 329–347.

Cooley, D., Nychka, D. and Naveau, P. (2007). Bayesian spatial modelling of extreme precipitation return levels, *J. Am. Statist. Ass.*, **102**, 820–840.

Cowpertwait, P.S.P. (1994). A generalized point process model for rainfall, *Proc. Roy. Soc.*, **A447**, 23–37.

Cowpertwait, P.S.P. (1995). A generalized spatial–temporal model for rainfall based on a clustered point process, *Proc. Roy. Soc.*, **A450**, 163–175.

Cowpertwait, P.S.P. (1998). A Poisson-cluster model of rainfall: some high-order moments and extreme values, *Proc. Roy. Soc.*, **A454**, 885–898.

Cowpertwait, P.S.P., Isham, V.S., Onof, C. (2007). Point process models of rainfall: developments for fine-scale structure, *Proc. Roy. Soc.*, **A463**, 2569–2587.

Cox, D.R. and Isham, V.S. (1994). "Stochastic models of precipitation" in: Barnett, V.D. and Turkman, K.F. (eds), *Statistics for the Environment, 2, Water Related Issues*, Wiley, Chichester, pp. 3–18.

Cox, D.R., Isham, V.S. and Northrop, P.J. (2002). Floods: some probabilistic and statistical approaches, *Phil. Trans. R. Soc. Lond.*, **A360**, 1389–1408.

Davison, A.C. (2003). *Statistical Models*, Cambridge University Press, Cambridge.

Davison, A.C. and Hinkley, D.V. (1997). *Bootstrap Methods and their Application*. Cambridge University Press, Cambridge.

Davison, A.C. and Smith, R.L. (1990). Models for exceedances over high thresholds, *J. R. Statist. Soc. B*, **52**, 393–442.

Deidda, R., Benzi, R. and Siccardi, F. (1999). Multifractal modeling of anomalous scaling laws in rainfall, *Water Resour. Res.*, **35**, 1853–1867.

Dunn, P.K. (2004). Occurrence and quantity of precipitation can be modelled simultaneously, *Int. J. Climatol.*, **24**, 1231–1239.

Eastoe, E.F. (2009). A hierarchical model for non-stationary multivariate extremes: a case study of surface-level ozone and NOx data in the UK, *Environmetrics*, **20**, 428–444.

Eastoe, E.F. and Tawn, J.A. (2009). Modelling non-stationary extremes with application to surface-level ozone, *J. R. Stat. Soc. C,* **58**, 25–45.

Evin, G. and Favre, A. -C. (2008). A new rainfall model based on the Neyman-Scott process using cubic copulas, *Water Resour. Res.*, **44**, doi: 10.1029/2007WR006054.

Fawcett, L. and Walshaw, D. (2006). A hierarchical model for extreme wind speeds, *J. R. Stat. Soc. C*, **55**, 631–646.

Fawcett, L. and Walshaw, D. (2007). Improved estimation for temporally clustered extremes, *Environmetrics*, **18**, 173–188.

Fealy, R. and Sweeney, J. (2007). Statistical downscaling of precipitation for a selection of sites in Ireland employing a generalised linear modelling approach, *Int. J. Climatol.*, **15**, 2083–2094.

Ferro, C.A.T. and Segers, J. (2003). Inference for clusters of extreme values, *J. R. Stat. Soc. B*, **65**, 545–556.

Finch, J.W., Bradford, R.B. and Hudson, J.A. (2004). The spatial distribution of groundwater flooding in a chalk catchment in southern England, *Hydrol. Process.*, **18**, 959–971.

Fisher, R.A. and Tippett, L.H.C. (1928). Limiting forms of the frequency distribution of the largest or smallest member of a sample, *P. Camb. Philol. Soc.*, **24**, 180–190.

Fowler, H.J., Blenkinsop, S. and Tebaldi, C. (2007). Linking climate change modelling to impacts studies: recent advances in downscaling techniques for hydrological modelling, *Int. J. Climatol.*, **27**, 1547–1578.

Fowler, H.J., Kilsby, C.G., O'Connell, P.E. *et al.* (2005). A weather-type conditioned multi-site stochastic rainfall model for the generation of scenarios of climate variability and change, *J. Hydrol.*, **308**, 50–66.

Furrer, E.M. and Katz, R.W. (2007). Generalized linear modeling approach to stochastic weather generators, *Clim. Res.*, **34**, 129–144.

Gabriel, K.R. and Neumann, J. (1962). A Markov chain model for daily rainfall in Tel Aviv, *Q. J. Roy. Meteor. Soc.*, **88**, 90–95.

Gaume, E., Mouhous, N. and Andrieu, H. (2007). Rainfall stochastic disaggregation models: calibration and validation of a multiplicative cascade model, *Adv. Water Resour.*, **30**, 1301–1319.

Giorgi, F. and Mearns, L.O. (2002). Calculation of average, uncertainty range, and reliability of regional climate changes from AOGCM simulations via the "Reliability Ensemble Averaging" (REA) method, *J. Climate*, **15**, 1141–1158.

Gnedenko, B. and Kolmogorov, A.N. (1968). *Limit Distributions for Sums of Independent Random Variables*, Addison-Wesley, New York.

Grunwald, G.K. and Jones, R.H. (2000). Markov models for time series with mixed distribution, *Environmetrics*, **11**, 327–339.

Gupta, V.K. and Waymire, E.C. (1993). A statistical analysis of mesoscale rainfall as a random cascade, *J. Appl. Meteorol.*, **32**, 251–267.

Hall, A.R. (2005). *Generalized Method of Moments*, Oxford University Press, Oxford.

Harte, D. (2001). *Multifractals: Theory and Applications*, Chapman & Hall, CRC, London.

Haylock, M.R., Cawley, G.C., Harpham, C. *et al.* (2006). Downscaling heavy precipitation over the United Kingdom: a comparison of dynamical and statistical methods and their future scenarios, *Int. J. Climatol.*, **26**,

1397–1415.

Heffernan, J. and Tawn, J.A. (2004). A conditional approach for multivariate extreme values (with discussion), *J. R. Stat. Soc.* B, **66**, 497–546.

Hosking, J.R.M., Wallis, J.R. and Wood, E.F. (1985). Estimation of the generalized extreme-value distribution by the method of probability-weighted moments, *Technometrics*, **27**, 251–261.

Hughes, J.P., Guttorp, P. and Charles, S.P. (1999). A non-homogeneous hidden Markov model for precipitation occurrence, *Applied Statistics*, **48**, 15–30.

Hulme, M., Jenkins, G.J., Lu, X. *et al.* (2002). *Climate Change Scenarios for the United Kingdom: The UKCIP02 Scientific Report*, Tyndall Centre for Climate Change Research, School of Environmental Sciences, University of East Anglia, Norwich, UK. Available from http://www.ukcip.org.uk/resources/publications/.

Hyndman, R.J. and Grunwald, G.K. (2000). Generalized additive modelling of mixed distribution Markov models with application to Melbourne's rainfall, *Aust. N.Z. J. Stat.*, **42**, 145–158.

Institute of Hydrology (1999). *Flood Estimation Handbook*, Institute of Hydrology, Wallingford, UK.

IPCC (2001). *IPCC Special Report on Emissions Scenarios*, Cambridge University Press, Cambridge.

Islam, S., Entekhabi, D., Bras, R.L. *et al.* (1990). Parameter estimation and sensitivity analysis for the modified Bartlett-Lewis rectangular pulses model of rainfall, *J. Geophys. Res.*, **95**, 2093–2100.

Jesus, J. and Chandler, R.E. (2011). Estimating functions and the generalized method of moments, *Interface Focus*, **1**, 871–885.

Kakou, A. (1998). A Point Process Model of Rainfall with a Dependent Duration and Intensity. *Research Report* 196, Department of Statistical Science, University College London. Available at http://www.ucl.ac.uk/stats/research/Resrprts/abs98.html#196 (Accessed on 05/04/2012).

Kalnay, E., Kanamitsu, M., Kistler, R. *et al.* (1996). The NCEP/NCAR 40-year reanalysis project, *B. Am. Meteorol. Soc.*, **77**, 437–471.

Katz, R.W. and Parlange, M.B. (1998). Overdispersion phenomenon of stochastic modeling of precipitation, *J. Climate,* **11**, 591–601.

Kilsby, C.G., Cowpertwait, P.S.P., O'Connell, P.E. *et al.* (1998). Predicting rainfall statistics in England and Wales using atmospheric circulation variables, *Int. J. Climatol.*, **18**, 523–539.

Koutsoyiannis, D. (2004). Statistics of extremes and estimation of extreme rainfall: II. Empirical investigation of long rainfall records, *Hydrolog. Sci. J.,* **49**, 591–10.

Koutsoyiannis D. and Onof, C. (2001). Rainfall disaggregation using adjusting procedures on a Poisson cluster model, *J. Hydrol.,* **246**, 109–122.

Leith, N.A. (2006). Point process models for sub-daily rainfall simulation, *Technical Report No.3, DEFRA Project FD2113.* Available at http://www.ucl.ac.uk/Stats/research/Rainfall/FD2113_rpt3.pdf (Accessed on 14/04/2012).

Leith, N.A. (2008). *Single-Site Rainfall Generation Under Scenarios of Climate Change*, PhD thesis, Department of Statistical Science, University College London.

Leith, N.A. and Chandler, R.E. (2010). A framework for interpreting climate model outputs, *J. R. Stat. Soc.* C, **59**, 279–296.

Lovejoy, S. and Schertzer, D. (1995). "Multifractals and Rain" in: Kundzewicz, Z.W. (ed.), *New Uncertainty Concepts in Hydrology and Water Resources*, Cambridge University Press, Cambridge, pp. 62–103.

Marani, M. (2003). On the correlation structure of continuous and discrete point rainfall, *Water Resour. Res.*, **39**, 1128.

Marani, M. and Zanetti, S. (2007). Downscaling rainfall temporal variability, *Water Resour. Res.*, **43**, W09415.

Maraun, D., Wetterhall, F., Ireson, A.M. *et al.* (2010). Precipitation downscaling under climate change — recent developments to bridge the gap between dynamical models and the end user, *Rev. Geophys.*, **48**, RG3003, doi: 10.1029/2009RG000314.

Marsan, D., Schertzer, D. and Lovejoy, S. (1996). Causal space–time multifractal processes: predictability and forecasting of rain fields, *J. Geophys. Res.*, **101**, 26333–26346.

Maurer, E. (2007). Uncertainty in hydrologic impacts of climate change in the Sierra Nevada, California under two emissions scenarios, *Climatic Change*, **82**, 309–325.

Menabde, M., Harris, D., Seed, A. *et al.* (1997). Multiscaling properties of rainfall and bounded random cascades, *Water Resour. Res.*, **33**, 2823–2830.

Met Office (2005). "Boscastle and North Cornwall Post Flood Event Study — Meteorological Analysis of the conditions leading to Flooding on 16 August 2004" in: Goulding, B. (ed.), *Forecasting Research Technical Report N. 459*, Met Office, Exeter.

Naden, P. (1992). Spatial variability in flood estimation for large catchments: the exploitation of channel network structure, *Hydrolog. Sci. J.*, **37**, 53–71.

NERC (1975). *Flood Studies Report*, Natural Environment Research Council, London.

Ngirane-Katashaya, G. and Wheater, H.S. (1985). Hydrograph sensitivity to storm kinematics, *Water Resour. Res.*, **21**, 337–345.

Northrop, P. (2006). Estimating the parameters of rainfall models using maximum marginal likelihood, *Student*, **5**, 173–183.

Onof, C. and Arnbjerg-Nielsen, K. (2009). Quantification of anticipated future changes in high resolution design rainfall for urban areas, *Atmos. Res.*, **92**, 350–363.

Onof, C., Chandler, R.E., Kakou, A. *et al.* (2000). Rainfall modelling using Poisson-cluster processes: a review of developments, *Stoch. Env. Res. Risk A.*, **14**, 384–411.

Onof, C., Faulkner, D. and Wheater, H.S. (1996). Design rainfall modelling in the Thames catchment, *Hydrolog. Sci. J.*, **41**, 715–733.

Onof, C. and Hege, L. (2009). Modifying a widely used rainfall model to correct for extreme value biases, *in preparation.*

Onof, C., Northrop, P., Wheater, H.S. *et al.* (1996). Spatiotemporal storm structure and scaling property analysis for modeling, *J. Geophy. Res.*, **101**, 26415–26425.

Onof, C., Townend, J. and Kee, R. (2005). Comparison of two hourly to 5-min rainfall disaggregators, *Atmos. Res.*, **77**, 176–187.

Onof, C. and Wheater, H.S. (1993). Modelling of British rainfall using a random parameter Bartlett-Lewis rectangular pulse model, *J. Hydrol.*, **149**, 67–95.

Onof, C. and Wheater, H.S. (1994). Improvements to the modelling of British rainfall using a modified random parameter Bartlett–Lewis rectangular pulse model, *J. Hydrol.*, **157**, 177–195.

Onof, C., Yameundjeu, B., Paoli, J.-P. *et al.* (2002). A Markov modulated Poisson process model for rainfall increments, *Water Sci. Technol.*, **45**, 91–97.

Over, T.M. and Gupta, V.K. (1996). A space–time theory of mesoscale rainfall using random cascades, *J. Geophys. Res.*, **101**, 26319–26331.

Paulson, K.S. (2004). Fractal interpolation of rain rate time-series, *J. Geophys. Res. — Atmosphere*, **109**, D22102.

Pickands, J. (1975). Statistical inference using extreme order statistics, *Ann. Stat.*, **3**, 119–131.

Pitt, M. (2008). *The Pitt Review. Learning the Lessons from the 2007 floods*, Cabinet Office, London.

Prudhomme, C., Reynard, N. and Crooks, S. (2002). Downscaling of global climate models for flood frequency analysis: where are we now? *Hydrological Processes*, **16**, 1137–1150.

Ramesh, N.I. and Davison, A.C. (2002). Local models for exploratory analysis of hydrological extremes, *J. Hydrol.*, **256**, 106–119.

Rodriguez-Iturbe, I., Cox, D.R. and Isham, V. (1987). Some models for rainfall based on stochastic point processes, *P. R. Soc. London A*, **410**, 269–288.

Rodriguez-Iturbe, I., Cox, D.R., Isham, V. (1988). A point process model for rainfall: further developments, *P. R. Soc. London A*, **417**, 283–298.

Sansó, B. and Guenni, L. (2000). A nonstationary multisite model for rainfall, *J. Am. Stat. Assoc.*, **95**, 1089–1100.

Schertzer, D. and Lovejoy, S. (1997). Universal multifractals do exist!: Comments on "A statistical analysis of mesoscale rainfall as a random cascade", *J. Appl. Meteor.*, **36**, 1296–1303.

Schmidli, J., Goodess, C.M., Frei, C. (2007). Statistical and dynamical downscaling of precipitation: an evaluation and comparison of scenarios for the European Alps, *J. Geophys. Res. — Atmosphere*, **112**, D04105–D04125.

Seed, A.W. (2003). A dynamic and spatial scaling approach to advection forecasting, *J. Appl. Meteor.*, **42**, 381–388.

Segond, M.-L., Onof, C. and Wheater, H.S. (2006). Spatial–temporal disaggregation of daily rainfall from a generalized linear model, *J. Hydrol.*, **331**, 674–689.

Segond, M.-L., Wheater, H.S. and Onof, C. (2007). The significance of spatial rainfall representation for flood runoff estimation: a numerical evaluation based on the Lee catchment, UK. *J. Hydrol.*, **347**, 116–131.

Singh, V. (1997). Effect of spatial and temporal variability in rainfall and watershed characteristics on stream flow hydrograph, *Hydrol. Process.*, **11**, 1649–1669.

Smith, R.L. (1985). Maximum likelihood estimation in a class of non-regular case, *Biometrika*, **72**, 67–90.

Smith, R.L. (1989). Extreme value analysis of environmental time series: an application to trend detection in ground-level ozone, *Stat. Sci.*, **4**, 367–377.

Smith, R.L. (1990). Regional estimation from spatially dependent data. Available from http://www.stat.unc.edu/postscript/rs/regest.pdf (Accessed 03/09/2012).

Smith, R.L. (1999a). "Bayesian and Frequentist Approaches to Parametric Predictive Inference" in: Bernardo, J.M., Berger, J.O., Dawid, A.P. *et al.* (eds), *Bayesian Statistics 6*, Oxford University Press, Oxford, pp. 589–612.

Smith, R.L. (1999b). Trends in Rainfall Extremes. Available from http://citeseerx.ist.psu.edu/viewdoc/download;jsessionid=897EF95856D54F74AE3DC29997BC146D?doi=10.1.1.40.9056&rep=rep1&type=pdf (Accessed 03/09/12).

Smith, L.A. (2002). What might we learn from climate forecasts? *PNAS*, **99**, 2487–2492.

Smith, P., Beven, K.J. and Tawn, J.A. (2008). Informal likelihood measures in model assessment: theoretic development and investigation, *Adv. Water Resour.*, **31**, 1087–1100.

Smith, R.L., Tawn, J.A. and Coles, S.G. (1997). Markov chain models for threshold exceedances, *Biometrika*, **84**, 249–268.

Smyth, G.K. (1989). Generalized linear-models with varying dispersion, *J. R. Stat. Soc.* B, **51**, 47–60.

Tebaldi, C. and Sansó, B. (2008). Joint projections of temperature and precipitation change from multiple climate models: a hierarchical Bayesian approach, *J. R. Stat. Soc.* A, **172**, 83–106.

Tebaldi, C., Smith, R.L., Nychka, D. *et al.* (2005). Quantifying uncertainty in projections of regional climate change: a Bayesian approach to the analysis of multi-model ensembles, *J. Climate*, **18**, 1524–1540.

Tessier, Y., Lovejoy, S. and Schertzer, D. (1993). Universal multifractals: theory and observations for rain and clouds, *J. App. Meteor.*, **32**, 223–250.

Timbal, B., Hope, P. and Charles, S. (2008). Evaluating the consistency between statistically downscaled and global dynamical model climate change projections, *J. Climate*, **21**, 6052–6059.

Underwood, F.M. (2008). Describing long-term trends in precipitation using generalized additive models, *J. Hydrol.*, **364**, 285–297.

Wagener, T., Wheater, H.S. and Gupta, H.V. (2004). *Rainfall-Runoff Modelling in Gauged and Ungauged Catchments*, Imperial College Press, London.

Walshaw, D. (2006). Modelling extreme wind speeds in regions prone to hurricanes. *J. R. Stat. Soc. C*, **49**, 51–62.

Wheater, H.S. (2008). "Modelling Hydrological Processes in Arid and Semi-Arid Areas" in: Wheater, H.S., Sorooshian, S. and Sharma, K.D. (eds), *Hydrological Modelling in Arid and Semi-Arid Areas*, Cambridge University Press, Cambridge.

Wheater, H.S., Chandler, R.E., Onof, C.J. *et al.* (2005). Spatial–temporal rainfall modelling for flood risk estimation, *Stoch. Env. Res. Risk A.*, **19**, 403–416.

Wilby, R.L. and Dawson, C.W. (2004). *Statistical Downscaling Model: SDSM version 3.1* (software and user guide). Available at http://co-public.lboro.ac.uk/cocwd/SDSM/ (Accessed on 25/04/2012).

Wilby, R.L. and Harris, I. (2006). A framework for assessing uncertainties in climate change impacts: low-flow scenarios for the River Thames, UK, *Water Resour. Res.*, **42**, W02419–W02429.

Wilby, R.L., Hassan, H. and Hanaki, K. (1998a). Statistical downscaling of hydrometerological variables using general circulation model output, *J. Hydrol*, **205**, 1–19.

Wilby, R.L. and Wigley, T.M.L. (2000). Precipitation predictors for downscaling: observed and general circulation model relationships, *Int. J. Climatol.*, **20**, 641–661.

Wilks, D.S. (1998). Multisite generalization of a daily stochastic precipitation generation model, *J. Hydrol.*, **210**, 178–191.

Wilks, D.S. (2009). A gridded multisite weather generator and synchronization to observed weather data, *Water Resourc. Res.*, **45**, doi: 10.1029/2009WR007902.

Wilks, D.S. and Wilby, R.L. (1999). The weather generation game: a review of stochastic weather models, *Prog. Phys. Geog.*, **23**, 329–357.

Woolhiser, D.A. and Pegram, G.G.S. (1979). Maximum likelihood estimation of Fourier coefficients to describe seasonal variations of parameters in stochastic daily precipitation models, *J. Appl. Meterol.*, **18**, 34–42.

Yang, C., Chandler, R.E., Isham, V. *et al.* (2005). Spatial–temporal rainfall simulation using generalized linear models, *Water Resour. Res.*, **41**, W11415–W11428.

Yang, C., Chandler, R.E., Isham, V.S. *et al.* (2006). Quality control for daily observational rainfall series in the UK, *Water Environ. J.*, **20**, 185–193.

CHAPTER 8

Uncertainty in Flood Frequency Analysis

Thomas R. Kjeldsen
Centre for Ecology & Hydrology, Wallingford, UK

Rob Lamb
JBA Consulting, Skipton, UK

Sarka D. Blazkova
T.G. Masaryk Water Research Institute, Prague, Czech Republic

8.1. Uncertainty Identification and Estimation

River flood frequency estimation is the process of determining the relationship between the size of a flood and its probability of occurrence. In engineering practice, "flood frequency" is often used to refer to the relationship between probability and river flow rate. Usually this expresses the probability that the flow in the river at a specified location exceeds a given threshold flow, or, conversely, the flow that is exceeded with a given probability. It is often convenient to interpret the probability scale in terms of a return period in years.

The flood frequency relationship is always an estimate, based on limited information. Hence there is always uncertainty about the value of flow or return period calculated. The root causes of this uncertainty are the limited record lengths for gauged flows (which give rise to sampling errors) and the inaccuracies in measured data. In addition, many estimates rely on the assumption of some type of model, for example a parametric distribution function for the flow rate or a more physically-based, conceptual model of river flow. If the model does not capture completely the real behaviour of the river catchment then this "model error" is also a source of uncertainty. Most modellers would acknowledge this to be the case most of the time.

This chapter deals with how we can characterise and analyse these uncertainties for different types of river flood frequency estimation methods. However, before discussing methods for the analysis of uncertainty, it is worth considering the relevance of such analysis from a practical point of view.

8.1.1. *What is the practical use of uncertainty of design floods?*

Flood frequency analysis is needed to estimate design flows or levels (for engineering works) or to provide a basis for planning and risk assessment, often at a broad spatial scale. These requirements lead to questions like "What is the 100-year peak water level?" or "What is the extent of a 100-year flood?". However, although these appear to be simple questions, answering them often involves difficult and complex problems, including analysis of flood frequency, river hydraulics and flood defence systems, with many uncertainties and interactions between the different component parts.

Until quite recently the treatment of uncertainty in flood estimation has been little more than descriptive. In the UK Flood Estimation Handbook (Institute of Hydrology, 1999) uncertainty was discussed in a chapter called "Looking ahead", where the authors urged researchers developing better methods for assessing uncertainty to ensure that these refinements, when they arrive, should be well used. Whilst uncertainty was seen as useful in helping to choose between alternative analysis methods, there was a concern to avoid uncertainty estimates being used to justify over-design by being interpreted as an additional "freeboard" allowance. This concern may have been well founded, for in the USA, freeboard had been interpreted in some US Army Corps of Engineers (USACE) documentation as an allowance for "the uncertainty inherent in the computation of a water surface profile" (Huffman and Eiker, 1991; National Research Council, 2000).

The call to ensure that refined uncertainty estimation methods should be "well-used" reflects a situation in the mid-1990s where uncertainty was widely acknowledged, but not deeply embedded in the techniques of flood analysis used in practice. An interpretation of uncertainty in terms of freeboard allowance stems from a flood management approach based on concepts of fixed standards of protection, e.g. seeking to defend communities against flooding up to a given level. Here, a typical design condition for a flood defence scheme might be the 100-year water level. Within this culture, uncertainty in the flood frequency estimate, although acknowledged, is a nuisance. Uncertainty has been recognised in the design or decision-making process by sensitivity analysis, although this would often involve calculating design levels for a T-year flow with a range of different values of hydraulic parameters, particularly roughness, but without quantifying the uncertainty in the underlying flow estimate.

At around the same time, a paradigm shift in flood management was beginning to taking place with a move away from fixed standards of protection towards the use of risk analysis techniques. In 1992, the USACE issued a draft Engineering Circular (EC) Risk-Based Analysis for Evaluation of Hydrology/Hydraulics and Economics in Flood Damage Reduction Studies (EC 1105-2-205), which was updated in 1994. A risk analysis approach uses probabilistic descriptions of the uncertainty in estimates of important variables, including flood frequency (and also stage–discharge, and depth–damage relationships) to compute probability distributions of potential flood damages. These computed estimates can be used to determine a design that provides a specified probability of containing a given flood, or of limiting risk exposure to a certain level. In the UK, similar concepts followed as the Department for Environment, Food and Rural Affairs (Defra) and Environment Agency developed their risk-based approaches for flood management, initially concentrating on strategic planning (Hall *et al.*, 2003).

With risk-based concepts for flood management, the analysis generally involves a simulation procedure in which the variables describing the source of risk are sampled from a suitable distribution and the generated values fed into models that calculate the resulting flood depth, damage or other required quantity. This provides a distribution for the resulting output variable. For example, a large number of flood flow values may be generated from a flood frequency distribution at a floodplain site, fed into a hydraulic modelling system which can calculate the number of properties flooded (which may also include some modelling of the performance of the flood defence system) and the outputs arranged in rank order to estimate the distribution of economic damage to property on that floodplain. Within this type of framework, uncertainty in the flood frequency estimation can be readily incorporated into the simulation (and need not even be presented explicitly to the decision maker).

How a risk-based analysis can best be used to support decision making is a separate question. Pappenberger and Beven (2006, "Decisions Are

Binary") note that although many decisions ultimately require specific thresholds, there is a rich literature on decision support systems that provides methods for decision making under uncertainty. In addition to building uncertainty into the calculation of risk, there are also methods that attempt to give a more sophisticated description of the probabilistic behaviour of risk. For example, Hashimoto *et al.* (1982a) considered operational estimators for measuring reliability (probability of system failure), resilience (probability of recovery once a failure has occurred) and vulnerability (likely magnitude of a failure). Also, Hashimoto *et al.* (1982b) introduced a measure of system robustness in the analysis of a water resources system, based on the probability that the actual economic costs of a system will not exceed some fraction of the minimum possible cost. A similar idea has been adopted more recently such as the information-gap approach (Ben-Haim, 2006) that seeks to examine how robust different decision options are as variables within the decision-making process deviate from assumed or "best estimate" values.

We have argued that the move towards a risk-based approach to flood management requires methods that can naturally accommodate uncertainty in flood estimation, but are there any benefits gained from the point of view of the analyst working on hydrology and flood estimation in practice? To some degree, the answer to this question rests with the agencies that drive the need for flood estimation. As flood management agencies adopt risk-based approaches, and if these approaches incorporate uncertainty, then of course it becomes relevant for those working in practice. It is also compelling to argue that transparency and honesty about uncertainties in a flood estimate are good things in principle!

One important practical reason to favour realistic communication of uncertainty to decision-makers is that "best estimates" of a flood frequency relationship can easily change as a result of changes in data, changes in accepted methodology or simply because of natural events. Two examples illustrate these issues. The first is the 1997 flood on the American River at Sacramento, California. This was the second major flood in 11 years out of a 93 year of hydrological record and had significant implications for the flood risk management decision process. The 1997 flood prompted two revisions of the previously accepted flood frequency relationship, which has been discussed in detail in National Research Council (1999). In this particular case, the revisions of the flood frequency analysis could have large planning and economic consequences for Sacramento, with the decisions resting on very thin margins. However, the uncertainty in the flood flow estimates was also large enough to suggest that a precautionary approach should be strongly considered. The second example is a recent update of recommended flood estimation methods for the UK Flood Estimation Handbook (Institute of Hydrology, 1999). Here, further scientific analysis since the original publication of the methods in 1999 has led to changes that could alter some 100-year flood flow estimates at ungauged catchments by as much as a factor of 2.25.

In cases like these, it may be seen as controversial when "best" flood flow estimates suddenly appear to change after some years, through a combination of additional data and improved methodology. It can be argued that analysis showing realistic uncertainty helps to insulate against the shock of such revisions, partly because it may be possible to demonstrate that additional data has in fact reduced uncertainty and partly because the original estimate may be less rigidly interpreted as a fixed value. Similarly, if we are to acknowledge the impact of future climate or catchment change on food frequency estimates, then it may be helpful in the long run to build uncertainty into the analysis.

8.1.2. *Sources of uncertainty in flood frequency estimation*

While Singh and Strupczewski (2002) expressed concerns that skills in statistics have become more important than hydrological knowledge in flood frequency estimation, the added focus on uncertainty will further emphasis the need of hydrologists for appreciation and understanding of statistical methods. However, for the purpose of discussing the more applied aspects of flood frequency estimation and the associated uncertainty analysis, it is probably worth considering the following statement by the distinguished statistician Sir Maurice Kendall (as quoted by David and Fuller, 2007) "*…work which achieves a slight gain in generality at the expense of being incomprehensible is defeating its own purpose*". While the definition of what is statistically incomprehensible may vary between hydrologists, this chapter will attempt to discuss uncertainty from a practical angle rather than presenting a comprehensive generic framework.

Any assessment of uncertainty in hydrological modelling will depend on what aspects of the model, its parameters and the forcing climatic time series are considered known and what aspects are considered random variables and in need of estimation based on available observations. A useful classification of error components might be achieved by considering the total error to consist of three distinct contributions from: (i) model errors;

(ii) sample errors; and (iii) error in the observed data. Further sub-division of the errors might be possible (see for example Ewen *et al.*, 2006; Gupta *et al.*, 2005) but for the purpose of this presentation the errors are considered aggregated into the three components listed above.

Start by considering a conceptual hydrological model, represented by the model operator M, which can simulate runoff from a catchment through a number of model parameters, represented by the parameter vector θ. Some of the model parameters are fixed while others are considered to be random variables which need to be estimated, $\hat{\theta}$, often by minimising the squared difference between simulated output and the observed data.

In the hypothetical case where the observed hydrological data were considered free of any errors and the true model parameters were known, it would still not be possible to exactly reproduce the observed runoff as the hydrological model is only a simple representation of all the complex processes and interactions taking place in the real catchment. Thus, by introducing the model error η_t component the relationship between the model M, the optimal model parameters, θ, and the observed (and error free) runoff Q_t can be represented as

$$Q_t = M_t(\theta) + \eta_t. \tag{8.1}$$

The optimal model parameters here refer to a set of values that would have been obtained if an infinitely long series of calibration data were available. Of course, in applied hydrology the model parameters are calibrated using data series of limited length and, thus, the estimated model parameters are only estimates of the true model parameters and the resulting simulated runoff,

$$\hat{Q}_t = M_t(\hat{\theta}), \tag{8.2}$$

is only a best estimate of the true runoff. By subtracting Equation (8.2) from Equation (8.1), the residual error is expressed as a sum of the two error components, model error and sampling (or calibration) error as

$$Q_t - \hat{Q}_t = M_t(\theta) - M_t(\hat{\theta}) + \eta_t = \varepsilon_t + \eta_t. \tag{8.3}$$

Of course, any observation of flow will be associated with a measurement error, ω_t, so that the final relationship is

$$Q_t - \hat{Q}_t = \varepsilon_t + \eta_t + \omega_t. \tag{8.4}$$

All error components have, by definition, a mean value of zero, are independent of each other and the uncertainty is specified through a covariance matrix. In some cases it might be useful to log-transform the flow variables to obtain normally distributed errors. In practice, only the difference between observed and modelled flow is known and, therefore, the exact values and covariance structures of the individual error components are not readily available. As discussed in Chapter 3, it is generally not possible to disaggregate the total error into the different error components when using more complex hydrological models. However, acknowledging the existence of the different error sources as well as understanding the error mechanism, the modeller can be more explicit about the assumptions behind any modelling results and the associated uncertainty estimates.

While the model and sample errors are specific to any particular modelling system under consideration, the errors in the observed flow values are identical for any modelling approach. Most standard textbooks on hydrology will cover the topics of how to, in principle, measure streamflow and any other hydrological quantity like, for example, precipitation, evaporation and soil moisture. However, they may not include, for example, explicit quantification of the range of uncertainties involved in deriving these quantities. As this chapter is mainly concerned with the estimation of extreme flow values, we shall limit the discussion here to uncertainty of observed flow. Issues related to uncertainty in observed rainfall and the process of converting point rainfall into areal rainfall will be addressed when discussing flood frequency analysis through rainfall-runoff modelling.

A comprehensive review of streamflow measurement techniques can be found in Herschy (1995) who also discussed the uncertainty involved in gauging flow and in the prediction of flow based on observed river stage, i.e. the rating curve. When considering flood flow, and in particular extreme flow of a magnitude rarely found in the observed systematic records, the uncertainty associated with extrapolation from the rating curve becomes an important error component to consider.

8.1.3. *Methods for flood frequency analysis*

This chapter will focus on two main approaches to flood frequency estimation: (i) direct statistical analysis of observed annual maximum series of instantaneous river flow; and (ii) rainfall-runoff model-based method where the probability of exceeding a specific flow value can be estimated by simulating the catchment response, to either a single design rainfall event or a continuous rainfall time series. The former rainfall-runoff model-based method is known as the event-based method or a derived distribution whereas the latter method is continuous simulation modelling. The two methods will be treated separately in the following sections.

While it might be possible to define other classifications of methods for flood frequency analysis, the division into three methods outlined above will form the basis of the subsequent discussion. The merits, applicability and caveats of each of the three methods mentioned above have been discussed extensively elsewhere (Lamb, 2005; Reed, 1999) and will not be repeated here. In the following sections the issues of uncertainty will be discussed in relation to flood frequency analysis through direct statistical analysis and rainfall-runoff modelling as well as providing actual examples of uncertainty estimation.

8.2. Statistical Analysis

Flood frequency analysis through direct statistical modelling of extreme peak flow data is the most widely used method for deriving design flood estimates. Most commonly the data series consist of the annual maximum peak flow, but also Peak-Over-Threshold (POT) data are sometimes used. Many very good texts providing a comprehensive introduction to the topic of frequency analysis of extreme events in hydrology are available and the reader is referred to these for a general introduction while this presentation will focus on the aspects of uncertainty.

Consider a series of annual maximum peak flow observations $x_1, \ldots, x_n$ which are considered to be realisations of a random variable X characterised by its Cumulative Distribution Function (cdf) $F(x)$. The design flood event with a return period T is denoted x_T and is obtained by inverting the cdf as

$$F(x_T) = 1 - \frac{1}{T}, \quad x_T = F^{-1}\left(1 - \frac{1}{T}\right). \tag{8.5}$$

The uncertainty of the design value x_T is often expressed as the variance of x_T, which for the simple case of a distribution fitted to a single data series is appropriate but for more complex regional procedures the issue of bias can be important.

Contribution to the total uncertainty of x_T can be attributed to the three general error types discussed in Section 8.1.2, i.e. model uncertainty, sampling uncertainty and uncertainty in the observations themselves. When considering regional methods for predicting the T-year event at both gauged and ungauged catchments, additional error components need to be considered, as will be discussed later.

8.2.1. *Model error*

The contribution to the total error arising from the particular choice of model, or cdf, is not often considered. By generating a large number of random samples from a specified distribution and comparing fits of various cdfs to these samples with the known cdf, it would be possible to quantify the error introduced by misspecifying the cdf. However, as the true underlying distribution is always unknown in practice it is not possible to directly estimate the effect of the choice of cdf on the error. It can be argued that a careful examination of the data with regards to goodness-of-fit of particular cdf will ensure this error contribution is minimised. The error contribution arising from model error, i.e. misspecification of cdf, will not be considered further here.

8.2.2. *Sampling error*

Traditionally, the uncertainties of flood quantiles derived from Equation (6.4) have been done so under the assumption that the chosen distribution is in fact the true distribution, i.e. no model error. By also ignoring the uncertainty of the individual data points, i.e. no measurement error, the confidence intervals of the T-year event is assumed to be only due to sampling error, i.e. limited data availability, and can be estimated using techniques available from the statistical literature. While these methods are generally well known, the complexity varies according to the number of parameters of a particular distribution and the chosen technique for estimating the model parameters, such as the method of moments, method of L-moments and the maximum likelihood method.

A more comprehensive review of methods for fitting statistical distributions to series of observed maximum data is provided in standard hydrological reference texts such as Kite (1977), Stedinger *et al.* (1993) and Rao and Hamed (2000), which contain examples of estimators of sampling uncertainty for many different distributions commonly used in flood frequency analysis as well as different methods of parameter estimation. As a foundation to the discussion of uncertainty in the more complex regional frequency methods, a short summary of uncertainty estimation is provided in the example below.

8.2.3. *Example*

This simple example illustrates the estimation of uncertainty of design flood derived by fitting a Gumbel distribution to a series of annual maximum instantaneous peak flow observations from a gauging station located on the River Lugg in Wales. The annual maximum series ($x_t, t = 1, \ldots, n$) has a record length of $n = 35$ years. The example considers three methods for estimating the sampling uncertainty: Taylor approximations (also referred to as the delta method); bootstrapping; and Monte Carlo simulations. Common for all three methods is that they assume the chosen distribution is the "true" distribution and ignore measurement error. In the following it is assumed that the two-parameter Gumbel distribution is the true distribution. The T-year event is estimated from a Gumbel distribution (Chow *et al.*, 1988) as

$$x_T = \xi + \alpha\, y_T \quad y_T = -\ln\left[-\ln\left(1 - \frac{1}{T}\right)\right], \tag{8.6}$$

where y_T is the Gumbel reduced variate and the Gumbel model parameters, ξ and α, can be estimated using a number of different methods including the Method of Moments (MOM), Maximum-Likelihood (ML) and the Method of L-moments (MOLM). In this example we consider only the simple MOM method. The first two product moments (mean and variance) of the Gumbel distribution are defined as

$$\mu = \xi + \alpha\gamma_e, \quad \sigma^2 = \frac{\pi^2}{6}\alpha^2, \tag{8.7}$$

where $\gamma_e = 0.5772\ldots$ is Euler's constant. By replacing the population moments in Equation (8.7) with the corresponding sample moments (sample mean and sample variance),

$$\bar{x} = \sum_{i=1}^{n} x_i, \quad s^2 = \frac{1}{n-1}\sum_{i=1}^{n}(x_i - \hat{\mu})^2, \tag{8.8}$$

the MOM parameter estimators are obtained as

$$\hat{\alpha} = \frac{\sqrt{6}s}{\pi}, \quad \hat{\xi} = \bar{x} - \gamma_e\alpha. \tag{8.9}$$

By combining Equations (8.9) and (8.7), the T-year event can be written as

$$\hat{x}_T = \bar{x} + K_T s = \bar{x}[1 + K_T C_v], \quad K_T = -\frac{\sqrt{6}}{\pi}(\gamma_e - y_T), \tag{8.10}$$

where K_T is a frequency factor that depends only on return period T. Uncertainty in the estimate can be assessed by assuming that the T-year event is normally distributed (unlike the annual maximum peak flow values themselves which are distributed according to the Gumbel distribution). The normality assumption is asymptotically true (increasingly so with increasing sample size) for most quantile estimators (Stedinger *et al.*, 1993). The uncertainty of the T-year event can then be expressed as the confidence interval

$$\left(\hat{x}_T - z_{1-p/2}\sqrt{\text{var}\{\hat{x}_T\}}; \hat{x}_T + z_{1-p/2}\sqrt{\text{var}\{\hat{x}_T\}}\right), \tag{8.11}$$

where z_p is p-th quantile of the standard normal distribution and $var\{\hat{x}_T\}$ is the variance of the T-year event $\hat{x}_T$. The variance of the T-year event will typically depend on the sample size and the degree of extrapolation (return period T) and can be estimated using a variety of methods. In the following sections, three techniques for estimation of the variance of the T-year event $var\{\hat{x}_T\}$ will be illustrated using: (i) analytical solution using the Taylor approximation; (ii) bootstrapping; and (iii) Monte Carlo simulations.

8.2.3.1. *Taylor approximations*

The Taylor method has often been used in hydrology to quantify the sampling variance of quantiles derived from statistical distributions fitted to annual maximum series of peak flow. The T-year event estimator is often a non-linear function of the model parameters which makes direct use of the variance operator impractical. By approximating the non-linear function with a linear function in the immediately vicinity of a known set of parameter values (e.g. the true parameter values), an approximation of the variance of the T-year estimator can be made. The performance of the approximation depends on the degree of non-linearity of the T-year estimator and how close the estimated model parameters are to the true model parameters. For an introduction to the theory behind the use of the Taylor approximation we refer to Kite (1977). Here we note that the Gumbel distribution is sufficiently simple so that using a variance operator directly on the expression of x_T in Equation (5.10) leads to the same results as using a Taylor approximation,

$$\begin{aligned}\mathrm{var}\{\hat{x}_T\} &= \mathrm{var}\{\bar{x}\} + K_T^2\mathrm{var}\{s\} + 2K_T\mathrm{cov}\{\bar{x}, s\} \\ &= \frac{\sigma^2}{n}(1 + 1.396K_T + 1.100K_T^2),\end{aligned} \tag{8.12}$$

where n is the record length, and the variance of the mean value and standard deviation, and the covariance between the two, are given by Kendall and Stuart (1977) but not reported here. Expressions similar to Equation (8.12) are available in the literature for a large number of distributions and estimation techniques.

8.2.3.2. *Bootstrapping*

The use of bootstrapping, and the related jackknife method have been found useful in frequency analysis of hydrological extremes (Zucchini and Adamson, 1989). The bootstrapping method is particularly useful in situations where exact or approximate analytical solutions (as discussed above) are difficult. For an in-depth introduction to both the bootstrap and jackknife methods, please refer to the comprehensive texts by Efron and Tibshirani (1993), Shao and Tu (1995) or Davidson and Hinkley (1997).

Estimation of the variance of a T-year event using bootstrapping involve the creation of a number of resampled data series of similar length to the original sample, but created from the original observed sample by random selection with replacement. Consider an observed series of annual maximum peak flow events $\mathbf{x} = (x_1, \ldots x_n)$ from which an estimate of the T-year event x_T is obtained using, in this example, the Gumbel distribution as in Equation (8.10). The new resampled series are denoted $\mathbf{x}_b(b = 1, \ldots, B)$, and can consist of any combination of elements from the original sample. For each of the resamples, the estimate of the T-year event can be estimated using Equation (8.10) and will be denoted $x_{T,b}$, $b = 1, \ldots B$.

In applied hydrology, the use of balanced resampling has become popular (Faulkner and Jones, 1999; Burn, 2003) where each element of the original sample is selected the same number of times. Burn (2003) suggested a practical balanced resampling procedure where B copies of the original sample are created and concatenated to create a new sample of length $n \times B$, where B is the number of resamples and n is the length of the original sample. The elements of the concatenated sample are randomly permutated and, subsequently, divided into B new samples.

Having obtained an estimate of the T-year event for each of the B samples, the variance of $\hat{x}_T$ is estimated as

$$\mathrm{var}\{\hat{x}_T\} = \frac{1}{B-1}\sum_{b=1}^{B}(\hat{x}_{T,b} - \bar{x}_T)^2, \quad \bar{x}_T = \frac{1}{B}\sum_{b=1}^{B}\hat{x}_{T,b}, \tag{8.13}$$

where $\hat{x}_{T,b}$ is the b-th estimate of x_T and $\bar{x}_T$ is the mean value across all resampled data sets. Note that both Faulkner and Jones (1999) and Burn (2003) were not interested in the variance of the T-year event as used here, but estimated a bootstrap confidence interval directly from the resampled values.

With regards to the required number of bootstrap samples, Faulkner and Jones (1999) used $B = 999$ and argued that using a limited number of resamples will introduce unnecessary random variation into the results. They recommended that a relatively large number of resamples should be used and that the sensitivity of the estimates should be checked by repeating the resampling procedure a few times using different values of B.

8.2.3.3. *Monte Carlo simulation*

As with bootstrapping, Monte Carlo simulations can be useful when direct analytical solutions are intractable. By assuming that the estimated parameters of the Gumbel distribution obtained using the MOM are the true parameter values, then M random samples can be generated each with a number of elements equal to the number of observations in the observed sample. Each element, x_p, in the stochastically generated sample is derived by first generating a realisation, u_p, from a uniform distribution defined on the interval $[0; 1]$, and then using the inverse of the cdf of the distribution as

$$F(x_p) = u_p, \quad x_p = F^{-1}(u_p), \tag{8.14}$$

which for the Gumbel distribution gives,

$$x_p = \xi - \alpha \ln[-\ln(u_p)]. \tag{8.15}$$

The record length of each of the M samples must be equal to the record length of the observed sample. For a discussion of how to generate random samples from particular statistical distributions commonly used in hydrology see for example Cheng *et al.* (2007). For each of the M random samples, the Gumbel parameters and the associated T-year event are estimated using Equation (8.8) to Equation (8.10). Having obtained an estimate of the T-year event for each of the M samples, the variance of $\hat{x}_T$ is estimated as

$$\text{var}\{\hat{x}_T\} = \frac{1}{M-1}\sum_{m=1}^{M}(\hat{x}_{T,m} - \bar{x}_T)^2, \quad \bar{x}_T = \frac{1}{M}\sum_{m=1}^{M}\hat{x}_{T,i}, \tag{8.16}$$

where $\hat{x}_{T,b}$ is the m-th estimate of x_T and $\bar{x}_T$ is the mean value of the M samples.

It can be argued that the procedure above will underestimate the variance because of the assumption that the estimated Gumbel parameters equal the true model parameters. An alternative two-stage Monte Carlo procedure can be used where the model parameters estimated for each of the M samples are themselves used for generating random samples. Alternatively, if the joint distribution of the model parameters could be specified, then M random parameter sets could be generated from where M random samples could be generated. However, this idea is not pursued further here.

8.2.3.4. *Results*

The mean and standard deviation of the 30-year long annual maximum series of peak flow at gauging station 55014 are $33.8\,\text{m}^3/\text{s}$ and $16.0\,\text{m}^3/\text{s}$, respectively. Estimating the parameters of the Gumbel distribution using MOM through Equations (8.8) and (8.9) gives

$$\hat{\alpha} = 12.51,$$
$$\hat{\xi} = 26.63.$$

Using these parameter values, the T-year flood can be estimated using Equation (8.10). Next, the standard deviation of this estimate is estimated using each of the three methods discussed above and the results shown for return periods of 10, 25, 50 and 100 years in Table 8.1.

In this particular case, the variance estimates from all three methods are of similar magnitude. However, this is not always the case. Based on application of a GEV distribution to observed series in Canada, Italy, and the UK, Burn (2003) observed that, in general, the bootstrap method gave estimates of uncertainty that were between 2%–47% lower than those obtained from an analytical solution. In a study investigating the performance of bootstrapping, Hall *et al.* (2004) concluded that the bootstrap

Table 8.1. Standard error (m^3/s) of T-year events for a range of return periods for each of the three different methods.

Method	T = 10	T = 25	T = 50	T = 100
$\hat{x}_T$ using Equation (6.10)	54.8	66.6	75.4	84.2
$\sqrt{\text{var}\{\hat{x}_T\}}$: Analytical	5.7	7.6	9.1	10.6
$\sqrt{\text{var}\{\hat{x}_T\}}$: Bootstrap	5.4	7.1	8.5	9.8
$\sqrt{\text{var}\{\hat{x}_T\}}$: Monte Carlo	6.7	8.6	10.0	11.5

method for some distributions and methods of parameter estimation often required record-lengths of about 100 years to provide reliable estimates. Therefore, choosing between the three methods based on these results is not a trivial task, and the fact is that the using the bootstrap method results in a smaller variance is not necessarily a valid reason for selecting this particular method, as this might be under estimating the real level of uncertainty. This simple example illustrates that even for the simple case of a single-site analysis, estimation of the uncertainty of predictions is not easy, and is only set to become more difficult as the complexity of the analysis and modelling system increases.

End of example

8.2.4. *Measurement error*

So far in this chapter the design flood estimates have been derived by fitting a statistical model directly to the sample of observed peak flow data. However, river flow is rarely observed but rather inferred from observations of stage (water level) through a stage–discharge relationship, i.e. a rating curve. A comprehensive review of streamflow measurement techniques can be found in Hersey (1995) who also discusses the uncertainty involved in flow gauging and the prediction of flow using rating curves. However, the effect of measurement error on the estimation of T-year floods is a topic that has attracted rather less attention.

We start this discussion by noting that a commonly used method for obtaining a rating curve is to fit a functional relationship directly to a set of N coherent observations of stage, h, and discharge, Q, (h_i, Q_i) $i = 1\ldots, N$ at a specific location on a river. The stage–discharge relationship is often defined by an equation in the form,

$$Q(h_i) = \gamma(h_i + \alpha)^\beta \quad i = 1\ldots N, \tag{8.17}$$

where γ, α and β are model parameters which can be fitted using regression techniques based on least square or maximum-likelihood principles (Clarke, 1999). If a change in river geometry at a stage h_m leads to a change in the stage–discharge relationship not captured by Equation (8.17), for example, when a river changes from being confined to the river channel to start flowing out-of-bank and into the surrounding floodplain, it might be necessary to have additional stage–discharge observations to develop this next leg of the rating curve. In practice, it is difficult to generalise rating curves as they are dependent on the local river geometry, measuring equipment and the number of available stage–discharge gaugings. Therefore, the following discussion of the effect of measurement uncertainty on flood frequency analysis is confined to the simple example of only one rating curve. It is assumed that there is sufficient hydraulic justification for this form of the rating curve, thus considering only a random error, η, of the observations scattered around the line

$$\ln[Q_i] = \ln[\gamma] + \beta \ln[h_i + \alpha] + \eta_i \quad i = 1 \ldots N. \tag{8.18}$$

For the case where a series of annual maximum flow is derived from a series of annual maximum levels through the use of a rating curve, two types of uncertainty has to be considered: (i) predicting discharge from Equation (8.18) based on observations of stage will introduce prediction uncertainty; and (ii) the correlation between annual maximum flow events introduced through the use of a common rating curve.

Considering first the prediction uncertainty of individual events, Clarke (1999) provided an example where the three parameters in Equation (6.18) are estimated using maximum-likelihood. In that case the prediction variance of the log-transformed flow, $\ln[Q_m]$ corresponding to the water level h_m is approximately given as

$$\begin{aligned}\operatorname{var}\{\ln[q_m] - \ln[Q_m]\} \approx{}& s_\eta^2 + \operatorname{var}\{\ln[\hat{\gamma}]\} + (\ln[h_m + \alpha])^2 \operatorname{var}\{\hat{\beta}\} \\ &+ \left(\frac{\beta}{h_m + \alpha}\right)^2 \operatorname{var}\{\hat{\alpha}\} + 2\ln[h_m + \alpha] \operatorname{cov}\{\ln[\hat{\gamma}], \hat{\beta}\} \\ &+ 2\left(\frac{\beta}{h_m + \alpha}\right) \operatorname{cov}\{\ln[\hat{\gamma}], \hat{\alpha}\} \\ &+ 2\ln[h_m + \alpha]\left(\frac{\beta}{h_m + \alpha}\right) \operatorname{cov}\left\{\hat{\beta}, \hat{\alpha}\right\},\end{aligned} \tag{8.19}$$

where q_m is the true (but unknown) value and the variance-covariance of the model parameters are obtained from the second-order derivatives of the log-likelihood function. The model error variance, s_η^2, is the sum of squared residual divided by $N - 3$.

To illustrate the effect of this type of measurement error on the actual flood frequency curve, consider the simple error model suggested by Potter and Walker (1981) and Rosso (1985) where, again, the true and measured discharge is denoted q and Q, respectively. By adopting a linear error model where a random measurement error is defined as

$$\sqrt{\mathrm{var}\{Q|q\}} = cq \tag{8.20}$$

it was shown by Potter and Walker (1981) and Rosso (1985) that the coefficient of variation, $C_v^{(Q)}$, of the measured series was larger than the $C_v^{(q)}$ of the corresponding true discharge values and given as

$$C_v^{(Q)} = \left[c^2 + (1 + c^2)(C_v^{(q)})^2\right]^{\frac{1}{2}} \geq C_v^{(q)}. \tag{8.21}$$

Considering again the Gumbel distribution, it is clear from Equations (8.21) and (8.10) that a higher value of C_v will automatically lead to higher values of T-year events. Thus, the existence of random error in the measurements result in more conservative estimates of the design flood events than would have been obtained if the true discharge values were observed.

The second, and perhaps less obvious, type of uncertainty is the introduction of correlation between estimates of annual maximum discharge, in a series when all values have been derived from water levels using the same rating curve (Clarke, 1999). The effect of correlation in a sample is generally to increase the variance of the estimated moments. Thus, the correlation will not affect the parameter values of the Gumbel distribution when estimated using the method of moments, but is likely to have an effect on the uncertainty of the T-year quantiles, though this effect remains unknown at this stage.

8.3. Regional Frequency Analysis

The relatively large uncertainties of T-year events estimated by extrapolation from statistical distributions fitted to limited at-site records prompted the development of models for regional frequency analysis. The regional models have an additional advantage allowing estimation at ungauged sites, though this issue is not pursued further here. The rationale for using a regional model is that more reliable estimates of T-year events can be obtained by augmenting the at-site record with data observed at neighbouring or hydrologically similar catchments. Often the regional approach is referred to as substituting space for time to distinguish it from other approaches of using information on historical flood events that occurred before systematic data collection was initiated at the site of interest or methods of record extensions, where the at-site record is extended by utilising the correlation between observed data at the site of interest, and a neighbouring observed series which extends further back (or forward) in time.

As regionalisation is an extension of the at-site analysis discussed above, many of the issues with regards to uncertainty estimation remain. While the inclusion of more data will decrease the sampling uncertainty, this gain should be offset against additional uncertainty due to departure from assumptions of homogeneity and the existence of cross-correlation between data at neighbouring sites.

Several regional frequency methods have been suggested in the literature, each relying on a different set of assumptions with regards to the degree of homogeneity and inter-site correlation present in the data from the considered region. In a study developing analytical estimators of sampling uncertainty for five different regional methods, Rosbjerg and Madsen (1995) argue that the choice of method should reflect the conditions observed in the data as closely as possible. They also found that the methods with the strictest assumptions tended to give estimates with lower uncertainties than estimates obtained from less restrictive models. Hosking and Wallis (1997) argued that, since the degree to which the underlying assumptions of a particular statistical model are justified are never truly known, they considered uncertainty estimates based on analytical approximations to be of little practical use — instead they advocated the use of Monte Carlo simulations. Of course, the utility of a Monte Carlo simulation procedure relies on a correct specification of the underlying, and still unknown, true structure of the underlying region, including correct specification of the type of distribution at each site, as well as the correlation between data observed at different sites. It seems that that the Monte Carlo simulation method would suffer from many of the same ills that Hosking and Wallis (1997) attribute to the more analytical methods. One clear advantage of Monte Carlo simulation method is that it allows for an assessment of the effect of any model ignoring structures in the data such as misspecification of distributions and the existence of heterogeneity. As an alternative to both the analytical approaches and the Monte Carlo simulations, some researchers, notably Faulkner and Jones (1999) and Burn (2003), have used a regional bootstrapping method for estimating the uncertainties of design events.

In the following each of the three methods discussed above will be illustrated for use with the index flood method for regional flood frequency analysis.

8.3.1. *The index flood method*

The index flood method has found widespread use in flood hydrology. The underlying key-assumption of the index flood method is that the distributions of flood data within a homogeneous region are identical except for a site-specific scale parameter, the index flood. Consider a region consisting of N different sites where the number of observations of annual maximum peak flow at each site is denoted n_i and the individual observation $x_{i,t}, t = 1, \ldots, n_i$. At each site within the region the regional T-year event is given as

$$x_{T,i}^{(R)} = F^{-1}\left(1 - \frac{1}{T}\right) = \mu_i \, z_T^{(R)}, \tag{8.22}$$

where μ_i is the site-specific index flood and $z_T^{(R)}$ is the regional dimensionless growth factor. The superscript R indicates a quantity based on regional analysis. The index flood is often defined as the mean or the median annual maximum flood. The growth factor z_T is the quantile function of a common distribution of $x_{i,t}/\mu_i$, which is assumed identical for all sites in the region. At each site in the region, the growth factor, $z_{T,i}$, is described by a number of model parameters θ_p, $p = 1, \ldots, P$ and each of these parameters can be combined across sites to form a regional average parameter as

$$\hat{\boldsymbol{\theta}}_p^{(R)} = \frac{1}{\Lambda}\sum_{i=1}^{N} \omega_i \hat{\boldsymbol{\theta}}_{p,i}, \quad \Lambda = \sum_{i=1}^{N} \omega_i, \tag{8.23}$$

where, again, the superscript R indicates a regional parameter. The weight ω_i can be specified as, for example, the record length so that $\omega_i = n_i$.

By considering the sampling variance of the regional parameter $\theta^{(R)}$ defined in Equation (8.23) for the simple case where there is only one such parameter, it is immediately clear that if all assumptions of the index-flood method are fulfilled, then the benefit of a regional approach over an at-site approach is a reduction in the variance of the estimated parameters of the distribution. For example, consider a hypothetical region consisting of data from five sites ($N = 5$) and that the variance of the parameter estimator $\hat{\theta}$

is known at each site in a region to be equal to one, i.e. $\text{var}\{\hat{\theta}\} = 1$. At each site it is assumed that $n_i = 10$. By using a variance operator directly on Equation (6.23) above the variance of the regional parameter is generally given as:

$$\text{var}\{\hat{\theta}^{(R)}\} = \frac{1}{\Lambda^2}\left(\sum_{i=1}^{N} n_i^2 \text{var}\{\hat{\theta}_i\} + 2\sum_{i=1}^{N-1}\sum_{j=i+1}^{N} n_i n_j \text{var}\{\hat{\theta}_i\}^{\frac{1}{2}} \text{var}\{\hat{\theta}_j\}^{\frac{1}{2}} \rho_{\theta,ij}\right). \tag{8.24}$$

Assuming the data from the different sites are independent, the correlation-term ρ becomes zero. This will result in a variance of the regional parameter equal to $5 \times (10^2 \times 1)/(5 \times 10)^2 = 0.2$, i.e. a significant reduction when compared to the variance of 1 at each individual site.

As a result of the model not fully complying with the underlying assumptions, the apparent gain in efficiency will not be fully realised. Firstly, it is well known that observations of annual maximum floods recorded at neighbouring sites tend to have some degree of positive correlation (e.g. Kjeldsen and Jones, 2006; Tasker and Stedinger, 1989), i.e. the covariance terms in Equation (8.24) are positive and larger than zero, resulting in a larger sampling error. Secondly, a model error might be introduced as a single distribution is specified for the entire region, which might not be fully justified. Finally, any departure from the assumption of complete homogeneity will introduce a degree of error into the analysis. The last two errors are classified as model errors, as they have been introduced by forcing a relatively simple model to represent a complex system. The model errors cannot easily be assessed directly from analysis of an observed data set, but their impact can be minimised by carefully analysing the data and ensuring compliance between data and model.

8.3.2. *Example*

Estimating the sampling variance of the T-year event was relatively simple for a single-site analysis based on the Gumbel distribution with parameters estimated using the MOM. However, the analysis becomes more complex when considering the regional T-year event estimator. The index flood method for the Gumbel distribution defines the regional T-year event estimator for the i-th site as

$$x_{T,i}^{(R)} = \bar{x}_i + K_T s_i = \bar{x}_i(1 + K_T C_v^{(R)}), \tag{8.25}$$

where C_v is the coefficient of variation $s/\bar{x}$ which is assumed to be a regional constant, whereas the index flood, $\bar{x}_i$ is site specific, and defined as the average annual maximum flood. Consider a region consisting of N catchments for which the underlying values of C_v are identical, then the record weighted mean regional mean value of C_v is

$$\hat{C}_v^{(R)} = \frac{1}{\Lambda}\sum_{i=1}^{N} n_i \hat{C}_{v,i}, \quad \Lambda = \sum_{i=1}^{n} n_i. \tag{8.26}$$

A very important part of regional frequency analysis is the identification of a region in which catchments can be considered to be homogeneous. For a more in-depth discussion of how to how to create homogeneous regions, please refer to Hosking and Wallis (1997).

Considering the same gauging station used in the single-site example (gauging station 55014), a regional estimate of the T-year flood was derived based on a region of 11 annual maximum series of peak flow from similar catchments located in the south-west of the UK have been defined. The catchments were chosen based on available record length and similarity with the subject site with regards to geographical location, catchment area, average annual rainfall and soil type. The locations of the eleven gauging stations are shown in Figure 8.1 and additional details shown in Table 8.2

As for the single-site analysis, we consider three methods for estimation of the variance of the regional T-year: (i) an analytical solution; (ii) bootstrapping; and (iii) Monte Carlo simulations.

8.3.3. *Analytical solutions*

Development of an analytical expression of the variance of the regional T-year event estimator requires a first-order Taylor approximation of Equation (5.25) centred at the true values of the population parameters of the mean, μ, variance, σ^2, and the coefficient of variation, C_v, i.e.

$$\begin{aligned} \mathrm{var}\{\hat{x}_{T,i}^{(R)}\} &= (1 + K_T C_r^{(R)})^2 \mathrm{var}\{\bar{x}_i\} + (\mu_i K_T)^2 \\ &\mathrm{var}\{\hat{C}_v^{(R)}\} + 2\mu_i K_T \mathrm{cov}\{\bar{x}_i, \hat{C}_v^{(R)}\}, \end{aligned} \tag{8.27}$$

where it is assumed that $\mathrm{cov}\{\bar{x}_i, \hat{C}_v^{(R)}\} = 0$ and

$$\mathrm{var}\{\bar{x}_i\} = \sigma^2/n_i. \tag{8.28}$$

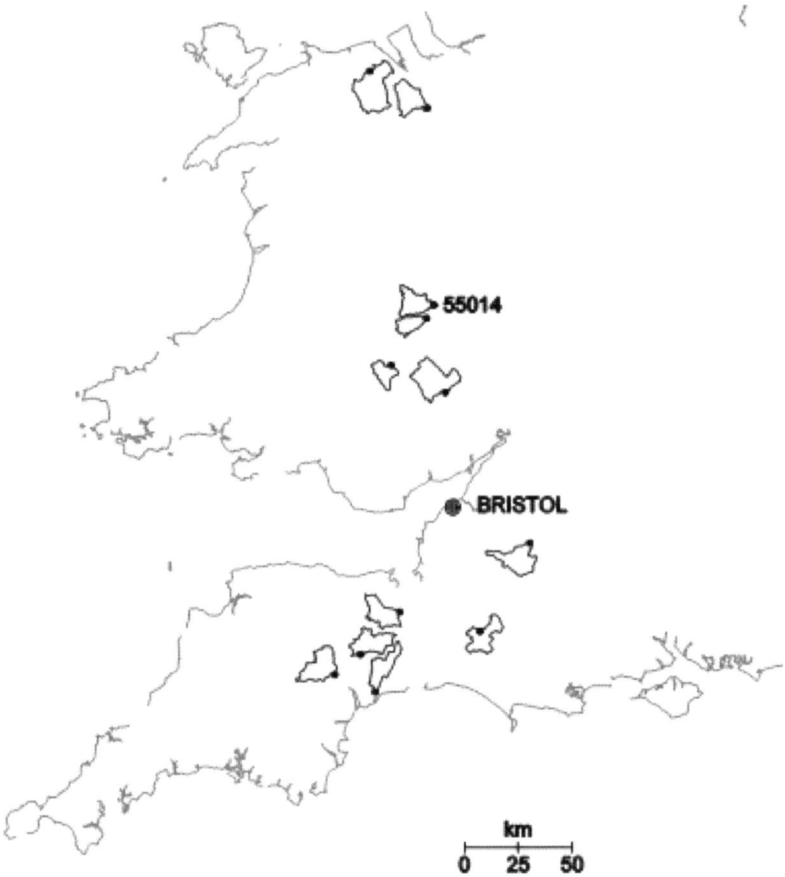

Figure 8.1. Map showing location of subject site (55014) and the additional 11 gauged catchments included in the regional frequency analysis.

The variance of the regional coefficient of variation is obtained by using the variance operator directly on Equation (8.26),

$$\operatorname{var}\{C_v^{(R)}\} = \frac{1}{\Lambda^2}\left(\sum_{i=1}^{N} n_i^2 \operatorname{var}\{\hat{C}_{v,i}\} + 2\sum_{i=1}^{N-1}\sum_{j=i+1}^{N} n_i n_j \operatorname{var}\{\hat{C}_{v,i}\}^{\frac{1}{2}} \operatorname{var}\{\hat{C}_{v,j}\}^{\frac{1}{2}} \rho_{C_v,ij}\right), \quad (8.29)$$

where $C_{v,i}$ and $C_{v,j}$ are the coefficient of variation at site i and j, and $\rho_{C_y,ij}$ is the correlation coefficient between $C_{v,i}$ and $C_{v,j}$. Kjeldsen and Rosbjerg

Table 8.2. Details of gauging stations included in regional analysis.

No.	River	Area [km^2]	n [no. years]	Mean [m^3/s]	Stdev [m^3/s]	CV
55014	Lugg	202.5	35	33.9	16.1	0.47
45003	Culm	228.9	41	78.8	40.8	0.52
67008	Alyn	225.8	38	23.3	8.82	0.38
52005	Tone	203.7	42	44.3	15.1	0.34
52006	Yeo	216.2	41	59.5	31.0	0.52
53007	Frome	263.7	42	61.2	19.2	0.31
45012	Creedy	263.6	39	80.9	39.3	0.49
55025	Llynfi	131.5	32	59.5	36.8	0.62
45005	Otter	202.8	41	79.4	49.7	0.63
55029	Monnow	355.1	30	153.5	37.3	0.24
55013	Arrow	125.9	35	29.0	10.8	0.37
66001	Clwyd	404.7	30	48.1	14.7	0.31

(2002) assumed that $\rho_{C_y,ij} = \rho_{ij}^2$ where ρ_{ij} is the correlation coefficient between the AMS at site i and j which can be estimated for pairs of sites with records overlapping in time. The variance of the $\hat{C}_{v,i}$ at each of the N sites is given by Rosbjerg and Madsen (1995) as

$$\mathrm{var}\{\hat{C}_{v,i}\} = \frac{C_{v,i}^2}{n_i}(1.1 - 1.14C_{v,i} + C_v^2), \qquad (8.30)$$

which should be substituted into Equation (8.29) to get the variance of the regional $C_v^{(R)}$ estimator, which again is used in Equation (8.27) to get the variance of the regional T-year event estimator.

8.3.4. *Bootstrapping*

The bootstrapping procedure used here is based on the procedure presented by Faulkner and Jones (1999) and adopted for regional flood frequency analysis by Burn (2003). Note that both of these studies were not interested in the variance of the T-year events but derived confidence intervals directly from the ranked bootstrapping results. However, for consistency, this presentation will focus on the estimation of the variance. When using bootstrapping to estimate the uncertainty of a T-year event obtained from a pooled frequency analysis Burn (2003) advocated the use of a vector bootstrapping approach to preserve the existing correlation between overlapping samples of data across sites in the pooling group. In vector bootstrapping, the procedure involves randomly selecting years (rather than observations) with replacement and then forming a vector containing

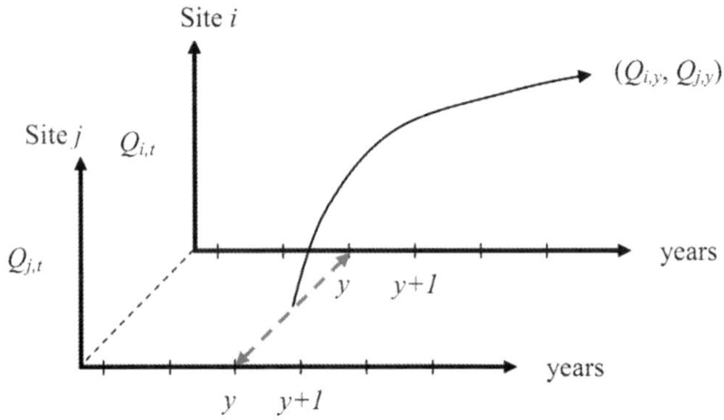

Figure 8.2. Bootstrap selection of all observations at all sites at a particular year.

data from each catchment in the pooling group for that particular year. Figure 8.2 illustrates the concept of selecting data across a region for a particular year for a simple bivariate case.

Similarly to the single-site procedure, the balanced resampling approach is adopted for the regional procedure, but ensuring that all years for which data are available are selected an equal number of times for the resampled data set (Burn, 2003). In practice, balanced resampling for regional analysis is implemented as described for the single-site case, but where the vector B in the single-site case contains individual data points, in the regional analysis it contains all the years where data are available at any site. For each of the B regional resamples, the regional estimate of the T-year event at the site of interest is obtained using the index-flood method and the variance of the T-year estimate calculated as

$$\mathrm{var}\{\hat{x}_{T,i}^{(R)}\} = \frac{1}{B-1}\sum_{b=1}^{B}\left(\hat{x}_{T,b}^{(R)} - \bar{x}_T\right)^2, \quad \bar{x}_T = \frac{1}{B}\sum_{b=1}^{B}\hat{x}_{T,b}^{(R)}, \tag{8.31}$$

where $\hat{x}_{T,b}^{(R)}$ is the b-th estimate of x_T and $\bar{x}_T$ is the mean value across all resampled data sets.

8.3.5. *Monte Carlo simulations*

A simple and practical algorithm for the use of Monte Carlo simulations for estimation of the variance was presented by Hosking and Wallis (1997). In the following, the number of Monte Carlo replications is denoted M.

(i) Consider again a region consisting of N gauged catchments where the records consist of $\{n_i, i = 1, \ldots, N\}$ year of observations.

(ii) For each catchment specify the marginal distribution $F_i(x) i = 1, \ldots, N$ and its parameters. The parameters can be sample values obtained from the observed records, or they can reflect a synthetic region specified by the analyst.

(iii) For each of the realisations generated by the Monte Carlo procedure, a set of sample data are generated for each catchment with a record-length equal to the observed record-length. From each replica, the T-year event is derived using the index-flood algorithm.

(iv) Having obtained M estimates of the T-year event, the variance can be estimated as

$$\mathrm{var}\left\{\hat{x}_T^{(R)}\right\} = \frac{1}{M-1}\sum_{m=1}^{M}\left(\hat{x}_{T,m}^{(R)} - \bar{x}_T\right)^2, \quad \bar{x}_T = \frac{1}{M}\sum_{m=1}^{M}\hat{x}_{T,m}^{(R)}, \tag{8.32}$$

where $\hat{x}_{T,m}^{(R)}$ is the m-th estimate of x_T and $\bar{x}_T$ is the mean value of the M samples.

The actual procedure for generating replica depends on whether inter-site dependence between the annual maximum flood series at different catchments is taken into consideration or not. If inter-site correlation is neglected, then the regional procedure is equivalent of generating samples from N individual catchments as discussed in the single-site case. If inter-site correlation is present, then the structure of the correlation should be specified in a $N \times N$ dimensional correlation matrix $\mathbf{C}$, where the diagonal elements are one and the off-diagonal elements correspond to the correlation ρ_{ij} between the annual maximum peak flow at catchment i and j.

$$\mathbf{C} = \begin{bmatrix} 1 & \rho_{12} & \rho_{13} & \cdots & \rho_{1N} \\ & 1 & \rho_{23} & \cdots & \rho_{2N} \\ & & 1 & \cdots & \rho_{3N} \\ & & & \ddots & \vdots \\ & & & & 1 \end{bmatrix}. \tag{8.33}$$

To generate a realisation, a total of $\max\{n_i, \mathrm{i}=1, \ldots, N\}$ vectors z_{ik}, $\mathrm{i}=1, \ldots, N$, $k = 1, \ldots, n_i$ containing realisations from a multivariate normal distribution with zero mean and a covariance-matrix $\mathbf{C}$ are generated. Please refer to standard texts for further details on how to generate multivariate normal vectors (Devroye, 1986). In practice it is often not possible to use the sample estimates of the cross-correlation in C as this can lead to a matrix which cannot be inverted (not symmetric positive semi-definite). Alternatively, Hosking and Wallis (1997) suggested using an average correlation for the entire region and Tasker and Stedinger (1989) related the correlation to the geographical distance between catchments.

Finally, each of the realisations z_{ik}, $i = 1, \ldots, N$, $k = 1, \ldots, n_i$ is transformed to the required marginal distribution (here Gumbel) as

$$x_{i,k} = F_i^{-1}(\Phi(z_{i,k})) = \xi_i - \alpha_i \ln\lfloor - \ln(\Phi\lfloor z_{i,k} \rfloor)\rfloor, \qquad (8.34)$$

where Φ is the cumulative distribution function of the standard normal distribution. It is common in Monte Carlo studies to generate $M = 10000$ or more samples.

8.3.6. *Results*

To estimate the T-year event for the annual maximum peak flow series at gauging station 55014 using the index-flood method, it is first necessary to estimate the regional coefficient of variation. Using Equation (8.26) with the data in Table 8.3, $\hat{C}_v^{(R)} = 0.25$. With a mean annual flood at the site of interest of $\bar{x} = 32.2\,\mathrm{m}^3/\mathrm{s}$ the T-year event can be estimated for any return period using Equation (8.25). The T-year events derived for $\mathrm{T} = 10, 25, 50$, and 100-years as well as the associated uncertainties estimated using each of the three methods are shown in Table 8.3.

The results in Table 8.3 can be compared to the corresponding results obtained for the single-site analysis in Table 8.1. While the magnitude of the T-year events themselves only changes slightly when comparing the results of the regional analysis to the single-site analysis, the associated reduction

Table 8.3. T-year flood and the associated standard error (m^3/s) estimated from the three regional procedures.

Method	T = 10	T = 25	T = 50	T = 100
$\hat{x}_T^{(R)}$ from Equation (6.25)	53.2	64.1	72.2	80.3
$\sqrt{\mathrm{var}\{\hat{x}_T^{(R)}\}}$: Analytical	4.4	5.4	6.2	7.0
$\sqrt{\mathrm{var}\{\hat{x}_T^{(R)}\}}$: Bootstrap	4.6	5.7	6.6	7.5
$\sqrt{\mathrm{var}\{\hat{x}_T^{(R)}\}}$: Monte Carlo	4.4	5.4	6.2	6.9

in uncertainty is more dramatic. A general reduction of between 20%–40% in standard deviation of the T-year event clearly illustrates the benefits of including additional regional information into flood frequency analysis.

It is worth noticing that the while the bootstrap method gave the lowest estimate of variance for the single-site case, it provides the highest estimate in the case of the regional analysis. However, there is no reason why this should always be the case. Also, note the close agreement between the estimates obtained for the analytical and the Monte Carlo methods.

8.4. Continuous Simulation Modelling

Continuous simulation methods for flood estimation have developed from process-based ideas, in which the runoff that produces flood flows is modelled, at least in principle, as part of a wider range of catchment responses. In this approach, a rainfall-runoff model that can simulate a long, unbroken flow record is used to generate synthetic flow data that can then be analysed using conventional statistical distribution-fitting methods or simple empirical estimates of probability. Alternatively, this long series of runoff can be used as input into another system for which the probabilistic behaviour is required, e.g. floodplain mapping. The simulation approach uses models that have some internal water balance accounting to track the state of the catchment over time and so allow the runoff response to vary with antecedent soil moisture. This integrates the variation in precipitation inputs over different timescales and hence avoids one of the most difficult problems with event-based models, which is the need to specify the joint distribution of "design inputs" (antecedent soil moisture and rainfall intensity; duration; and profile or snowmelt volume). A price to pay for avoiding this difficult joint distribution is that a complete precipitation series is needed to drive the runoff model. The aim is generally to have a synthetic flow record that is long enough to make the uncertainty arising from random sampling of the flow data very small. There may be situations where long gauged precipitation records exist and the required design flow is at a relatively short return period, in which case the gauged data could be used directly to drive the runoff modelling. But often it will be necessary to use a model to generate long precipitation sequences, which of course introduces uncertainty through the same combination of data, sample and model-specification errors that occur for flow models.

One of the motivations for continuous simulation, as noted by Bras *et al.* (1985), is that rain-gauge networks are often more extensive and longer established than flow gauges. Hence more information may be available to represent the rainfall inputs than exists for the river flows. Data are also required to support snow accumulation and melt models in some regions and for the calculation of potential evaporation. Long-term variations in evaporation may affect the flood regime in some catchments by controlling soil moisture deficits prior to rain storms. However, it is most likely that rainfall or snowmelt inputs control the probability distribution of flood flows, at least for large events and in humid temperate environments.

Where rainfall-runoff models are used for flood estimation, estimation of uncertainty is complicated by a number of factors. Firstly, rainfall-runoff models applied in continuous simulation tend to be highly non-linear and including thresholds in their responses, which makes estimation of the parameter values difficult. Secondly, the probability distributions of the model parameters are unlikely to be independent because most rainfall-runoff models are designed from the principle of trying to reflect process understanding of catchment hydrology, rather than to provide a "best" explanation for the variation in measured data in a statistical sense. The models may therefore have multiple pathways that can give rise to equivalent outputs; for example there may be several different attenuation functions representing surface, soil and groundwater storage. One exception is the class of models based on transfer functions that are inferred directly from data and subsequently interpreted conceptually. Finally, multiple output variables from the rainfall-runoff model (e.g. flood peaks, low flows and soil moisture) can be defined against which observational data may be compared and different measures of performance evaluated.

Due to these complications, measures of uncertainty have tended to be based on importance sampling using Monte Carlo methods, or concepts of Pareto optimality (Gupta *et al.*, 2002). For example, Cameron *et al.* (1999, 2000b) and Blazkova and Beven (2002, 2004) used the GLUE (Generalised Likelihood Uncertainty Estimation) method of Binley and Beven (1992), a generalised Monte Carlo approach (which can include both formal and informal Bayes approaches and fuzzy set approach, see 2.1.4.2), to compute uncertainty bounds for flood frequency curves.

Although it is an elegant approach in principle, there are two important problems that have to be solved to make practical use of the continuous simulation method. Firstly, it will often be necessary to extend simulations in time so as to generate a sufficiently long synthetic flow series to reduce sampling uncertainty. Secondly, it will often be necessary to simulate ungauged locations. Both situations can be seen as extrapolations of the method (in time and space, respectively). In both cases, the extrapolation effectively introduces additional models and uncertainty. Temporal extension requires models for precipitation (and other climatic forcing variables). Spatial extrapolation requires models for parameterisation of the runoff and precipitation simulators. The combination of two or three model components means the uncertainty analysis can become rather complex.

The following sections introduce uncertainty analysis methods that have been developed separately for the temporal and spatial extrapolation of continuous simulation flood frequency estimates. Estimation of uncertainty for combined precipitation and runoff modelling is illustrated with examples using the GLUE method (for introduction in GLUE, see Chapter 6). Spatial extrapolation is then discussed with reference to a national study in the UK where a theoretical analysis of uncertainty in runoff modelling has been developed to include transfer of information about parameter estimates to ungauged locations.

8.4.1. *Simulation uncertainty*

For continuous simulation of long series, two models are in fact used: a precipitation and temperature simulator and a precipitation-runoff model. Equation (3.1) could be used for each of them separately or the parameter set θ could contain parameters of both. If there are rainfall and snow data available in sufficient quantity, it could be possible to try to estimate the input and commensurability errors, and in case of the availability of rating data, the observation error could be estimated using the acceptability approach of GLUE (Beven, 2006). The usual situation, however, is that those data are not available. In such cases, the only uncertainty we are dealing with explicitly is the parameter uncertainty, the estimate of which we are getting by sampling parameters independently from physically reasonable ranges. The parameters then compensate for all the other errors of Equation (3.1). In the following text we will present some examples of precipitation simulators and of likelihood measures used in connection with producing long series of annual peaks for catchments in Wales and in the Czech Republic.

8.4.2. *Precipitation simulators*

Eagleson (1972) developed derived flood flow distributions based on a statistical rainfall model combined with a kinematic wave hillslope flow model. This early work integrated the rainfall and flow models analytically.

With the subsequent development of computer codes to run water balance accounting models, the derived distribution idea could be generalised to use rainfall and runoff models that would have been too complex to integrate analytically. Some of the early developments in this field were reported by Beven (1986a,b, 1987) who adopted and then extended the Eagleson rainfall model. The parameters of the simulator were the parameters of independent exponential distributions for average rainfall intensity, and the average duration of storm events. For continuous simulation, a parameter controlling an exponential distribution of storm arrival time was added.

A number of studies found that it was necessary to model separate low and high intensity event types, each of which had three independent parameters. Cameron *et al.* (1999, 2000a) found dependence between average duration and intensity on some UK catchments and as a result modelled events in seven duration classes and extrapolated the upper tail of each distribution by fitting a generalised Pareto distribution. They were able to evaluate uncertainty separately for the rainfall simulator and for the catchment rainfall-runoff model "TOPMODEL" and then created coupled simulations by random sampling catchment model parameters, driven by realisations of sequences of rainstorms.

It seems to be intuitively physically more reasonable to simulate rain cells directly. The pulse-based models often reproduce many of the properties of data series adequately (Cowpertwait *et al.*, 1996; Onof and Wheater, 1993) but there is a large uncertainty due to limited representativeness of the observed rainfall data sample for short durations of the order of one hour and problems about defining the upper limits of the statistical distributions of rainfall intensities assumed (see Cameron *et al.*, 2001a and the citations therein).

A precipitation simulator or weather generator (if temperatures are also modelled) could be used for the estimation of the climate change impact on floods and the hydrological regime in general (Fowler *et al.*, 2005). It is important to realise, however, that the parameter sets which are identified as being behavioural (i.e. in reasonable agreement with the observed data) under current climate conditions are also assumed to be behavioural under climate change (Cameron, 2006).

8.5. Likelihood Measures

There are many possible likelihood measures suggested for use within the GLUE methodology, including formal Bayes likelihood functions

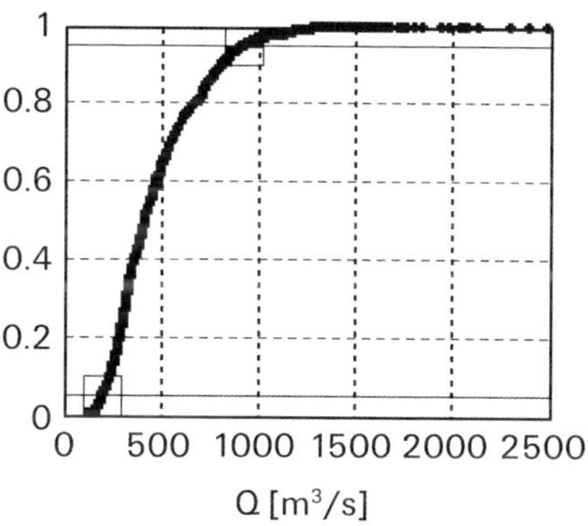

Figure 8.3. Cumulative frequency curve for 1000 years return period at Dolni Kralovice; horizontal lines indicate 5 and 95% prediction bounds. (Blazkova and Beven, 2004, reproduced with permission of *J. Hydrol.*)

(based on statistical models for the error structures), simpler coefficients of determination for hydrograph simulation (such as the Nash–Sutcliffe criterion), sums of absolute errors between modelled and observed ordinates or flood frequency quantiles or fuzzy measures. The computed likelihoods are first used for rejection of non-behavioural (i.e. totally implausible) model structures and/or parameter sets, and then they are used as weights for the construction of the cumulative likelihood weighted distributions from the behavioural simulations (Figure 8.3), which gives an expression of uncertainty in a flood quantile (see the example below).

The construction of the likelihood measure depends on what data is available. If there are observed series of precipitation and flow covering the same period, then it is possible to find parameter sets of the rainfall-runoff model producing simulations which agree with the observed flow record (behavioural simulations) and to reject those simulations (parameter sets) which do not. This was done by Cameron *et al.* (1999) on the data-rich Wye catchment. The more usual situation is an existing record of annual maxima and daily flow, in which case it is possible to constrain the simulation on the flood frequency curve estimated from the observed data and on the observed flow duration curve. For flood studies it is preferable in most cases to use data with a sub-daily resolution to capture the real peaks rather than the daily average flow values.

Depending on the data availability, the likelihoods measures used in the uncertainty evaluation can be combined in various ways. For example, Cameron *et al.* (1999) defined one measure for a rainfall model, another for the rainfall-runoff model, and from these measures constructed a Combined Measure, *CM*, which assumed equal weighting for each measure and is expressed analytically as

$$CM = \exp\left\{l(\rho) + \frac{1}{nd}\sum_{i=1}^{nd}[l_s(\Theta_i)]\right\}, \tag{8.35}$$

where nd is the number of rainfall duration classes, and $\mathrm{l_s}(\Theta)$ is the rescaled version of each likelihood function, $\mathrm{l}(\Theta)$, for a range of rainfall duration defined as

$$l(\Theta)_{du} = \sum_{i=1}^{np}\{-\log\sigma + (-1/\kappa - 1)\times\log[1+\kappa(x_i - u)/\sigma]\}, \tag{8.36}$$

where du is the duration class, x_i is a particular event in the corresponding upper tail, np is the number of events in that tail, and κ, u and σ are shape, location (or threshold) and scale parameters. The rainfall likelihood functions are rescaled such that they share a common scale with the likelihood function defined for the peak flow series as

$$l(\rho) = \sum_{i=1}^{19}\Big\{-\log\alpha_s + (-1/k_s - 1)\log[1+k_s(y_i - u_s)/\alpha_s] - [1+k_s(y_i-u_s)/\alpha_s]^{-1/k_s}\Big\}, \tag{8.37}$$

where α_s, k_s and u_s are the scale, shape and location parameters of a GEV distribution, fitted to the annual maximum peak flows of the simulated series, and y_i is the corresponding value of the peak flow estimated from a GEV distribution fitted to the observed series of annual maximum peak flows. Note that in the study by Cameron *et al.* (1999) the summation goes from 1 to 19, because 19 of the 21 observed peaks had a non-exceedance probability of less than or equal to 0.9 (ten-year return period).

Another possibility is to incorporate the parameters of the precipitation model into the uncertainty analysis together with the runoff model parameters, hence providing an estimate of the uncertainty of the combined simulation as has been done in the example below.

In the applications of the GLUE methodology reported by Blazkova and Beven (2002, 2004), fuzzy set theory has been used for the formulation of likelihood measures (see Section 3.4.1). In evaluating the modelled flood frequency curves against frequency curves estimated from observations, three fuzzy sets (defined by three membership functions) have been created, linguistically described as "wrong", "reasonable" and "good" based on a measure defined as 1/SUMF where SUMF is sum of absolute errors between the modelled and "observed" flood frequency curves, rescaled to the range between 0 and 1 (on the x-axes in the plots in columns a and d in Figure 8.4). A modelled frequency curve can belong to two of the sets at the same time to various degrees. The vertical line in column a shows that the simulation with $1/\text{SUMF} = 0.307$ belongs to a large degree to the set "reasonable" but to some degree also to the set "wrong".

This can be further used as an input into a fuzzy system combining the flood frequency goodness of fit with other goodness-of-fit criteria, here by considering flow duration curve. There are two membership functions for flow duration in Figure 8.4, "wrong" and "good" in columns b and e. The criterion here is 1/SUMD (where SUMD is sum of the absolute errors between simulated flow duration curve and the curve from observations). The combining is done in this case by constructing six linguistic rules, which can be expressed, for example, as (see row 4 in columns a, b and c).

If "floods" is "reasonable" and "duration" is "good" then "likelihood" is "medium".

The outputs of the rules are also fuzzy-membership functions (in the Mamdani system, see column c). The outputs of all the rules are then aggregated using e.g. summation (see dark shaded area in the last plot in column c) and then defuzzified using e.g. centroid (pointed out by the short vertical line in this plot). The result is a crisp number representing a likelihood weight for the simulation (that is not a particularly good one, 0.313). The Sugeno system does not have membership functions as output as the Mamdani system, but crisp numbers (column f in Figure 8.4). The defuzzification is then achieved by weighted average. In the Sugeno plots we have chosen a relatively good simulation (see the vertical lines in columns d and e). The combined likelihood is 0.617. Matlab Fuzzy Toolbox offers more opportunities for combining functions, and of defuzzification, and is easy to use.

If neither precipitation runoff nor flood frequency data are available on the catchment in question it is possible to use a regional estimate of flood frequency curve and flow duration curve for selecting the behavioural parameter sets as was done in a study of Blazkova and Beven (2002) where a gauged basin was treated as ungauged.

Flood frequency can also be computed within the limits of acceptability GLUE framework where limits (based e.g. on rating curves data) are set before running the simulations (Blazkova and Beven, 2009b). This study

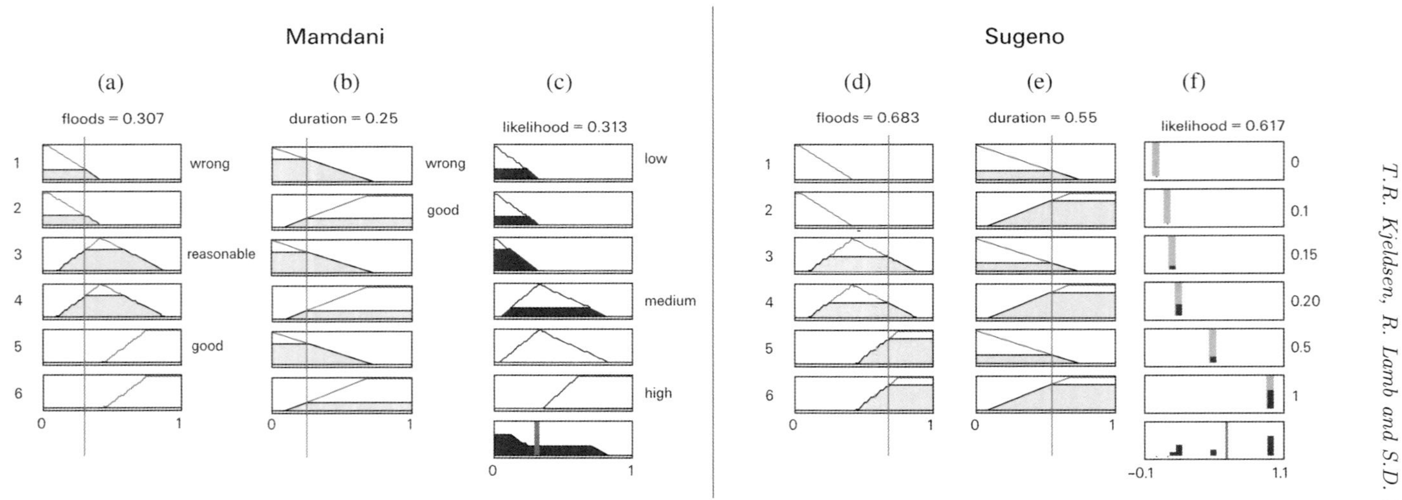

Figure 8.4. An example of the Mamdani and Sugeno fuzzy systems for the same case of computing fuzzy likelihood from two criteria: flood frequency with three membership functions (described linguistically) and flow duration with two. The system has six fuzzy rules (one is given as an example in the text). The long vertical lines over the columns a, b, d and e give the value of the two criteria for a simulation in question 1/SUMF and 1/SUMD for the flood frequency and flow duration, respectively. Shaded areas visualise to which extent each membership function is in use; in the output functions of the Mamdani system (column c) the shaded areas of the individual rows (rules) are aggregated giving the shape in the last row where then the centroid is found. In the Sugeno system the resulting likelihood is found as weighted average (column f) (Software Fuzzy Toolbox of Matlab).

also shows procedures of relaxing the limits if the number of acceptable simulations is small in order to be able to construct prediction limits.

8.5.1. *Example*

The Zelivka catchment (1188.6 km^2) drains part of the Czech-Moravian highlands, located between Bohemia and Moravia in the Czech Republic. The Švihov reservoir located at the outlet of the catchment is the most important source of drinking water for Prague. Most of the catchment lies on relatively impermeable crystalline bedrock. Deeper flow pathways occur in bedrock fractures. Land use in the catchment is dominated by agriculture and forestry. The maximum altitude of the catchment is 765 m above sea level, and the altitude at the Švihov dam is 400 m above sea level. The catchment average annual precipitation is 670 mm. In the past floods in the catchment have been caused by both summer rainfall and snowmelt.

For modelling purposes the catchment has been sub-divided into seven sub-catchments. Gauged flow information was available for four of these (Figure 8.5). The stochastic precipitation model distinguishes two types of rainfall — with low and high intensity, with parameters that vary for four characteristic seasons of different lengths. The model allows for a probability that storms arrive from different directions and for correlation in sub-catchment averaged rainfall volumes on different sub-catchments at different time steps. The temperature model has Fourier and autoregressive stochastic components for both seasonal and daily temperature variations. A simple degree-day routine is employed for snowmelt. A detailed description can be found in Blazkova and Beven (2004).

This study adopted the TOPMODEL (Beven, 1986a,b, 1987, 2001; Blazkova and Beven, 2002, 2004) as the rainfall-runoff model. The TOPMODEL has a number of parameters, some of which were considered fixed and others considered to be random variables. Using the GLUE methodology the following TOPMODEL parameters were sampled from uniform distributions with physically reasonable ranges: depletion and rootzone parameters, soil transmissivity, routing velocities and a parameter for recharge to the saturated zone. For the weather generator, the winter season storm parameters, degree day factor for snowmelt and reductions of point rainfall with the areas of the subcatchments, have been considered to be uncertain.

For each of the four sub-catchments with observed data, a fuzzy model has been set up combining three criteria: goodness of fit (1/sum of the

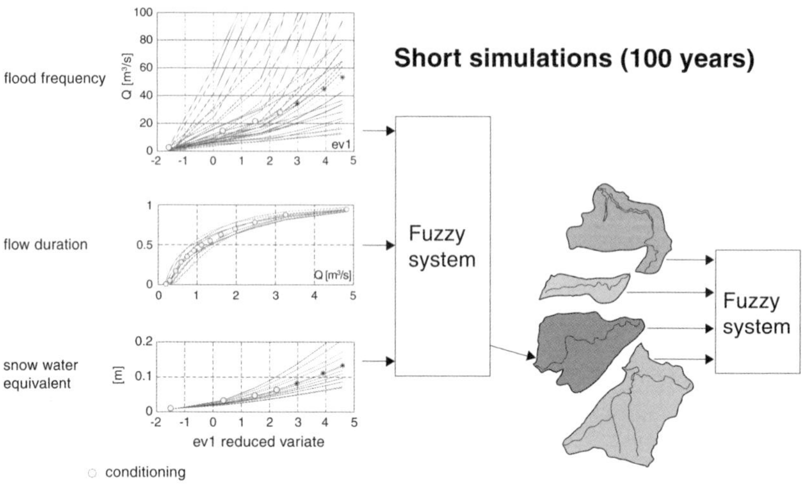

Figure 8.5. Computation of likelihoods from short (100 years) simulations. Flood frequency — the exceedance curves of annual flow maxima, ev1 is Gumbel distribution, ev1 = 4.6 is for 100-years flood, snow water equivalent — exceedance curve of annual maxima of snow water equivalent, conditioning — points on which the likelihood has been computed; (Blazkova and Beven, 2009a, reproduced with permission of Structure and Infrastructure Engineering).

absolute errors) of the flood frequency curve, of the flow duration curve and of the maximum annual snow water equivalent (schematically shown in Figure 8.5).

The result is the fuzzy-likelihood of the simulation on each sub-catchment in question. A combined measure expressing the likelihood of each simulation on the whole catchment was computed with another fuzzy system, which has been used for the evaluation of short (100-year) simulations. With the behavioural parameter sets 10,000-year time series have been simulated to get better estimates of the longer-term frequency statistics. A final evaluation, again in a fuzzy form, has been used — the relation of the simulated rainfall to the Probable Maximum Precipitation (PMP). PMP has been derived by the Institute of Atmospheric Physics, Rezacova *et al.*, 2005. Prediction quantiles of peak flow are then obtained by weighting the predictions from all the behavioural models with their associated final fuzzy likelihood measure and constructing a cumulative frequency curve (see Figure 8.3). On Figure 8.6 the prediction bounds of flood frequency at Dolni Kralovice together with the observed data are shown.

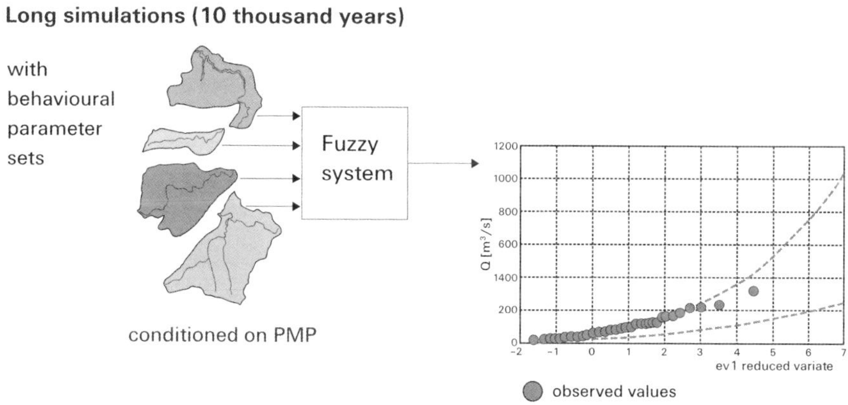

Figure 8.6. Computation of likelihoods from long (10,000 years) simulations, behavioural parameter sets — parameter sets found acceptable in 100 years simulations, PMP — probable maximum precipitation (Blazkova and Beven, 2009a, reproduced with permission of Structure and Infrastructure Engineering).

8.6. Generalised Model Uncertainty

It is common in practice to compute design flood flows for ungauged sites where there is no local data to help with calibration or estimation of uncertainty. In this situation, models used for continuous simulation have to be parameterised based on transfer of information from gauged locations (as is the case for other flood frequency estimation methods applied at an ungauged catchment). This transfer of information involves estimating the parameters of the runoff model either by taking an average of neighbouring estimates at gauged sites or by modelling the parameters themselves, typically using regression against catchment properties such as area, mean elevation or mean slope. The runoff model parameter estimation therefore introduces its own uncertainty.

The first analysis of the uncertainty of ungauged site continuous simulation was by Lamb and Kay (2004), using Monte Carlo simulation to generate confidence intervals for flood frequency curves at ungauged catchments. They fitted regression relationships between hydrological model parameters and catchment properties based on calibrated estimates for 40 gauged sites in the UK. Random samples were then generated from the distribution of the residuals surrounding each regression equation and supplied to the hydrological model to produce simulated flow series. Flood

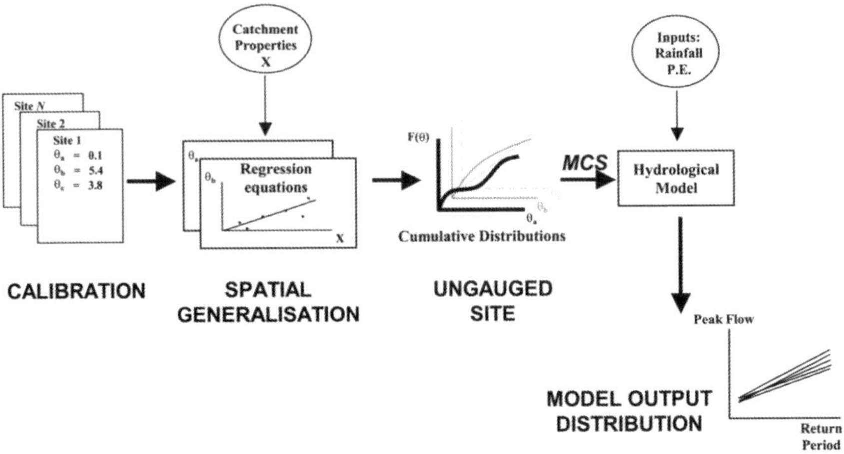

Figure 8.7. Schematic of the method used to calculate approximate confidence intervals using Monte Carlo Simulations (MCS) for catchments treated as ungauged (Reproduced from Lamb and Kay, 2004).

quantiles for specified return periods were calculated for every simulation and used to construct an empirical distribution for the design flows, and hence to calculate confidence intervals; the procedure is illustrated in Figure 8.7.

The confidence intervals calculated for the "ungauged" continuous simulation were compared with intervals derived for a Generalised Pareto Distribution (GPD) fitted directly by maximum likelihood to gauged data that had been withheld from the continuous simulation. The degree of uncertainty in the spatially generalised continuous simulation varied between sites but was often no greater than for the site-specific frequency curve fitted directly to the gauged flows. In particular the study showed that the estimation of parameters from catchment properties did not lead to the rapid inflation of uncertainty at higher design flows that was seen for the GPD, presumably as a result of sampling error.

The Lamb and Kay analysis considered uncertainty in estimates of runoff model parameter values that would be obtained for a catchment given a limited calibration record, rather than estimates of a hypothetical "true" parameter. A more complete analysis of uncertainty for a spatially generalised flood simulation model has been developed by Calver *et al.* (2005) and Jones and Kay (2006) where a framework was proposed to assess the likely error in estimating a notional "true" parameterisation at the ungauged site rather than a parameterisation that would be obtained given a limited sample of calibration data. Here, the calibration error

$$\varepsilon = Y - T \tag{8.38}$$

was defined as the difference between the calibrated estimate Y of a parameter at a gauged site and a hypothetical underlying "true" value, T, which represents the value to which a calibration procedure would converge for a very long hydrological record. The generalisation error

$$\eta = T - \mu \tag{8.39}$$

represents the difference between the notional true value T for the catchment, and the estimated value μ obtained from a generalisation rule (e.g. a regression model linking model parameters to catchment properties). The proposed uncertainty analysis distinguished between a notional "true generalisation" μ that would be based on an arbitrarily large sample of catchments, and the "sample generalisation" estimate $T^* = \mu^*$ derived from the generalisation procedure applied to a real, limited set of catchments.

Uncertainty arising from the parameter generalisation is captured in the above model by the variance of the generalisation error $\boldsymbol{\sigma}_\eta^2$ whilst the calibration uncertainty is associated with the variance of the calibration error $\boldsymbol{\sigma}_\varepsilon^2$. The calibration error variance was estimated using a jackknife procedure in which a re-calibration was carried out multiple times for each gauged catchment, leaving out a year of record on each pass. The calibration error variance was estimated separately for each gauged catchment i as $\boldsymbol{\sigma}_{\varepsilon,i}^2$. The generalisation error variance was then computed for the whole set of catchments using an iterative procedure that simultaneously found the best generalised parameter estimates Y_i^* consistent with the calibration uncertainty at each catchment. If the generalisation model is a least squares regression between model parameters and catchment properties, then the above process effectively estimates the generalisation uncertainty as a weighted sum of squares of the residuals of the regression model.

Total uncertainty in the generalised estimate of a model parameter for an ungauged site (denoted by a subscript u) can then be expressed in terms of the variance of the total estimation error

$$\mathrm{var}(T_u - \hat{T}_u) = \sigma_\eta^2 + \sum_{j=1}^{n} w_{u,j}^2(\sigma_\eta^2 + \sigma_{j,\varepsilon}^2), \tag{8.40}$$

where $w_{u,j}$ is a weight applied to the calibrated parameter estimate from gauged catchment j when constructing the estimated parameter value at ungauged catchment u. The terms on the right-hand side represent the generalisation uncertainty, plus the transferred effects of both generalisation uncertainty, and calibration uncertainty, from n gauged catchments to the target ungauged site.

The above analysis is presented for a single model parameter, but as we have noted, rainfall-runoff models used for continuous simulation typically have several parameters with some dependence between them. The study by Calver *et al.* (2005) extended the analysis of uncertainty to include the covariance between the calibration and generalisation errors as well as the variances discussed above. By using the derived covariance matrix, random sets of parameters were generated and supplied to the runoff model to produce uncertainty bounds. Examples of the results are shown in Figure 8.8.

Uncertainty bounds were computed for 112 catchments across the UK and the results summarised in terms of the mean width of the 90%,

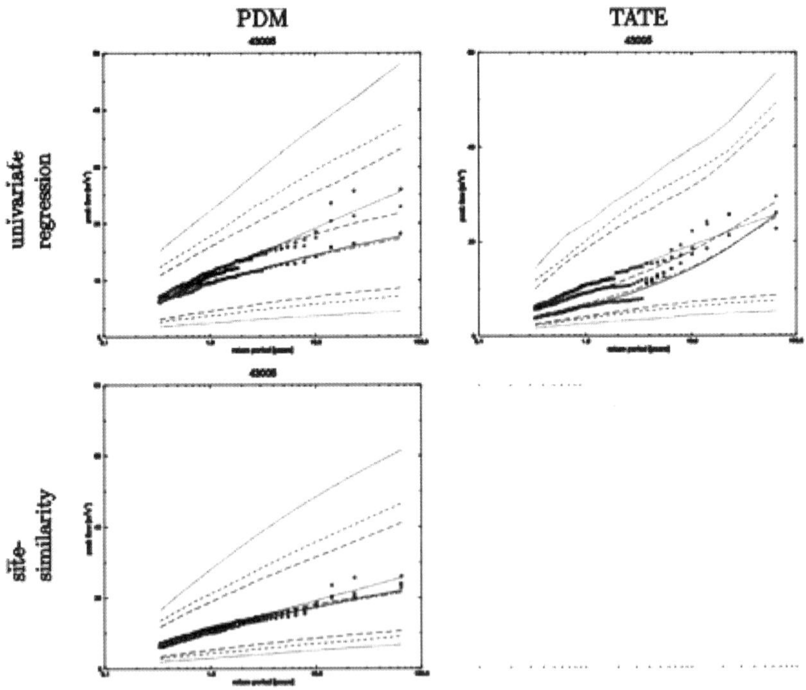

Figure 8.8. Example of uncertainty bound (blue; 50% - dot/dash, 90% - long dashed, 95% - short dashed, 99% - dotted) on generalised flood frequency curves (black solid), for two catchments (Reproduced from Calver *et al.*, 2005).

95% and 99% bounds for estimates of the 10-year and 50-year flood flow, where this width was standardised by the median simulation, to permit comparison across all of the catchments. In part, the uncertainty analysis was used examine the performance of the continuous simulation method as a "national" methods for flood flow estimation when using either of two different runoff models and two alternative procedures for estimation in ungauged catchments. One model produced uncertainty bounds for the design flows where the upper bounds were generally higher relative to the "best estimate", whereas the other model produced lower bounds that were generally wider. The study did not conclude firmly that one particular runoff model and parameter generalisation procedure was a clear winner. But a combination of the PDM rainfall-runoff model with a weighted average of calibration parameters based on catchment similarity seemed to be favoured, based on comparison of the generalised model uncertainty bounds with gauged data.

8.7. Conclusions

As illustrated in this chapter, there are many methods available for the analysis of uncertainty in flood estimation. It is not always obvious what method should be preferred, and the exact nature of the underlying assumption might not be apparent (or even palatable) to an individual hydrologist. It is important to remember that the estimated uncertainty will, to a large degree, depend on the assumptions made in the course of a modelling exercise. The more the hydrologist is prepared to constrain the analysis by accepting the assumptions made in a particular model, the less uncertain the outcome will appear. For example, by assuming that a simple lumped rainfall-runoff model is an accurate representation of a catchment, the model error might be left out of the analysis. As a result, the uncertainty of predictions made using the model will be reduced when compared to a situation where this model error was included because the analyst considered the assumptions made in formulating the model to be sufficiently inaccurate to be questionable. As is often argued, an estimate (say, the 100-year peak flow) should be accompanied by a statement of the uncertainty of that estimate. Similarly, we advocate here that an estimate of uncertainty should be accompanied by a statement of the assumptions underlying this analysis specifying what parts of the modelling system are considered certain and what parts are assumed to contribute to the uncertainty. These assumptions should as far as possible be supported by evidence derived from the observed data. Once the hydrological model has been constructed and the assumptions listed, it should be clear that the only way to reduce uncertainty is to: (i) develop a better model (reduce model uncertainty); or (ii) include more information (data) in the modelling exercise, which will reduce the sampling uncertainty. Of course, these two activities are not mutually exclusive, and often the availability of more data will allow the analyst to make inference about how to improve a model.

What method is used for estimating the uncertainty depends on the complexity of the modelling system. For simple models, analytical methods can be used but for more complex and coupled modelling systems, such methods are not feasible. However, for all methods it is clear that a thorough understanding of the modelling system is required by the analyst to ensure that whatever assumptions are made, they are based on solid foundations. The added complexity of uncertainty assessments adds to the well-known mantra that modelling should only be attempted by properly trained and well-informed hydrologists.

Acknowledgement

The third author wishes to acknowledge the support of the COST EU grant ES0901 and the grants of the Czech Ministeries of Environment (MZP0002071101) and Education (OC10024) and Czech Science Foundation (P209/11/2045).

References

Beven, K. (1986a). "Runoff Production and Flood Frequency in Catchments of Order n: An Alternative Approach" in: Gupta, V.K., Rodriguez-Iturbe, I. and Wood, E.F. (eds), *Scale Problems in Hydrology*, D. Reidel Publishing Company, Dordrecht, the Netherlands, pp. 107–131.

Beven, K. (1986b). "Hillslope Runoff Processes and Flood Frequency Characteristics", in: Abrahams, A.D. (ed.), *Hillslope Processes*, Allen and Unwin, London, pp. 181–202.

Beven, K. (1987). Towards the use of catchment geomorphology in flood frequency predictions, *Earth Surf. Proc. Land.*, **12**, 69–82.

Beven, K. (2001). *Rainfull-Runoff Modelling: the Primer 1st Edition*, Wiley, Chichester.

Beven, K. (2006). A manifesto for the equifinality thesis, *J. Hydrology*, **320**, 18–36.

Blazkova, S. and Beven, K. (1997). Flood frequency prediction for data limited catchments in the Czech Republic using a stochastic rainfall model and TOPMODEL, *J. Hydrol.*, **195**, 256–279.

Blazkova, S. and Beven K. (2002). Flood frequency estimation by continuous simulation for a catchment treated as ungauged (with uncertainty), *Water Resour. Res.*, **38**, WR000500.

Blazkova, S. and Beven, K. (2004). Flood frequency estimation by continuous simulation of subcatchment rainfalls and discharges with the aim of improving dam safety assessment in a large basin in the Czech Republic, *J. Hydrol.*, **292**, 153–172.

Blazkova, S. and Beven, K. (2009a). Uncertainty in flood estimation, *Struct. Infrastruct. E.*, **5**, 325–332.

Blazkova, S. and Beven, K. (2009b). A limits of acceptability approach to model evaluation and uncertainty estimation in flood frequency estimation by continuous simulation: Skalka catchment, Czech Republic, *Water Resour. Res.*, **45**, 1–12.

Bras R.L., Gaboury, D.R., Grossman, D.S. *et al.* (1985). Spatially varying rainfall and flood risk analysis, *J. Hydraul. Eng.*, **111**, 754–773.

Burn, D.H. (2003). The use of resampling for estimating confidence intervals for single site and pooled frequency analysis, *Hydrol. Sci. J.*, **48**, 25–38.

Calver, A., Crooks, S., Jones, D. *et al.* (2005). *National River Catchment Flood Frequency Method using Continuous Simulation Modelling*, R&D Technical Report FD2106/TR, Defra, London.

Cameron, D. (2006). An application of the UKCIP02 climate change scenarios to flood estimation by continuous simulation for a gauged catchment in the northeast of Scotland, UK (with uncertainty), *J. Hydrol.*, **328**, 212–226.

Cameron, D., Beven, K.J., Tawn, J. *et al.* (1999). Flood frequency estimation by continuous simulation for a gauged upland catchment (with uncertainty), *J. Hydrol.*, **219**, 169–187.

Cameron, D., Beven, K.J., Tawn, J. *et al.* (2000a). Flood frequency estimation by continuous simulation (with likelihood based uncertainty estimation), *Hydrol. Earth Syst. Sc.*, **4**, 23–34.

Cameron, D., Beven, K.J. and Tawn, J. (2000b). An evaluation of three stochastic rainfall models, *J. Hydrol.*, **228**, 130–149.

Cameron, D., Beven, K.J. and Tawn, J. (2001a). Modelling extreme rainfalls using a modified random pulse Bartlett-Lewis stochastic rainfall model (with uncertainty), *Adv. Water Resour.*, **24**, 203–211.

Cameron, D., Beven, K. and Naden, P. (2001b). Flood frequency estimation under climate change (with uncertainty), *Hydrol. Earth Syst. Sci.*, **4**, 393–405.

Cheng, K.S., Chiang, J.L. and Hsu, C.W. (2007). Simulation of probability distributions commonly used in hydrological frequency analysis, *Hydrol. Process.*, **21**, 51–60.

Chow, V.T., Maidment, D.R. and Mays, L.W. (1988). *Applied Hydrology*, McGraw-Hill, Singapore.

Clarke, R.T. (1999). Uncertainty in the estimation of mean annual flood due to rating-curve indefinition, *J. Hydrol.*, **222**, 185–190.

Cowpertwait, P.S.P., O'Connell P.E., Metcalfe A.V. *et al.* (1996). Stochastic point process modelling of rainfall. 1. Single-site fitting and validation, *J. Hydrol.*, **175**, 17–46.

David, H.A. and Fuller, W.A. (2007). Sir Maurice Kendall (1907–1983): a centenary appreciation, *Am. Stat.*, **61**, 41–46.

Davidson, A.C. and Hinkley, D.V. (1997). *Bootstrap Methods and Their Application*, Cambridge University Press, Cambridge.

Devroye, L. (1986). *Non-Uniform Random Variate Generation*, Springer-Verlag, New York.

Eagleson, P.S. (1972). Dynamics of flood frequency, *Water Resour. Res.*, **8**, 878–898.

Efron, B. and Tibshirani, R.J. (1993). *An introduction to the Bootstrap, Monogram on Statistics and Applied Probability*, 57, CRC, Boca Raton.

Ewen, J., O'Donnell, G., Burton, A. *et al.* (2006). Errors and uncertainty in physically-based rainfall-runoff modelling of catchment change effects, *J. Hydrol.*, **330**, 641–650.

Faulkner, D.S. and Jones, D.A. (1999). The FORGEX method of rainfall growth estimation, III: Examples and confidence intervals, *Hydrol. Earth Syst. Sc.*, **3**, 205–212.

Fowler, H.J., Kilsby, C.G., O'Connell, P.E. *et al.* (2005). A weather-type conditioned multi-site stochastic rainfall model for the generation of scenarios of climatic variability and change, *J. Hydrol.*, **308**, 50–66.

Gupta, H.V., Beven, K.J. and Wagener, T. (2005). "Model Calibration and Uncertainty Estimation" in: Anderson, M.G. and McDonnell, J.J. (eds), *Encyclopedia of Hydrological Sciences*, Vol. 3, Wiley, London.

Hall, J.W., Dawson, R.J., Sayers, P.B. *et al.* (2003). A methodology for national-scale flood risk assessment, *P. I. Civil Eng.*, (Water & Maritime Engineering), **156**, 235–247.

Hall, M.J., van den Boogaard, H.F.P., Fernando, R.C. *et al.* (2004). The construction of confidence intervals for frequency analysis using resampling techniques, *Hydrol. Earth Syst. Sc.*, **8**, 235–246.

Hashimoto, T., Loucks, D.P. and Stedinger, J.R. (1982a). Reliability, resilience and vulnerability criteria for water resource system performance, *Water Resour. Res.*, **18**, 14–20.

Hashimoto, T., Loucks, D.P. and Stedinger, J.R. (1982b). Robustness of water resources systems, *Water Resour. Res.*, **18**, 21–26.

Herschy, R.W. (1995). *Streamflow Measurement 2nd Edition*, E & FN Spon, an imprint of Chapman and Hall, London.

Hosking, J.R.M. and Wallis, J.R. (1997). *Regional Frequency Analysis: An Approach Based On L-Moments*, Cambridge University Press, Cambridge.

Huffman, R.G. and Eiker, E. (1991). "Freeboard Design for Urban Levees and Floodwalls" in: Proceedings of a Hydrology and Hydraulics Workshop on Riverine Levee Freeboard, Monticello, Minnesota, 27–29 August, 1991. US Army Corps of Engineering Hydrologic Engineering Center, Davis, California, pp. 5–11.

Institute of Hydrology (1999). *Flood Estimation Handbook*, Institute of Hydrology, Wallingford, UK.

Jones, D.A. and Kay, A.L. (2006). Uncertainty analysis for estimating flood frequencies for ungauged catchments using rainfall-runoff models, *Adv. Water Resour.*, **30**, 1190–1204.

Kendal, M.G. and Stuart, M.A. (1977). *The Advanced Theory of Statistics: Distribution Theory 4th Edition*, Griffin, London.

Kite, G.W. (1977). *Frequency and Risk Analysis in Hydrology*, Water Resources Publications, Colorado, USA.

Kjeldsen, T.R. and Jones, D.A. (2006). Prediction uncertainty in a median-based index flood method using L moments, *Water Resour. Res.*, **42**, W07414, doi:10.1029/2005WR004069.

Kjeldsen, T.R. and Rosbjerg, D. (2002). Comparison of regional index flood estimation procedures based on the extreme value type I distribution, *Stoch. Env. Res. Risk A.*, **16**, 358–373.

Lamb, R. (2005). "Rainfall-Runoff Modelling for Flood Frequency Estimation" in: Anderson, M.G. and McDonnell J.J. (eds), *Encyclopedia of Hydrological Sciences*, Wiley, London.

National Research Council (1999). *Improving American River Flood Frequency Analyses, Committee on American River Flood Frequencies Water Science and Technology Board, Commission on Geosciences, Environment, and Resources*, National Academy Press, Washington, DC.

National Research Council (2000). *Risk Analysis and Uncertainty in Flood Damage Reduction Studies, Committee on Risk-Based Analysis for Flood Damage Reduction, Water Science and Technology Board*, National Academy Press, Washington, DC.

Onof, C. and Wheater, H.S. (1993). Modelling of British rainfall using a random parameter Bartlett-Lewis rectangular pulse model, *J. Hydrol.*, **149**, 67–95.

Potter, K.W. and Walker, J.F. (1981). A model of discontinuous measurement error and its effect on the probability distribution of flood discharge measurements, *Water Resour. Res.*, **17**, 1505–1509.

Rao, A.R. and Hamed, K.H. (2000). *Flood Frequency Analysis*, CRC Press LLC, Florida.

Reed, D.W. (2002). "Reinforcing Flood-Risk Estimation", *Philosophical Transactions: Mathematical, Physical and Engineering Sciences*, Vol. 360, No. 1796, Flood Risk in a Changing Climate (Jul. 15, 2002), pp. 1373–1387.

Rezacova, D., Pesice, P. and Sokol, Z. (2005). An estimation of the probable maximum precipitation for river basins in the Czech Republic, *Atmos. Res.*, **77**, 407–421.

Rosbjerg, D. and Madsen, H. (1995). Uncertainty measures of regional flood frequency estimators, *J. Hydrol.*, **167**, 209–224.

Rosso, R. (1985). A linear approach to the influence of discharge measurement error on flood estimates, *Hydrol. Sci. J.* **30**, 137–149.

Shao, J. and Tu, D. (1995). *The Jackknife and Bootstrap*, Springer, New York.

Sing, V.P. and Strupczewski, W.G. (2002). On the status of flood frequency analysis. *Hydrol. Process.*, **16**, 3737–3740.

Stedinger, J.R., Vogel, R.M. and Foufoula-Georgiou, E. (1993). "Frequency Analysis of Extreme Events" in: Maidment, D. (ed.), *Handbook of Hydrology*, McGraw-Hill, New York.

Tasker, G.D. and Stedinger, J.R. (1989). An operational GLS model for hydrological regression, *J. Hydrol.*, **111**, 361–375.

Zucchini, W. and Adamson, P.T. (1989). Bootstrap confidence intervals for design storms from exceedance series, *Hydrolog. Sci. J.*, **34**, 41–48.

CHAPTER 9

Minimising Uncertainty in Statistical Analysis of Extreme Values

C. Keef
JBA Consulting, UK

9.1. Introduction

There are two main sources of uncertainty due to statistical analysis of a flood frequency curve or extreme sea levels. The first is the statistical uncertainty due to the fact that we observe only a finite number of years of data, and so we have only a sample of all possible observations. The second source of uncertainty is due to the choice of statistical distribution fitted to the data.

Statistical methods to estimate the uncertainty due to sampling (only having observations from a fixed time span) are well developed and all standard statistical inference procedures have methods for producing confidence, or credibility intervals. For example, confidence intervals for parameters estimated by maximum likelihood or L-moments can be obtained using the fact that, as the sample size increases, the distribution of the estimator tends to a normal distribution. In addition to this, the bootstrap method can be used to obtain confidence intervals which do not rely on assumption of normality. Credibility intervals for parameter estimates obtained from a Bayesian analysis can be obtained directly from the estimated posterior distribution for that parameter.

A problem that is less well covered than sample uncertainty is the selection of an appropriate distribution for the data. All statistical distributions are mathematical constructs that describe the probability of observing data values and many are derived as limiting distributions and, as such, rely on a convergence law.

A simple example of a convergence law is the Poisson approximation to the binomial distribution. The binomial distribution describes the probability of observing a certain number of successes from a fixed number of trials where the probability of success is the same in each trial. It can be proved mathematically that if the number of trials is large and the probability of success small, then the distribution of the number of successes can be closely approximated by a Poisson distribution.

An example of the application of this approximation is the use of the Poisson distribution to model the number of days in a year where the total rainfall is above a high threshold (e.g. the 0.99 probability threshold). Underlying this choice of distribution is the implication that the probability of exceeding a certain threshold is the same on each day of the year, or at least the number of days where this is possible is constant from year to year, and that observations from one day to the next are independent. This assumption of independence is likely to be safe for rainfall data. If we did not believe that the probability of exceeding a threshold is constant then we should look for an alternative distribution that does not rely on us making the assumption.

The assumptions on which we base our choice of statistical distribution for flood flows, extreme sea levels or rainfall are usually more important than in most other applications of statistics. The reason for this is that by definition floods are extreme, rare events and typically the events of most particular interest are those larger than any observed events. This means that it is usually not possible to check our assumptions against real data in the region in which we are most interested. This is particularly difficult because rainfall and river flows typically have a heavy tailed distribution; in other words, very high values can be observed.

Figure 9.1 shows an example of two distributions commonly fitted annual maxima flow data, the Generalised Extreme Value (GEV) and Generalised Logistic (GL), fitted to the observed annual maxima data (48 years) from gauging station Nith at Friars Carse in south-west Scotland.

We can see that in the range of the observed data there is very little to choose between the two model fits. However, when each fitted distribution is extrapolated beyond the range of the data, the fits are very different. For instance, the difference in estimated flow for the 1000-year return period is over $100\,\mathrm{m}^3\mathrm{s}^{-1}$. As we have no data in this region it is difficult to tell which distribution we should choose.

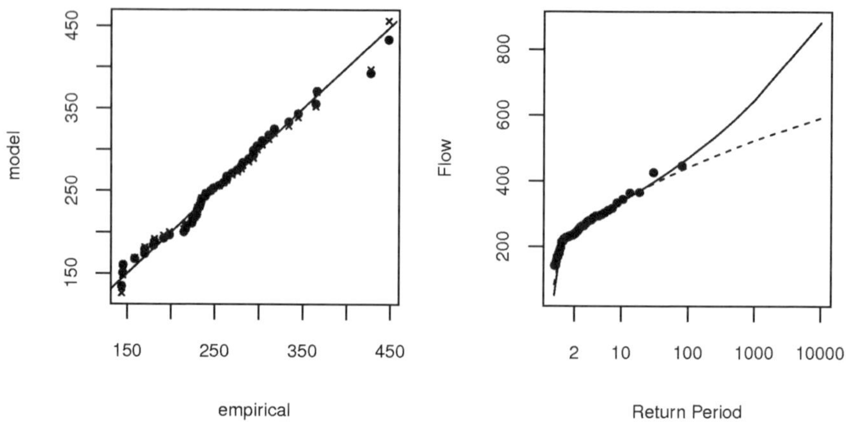

Figure 9.1. Left plot Q–Q plot of fitted annual maxima distribution for the River Nith, dots GEV, crosses GL. Right plot extrapolated dashed line GEV distribution solid line GL distribution; dots show empirical distribution of observed data.

This difficulty in choosing an appropriate fit to flow data has been known for a number of years; Cunnane (1987) stated that:

> "Choosing between such candidates is by no means trouble free as the sampling variations inherent in small samples from skewed, heavily tailed populations tend to mask true between-distribution differences."

The same source also states that goodness-of-fit indices all tend to be inconclusive. However, from a practical point of view choices do have to be made and so a number of studies into the best choice of statistical distribution for flood flows have been published (e.g. Benson, 1968 and NERC, 1975).

A more recent example of a practical approach to the problem of choosing an appropriate distribution is given by Robson and Reed (1999) in the Flood Estimation Handbook (FEH). In developing guidance for a suitable default distribution for annual maxima of river flows in the UK they fitted the GEV distribution, GL distribution, the log-normal distribution and the Pearson Type III distribution to annual maxima from 698 rural sites. When the fits of each of these were compared to the data they found that overall the best fitting distribution was the GL followed by the GEV. In this chapter we present statistical theory that provides evidence as to why this might be the case.

Although this analysis in the FEH gives useful empirical evidence that the GL may provide a good estimate of flood flows, the FEH does not provide a mathematical explanation of why the GL distribution may be used for extrapolation. Distributions which do have a mathematical basis for extrapolation are the domain of Extreme Value statistics. Extreme Value statistics is a collection of mathematical theory and statistical models to describe the statistical distribution of very large (or small) values. It is a relatively new area of statistics, with the PhD thesis of de Haan (1970) generally being accepted as the first modern work on extreme values, although the basic results pre-date this. As such, theoretical development is rapid which suggests that a review of current statistical extreme value theory is likely to result in useful advances for flood risk estimation.

An additional point to note in any review of recent statistical developments is that, as with many areas in flood risk, availability of computational resources have made massive changes to the field of statistics, including extremes. Consequently, methods that would have been out of reach for most practitioners not long ago are now commonplace, for example, bootstrapping, simulation studies and Markov Chain Monte Carlo (MCMC) methods.

Most studies in flood risk management look to estimate the level that will be exceeded once every X-years. If a distribution is fitted to the Annual Maxima (AMAX) of a data record then this X-year level can be obtained by simply inverting the fitted distribution. As a result, it has been argued that it is simpler to estimate this value using annual maxima data, and certainly such methods are more widespread. It is also true that in some situations it is easier to obtain AMAX data than continuous observations because the continuous observations have not been digitised. Other reasons given for using AMAX data are: there is less data to handle, which given modern computing capabilities is no longer a problem; and the fact that independent events must be identified, a problem that can be overcome by using the automated procedures described in this chapter.

This use of AMAX data is in contrast to the majority of recent applications by extreme value statisticians. There are a number of reasons why Peaks-Over-Threshold (POT) methods can be considered to have an advantage. The main reason a POT analysis is preferable to an AMAX approach is that it makes a much more efficient use of data. The number of observations available for a POT analysis is much larger than for an AMAX analysis, and we can also be sure that all the observations are "extreme". Due to annual variability, it is not uncommon for some years to have no "extreme" observations and for other years to have many. This means that the number of threshold exceedances (events) per year varies from year to year. For a POT series with an average number of events per year of two, where the number of events per year followed a Poisson distribution, then the expected number of years in a 50-year series without an event would be 6.7.

The other differences between an AMAX approach and a POT approach are that in modelling peaks above a threshold we must first choose the threshold, and to obtain estimates of the flows that will be exceeded once every X-number of years we must also explicitly choose a distribution for the number of events per year. These choices give an additional level of complexity to the modelling process. However, as we will demonstrate in this chapter, threshold selection is a problem for which solutions are well developed. Additionally, by explicitly choosing a distribution for event rate, we also add an additional level of flexibility to the modelling process.

The final area of extreme value statistics that we cover in this chapter are methods available for pooling data from different locations. Currently most hydrological studies (e.g. Guse *et al.*, 2009, Li *et al.*, 2010 and Robson and Reed, 1999) for pooling data from a set of sites use methods based on the index flood method of Dalrymple (1960). In this method it is assumed that apart from a scaling parameter (the index flood) which varies between sites, the flood frequency distribution is identical for all sites within the set. This can be thought of as letting one distributional parameter vary between sites. Relatively recent developments (the earliest references are Davison and Smith, 1990, and Smith, 1990a) show how it is possible to pool data from a number of sites, thus allowing more than one parameter to vary between the sites. The main benefit of these methods over the earlier methods is that they provide a more flexible fit to multiple sites. By allowing a more flexible fit the resulting estimates will be more robust to inhomogeneous pooling groups, and also potentially allow a better fit to each site.

In this chapter we briefly describe the development and main elements of classical extreme value statistics, with an emphasis upon the treatment of uncertainty in these methods. We then discuss methods of modelling spatial and temporal dependence in extreme values. The next section gives examples of pooling data from different locations. We then describe some problems specifically related to river flows and show how these can be overcome. The final section gives an example of a method to estimate the return flow using POT methods. We explore the uncertainties associated with these estimates using profile likelihood and block bootstrap methods.

In this chapter we have focused our attention mainly on flow data. Generally, river flows exhibit a greater level of temporal dependence and seasonality and so, at a single site, flow data presents more modelling problems than either rainfall or sea surge data. However, the methods presented in this chapter are applicable to both rainfall and surge data.

9.2. Univariate Extreme Value Statistics

9.2.1. *Block maxima*

The very first statistical result for extreme values was derived by Fisher and Tippet (1928) and Gredenko (1943) and described the possible limits laws for the distribution of the sample maximum of Independent and Identically Distributed (i.i.d.) data. This is the famous Extremal Types Theorem and states that:

> "If for a variable X with distribution $F(x)$ we denote the maximum of a sample from X of size n then there exist sequences of constants $a_n > 0$ and b_n such that, as n tends to infinity
>
> $$\Pr\left(\frac{M_n - b_n}{a_n} \leq x\right) = [F(a_n x + b_n)]^n \to G(x), \tag{9.1}$$
>
> for some non-degenerate distribution G, then G is of the same type as the Gumbel (EVI), Fréchet (EVII) or Negative Weibull (EVIII) distribution".

An extension to this result was the derivation of the Generalised Extreme Value (GEV) distribution, which simply combines the three types of extreme value distribution. The GEV was derived by von Mises (1954) and Jenkinson (1955) and is defined as follows:

$$G(x) = \exp\left\{-\left[1 + \xi\left(\frac{x-\mu}{\sigma}\right)\right]_+^{-\frac{1}{\xi}}\right\}, \tag{9.2}$$

where $x_+ = \max(x, 0)$ and $\sigma > 0$. This collapses to

$$G(x) = \exp\left\{-\exp\left[1 - \left(\frac{x-\mu}{\sigma}\right)\right]\right\}, \tag{9.3}$$

when $\xi = 0$.

The three types of extreme value distribution fall out of the GEV, depending on the value of ξ. When $\xi < 0$ the GEV is the same as the Negative Weibull distribution, when $\xi > 0$ the GEV is the same as the Fréchet distribution and when $\xi = 0$ the GEV is the same as the Gumbel distribution.

The implication of the Extremal Types Theorem and subsequent derivation of the GEV is that if a data set is split into equally sized blocks (or samples) then if a limit distribution exists the GEV will be a good approximation to the distribution of the maxima of these blocks, provided that we can make the assumption that the data are i.i.d. and that the size of the blocks are sufficiently large.

9.2.2. *Threshold exceedances*

The limit distribution for threshold exceedances is often derived from the GEV and was first derived in Pickands (1975). This is significantly more recent than the equivalent distribution for block maxima although it should be noted that threshold methods were in use before this date. If the Extremal Types Theorem holds for a variable X, then we have that for a sufficiently high threshold u the exceedances $Z = X - u$ of this threshold have the Generalised Pareto (σ_u, ξ) (GP) distribution function

$$\Pr(\mathrm{Z} \leq \mathrm{z}|\mathrm{X} > \mathrm{u}) = \begin{cases} 1 - \left[1 + \xi\left(\dfrac{z}{\sigma_u}\right)\right]_+^{-\frac{1}{\xi}} & \text{for } \xi \neq 0 \\ 1 - \exp\left(-\dfrac{z}{\sigma_u}\right) & \text{for } \xi = 0, \end{cases} \tag{9.4}$$

where $\sigma_u = \sigma + \xi(u - \mu)$.

9.3. Inference Methods

In fitting extreme value distributions, there are two statistical inference methods which are most commonly used: maximum likelihood and L-moments; both of which can be used to provide confidence intervals for parameter estimates and return levels (Bierlant *et al.*, 2004). In general, maximum likelihood is more commonly used in the statistical community and L-moments in the hydrological community. In addition, the newer technique of MCMC method is also used in extreme value modelling.

9.3.1. *L-moments*

The method of L-moments is based on estimating linear combinations of order statistics, L-moments, in the observed data, and setting the values of the parameters of the distribution to be equal to those which give rise to the observed L-moments. L-moments are an extension of probability weighted moments, first derived by Greenwood *et al.* (1979). An extensive derivation of L-moments, is given by Hosking (1990). The theory behind L-moments is based on the fact that for any distribution with finite mean, theoretical L-moments can be derived. Uncertainty around the L-moment estimators is estimated based on asymptotic results for order statistics which give that for large sample sizes, estimated L-moments follow a normal distribution, with the exact details given by Hosking (1990).

9.3.2. *Likelihood methods*

The method of maximum likelihood is based on the principle of estimating which values of the parameters give the observed data the highest probability of occurring, so which parameter values maximise the likelihood, or probability, or occurrence. In maximum likelihood estimation, the likelihood of observing a certain data set, $\boldsymbol{x}$, given a set of parameters, θ, is given by

$$L(x|\theta) = \prod_{i=1}^{n} f(x_i|\theta), \tag{9.5}$$

where $f(x_i|\theta)$ is the density for of the *ith* observation of $\boldsymbol{x}$ conditional on the parameter set θ (the probability of observation), and n is the number of observations in the sample. The maximum likelihood estimate is the value of θ where $L(\boldsymbol{x}|\theta)$ is the maximum. For computational reasons this maximisation is usually carried out on the log of $L(\boldsymbol{x}|\theta)$. By making certain assumptions about the form of the distribution function it is possible to obtain confidence intervals for the parameter estimates using the Fisher information matrix, $I(\hat{\theta})$. Under these assumptions, for large samples, the distribution of the maximum likelihood estimate of θ, $\hat{\theta}$ is normally distributed with mean θ, and variance $I(\hat{\theta})^{-1}$ (Cox and Hinckley, 1974). Although the GEV distribution does not meet the criteria usually needed for the asymptotic normality of the maximum likelihood estimate, Smith (1985) showed that as long as the shape parameter is greater than -0.5, the standard result can still be used. For many distributions it is possible to obtain the maximum likelihood estimate using analytical methods; however, for some distributions, including extreme value distributions, it is necessary to use numerical methods.

Likelihood methods can also be used to derive the profile likelihood. This is a commonly used method of obtaining confidence intervals around parameters estimates that is particularly useful when the parameter distribution is skewed, i.e. the parameter is more likely to be under or overestimated so symmetric confidence intervals would not be appropriate. Profile methods can also be used to obtain maximum likelihood parameter estimates. Profile likelihood plots are obtained by fixing one parameter of the distribution and maximising the likelihood over all other parameters; for full details see Coles (2001). The resulting log likelihood values are then plotted against the value of the parameter we have fixed. Confidence intervals using the profile likelihood method can be obtained using distributional properties of the likelihood. For a generic parameter θ, if $\boldsymbol{l}(\hat{\theta})$ is the maximum log likelihood, and $\boldsymbol{l}(\theta)$ is the likelihood at a different value of θ, then $D(\theta) = 2\{\boldsymbol{l}(\hat{\theta}) - \boldsymbol{l}(\theta)\}$ follows a chi-squared distribution with degrees of freedom 1 if $\hat{\theta}$ is the true value of θ. A 95% confidence interval is obtained as being all θ such that $D(\theta)$ is less than or equal to the 0.95 quantile of the chi-squared(1) distribution (3.841).

9.3.3. *Bootstrap procedures*

For situations where it is not possible to make the assumptions on which uncertainty estimates for L-moments or likelihood methods are based a bootstrapping procedure is often used, which is described fully in Davison and Hinkley (1997). The basic non-parametric bootstrap is a technique to resample the data to obtain information about the variation in the data. Let X be a univariate sample of i.i.d. data of size n. If we are estimating the value of a parameter θ of the distribution from which the sample X is taken and we wish to estimate the uncertainty in the estimate $\hat{\theta}$ of θ then the basic bootstrap method is as follows.

(1) Re-sample X with replacement to obtain a bootstrapped sample X^* of size n.
(2) Calculate $\hat{\theta}$ for X^*.

Steps 1 and 2 are repeated B, B large, times to obtain a sample $\hat{\theta}^B$, of estimates $\hat{\theta}$ of of size B. The variation in this sample can then be assessed and used as the estimate of uncertainty in the parameter estimation. In particular, if we wish to obtain a 95% confidence interval then we can take the end points of this interval to be the 0.025 and 0.975 quantiles of $\hat{\theta}^B$.

One application in which a bootstrap procedure is often used is when temporal dependence must be taken into account. In the block bootstrap the original time series X_t is divided into blocks. These blocks are then resampled with replacement to create the bootstrapped sample X_t^*. The blocks are chosen to be large enough to preserve the temporal dependence in the time series but small enough to allow a large number of possible combinations in each resample. When modelling flow data, due to the seasonality of the data it is sensible to choose blocks that correspond to a whole year. In making this choice we make the assumption that floods in one year are independent of floods in the previous year. If we chose blocks that corresponded to calendar years, then this assumption would be invalid. This is because the start date of the calendar year falls in the middle of the flood season and what happens in the second half of a flood season is not independent of what happens in the first half. The start day of the block year should be chosen so that it is in the middle of the lower seasonal flow periods.

9.3.4. *Discussion of L-moments and likelihood methods*

The main argument for using L-moments as a preference over maximum likelihood in fitting extreme value distributions is that for small samples maximum likelihood estimates have been shown to have more bias. In using L-moments to fit the GEV to data we force the shape parameter to be less than one ($\xi < 1$). This forcing is necessary because moments only exist if the mean is finite, which is only the case if $\xi < 1$. A method of making maximum likelihood estimates replicate this forcing was developed by Coles and Dixon (1999). By penalising the likelihood function, they found that it was possible to obtain parameter estimates using maximum likelihood techniques that had similar bias to those obtained by L-moments, and smaller mean squared errors. The penalty function introduced was

$$P(\xi) = \exp\left[-\left(\frac{1}{1-\xi} - 1\right)\right] \tag{9.6}$$

for $0 < \xi < 1$, and $\mathrm{P}(\xi) = 1$ for $\xi \leq 0$ and $\mathrm{P}(\xi) = 0$ for $\xi \geq 1$. The corresponding penalised likelihood function then becomes $L_{pen}(x; \mu, \sigma, \xi) = L(x; \mu, \sigma, \xi)P(\xi)$.

Therefore, because it is possible to reduce the bias when using maximum likelihood to fit the GEV distribution, the main difference in usefulness between maximum likelihood and L-moments is in the ability to extend the method to include covariates in the distributional parameters. However, L-moments do still have the additional feature that they are more robust to "outliers", although in the study of extremes this is both a good and bad feature. Often it is these very large (or small, when studying minima) observations that provide most information about the extremes of the process being modelled.

9.3.5. *Markov Chain Monte Carlo (MCMC)*

An alternative approach to fitting models to data is the Bayesian technique of MCMC; for full details see Gamerman and Lopes (2006). This is a collection of Bayesian methods that rely on stochastic simulation. Bayesian inference is based on the assumption that an improved estimate can be obtained by including prior information about the parameters. In Bayesian inference uncertainty estimates are thought about in a very different way when compared with frequentist inference. In frequentist inference the distribution from which the parameter estimates are sampled is estimated, and uncertainty estimates are based on this distribution. In Bayesian inference the fact that the parameters themselves have distributions is stated up front; this full distribution (posterior density) is estimated and so estimates of uncertainty are given by this distribution.

In Bayesian analysis prior information is included using Bayes Theorem which gives that for a parameter θ the posterior density of θ conditional on data observations $\boldsymbol{x}$ is given by

$$f(\theta|x) \propto f(\theta)L(x|\theta), \tag{9.7}$$

where $f(\theta)$ is the prior distribution of parameter θ and $L(\boldsymbol{x}|\theta)$ is the likelihood of observing the data given the parameter θ. If the prior distribution given is non-informative, i.e. the same for all values of θ, we can see that this reduces simply to the likelihood. In conducting Bayesian inference it is important that the prior distribution should be chosen independently of observing the data, otherwise the information provided by the data is effectively double-counted and so the resulting estimates of uncertainty will be too small.

MCMC techniques rely on the fact that as long as it is possible to write down the prior distribution $f(\theta)$ and likelihood function $L(\boldsymbol{x}|\theta)$ it is possible to simulate a Markov chain that has a stationary distribution equal to the posterior distribution $f(\theta|\boldsymbol{x})$. In MCMC, stationary distribution of the Markov chain is approximated by the empirical distribution of a large number of simulated observations from this chain.

A Markov chain is a random process $\theta_1, \theta_2, \ldots$ in which given a single (present) observation θ_i, future observations $\theta_{i+1}, \theta_{i+2}, \ldots$ are independent of past observations $\theta_{i-1}, \theta_{i-2}, \ldots$ In MCMC the method of getting from one observation to the next is as follows:

1. Simulate a proposal value θ^* from proposal density $q(\theta^*|\theta_i)$.
2. Obtain acceptance probability α where

$$\alpha = \min\left\{1, \frac{L(x|\theta^*)f(\theta^*)q(\theta_i|\theta^*)}{L(x|\theta_i)f(\theta_i)q(\theta^*|\theta_i)}\right\}.$$

3. Set

$$\theta_{i+1} = \begin{cases} \theta^* \text{ with probability } \alpha \\ \theta_i \text{ with probability } 1-\alpha \end{cases}.$$

Theoretically the proposal density can be arbitrarily chosen. However, in practical terms the choice can have a large effect on how well the values in the chain cover the full posterior distribution. In the basic Metropolis–Hastings algorithm the proposal density is symmetric, so $q(\theta_i|\theta^*) = q(\theta^*|\theta_i)$. In Gibbs sampling the proposal density is formed in such a way that all proposal values are accepted.

As well as being used in a true Bayesian context, MCMC is also often used with non-informative priors. This is because it can be used to fit models that would otherwise be intractable using frequentist maximum likelihood techniques.

9.4. Spatial and Temporal Dependence

9.4.1. *Overview*

There are two reasons to model spatial and temporal dependence in flow data. The first is to account for it in estimation of extreme return levels and the second is to assess the joint behaviour of extremes at multiple sites.

In terms of statistical extreme value theory, limited temporal dependence can be handled by the theorem of Leadbetter (1983). Informally, this states that if a time series is locally dependent, but that observations separated by a sufficient period of time are effectively independent, then the GEV distribution is still the limiting form for the block maxima of the time series; for a fuller description see Chapter 10 of Beirlant *et al.* (2004). The relationship between the distribution if the data were independent and the dependent data is governed by the extremal index. If above a certain (high) threshold the mean cluster (or event) length is constant we can term this the limiting cluster length. The extremal index is the inverse of this limiting cluster length.

An example of a flow record that would satisfy this condition is one in which the flow on one day is useful in predicting the flow the next day, but that the flow in September had no value in predicting the likely flow in November. Although for some regions where the rivers respond very quickly to rainfall this assumption is likely to be valid, for larger or slower responding rivers it is difficult to justify this assumption. For some baseflow dominated rivers, the flows respond so slowly to rainfall that they exhibit dependence over timescales of months.

Another statistical result derived by Leadbetter (1991) describes the asymptotic relationship between the distribution of cluster maxima and cluster exceedances. It is usually taken to mean that the cluster maxima have an identical distribution to that of marginal exceedances, although as we will see later this theorem does not necessarily hold for the threshold at which POT analysis is typically carried out.

When assessing the joint behaviour of extremes at multiple sites the difference between block maxima and POT methods is more profound than in a single-site analysis. The question of which approach should be taken is no longer as simple as choosing which will give the best point estimates; instead we must think more carefully about the question of interest. Typically the questions of interest take the form of, "How much flooding will we have to cope with at once?". To answer these questions we need an approach that can examine the extremes of simultaneous flows at multiple locations. If we were to examine block maxima there is no guarantee that within each block the days on which the maximum is observed would be the same for each station. Therefore it is sensible to look to a method that is equivalent to a POT approach which limits our choice.

Currently much statistical research is dedicated to max stable processes (de Haan, 1984; Padoan *et al.* 2010; Smith, 1990b). Max stable processes have similar asymptotic motivation to the GEV and as such have the advantage that they are underpinned by mathematical theory, but they are only suitable for block maxima. Additionally, they rely on strong assumptions about the form of extremal dependence, specifically the only forms possible are asymptotic dependence and complete independence. The results of Keef *et al.* (2009a) suggest that neither of these assumptions is appropriate for the dependence between UK flow or rainfall extremes at different locations.

The dependence between the extremes of two (or more) variables can take two forms, asymptotic dependence or asymptotic independence. Two variables are asymptotically dependent if the largest observations in both variables occur together with a positive probability. If two variables are asymptotically independent then the probability of the largest observations of each variable occurring together is zero (Ledford and Tawn, 1996). There are three sub-classes of asymptotic independence, positive extremal association, near independence, and negative extremal association. These three classes correspond respectively to joint extremes of two variables occurring more often than, approximately as often as, or less often than joint extremes, if all components of the variable were independent. Variables that have a bivariate normal dependence structure with correlation, ρ, greater than zero are examples of asymptotic independence with positive extremal association.

Similarly to extreme value modelling of single site data, correct specification of the statistical model used for model dependence between sites is vital to ensure correct estimation of the probability of multiple floods. Examples of the consequences of incorrect specification are given in Keef *et al.* (2010a) and Lamb *et al.* (2010).

9.4.2. *Temporal event identification*

If POT methods are to be used to estimate the X-year flow we need a model for the number of events per year. This means we must first de-cluster the flow time series into independent clusters. One of the independence criteria used in the FEH was to say that two peaks are independent if "the minimum discharge in the trough between two peaks must be less than two-thirds of the discharge of the *first* of the two peaks"; the other was a specification on time between peaks.

Currently there are three main de-clustering methods of separating out exceedance of a fixed threshold, u, into independent clusters that are in use in the statistical community. The first is the runs estimator of Smith and Weissman (1994). This estimator simply states that two exceedances of threshold u are independent if they are separated by a fixed number, m, of non-exceedances.

An extension to the runs estimator (Laurini and Tawn, 2003) introduces a second threshold, it defines two exceedances of u as independent if they are separated by m non-exceedances or if any of the intervening values is below a second, lower threshold u_2. The third method is the intervals method of Ferro and Segers (2003). This method can be used to split the exceedances into clusters according to the inter-exceedance times.

Out of these methods, the one most similar to that used in hydrology is the two thresholds extension to the runs estimator, so we would expect it to perform "best" at identifying clusters. In a statistical analysis, a common reason for de-clustering a time series is to estimate the mean cluster size and so the extremal index. Keef (2007) compared the three methods for eight daily mean flow gauging stations. Varying thresholds u were examined, and the second threshold u_2 in the two thresholds estimator was set to the 0.65 quantile. Keef found very little difference in the estimates of mean cluster size obtained using the runs estimator and the two threshold estimator for values of m in the range 5–20 days, with the second threshold only affecting the estimator for longer run lengths. The intervals method did not perform as well, particularly for slowly responding rivers. The reason for this is likely to be that it relies on a stronger assumption of stationarity than the other two methods, and so the seasonality of river flows is likely to have a larger effect on the resulting estimates.

9.4.3. *Dependence modelling*

In this section we discuss three models for extremal dependence of asymptotically independent data that have been used for flood risk variables.

In order to apply any of these methods we must use the theory of copulas (Sklar, 1959). This states that for a d-dimensional multivariate random vector with joint distribution $F(\boldsymbol{x}) = F(x_1, \ldots, x_d)$ and marginal distributions $F_i, i = 1, \ldots, d$, then it is possible to write $F(\boldsymbol{x})$ as

$$F(x) = C\{F_1(x_1), \ldots, F_d(x_d)\}.$$

The function C is called the copula, it has domain $[0, 1]^{\mathrm{d}}$ and contains all the information about the dependence structure of $F(\boldsymbol{x})$.

In simple terms all this theory says is that it is possible to estimate the joint distribution function of a set of variables into the separate marginal distribution functions of each of the variables and the dependence structure. This is useful because it allows us to analyse the marginal distributions and dependence structure separately. It also allows us to estimate the dependence between a set of variables on any marginal distribution we choose. For temporal modelling the variables $X_1, \ldots, X_d$ are simply the time series at lags $0, \ldots, d-1$.

The models we discuss are those of Ledford and Tawn (1996, 2003), Heffernan and Tawn (2004) and Bortot *et al.* (2000). Both the Ledford and Tawn and Heffernan and Tawn dependence models were initially defined for variables with a specified marginal distribution, the Ledford and Tawn model uses Fréchet margins and the Heffernan and Tawn uses Gumbel margins. The transformation to these margins can be achieved using the probability integral transform. In simple terms, the probability integral transform uses the result that for a sample $x_i, i = 1, \ldots, n$ of a random variable X, $F_i(x_i)$ has a uniform distribution where F is the distribution function of X. For more details see Grimmett and Stirzaker (1992).

The Ledford and Tawn model can be described as follows. For a pair of Fréchet distributed variables (S, T) the joint survivor function $\bar{F}(s, s) = Pr(S > s, T > s)$ can be modelled as $\bar{F}(s, s) = \mathcal{L}(s)s^{-\frac{1}{\eta}}$ where $\eta \in (0, 1]$ is called the coefficient of tail dependence and $\mathcal{L}()$ is a function that becomes approximately constant as s gets large (a slowly varying function). It is then possible to model the conditional probability $\Pr(T > s|S > s)$ by as s tends to infinity $\Pr(T > s|S > s) \approx \mathcal{L}(s)s^{1-1/\eta}$.

This result is used for a time series X_t, by simply substituting S and T for X_t and $X_{t+\tau}$.

In contrast with the Ledford and Tawn model, the Heffernan and Tawn model is capable of modelling truly multivariate data sets. However, because it is described semi-parametrically, some inference procedures, such as extensions to include covariates, are hard. For a single variable S and a set of variables $\boldsymbol{T} = \{T_1, \ldots, T_d\}$, all with Gumbel marginal distributions, the Heffernan and Tawn method is based on the relatively weak assumption that there exist normalising vectors $\boldsymbol{a} = \{a_1, \ldots, a_d\}$ and $\boldsymbol{b} = \{b_1, \ldots, b_d\}$ such that

$$\Pr\left(\frac{T - \boldsymbol{a}s}{s^{\boldsymbol{b}}} < z \,\middle|\, S = s\right) \to H(z) \quad \text{as } s \to \infty$$

where all marginal distributions of $H(\boldsymbol{z})$ are non-degenerate, i.e. not a constant value.

The model is based on the approximation that the above limiting relationship holds exactly for all values of $S > v_p$ for a suitable high threshold v_p, which has probability p of being exceeded. A consequence of this assumption is that when $S = s$, with $s > v_p$ the random variable $\boldsymbol{Z}$, defined by

$$\mathbf{Z} = \frac{T - \boldsymbol{a}s}{s^{\boldsymbol{b}}}$$

is independent of S and has distribution function H. It is this assumption that allows us to extrapolate beyond the range of the data. For a time series X_t we simply substitute the variable X at time t (X_t) for S and the lagged variables $X_{t+\tau}, \tau = 1, \ldots, \tau_{\max}$ for $\boldsymbol{T}$. Keef *et al.* (2009a) showed how this model can be extended to model both temporal and spatial dependence, and to handle missing data.

A large-scale application of spatial and temporal dependence modelling was undertaken by Keef *et al.* (2009b) who mapped the level of spatial dependence of extreme river flows and rainfall over the UK using the Heffernan and Tawn model. The Heffernan and Tawn model has also been used to assess flood risk over large regions or multiple locations by Lamb *et al.* (2010), Keef *et al.* (2010b) and Speight *et al.* (2010).

The model used in Bortot *et al.* (2000) is to assume that the joint tail of multivariate variables can be accurately modelled by a Gaussian distribution. In terms of assumptions made about the dependence between variables, the use of the Gaussian copula necessitates much stronger assumptions than either the Ledford and Tawn or Heffernan and Tawn models. The assumptions made are that the level of dependence between a pair of variables does not change as the variables become more extreme. It also makes the assumption that the variables are always asymptotically independent. In contrast, the other models discussed here allow the dependence to vary with extremeness and also do not need any assumptions about the form of dependence to be made.

9.5. Pooling Data

In pooling data from different sites to obtain better estimates of return levels there are two issues to address: the first is the spatial dependence between sites; and the second is accounting for covariate effects such as catchment area, soil type, etc. Examples of applications of extreme value methods including covariate effects are numerous and diverse. A selection of areas of application includes sports records (Adam, 2007), lichenometry (Cooley *et al.*, 2006), radioactive matter dispersal (Davison and Smith, 1990), wind speed (Fawcett and Walshaw, 2006), minimum temperature (Chavez-Demoulin and Davison, 2005) and precipitation (Cooley *et al.*, 2007).

Maximum likelihood methods to account for dependence in the data have been known since Smith (1990a), with more recent references being Chandler and Bate (2007) and Fawcett and Walshaw (2007). Generally these methods involve first estimating the parameter uncertainty, and subsequently adjusting for dependence.

Similarly, methods to include covariate effects in extreme value distribution parameters are not new (Davison and Smith, 1990 and Smith, 1990a). The earliest and simplest approaches were to simply model the location, scale and shape parameters as functions (f, g, h) of linear combinations of a set of covariates $v = \{v_1, v_2, \ldots, v_m\}$, so for a GEV distribution set $\mu = f(\boldsymbol{v\beta})$, $\sigma = g(\boldsymbol{v\gamma})$, $\xi = h(\boldsymbol{v\delta})$ where $\boldsymbol{\beta}$, $\boldsymbol{\gamma}$, $\boldsymbol{\delta}$ are the multipliers of the covariates in the linear combinations. The use of a function can ensure that an estimated parameter is within the possible range, for example, greater than zero for the scale parameter (c.f. generalised linear modelling, Dobson, 2001). Because the shape parameter is difficult to estimate it is common not to include covariates in this.

Examples of extreme value models that include covariates vary in complexity and procedure used to fit the model. A relatively simple example of extreme value modelling with covariates is that of Fawcett and Walshaw (2006), who modelled extremes of hourly maximum wind speed using a GP distribution taking into account site and seasonal effects using MCMC techniques; the extension to the original methods in this case is by grouping the observations by site.

Examples of peaks-over-threshold models taking a spatial covariate into account are those of minimum winter temperature using maximum penalised likelihood estimation of Chavez-Demoulin and Davison (2005) and maximum daily precipitation using MCMC of Cooley *et al.* (2007). However, neither of these explicitly accounted for any remaining spatial dependence in the observations. In other words, they assume that given the overall spatial effect, observations were independent.

An application where spatial dependence is taken into account in modelling is that of Northrop and Jonathan (2010), who modelled extreme hurricane induced wave heights using quantile regression methods with a GP distribution and maximum likelihood.

9.6. Problems and Solutions Specific to Fluvial Estimation

The Extremal Types Theorem gives the Generalised Extreme Value (GEV) distribution as the limiting distribution of the maxima of samples where each observation, within each sample and for different samples, is i.i.d. For river flow data from a single site hourly, daily, or even monthly observations are neither independent nor identically distributed. River flow observations are time series which exhibit both temporal dependence, seasonality, and sometimes step changes. These facts led the authors of the Flood Studies Report to conclude that classical extreme value theory was of little use in flood frequency analysis. However, recent developments in extreme value theory mean that this conclusion ought to be re-visited.

Longer-term trends and step changes can be handled by the inclusion of covariates by the methods described in the previous section.

The fact that seasonality means that flow data is not identically distributed is often ignored in fitting a distribution. A possible method to account for this non-stationarity would be to derive a model for the underlying seasonality in the series, for instance by modelling the expected flows on each day of the year, and then to carry out an extreme value analysis on the deviations from this model. However, it can be debated whether or not this type of approach would work for river flow data. The first problem is that of characterisation of the seasonal process, which is unlikely to be a simple sinusoidal variation. The second problem is that it is not unlikely that summer deviations would have different characteristics to winter deviations; the flow processes from rainfall on dry ground with a high level of vegetation are different to those when rain falls on wet ground when many plants are dormant. This means that fitting a suitable distribution to these deviations is likely to be no less complicated than fitting a distribution to the raw flow observations. It also means that peak observations in summer (which are unlikely to result in flooding) may provide little information about the distribution of peak observations in winter. An approach for this type of modelling was developed by Eastoe and Tawn (2009) and demonstrated using extremes of air pollution data.

An alternative to the strong assumption of seasonality is to assume that once the ground conditions are such that flooding is possible, each flood event is independent and has an identical distribution. This assumption is implicit in using POT methods. As we have seen in the introduction, if in addition the number of days in which ground conditions were such that flooding is possible were the same in each year, then the block size for each year would be identical and so the GEV distribution would still be observed as the correct limit distribution for annual maxima. In reality there is between year variation in ground conditions and so the number of days for which flooding is possible varies greatly from year to year.

There is much empirical evidence (Lang, 1999; Robson and Reed, 1999) that a negative binomial distribution provides a better fit to the number of flood events per year than the Poisson distribution. Compared to the

Poisson distribution, the negative binomial is over dispersed i.e. the variance is greater than the mean. The negative binomial is an attractive distribution to use in modelling the number of events per year for a number of reasons. First a special case of the negative binomial when combined with the GP distribution for the size of flood events leads to the GL distribution for AMAX data. The second main reason is that it can be formulated as a combination of the gamma and Poisson distributions, in that the number of events in any particular year follows a Poisson distribution with parameter λ where λ follows a gamma distribution. An important point to note here is that although the gamma distribution is sufficiently flexible that it is likely to provide a suitable fit to most data sets, there is no theoretical justification for its use.

To overcome some of the limitations of classical extreme value theory we present two alternative models developed by Eastoe and Tawn (2010, 2012). The first new statistical model we present is that of Eastoe and Tawn (2010). They use the Leadbetter (1983) theorem and assume that event maxima follow a GP(σ, ξ) distribution, but that the number of events per year follow a specific formulation of the negative binomial $(1/\alpha, 1/(1+\lambda\alpha))$ distribution. With these assumptions, the annual maxima of flows (x) when x is above the GP threshold u have the following distribution:

$$G(x) = \begin{cases} p^r \left\{1 - (1-p)\left(1 - \left[1 + \xi\left(\dfrac{x-u}{\sigma}\right)\right]_+^{-\frac{1}{\xi}}\right)\right\}^{-r} & \text{for } \xi \neq 0 \\ p^r \left\{1 - (1-p)\left[1 - \exp\left(-\dfrac{x-u}{\sigma}\right)\right]\right\}^{-r} & \text{for } \xi = 0 \end{cases}$$

where $p = 1/(1+\lambda\alpha)$ and $r = 1/\alpha$. The standard GL distribution is the special case when $\alpha = 1$. Eastoe and Tawn successfully fitted this distribution to annual maxima from the Thames at Kingston, which has a record length of 123 years, using Bayesian MCMC techniques, and introducing covariates for the between year differences.

The second model (Eastoe and Tawn, 2012) is for the cluster maxima of exceedances of sub-asymptotic thresholds. The model is motivated by the observation that a GP distribution fitted to cluster maxima is often different to a GP distribution fitted to all exceedances of a threshold, which suggests that the asymptotic argument supported by the Leadbetter (1991) theorem is invalid. This is consistent with the findings of Fawcett and Walshaw (2007) who found evidence of bias in estimates of σ_u and ξ when these estimates were obtained using cluster maxima only. Eastoe and Tawn's solution is to introduce a modifier to the GP distribution that describes how the degree of clustering varies with the level of extremness. If the distribution of peaks over a threshold is GP(σ_u, ξ) the resulting model for the exceedance probability of x takes the form

$$\Pr(X > x|X \geq u) = 1 - \frac{\theta(x,m)}{\theta(u,m)}\left[1 + \xi\frac{(x-u)}{\sigma_u}\right]_+^{-1/\xi}, \tag{9.1}$$

where $\theta(x, m)$ is given by

$$\theta(x,m) = \Pr[\max(X_2,\ldots,X_m) < x|X_1 > x],$$

with m being the run length used to identify independent clusters. The implication of this model is that the GP is only an appropriate distribution for cluster maxima if $\theta(x, m) = \theta(u, m)$ for all $x > u$. In other words, if the length of time a river can be expected to be above a threshold u in a single cluster is the same as the length of time a river can be expected to be above a threshold x at one time. In the range of the data, $\theta(x, m)$ can be estimated empirically using the runs estimator (Smith and Weissman, 1994). However, for the range of data for which there are no observations the temporal dependence must be explicitly modelled. Eastoe and Tawn examine two dependence models, those of Ledford and Tawn (1996) and Heffernan and Tawn (2004), and found that both resulted in models that fitted cluster maxima from the River Lune at Caton, but that the Heffernan and Tawn model has the advantage of a much lighter computational burden.

9.7. New Approach

In this section we show how it is possible to use the recent statistical developments to accurately estimate the return period of river flows. We use the distribution of cluster maxima given by model (Equation 9.1) with the ideas of modelling the numbers of events per year to produce estimates of high return period flows. The main assumption made in using this approach is that the GP distribution is a suitable limit distribution for flows above a certain threshold.

In this process there are a number of modelling decisions to be made. The first is what distribution to use for the number of events per year. From empirical evidence it appears that the form of negative binomial used in Eastoe and Tawn (2010) does not provide a uniformly good fit for all stations. Instead we simply use the observed number of events per year. To increase sample size we pool stations with similar mean numbers of events per year together.

The second modelling decision is which dependence model to use for temporal dependence, and also which run length to use in estimating θ. The dependence model we choose is the Heffernan and Tawn model, when compared with the Ledford and Tawn model it is simpler to use and the results of Eastoe and Tawn (2012). The run length we choose is 14 days, i.e. we assume that exceedances separated by at least two weeks of consecutive non-exceedances are independent. This choice is in accordance with results from Keef *et al.* (2010b) who found that extremal dependence drops to approximately zero within this time frame for most sites in the UK.

To obtain the distribution function for annual maxima data from the distributions of cluster maxima and numbers of events per year we use a simulation procedure. To minimise simulation noise we simulate 10,000 years of cluster maxima 100 times. For each of these 100 simulations of 10,000 years we calculate the empirical distribution function. For simulated years when the number of clusters is equal to zero, we use the empirical distribution function of the observed annual maxima series conditional on the annual maxima being below the GP conditioning threshold. Our estimated annual maxima distribution function is the median of these 100 simulations.

One of the implications of using this method of estimating the return period of river floods is that we must use a bootstrap procedure to produce confidence intervals.

9.8. Data Analysis

9.8.1. *Data*

The data set we use comprises 31 sites with complete records from 1st January 1960 to 31st December 2008, the positions of which are shown in Figure 9.2. We only directly present results from a small number of these stations.

9.8.2. *Fitting GP distribution*

We illustrate the method of fitting a GP distribution using data from the Nith at Friar's Carse. The first step in this process is to choose a suitable threshold u, this should be high enough for the asymptotic properties to hold, and low enough that there is still sufficient data to obtain an

Figure 9.2. Locations of gauges used in study.

acceptable amount of variance in the estimators. There are three main diagnostic plots to help threshold selection. The first is the mean excess plot; the mean threshold exceedance $(X - u)$ should be constant if with respect to the threshold u if u is high enough for the asymptotic properties to hold.

The second and third checks relate to the threshold stability of the estimators. If the asymptotic limits hold for threshold u, then the modified scale parameter $\sigma_v^* = \sigma_v - \xi(v - u)$ for all thresholds v where $v > u$ should be the same. Similarly, the estimated shape parameter for all thresholds greater than the lowest for which the asymptotic properties hold should also be constant. Figure 9.3 shows examples of all three plots. The wild variation for increasing thresholds is usual. It is a feature of the threshold being so high that there is not enough data above it to accurately estimate the

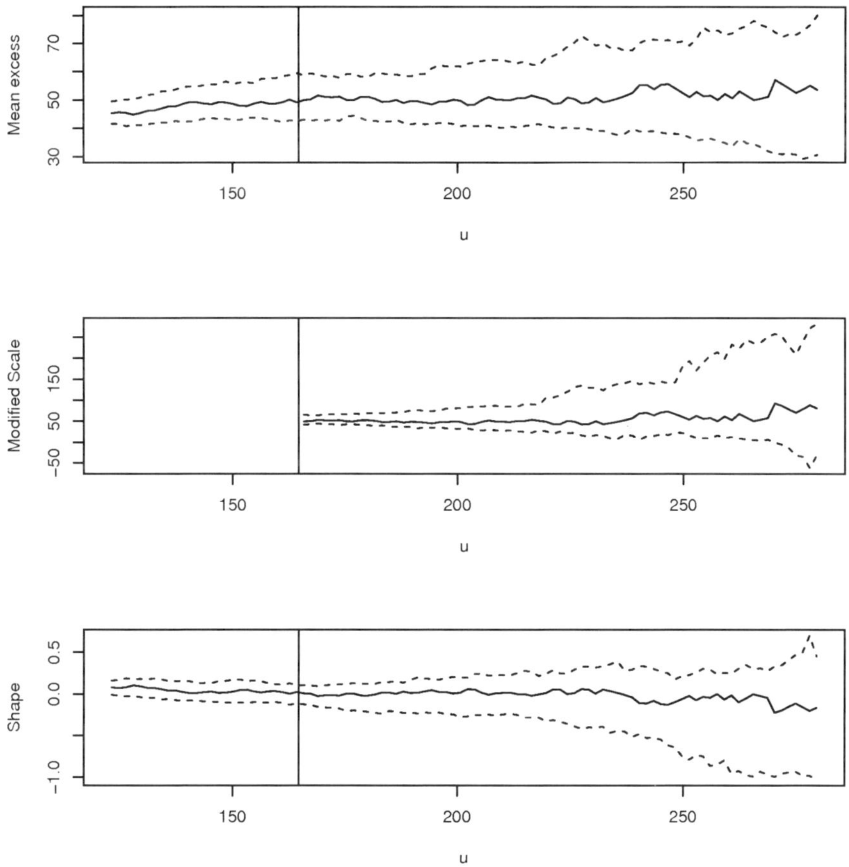

Figure 9.3. Diagnostic plots for threshold selection for Nith at Friars Carse, vertical line shows chosen threshold. Top mean excess plot, middle modified scale parameter, bottom, shape parameter. In all plots, the solid line shows best estimate, dashed lines show 95% block bootstrapped confidence intervals.

parameters. From Figure 9.3 we can see that a suitable threshold is likely to be about $160\,\mathrm{m}^3\mathrm{s}^{-1}$, which is equal to the 0.99 probability threshold.

To fit the GP distribution we use penalised maximum likelihood, the penalty imposed is that suggested by Coles and Dixon (1999) for the GEV distribution. This is likely to be a sensible choice because of the links between the GEV and GP distributions, although we have not checked this assumption.

Because we have used all exceedances and have not first declustered the time series there is dependence within the data. Although this does not

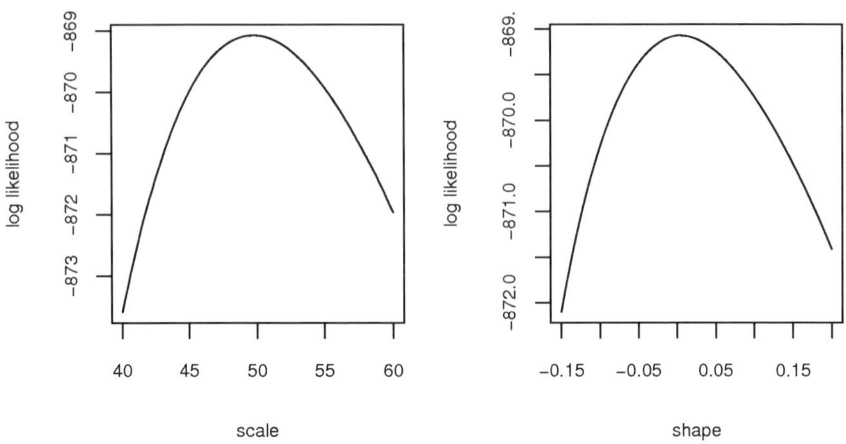

Figure 9.4. Profile likelihood plots for GP parameter estimates, horizontal line shows approximately 95% confidence intervals.

affect the parameter estimates it does affect the variance of the estimation. This means that many of the standard techniques for estimating uncertainty will under estimate the true level. To take this into account we use a yearly block bootstrap, described in Section 7.3.3. In Keef (2007) it was found that a suitable date to choose as the start date of a year for the purposes of block bootstrapping river flows was 1st August as most UK rivers are in the middle of seasonal low flows on this date.

For illustrative purposes we show confidence intervals obtained using the technique of profile likelihood. Figure 9.4 shows profile likelihood plots for the shape and scale parameters using a threshold of $160\,\mathrm{m}^3\mathrm{s}^{-1}$. The confidence intervals we show here are approximate for two reasons; firstly, we do not take temporal dependence into account, and secondly, the introduction of the penalty function will make the estimates imprecise (Coles and Dixon, 1999).

For the scale parameter σ_u the 95% confidence interval using profile likelihood methods is equal to (43.2, 57.8) and for the shape parameter ξ (−0.12, 0.18).

In carrying out the block bootstrap procedure we re-estimated the threshold u for each bootstrapped sample to be the 0.99 probability quantile and carried out 100 replications. The re-estimation of the threshold will give us a better description of the uncertainty, however, because of the dependence between the threshold and scale parameters it does mean we are not able to produce confidence intervals for the scale parameter. The 95% confidence interval for the shape parameter is (−0.11, 0.12).

The discrepancy between the shape parameter confidence intervals using the two different methods is not uncommon. The intervals obtained using profile likelihood use a strict distributional assumption, whereas those using the bootstrap method simply reflect the variation observed in the data. Which is the most sensible is more difficult to decide, the profile likelihood estimates rely on a strict assumption of i.i.d. data, whereas the bootstrapped estimates rely on our sample of data being a very good representation of the uncertainty in the full population of data, for a long data record such as ours, this second assumption is likely to be valid, however, for shorter records it may be more difficult to defend. In practical terms nonetheless both are similar enough that this difference will not make a large impact on any subsequent modelling.

Figure 9.5 shows plots of the fitted distribution. The QQ-plot shows that the distribution fits well within the range of the data. The wide confidence intervals for the fitted distribution reflect the large amount of uncertainty in the extrapolation. In this case, a contributing factor is likely to be the fact that the confidence interval for the shape parameter spans both negative and positive values. The bootstrapped instances when the fitted shape parameter is negative will have short tails, and so low exceedance probability flows will not be very much larger than high exceedance probability flows.

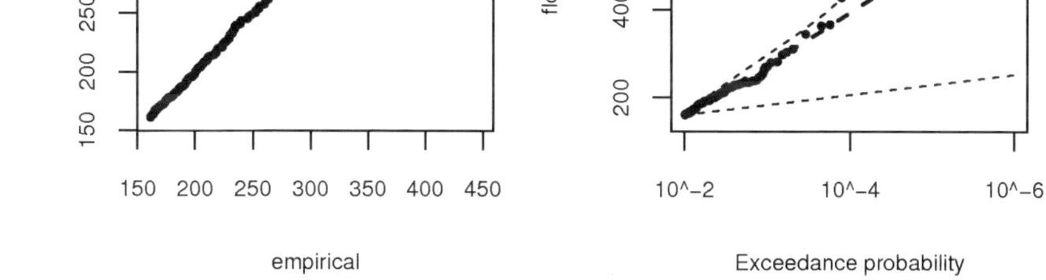

Figure 9.5. Diagnostic plots for GP distribution fit. Left plot QQ-plot, right plot fitted distribution with 95% confidence intervals.

9.8.3. *Fit to annual maxima data*

Before we proceed there are a number of modelling choices to discuss. Firstly, the choice of threshold above which to fit the GP distribution, when diagnostic plots such as those in Figure 9.3 were examined, we found that in general the 0.99 probability threshold was suitable for nearly all sites. The second choice was how many groups to use for obtaining sensible estimates of the number of events per year; we chose five groups, each with an (approximately) equal number of sites in each. The final decision was to use 100 bootstrap replications.

To compare our new method of estimating the distribution of annual maxima we also fit the GEV and GL distributions. To ensure that we make a fair comparison of the relative benefits of our method, we use the same

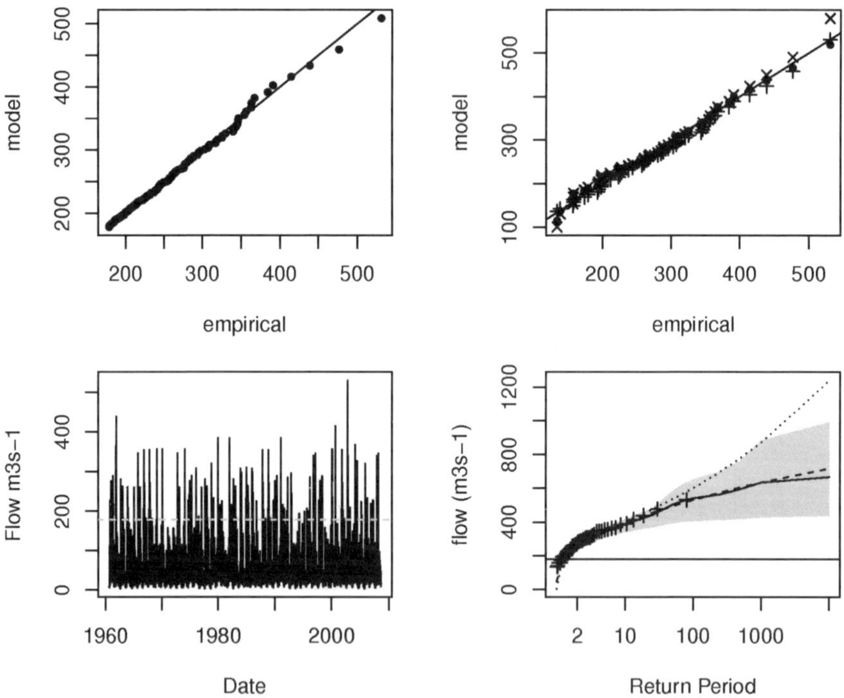

Figure 9.6. Diagnostic plots for Dee at Woodend. Clockwise from top left; Q-Q plot for GP fit; Q-Q plot for AMAX fits, dots show GEV, diagonal crosses show GL, vertical crosses show new method; flow-return period plots, dashed line shows GEV, dotted line shows GL, solid line new method with 95% confidence intervals shaded, GP threshold shown in grey, observations are crosses; times series plot of the data, black dotted line shows GP threshold.

block bootstrapping procedure with the same bootstrap samples to obtain confidence intervals for all methods.

We show results of the model fit for two sites which represent both ends of the goodness of fit spectrum. These sites are the Dee at Woodend in North East Scotland, and the Itchen at Highbridge and Allbrook in Hampshire. For the whole data set only five sites had a fit noticeably worse than distributions fitted directly to the annual maxima data and only two had a poor fit. These two sites are baseflow dominated and so have a very high level of temporal dependence in the series.

The first check in the modelling process is the fit of the GP distribution to the data. For most sites the GP appears to fit well, for instance the Dee at Woodend (Figure 9.6), however, for the Itchen at Highbridge and Allbrook this is not the case (Figure 9.7). This site has a particularly high level of

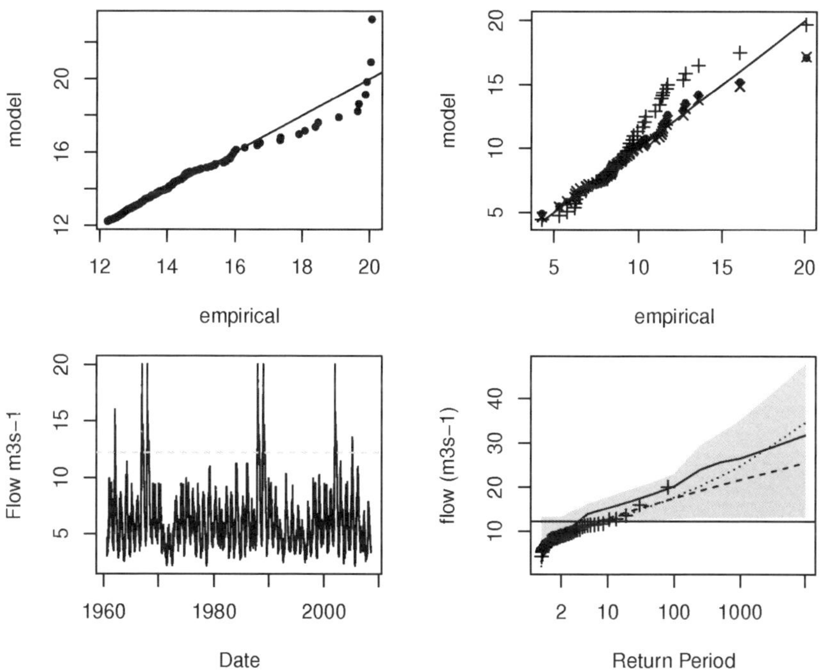

Figure 9.7. Diagnostic plots for Itchen at Highbridge and Allbrook. Clockwise from top left; Q-Q plot for GP fit; Q-Q plot for AMAX fits, dots show GEV, diagonal crosses show GL, vertical crosses show new method; flow-return period plots, dashed line shows GEV, dotted line shows GL, solid line new method with 95% confidence intervals shaded, GP threshold shown in grey, observations are crosses; times series plot of the data, black dotted line shows GP threshold.

temporal dependence, and when it experiences high flows the river tends to stay high for a long period of time. This means that the number of clusters is very small, and a large number of years have no exceedances. In this situation we can also see that the highest observations, above $17\,\mathrm{m}^3\mathrm{s}^{-1}$, are all from the same event. This is not an uncommon feature of flow data from slowly responding rivers.

There are two possible reasons for the poor fit of the GP distribution, first it is not a suitable limiting distribution for very slowly responding catchments, second the threshold above which we fitted the distribution is too low, i.e. the asymptotic limit is not reached. Without a much longer data record it is difficult to verify which of these reasons applies here.

Figures 9.6 and 9.7 also show diagnostic plots for the annual maxima distributions. For the Dee at Woodend we can see that all three distributions fit the data well within the range of the data, however, when the fits are extrapolated the GL gives flows for high return periods that are much higher than those estimated using the other two methods. For the Itchen at Highbridge and Allbrook, unsurprisingly, because our new method is based on the fitted GP distribution, the fit within the range of the data is poorer for our new method than for the other two. All three fitted distributions, however, have very similar properties beyond the range of the fitted data.

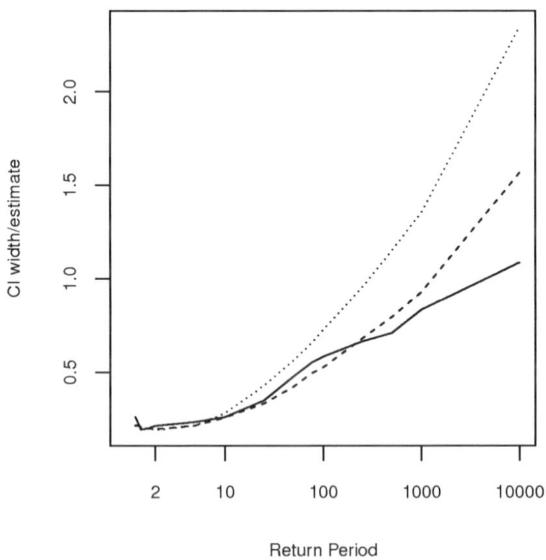

Figure 9.8. Confidence interval widths scaled by estimate. Solid line median for new method, dashed lines median for GEV, dotted line median for GL.

Our final plot shows the benefit in terms of reduced uncertainty of using all the information for extremes available in the data. Figure 9.8 shows the 95% confidence intervals widths of the estimated return flow, scaled by the estimate of return flow. We can see that for high return periods our new estimate gives narrower confidence intervals overall. We can also see that the GL distribution gives much wider confidence intervals.

9.9. Conclusion

In this chapter we have presented old and new statistical approaches to estimating the extreme values of data. We have shown that many of the features of river flow data that complicate statistical inference can be handled by modern statistical models.

The aim of the data analysis in this chapter has been to emphasise two things: first that estimates of uncertainty vary depending on how they are obtained; second that the uncertainty associated with different methods of estimation also differs.

The methods of uncertainty estimation we compared were profile likelihood, which is based on the assumption of asymptotic normality of maximum likelihood estimates, and non-parametric block bootstrapping. If the assumptions on which the parametric modelling is based are sound then generally the parametric methods of estimation are more efficient, however, where these assumptions are invalid better estimates can be obtained using non-parametric methods.

The methods of obtaining return level estimates we compared are simply fitting the GL and GEV distributions to annual maxima data and a new approach based on separate models for the event maxima and the number of events per year. The new method in this chapter could be extended in many ways, for example, the fixed probability threshold could be allowed to vary or a more sophisticated model could be used for the number of events per year. However, in its current form it does demonstrate the benefits in terms of reduced uncertainty for high return periods gained by using a larger proportion of the data.

When the overall fits of the GL and GEV distributions are compared the biggest difference is in the level of uncertainty. The confidence intervals for return levels estimated by fitting a GL distribution to AMAX data were much wider than those estimated by fitting a GEV distribution to AMAX data. This is probably due to the increased level of flexibility offered by the GL distribution which allows the number of events per year to be over-dispersed compared to the GEV distribution. This means that the chances of obtaining very high return level estimates are likely to be higher. This feature is also likely to be the reason why in Robson and Reed (1999) the GL was found to be a slightly better distribution for flow data than the GEV. However, as we have shown in this paper, an even better approach, with a safer basis for extrapolation, would be obtained by explicitly modelling the distribution of event maxima and the numbers of events per year.

Although we have not provided an example of pooling flow records to directly estimate extreme value distribution parameters, such a wealth of applications to other data sets exist that this must surely be possible. The benefits of joint estimation in this way are that uncertainty bounds for all parts of the process can be estimated using standard statistical inference techniques.

Another problem highlighted in this chapter is the difficulty in defining suitable distributions for very slowly responding catchments. Simply fitting an extreme value distribution to this data is unlikely to provide safe extrapolation beyond the range of the data. This is because the only available distributions that have underlying mathematical theory to safely extrapolate beyond the range of the data rely on asymptotic assumptions that are not valid for flow data from these catchments. Although the distributional parameters can be coerced to fit, the resulting distribution will not have a mathematical basis for extrapolation. For sites such as these a pooled analysis, or estimation based on continuous simulation rainfall-runoff modelling or a more sophisticated statistical modelling of the underlying process may be the best answer.

Acknowledgements

A number of people have helped with this chapter, Emma Eastoe, Jon Tawn, Rob Lamb and the editors, provided useful comments on this paper and its structure. The data were provided by the National River Flow Archive and computing was carried out using R (R Development Core Team, 2009).

References

Adam, M.B. (2007). *Extreme Value Modelling of Sports Data*, PhD Thesis, Lancaster University.

Bierlant, J., Goegebeur, Y., Teugels, J. *et al.* (2004). *Statistics of Extremes Theory and Applications*, John Wiley & Sons, London.

Bortot, P., Coles, S.G. and Tawn, J.A. (2000). The multivariate Gaussian tail model: an application to oceanographic data, *Appl. Stat.*, **49**, 31–49.

Chandler, R.E. and Bate, S. (2007). Inference for clustered data using the independence log likelihood, *Biometrika*, **94**, 167–183.

Coles, S. (2001). *An Introduction to Statistical Modeling of Extremes*, Springer-Verlag, London.

Coles, S.G. and Dixon, M.J. (1999). Likelihood-based inference for extreme value models, *Extremes*, **2**, 5–23.

Cooley, D, Naveau, P. Jomelli, V. *et al.* (2006). A Bayesian hierarchical extreme value model for lichenometry, *Environmetrics*, **17**, 555–574.

Cooley, D., Nychka, D. and Naveau, P. (2007). Bayesian spatial modelling of extreme precipiration return levels, *J. Am. Stat. Assoc.*, **102**, 824–840.

Cox, D.R. and Hinckley, D.V. (1974). *Theoretical Statistics*, Chapman & Hall, Boca Raton.

Cunnane, C. (1987). "Review of statistical models for flood frequency estimation" in: Singh, V.P. (ed.), *Hydrologic Frequency Modeling: Proceedings of the International Symposium on Flood Frequency and Risk Analyses*, 14–17 May 1986, Louisiana State University, Baton Rouge, USA, Springer.

Dalyrymple, T. (1960). *Flood Frequency Analysis*, USGS Water-Supply Paper, US, 1543A.

Davison, A.C. and Hinckley, D.V. (1997). *Bootstrap Methods and their Application*, Cambridge University Press, Cambridge.

Davison, A.C. and Smith, R.L. (1990). Model for exceedances over high threshold, *J. R. Stat. Soc. B*, **52**, 393–442.

de Haan, L. (1970). *On Regular Variation and its Applications to the Weak Convergence of Sample Extremes*, Mathematical Centre Tract, Amsterdam.

de Haan, L. (1984). A spectral representation for max-stable processes, *Ann. Probab.*, **12**, 1194–1204.

Dobson, A.J. (2001). *An Introduction to Generalized Linear Models 2nd Edition*, Chapman & Hall, CRC, Boca Raton.

Eastoe, E.F. and Tawn, J.A. (2009). Modelling non-stationary extremes with application to surface level ozone, *Appl. Statist.*, **59**, 25–45.

Eastoe, E.F. and Tawn, J.A. (2010). Statistical models for overdispersion in the frequency of peaks over threshold data for a flow series, *Water Resour. Res.*, **46**, W02510, doi:10.1029/2009WR007757.

Eastoe, E.F. and Tawn, J.A. (2012). The distribution for the cluster maxima of exceedances of sub-asymptotic thresholds, *Biometrika*, **99**, 43–55.

Fawcett, L. and Walshaw, D. (2006). A hierarchical model for extreme wind speeds, *App. Stat.*, **55**, 631–646.

Fawcett, L. and Walshaw, D. (2007). Improved estimation for temporally clustered extremes, *Environmetrics*, **18**, 173–188.

Ferro, C.A.T. and Segers, J. (2003). Inference for clusters of extreme values, *J. R. Stat. Soc. B*, **65**, 545–556.

Fisher, R.A. and Tippet, L.H.C. (1928). On the estimation of the frequency distributions of the largest or smallest member of a sample, *P. Cam. Philol. S.*, **24**, 180–190.

Gamerman, D. and Lopes, H.F. (2006). *Markov Chain Monte Carloe: Stochastic Simulation for Bayesian Inference 2nd Edition*, Chapman & Hall, CRC, Boca Raton.

Gnedenko, B.V. (1943). Sur la distribution limte du terme maximum d'une série aléatoire, *Ann. Math.*, **44**, 423–453.

Greenwood, J.A., Landwehr, J.M., Matalas, N.C. *et al.* (1979). Probability weighted moments: definition and relation to parameters of several distributions expressable in inverse form, *Water Resour. Res.*, **15**, 1049–1054.

Grimmet, G.R. and Stirzaker, D.R. (1992). *Probability and Random Processes 2nd Edition*, Oxford University Press, Oxford.

Guse, B., Castellarin, A., Thieken, A.H. *et al.* (2009). Effects of intersite dependence of nested catchment structures on probabilistic regional envelope curves, *Hydrol. Earth Syst. Sc.*, **13**, 1699–1712.

Heffernan, J.E. and Tawn, J.A. (2004). A conditional approach to modelling multivariate extreme values (with discussion), *J. R. Stat. Soc. B*, **66**, 497–547.

Hosking, J.R.M. (1990). L-Moments: analysis and estimation of distributions using linear combinations of order statistics, *J. R. Stat. Soc.*, **52**, 105–124.

Jenkinson, A.F. (1955). The frequency distribution of the annual maximum (or minimum) of meteorological elements, *Q. J. Roy. Meteor. Soc.*, **81**, 158–171.

Keef, C. (2007). *Spatial Dependence in River Flooding and Extreme Rainfall*, PhD Thesis, Lancaster University.

Keef, C., Lamb, R., Tawn, J.A. *et al.* (2010a). "The Risk of Widespread Flooding — Capturing Spatial Patterns in Flood Risk from Rivers and Coast", *Spatial Coherence of Flood Risk*, SC060008/SR, Environment Agency.

Keef, C., Lamb, R. Tawn, J.A. *et al.* (2010b). "A Multivariate Model For The Broad Scale Spatial Assessment Of Flood Risk" in: *Proceedings of the British Hydrology Society 2010 International Symposium*, Newcastle, July 2010.

Keef, C., Svensson, C. and Tawn, J.A. (2009a). Spatial dependence of extreme river flows and precipitation in Britain, *J. Hydrol.*, **378**, 240–252.

Keef, C., Tawn, J.A. and Svensson, C. (2009b). Spatial risk assessment for extreme river flows, *App. Stat.*, **58**, 601–618.

Lang, M. (1999). Theoretical discussion and Monte-Carlo simulations for a negative binomial process paradox, *Stoch. Env. Res. Risk A.* **13**, 183–200.

Laurini, F. and Tawn, J.A. (2003). New estimators for the extremal index and other cluster characteristics. *Extremes*, **6**, 189–211.

Leadbetter, M.R. (1983). Extremes and local dependence in stationary sequences, *Zeitschrift für Wahrscheinlichkeitstheorie und Verwandte Gebeite* **65**, 291–306.

Leadbetter, M.R. (1991). On a basis for "peaks over threshold" modelling, *Stat. Probabil. Lett.* **12**, 357–362.

Ledford, A.W. and Tawn, J.A. (1996). Statistics for near independence in multivariate extreme values, *Biometrika*, **83**, 169–187.

Ledford, A.W. and Tawn, J.A. (2003). Diagnostics for dependence within time series extremes, *J. R. Stat. Soc. B*, **65**, 521–543.

Li, M., Shao, Q., Zhang, L. *et al.* (2010). A new regionalisation approach and its application to predict flow duration curve in ungauged basin, *J. Hydrol.*, **389**, 137–145.

NERC (1975). *Flood Studies Report*, Natural Environment Research Council.

Northrop, P. and Jonathan, P. (2010). *Modelling Spatially Dependent Non-Stationary Extremes with Application to Hurricane Induced Wave Heights*, University College London Research, Department of Statistical Science, Report 306, University College London, London.

Padoan, S.A. Ribatet, M. and Sisson, S.A. (2010). Likelihood-based inference for max-stable processes, *J. Am. Stat. Assoc.*, **105**, 263–277.

Pickands, J. (1975). Statistical inference using extreme order statistics, *Ann. Stat.* **3**, 119–131.

R Development Core Team (2009). *R: A Language and Environment for Statistical Computing*, R Foundation for Statistical Computing, Vienna, Austria.

Smith, R.L. (1985). Maximum likelihood estimation in a class of non-regular cases, *Biometrika*, **72**, 67–90.

Smith, R.L. (1990a). *Regional Estimation from Spatially Dependent Data*, University of Surrey, Guildford. Available at http://www.stat.unc.edu/postscript/rs/regest.pdf (Accessed on 20/04/12).

Smith, R.L. (1990b). *Max-Stable Processes and Spatial Extremes*, University of Surrey, Guildford. Available at http://www.stat.unc.edu/postscript/rs/spatex.pdf (Accessed on 20/04/12).

Smith, R.L. and Weissman, I. (1994). Estimating the extremal index, *J. R. Stat. Soc. B*, **56**, 515–528.

Speight, L., Hall, J., Kilsby, C. *et al.* (2010). "Adding Value to Catastrophe Models: A Multi-Site Approach to Risk Assessment for the Insurance Industry" in: *Proceedings of the British Hydrology Society 2010 International Symposium*, Newcastle, July 2010.

Svensson, C. and Jones, D.A. (2002). Dependence between extreme sea surge, river flow and precipitation in eastern Britain, *Int. J. Climatol.*, **22**, 1149–1168.

Svensson, C. and Jones, D.A. (2004). Dependence between extreme sea surge, river flow and precipitation in south and west Britain, *Hydrol. Earth Syst. Sc.*, **8**, 973–992.

von Mises, R. (1954). "La distribution de la plus grande de n valuers" in: *Selected Papers*, Volume II, American Mathematical Society, Providence, pp. 271–294.

CHAPTER 10

Uncertainty in Flood Inundation Modelling

Paul D. Bates
School of Geographical Sciences,
University of Bristol, UK

Florian Pappenberger
European Centre for Medium-Range
Weather Forecasts, UK

Renata J. Romanowicz
Institute of Geophysics, Polish Academy of Sciences,
Poland

10.1. Introduction: Floodplain Flow Processes, Inundation Models and Effective Parameters

Flood inundation models are central components in any flood risk analysis system as they transform the bulk discharge outputs from flood-frequency analyses or rainfall-runoff models into distributed predictions of flood hazard in terms of water depth, inundation extent and flow velocity. Predictions may be dynamic in time and can be derived from a range of codes which vary in complexity from non-model approaches, such as fitting a planar surface to digital elevation data (see Matgen *et al.*, 2007; Puesch and Raclot, 2002), through to numerical solutions of fluid dynamics equations derived from considerations of mass and momentum conservation. Whilst such models are parsimonious in terms of their data requirements and number of unconstrained parameters relative to other environmental simulation software, their underlying equations may be highly non-linear.

Moreover, the data sets that they do require may be subject to complex, but poorly known errors that may vary markedly in time and space. As a consequence, considerable research has, in recent years, sought to understand and better estimate these uncertainties in order to improve flood risk analyses.

The physics of floodplain inundation is still the subject of considerable debate, yet in recent years a combination of laboratory experimentation (e.g. Knight and Shiono, 1996; Sellin and Willetts, 1996), field (e.g. Babaeyan-Koopaei *et al.*, 2002; Nicholas and Mitchell, 2003) and remote sensing observations (e.g. Alsdorf *et al.*, 2000, 2007; Bates *et al.*, 2006; Schumann *et al.*, 2007, 2008a,c; Smith and Pavelsky, 2008) has resulted in significant advances in understanding. These studies have yielded detailed observations of turbulent flow structures during overbank flood flows and the first observations of the dynamics of shallow plain flooding over complex topography. Whilst there is still much to learn, a conceptual model of floodplain inundation is evolving that suggests that whilst in-channel routing may be treated as a one-dimensional flow in the streamwise direction, floodplain flow is clearly at least a two-dimensional processes as the flow paths cannot be predicted *a priori*. Moreover, in localised zones, such as at the interface between the main channel and floodplain, strongly three-dimensional and turbulent flow is likely to exist.

A "brute force" approach to this problem would be to model both channel and floodplain flows as fully three-dimensional and turbulent through the solution of the full Navier–Stokes equations. However, for typical flood wave flows (i.e. unsteady, non-uniform flows of high Reynolds number in a complex geometry) the direct numerical simulation of the Navier–Stokes equations is computationally prohibitive. Modellers have therefore sought to isolate, from the complex assemblage of hydraulic processes known to occur during floods, those that are central to the problem of flood routing and inundation prediction. An obvious first step is to assume that it is unnecessary to compute the details of the instantaneous turbulence field and that one can approximate this using the three-dimensional Reynolds Averaged Navier-Stokes (RANS) equations and some appropriate turbulence closure model. Such models have been used to simulate steady state flows in reach scale compound channels (see for example Stoesser *et al.*, 2003), but simulations of dynamic inundation are currently rare due to computational cost. Three-dimensional modelling through the whole domain may also be inappropriate for floodplain inundation where the majority of the flow can be adequately described using the two-dimensional shallow water equations. These equations are derived from the three-dimensional RANS equations by integrating over the flow depth, and apply where: (i) the horizontal length scale is much greater than the vertical length scale; and (ii) the vertical pressure gradient is hydrostatic. This is often sufficient for the majority of compound channel flows, and by invoking the shallow water approximation we assume that the effect of three-dimensional and turbulent processes is either small, or can be parameterised in terms of their contribution to energy losses or through an effective eddy viscosity coefficient (see for example Bates *et al.*, 1998).

If we consider in-channel flow over reach scales appropriate to flood routing problems (i.e. >1 km) then it is clear that flow varies much more in the downstream direction than in the cross-stream or vertical directions. In this situation it is common (Knight and Shiono, 1996) to treat flow using some form of the one-dimensional Saint Venant equations (see for example, Fread, 1993). Whilst strictly only applicable to in-channel flow, many authors have applied one-dimensional models to compound channel flood flow problems. For example, Horritt and Bates (2002) demonstrate for a 60 km reach of the UK's River Severn that one-dimensional, simplified two-dimensional and full two-dimensional models perform equally well in simulating flow routing and inundation extent. This suggests that although gross assumptions are made regarding the flow physics incorporated in a one-dimensional model applied to out-of-bank flows, the additional energy losses can be compensated for using a calibrated effective friction coefficient. Here one assumes that the additional approximations involved in continuing to treat out-of-bank flow as if it were one-dimensional are small compared to other uncertainties (for a discussion, see Ali and Goodwin, 2002). Many one-dimensional codes now also include a representation of floodplain storage using a series of user-defined polygonal compartments into which overbank flow can spill. Floodplain conveyance between compartments can be represented using weir equations as first described by Cunge *et al.* (1980). However, the definition of both the geometry and connections between storage areas requires considerable operator intervention and an *a priori* knowledge of floodplain flow paths. In using such models there is, subsequently, a strong danger that the resulting simulations merely reflect operator biases and that one simply gets the result that one was expecting.

A final class of models that is frequently applied to floodplain inundation simulation is coupled one-dimensional/two-dimensional codes which seek to combine the best features of each model class. Such models typically treat in-channel flow with some form of the one-dimensional Saint Venant equations, but treat floodplain flows as two-dimensional using either the full shallow water equations or storage cells linked using uniform flow formulae applied over a structured grid. Such a treatment alleviates the need to use a very fine spatial resolution grid in order to represent channel geometry and hence can result in a considerable reduction in computational cost. Using a structured grid to form the storage cells also avoids the need for operator intervention to define floodplain compartments and the linkages between them, and hence eliminates the potential for bias.

In developing a model for a particular site, choices also need to be made about the discretisation of space and time. These will clearly depend on the resolution of available terrain data, the length scales of terrain features in the domain and the length and timescales of relevant flow processes. With the development over the last decade of high resolution mapping technologies, such as airborne scanning laser altimetry, terrain data are usually available at scales much finer than it is computationally possible to model over wide areas. However, deciding which terrain and flow length scales need to be incorporated in a model is a much more subjective choice. Clearly, as spatial resolution is increased, particular terrain and flow features will no longer be adequately represented, and the impact of these sub-grid scale effects on the model predictions will need to be parameterised.

As a result, for any given situation there are a variety of modelling tools that could be used to compute floodplain inundation and a variety of space and time resolutions at which these codes could be applied. All codes make simplifying assumptions and only consider a reduced set of the processes known to occur during a flood event. Hence, all models are subject to a degree of structural error that is typically compensated for by calibration of the friction parameters. Calibrated parameter values are not, therefore, physically realistic, as in estimating them we also make allowance for a number of distinctly non-physical effects, such as model structural error and any energy losses or flow processes which occur at sub-grid scales. Calibrated model parameters are therefore area-effective, scale-dependent values which are not drawn from the same underlying statistical distribution as the equivalent at-a-point parameter of the same name. Thus, whilst we may denote the resistance coefficient in a wide variety of hydraulic models as "Manning's n", in reality the precise meaning of this resistance term changes as we change the model physical basis, grid resolution and time step. For example, a one-dimensional code will not include frictional losses due to channel meandering in the same way as a two-dimensional code.

In the one-dimensional code these frictional losses need to be incorporated into the hydraulic resistance term. Similarly, a high-resolution discretisation will explicitly represent a greater proportion of the form drag component than a low-resolution discretisation using the same model. Little guidance exists on the magnitude of such effects and some of the differences generated in this way may be subtle. However, complex questions of scaling and dimensionality do arise which may be difficult to disentangle. In general, as the dimensionality increases and grid scale is reduced we require the resistance term to compensate for fewer unrepresented processes and: (i) the model sensitivity to parameter variation reduces; and (ii) the calibrated value of the resistance term should converge towards the appropriate skin friction value.

The major sources of uncertainty in flood inundation modelling are therefore: (i) errors in the model input data (principally hydrometric or rainfall-runoff model data to set boundary and initial conditions, topography data, friction parameters and details of any hydraulic structures present along the reach); (ii) errors in the independent observed data used to estimate model parameters or simulation likelihoods; (iii) model structural errors; and (iv) conceptual model uncertainty as different classes of model may fit sparse validation data equally well, yet give substantially different results in prediction. In this chapter we review the sources of uncertainty in model input data in Section 10.2, before considering errors in the observations available to help estimate and constrain predictive uncertainty in such codes in Section 10.3. In all such studies there is an inevitable trade-off between computational costs, the physical realism of the inundation model (i.e. the structural errors and conceptual model uncertainty) and the number of simulations that can be used to sample the space of possible model behaviours. Such trade-offs are discussed in Section 10.4 before a discussion of possible error models and uncertainty analysis techniques for use in flood inundation prediction in Section 10.5.

10.2. Uncertainty in Flood Inundation Modelling Input Data

The input data required by any hydraulic model are principally: (i) hydrometric or rainfall-runoff model data to set boundary and initial conditions; (ii) topography data; (iii) information on the geometry and operation rules of any hydraulic structures present along the reach; and (iv) friction coefficients to represent all the energy loss mechanisms not explicitly included in the model physics. In this section we review our current understanding of the uncertainties associated with each of these input data sources.

10.2.1. *Boundary and initial condition data*

All numerical models require boundary and initial condition information as starting points for their computational procedures. Hydraulic models then interpolate these data in space and time using the model physical equations to constrain this process. The result is a temporally dynamic and spatially distributed prediction of the model deterministic variables.

Boundary condition data consists of values of each model dependent variable at each boundary node, and for unsteady simulations the modeller needs to provide these values for each time step. From these the model is then able to compute the fluxes of mass (and perhaps turbulent energy) entering and leaving the model through each open boundary to give a well-posed problem that can be solved by the numerical scheme. As model dimensionality changes so does the number of dependent variables, and hence the required boundary condition data may vary depending on the physics being solved. Moreover, required boundary conditions may also change with reach hydraulics as we shall see below. For most one- and two-dimensional inundation models well posed boundary conditions consist of the fluxes into the model through each inflow boundary and the water surface elevation at each outflow boundary to allow backwater effects to be taken into account. These requirements reduce to merely the inflow flux rates for super-critical or kinematic flow problems, as when the Froude number $Fr > 1$ or when a kinematic version of the momentum equation is used, information cannot propagate in an upstream direction. In addition, three-dimensional codes require the specification of the velocity distribution at the inlet boundary and values for the turbulent kinetic energy. In most cases, hydrological fluxes outside the channel network, e.g. surface and subsurface flows from hillslopes adjacent to the floodplain and infiltration of flood waters into alluvial sediments, are ignored (for a more detailed discussion see Stewart *et al.*, 1999).

Initial conditions for a hydraulic model consist of values for each model dependent variable at each computational node at time $t = 0$. In practice, these will be incompletely known, if at all, and some additional assumptions will therefore be necessary. For steady-state (i.e. non-transient) simulations any reasonable guess at the initial conditions is usually sufficient, as the simulation can be run until the solution is in equilibrium with the boundary conditions and the initial conditions have ceased to have an influence. However, for dynamic simulations this will not be the case and whilst care can be taken to make the initial conditions as realistic as possible, a "spin up" period during which model performance is impaired will always exist. For example, initial conditions for a flood simulation in a compound channel are often taken as the water depths and flow velocities predicted by a steady-state simulation with inflow and outflow boundary conditions at the same value as those used to commence the dynamic run. Whilst most natural systems are rarely in steady state, careful selection of simulation periods to coincide their start with near steady-state conditions can minimise the impact of this assumption.

Data to assign these values can come either from hydrometric networks (or from additional relationships such as flood frequency analyses derived from hydrometric data) or from rainfall-runoff modelling of the upstream catchment. Either source of boundary and initial condition data is subject to considerable uncertainty and these will be discussed in turn. First, hydrometric data are affected by measurement error, particularly for discharge, and uncertainties deriving from uncharacterised fluxes as a result of low density or short duration records. Hydrometric measurement errors are relatively well understood in general terms, but may not be well characterised at any given gauging site. Typically, water level is the measured quantity, and this can usually be determined to very high ($\sim$1 cm) accuracy. Once water level has been determined the discharge flux is usually determined via a rating curve. These are constructed by undertaking a number of gauging (area $\times$ velocity) measurements over time and producing a least squares relationship relating discharge to water level that can then be extrapolated for large flows not recorded during the measurement campaign. Where flow is contained within the river, channel discharge can likely be measured to within $\pm$5% by a skilled operator. However, measuring velocity across wide vegetated floodplains when flow is in an out-of-bank condition is much more difficult and in this situation errors of $\pm$10–15% are probably more typical (see Figure 10.1). Errors in extrapolating rating curves to flows higher than those measured can also add to this, particularly if the gauge location is bypassed when flow reaches a particular water level, and can result in very substantial errors (up to 50% at some sites the authors are aware of). One should also note that rating relationships change over time as a result of local morphological change and do not account for flow hysteresis which may be a significant factor at particular sites.

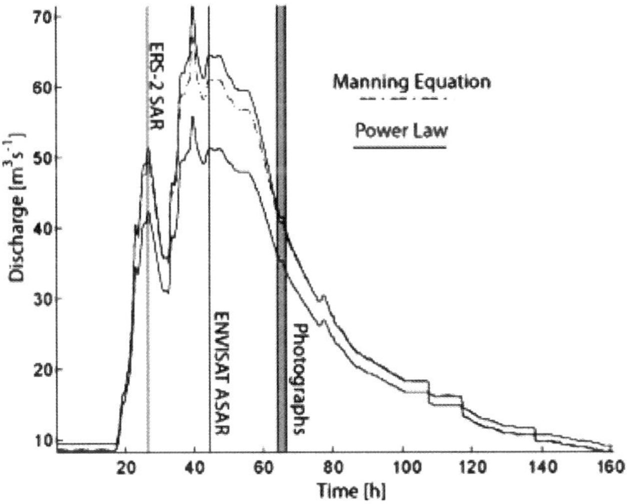

Figure 10.1. 5% and 95% percentile of a hydrograph for the Manning equation and the power law (used as upstream boundary condition of a flood inundation model). The figure illustrates the range of inflow created through the uncertainty of the rating curve (taken from Pappenberger *et al.* 2006a).

Depending on the installation, gauging station data may be available at anything from 15 minute to daily time intervals, and in developed countries in Europe and North America gauges are typically spaced 10–60 km apart for flood warning and forecasting purposes. In other parts of the world gauge networks are much less dense and in many areas are in severe decline as a result of falling investment levels (see Vörösmarty, 2002 for an extended discussion) such that many records are of short duration. Moreover, certain fluxes cannot be measured operationally using current technology, for example fluxes from hillslope sections to the channel. As a consequence, for many flood modelling applications sufficient gauge data may not exist to characterise all significant fluxes into and out of the domain. For example, Stewart *et al.* (1999) show for a 60 km reach of the River Severn in the UK that as the catchment wets up over a winter flood season the ungauged flow contributions increase from 7% to over 30% of total downstream discharge. In such situations the modeller needs to estimate the ungauged flows in some reasonable manner either using a hydrological model calibrated on nearby gauged catchments with similar properties (e.g. Stewart *et al.*, 1999) or by scaling the fluxes according to the ungauged catchment area and some average rate of runoff production.

For ungauged catchments or for forecasting studies, gauged data are unavailable and therefore boundary condition data for hydraulic modelling must come from rainfall-runoff models of the upstream catchment, which themselves may use ensemble numerical weather prediction models as their input (see for example Pappenberger *et al.*, 2005b). Here the hydraulic model input data are subject to all the uncertainties present in the upstream chain of simulation models and these can be considerable (see Chandler *et al.* (Chapter 7 of this volume) for a discussion of uncertainties in rainfall data). Formal estimation of how uncertainty cascades through such systems may therefore require large numbers of model runs and efficient sampling schemes to reduce the dimensionality of the model space to be searched.

Lastly, for many flood risk analyses we need to consider low probability events (annual probability of 0.01 or less) that have not been observed in the hydrometric record. Few gauging stations have record durations longer than 40 years and therefore are unlikely to have recorded examples of the high magnitude, low frequency events typically considered in design and planning studies. Even if large events have been recorded it is still necessary to conduct analyses to correctly estimate their probability in order to set design levels. Such analyses use extreme value statistical theory (see Coles, 2001 and Walshaw, this volume for an extended discussion) to estimate magnitude–frequency relationships for observed flows which can then be extrapolated to low frequency events. As with other forms of extrapolation this gives the potential for considerable error at high recurrence intervals. Moreover, whilst extending the data record through regionalisation and the procedures used in the UK Flood Estimation Handbook (Institute of Hydrology, 1999) or continuous simulation (Blazkova and Beven, 1997; Cameron *et al.*, 1999) have both been proposed as a solution here, neither is without its problems and estimation of design flood magnitude will undoubtedly continue to be a major source of uncertainty in flood risk analyses.

10.2.2. *Digital elevation models and channel bathymetry*

High resolution, high accuracy topographic data are essential to shallow water flooding simulations over low slope floodplains with complex micro-topography, and such data sets are increasingly available from a variety of remotely mounted sensors (see for example Sanders, 2007 for an extended discussion). Traditionally, hydraulic models have been parameterised using ground surveys of cross-sections perpendicular to the channel at spacings of between 100 and 1000 m. Such data are accurate to within a few millimetres in both the horizontal and vertical, and integrate well with one-dimensional hydraulic models. However, ground survey data are expensive and time consuming to collect and of relatively low spatial resolution. Hence, they require significant interpolation to enable their use in typical two- and three-dimensional models, whilst in one-dimensional models the topography between cross-sections is effectively ignored and results are sensitive to cross-section spacing (e.g. Samuels, 1990). Moreover, topographic data available on national survey maps tends to be of low accuracy with poor spatial resolution in floodplain areas. For example in the UK, nationally available contour data are only recorded at 5 m spacing to a height accuracy of ± 1.25 m and for a hydraulic modelling study of typical river reach, Marks and Bates (2000) report found only three contours and 40 unmaintained spot heights within a $\sim 6\,km^2$ floodplain area. When converted into a Digital Elevation Model (DEM) such data led to relatively low levels of inundation prediction accuracy in hydraulic models (see Wilson and Atkinson, 2005).

Considerable potential therefore exists for more automated, broad-area mapping of topography from satellite and, more importantly, airborne platforms. Three techniques which have shown potential for flood modelling are aerial stereo-photogrammetry (Baltsavias, 1999; Lane, 2000; Westaway *et al.*, 2003), airborne laser altimetry or LiDAR (Gomes Pereira and Wicherson, 1999; Krabill *et al.*, 1984) and airborne Synthetic Aperture Radar (SAR) interferometry (Hodgson *et al.*, 2003). Radar interferometry from sensors mounted on space-borne platforms, and in particular the Shuttle Radar Topography Mission (SRTM) data (Rabus *et al.*, 2003), can also provide a viable topographic data source for hydraulic modelling in large, remote river basins where the flood amplitude is large compared to the topographic data error (see for example Wilson *et al.*, 2007). At the other end of the scale range, ground-based scanning lasers are also beginning to be used to define three-dimensional topography at very high resolution and accuracy, particularly in urban areas where such systems can be vehicle-mounted in order to allow wide area acquisition.

Remotely sensed terrain data typically comes in the form of a Digital Surface Model (DSM) which includes surface artefacts (buildings, vegetation, etc). A key step in using these data in flood inundation models is therefore to remove these artefacts to leave a "bare earth" Digital Elevation Model (DEM). How this is done most effectively depends on the nature of the sensor and how the signal interacts with the target. Thus processing algorithms for scanning laser or interferometric radar systems differ considerably and an extended discussion of these is beyond the scope of this chapter. Uncertainty in terrain data is therefore a function of sensor technology, sensor flight altitude, data ground resolution and the quality of any post-processing. Considerable potential exists for reductions in terrain data uncertainties as a result of continued improvements in sensor technology and fusion of data sources to aid removal of surface features, particularly for complex urban environments. For example, Mason *et al.* (2007) show that using surface characteristic information contained in digital map data can improve substantially our ability to discriminate between surface features and bare earth in urban LiDAR scenes or to distinguish between vegetation and buildings. This latter development raises the possibility of using the surface feature information in remotely sensed terrain data in the model building process to, for example, parameterise vegetation resistance (e.g. Cobby *et al.*, 2003; Mason *et al.*, 2003) or define building shapes and locations (e.g. Hunter *et al.*, 2008; Schubert *et al.*, 2008). How this information is then treated by the model is dependent on the scheme's dimensionality and how it discretises space. Moreover, whether it is appropriate to include buildings as topographic barriers or to use a bare earth DEM with the blockage and storage effect of buildings parameterised through an increased friction coefficient or through a porosity parameter is strongly scale-dependent and again uncertainty can arise here.

10.2.3. *Hydraulic structures*

Flow in many rivers is often strictly controlled by hydraulic structures such as weirs, gates, bridges and locks. The representation of these structures can significantly influence flood inundation predictions. For example, different representations of bridge structures can lead to significantly different water level predictions (see Figure 10.2, Pappenberger *et al.*, 2006b).

Moreover, flood defence structures such as embankments, walls, diversions and temporary storage areas provide mechanisms for managing flood risk, but will have a small yet poorly known probability of failure. Fortunately, modern terrain data capture systems such as LiDAR are increasingly able to determine important parameters such as defence crest elevations over large areas. However, representing hydraulic structures can still be problematic in certain classes of hydraulic model. For example, representing the operation of gates in a full two-dimensional model may pose serious challenges and in all cases hydraulic structure operating rules may not be known by the modeller or, even if they are, may not be implemented correctly

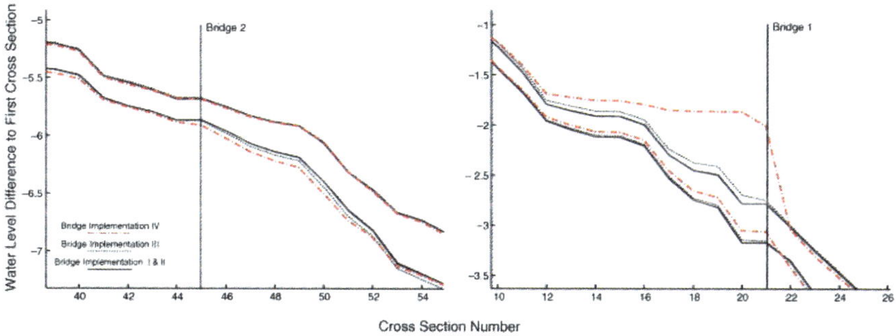

Figure 10.2. Range of predictions of difference in water levels to first cross-section over all behavioural models at the time of an ERS-2 SAR image. The difference between the water levels predicted at each cross-section (numbered on abscise from upstream to downstream) and the first cross-section is computed for each model realisation. The 5% and 95% envelope curves of all these computations are plotted on this figure.

during an event. Failure probabilities can be represented statistically using a fragility curve methodology (see, for example, Hall *et al.*, 2003), which gives the likelihood that a structure will fail for a given external load, but exploration of the full parameter space implied by such a treatment may require a large number of model simulations. For example, Dawson *et al.* (2005) demonstrated that a probabilistic analysis to determine flood risk due to possible failures of a system of defences around the town of Burton-on-Trent in the UK required upwards of 5000 simulations, even if all other model variables (inflow rate, friction factors, terrain) were treated deterministically.

10.2.4. *Roughness parameterisation*

As noted above, friction is usually an important unconstrained parameter in a hydraulic model. Two- and three-dimensional codes which use a zero equation turbulence closure may additionally require specification of an "eddy viscosity" parameter which describes the transport of momentum within the flow by turbulent dispersion. However, this prerequisite disappears for most higher-order turbulence models of practical interest. Hydraulic resistance is a lumped term that represents the sum of a number of effects: skin friction, form drag and the impact of acceleration and deceleration of the flow. These combine to give an overall drag force *Cd*, that in hydraulics is usually expressed in terms of resistance coefficients such as Manning's n and Chezy's C, which are derived from uniform flow theory. This approach then assumes that the rate of energy dissipation for non-uniform flows is the same as it would be for uniform flow at the same water surface (friction) slope. As noted in Section 10.1, the precise effects represented by the friction coefficient for a particular model depend on the model's dimensionality, as the parameterisation compensates for energy losses due to unrepresented processes, and the grid resolution. We currently do not possess a comprehensive theory of "effective roughness" that accounts for variations in model structure and scale, and as a consequence friction parameters in hydraulic models are typically calibrated. Recent work (Hunter *et al.*, 2005, 2007) has shown that the data against which a model is calibrated, and the specific objective function which the modeller seeks to optimise, can make a significant difference to values of "best fit" parameters determined by calibration. Moreover, different models show different degrees of sensitivity to friction parameters, which may be more or less non-linear (see for example Horritt and Bates, 2001 and Pappenberger and Ratto, this volume) depending on the underlying controlling equations solved by the code.

Certain components of the hydraulic resistance, such as bed material grain size (e.g. Hey, 1979) and form drag due to vegetation (e.g. Kouwen, 2000) can be constrained from field data, and increasingly remote sensing is being used to undertake this. For example, photogrammetric techniques to extract grain size information from ground-based or airborne photography (Butler *et al.*, 2001) and methods to determine specific plant biophysical parameters from LiDAR data are currently under development (e.g. Cobby *et al.*, 2001). Using such techniques, Mason *et al.* (2003) and Straatsma and Baptist (2008) have presented methods to calculate time and space distributed friction coefficients for flood inundation models directly from airborne altimetry data. Furthermore, the development of an ability to determine realistic spatial distributions of friction across a floodplain has also led Werner *et al.* (2005) to examine the identifiability of roughness classes in typically available model validation data. This showed that very few (1 or 2) floodplain roughness classes are required to match current data sources to within error and this suggests that application of complex formulae to establish roughness values for changed floodplain land use would seem inappropriate until model validation data are improved significantly. Much further work is required in this area; nonetheless such studies are beginning to provide methods to explicitly calculate important elements of frictional resistance for particular flow routing problems. This leads to the prospect of a much reduced need for calibration of hydraulic models and therefore a reduction in predictive uncertainty.

10.3. Data for Uncertainty Estimation in Flood Inundation Models

In all but the simplest cases (for example, the planar free-surface "lid" approach of Puesch and Raclot, 2002), some form of calibration is required to successfully apply a floodplain flow model to a particular reach for a given flood event. Calibration is undertaken in order to identify appropriate values for parameters such that the model is able to reproduce observed data and, as previously mentioned, typically considers roughness coefficients assigned to the main channel and floodplain and values for turbulent eddy viscosity if a zero-equation turbulence closure is used. Though it has been claimed that these values can be estimated in the field with a high degree of precision (Cunge, 2003), it has proven very difficult to demonstrate that such "physically-based" parameterisations are capable of providing accurate predictions from single realisations for reasons discussed in the critiques of Beven (1989, 2002, 2006) and Grayson *et al.* (1992). As such, values of parameters calculated by the calibration of models should be recognised as effective values that may not have a physical interpretation outside of the model structure within which they were calibrated. In addition, the process of estimating effective parameter values through calibration is further convoluted by the uncertainties inherent in the inundation modelling process discussed in Section 10.2 which cast some doubt on the certainty of calibrated parameters (Aronica *et al.*, 1998; Bates *et al.*, 2004; Horritt, 2000). Principally, these errors relate to the inadequacies of data used to represent heterogeneous river reaches (i.e. geometric integrity of floodplain topography and flow fluxes along the domain boundaries), but also extends to the observations with which the model is compared during calibration and the numerical approximations associated with the discrete solution of the controlling flow equations. The model will therefore require the estimation of effective parameter values that will, in part, compensate for these sources of error (Romanowicz and Beven, 2003).

Given that the number of degrees of freedom in even the simplest of numerical models is relatively large, it is no surprise that many different spatially distributed combinations of effective parameter values may fit sparse validation data equally well. Such equifinality in floodplain flow modelling has been well documented (see, for example, Aronica *et al.*, 1998, 2002; Bates *et al.*, 2004; Hankin *et al.*, 2001; Romanowicz *et al.*, 1996; Romanowicz and Beven, 2003) and uncertainty analysis techniques have been developed and applied in response. For example, the Generalised Likelihood Uncertainty Estimation (GLUE) of Beven and Binley (1992) has been used by a number of authors to make an assessment of the likelihood of a set of effective parameter values being an acceptable simulator of a system when model predictions are compared to observed field data.

Typical data available to compare to model predictions include: (i) point-scale measurements of water stage, discharge or velocity made at established gauging stations, taken during field campaigns (e.g. Lane *et al.*, 1999), determined from satellite altimetry (e.g. Wilson *et al.*, 2007) or estimated by intersecting flood extent maps with high resolution DEMs (Mason *et al.*, 2007; Schumann *et al.*, 2008b); (ii) vector data on flood inundation extent determined by ground surveys either during or after an event or from remotely sensed images of flooding (e.g. Bates *et al.*, 2006; Horritt, 1999; Pappenberger *et al.*, 2005a; Romanowicz and Beven, 2003); (iii) interferometric measurements of water surface height change determined from repeat-pass satellite radar images (e.g. Alsdorf *et al.*, 2007); and (iv) anecdotal evidence collected from eyewitness to flood events. As with all other types of hydraulic modelling data, errors in validation data may be very poorly characterised, and individual data sources may only test very limited aspects of a given model's performance (e.g. Hunter *et al.*, 2006). For example, stage data from gauging stations are typically available at sites 10–60 km apart and only test the bulk routing performance of hydraulic models and not their ability to produce spatial predictions. Neither can such bulk flow data be used to identify spatially variable parameter fields. In a similar vein, Hunter *et al.* (2006) show that using maps of flood extent to calibrate separate values of channel and floodplain friction leads to the conclusion that both high channel friction/low floodplain friction and high channel friction/low floodplain friction combinations produce equally plausible inundation extent simulations, despite the fact that in reality channels will be considerably smoother than rough vegetated floodplains. In effect, inundation extent data are not able to discriminate between different spatial patterns of friction, despite the fact that our physical understanding shows that some parameter fields are very unlikely to occur in practice. Hunter *et al.* (2006) showed that this equifinal behaviour was due to trade-off effects between the two parameters and that implausibly high channel friction values could be compensated for by implausibly low floodplain friction values such that the model still performed well in terms of flood extent prediction. Hunter *et al.* (2006) went on to show that it was only by including flood wave travel time data as an additional constraint to the parameter identification process that one was able to eliminate this pathological behaviour. In this case, flood wave travel time could only be replicated by the model if the channel friction was (correctly) lower than that on the floodplain, and hence the only calibrations that could replicate both extent and travel time data simultaneously were those that had an intuitively correct arrangement of roughness elements. Importantly, using wave travel time data alone led to a wider spread of acceptable parameter values than if both data sets were used to constrain the model. Such compensating effects are probably very common and can only really be eliminated by using multiple model validation data sources. However, as one builds more data into the calibration process the more likely it becomes that a single model or parameterisation cannot optimally match all of the data all of the time as a result of model structural errors. The choice of the objective function that is to be optimised will also strongly affect conclusions regarding the sensitivity and uncertainty behaviour of a particular model.

In general, water surface elevation measurements obtained from gauging stations are accurate to approximately ± 1 cm, and often absolute elevations are also known as such permanent installations can be levelled very precisely. Stages derived from bespoke field campaigns can have similar accuracy, although absolute elevations may be more difficult to establish at remote field sites. Elevations derived from post-event wrack mark surveys will have higher errors as it is more difficult to be certain that the elevation recorded actually was generated by the flood peak and not by some minor peak during the flood recession. As well as being difficult to interpret because of such biases, wrack marks composed of trash lines or water stains also are typically spread over an elevation range and are thus at best accurate to ± 10–15 cm. Elevations from satellite radar altimetry are typically obtained using instruments designed from oceanic studies, which have a footprint of ~ 2 km and vertical accuracy of decimetres to metres. Over the continental land surface such instruments therefore only record elevations over the very largest rivers, however, sophisticated re-tracking algorithms have recently been developed (e.g. Berry *et al.*, 2005) which allow separation of water and other signals in mixed pixels. In this way the elevation of smaller water bodies (~ 100 s of metres across) can be obtained and used for model validation (e.g. Wilson *et al.*, 2007).

Errors in discharge measurement and rating curve construction are already discussed in Section 10.2.1 so are not considered further here, but it is appropriate to look briefly at the accuracy of velocity data. Such information is usually obtained from handheld current meters or acoustic doppler current profilers mounted on boats and linked to a GPS for position determination. Such data have typical errors of ±5–10% for the primary downstream velocity field, although lower velocity cross-stream and vertical flows may be more difficult to measure accurately and precisely. In addition, research is ongoing into techniques to remotely sense velocities either from the ground or from airborne platforms. Ground-based methods include using a large-scale variant of the particle image velocimetry technique increasingly used in laboratory studies (e.g. Kim *et al.*, 2008) or microwave doppler radar systems (e.g. Costa *et al.*, 2000), whilst from airborne platforms velocities can be determined using along-track interferometry (e.g. Romeiser *et al.*, 2007). In the future such systems may revolutionise how we constrain inundation models; however, all such remote sensing methods should currently be regarded as experimental. For example, Mason *et al.* (2003) demonstrate a situation where wide area velocity data would have been able to discriminate between two competing model parameterisations, although data were not available at the time of this study, each parameterisation had to be considered equally likely.

Vector data on flood extent can be obtained from either ground survey during or after flood events, or from visible band imagers or radars flown onboard aircraft or satellites. Ground survey during floods can potentially yield excellent data, but a wide area synoptic view may be very difficult to obtain, and there can also be problems of access or safety. After flood events it is often possible to identify wrack marks although such data will suffer from the errors noted above in relation to point scale data, and in addition may only provide a fragmented view. Synoptic images are therefore advantageous in this latter respect and a variety of data sources can be used. Aerial photographs during floods are a common source of such data (e.g. Biggin and Blyth, 1996), provided weather conditions and emergency operations do not restrict flying. Aerial data have the advantage of being collected (relatively) close to the target so effective pixel sizes can be of the order metres. However, determining shorelines from aerial photographs is a manual process that is rather labour intensive, and (especially with oblique photographs) one which may actually be subject to a degree of subjectivity (e.g. Romanowicz and Beven, 2003). Visible band satellite imagery (e.g. 30 m resolution Landsat or coarser 250 m resolution MODIS data) can detect floods (e.g. Bates *et al.*, 1997), although cloud cover and restriction to daytime only operation can limit the utility of these data. For these reasons SAR data are often preferred for flood remote sensing.

SARs are active systems which emit microwave pulses at an oblique angle towards the target. Open water acts as a specular reflector and the microwave energy is reflected away from the sensor so such areas appear as smooth areas of low backscatter in the resulting imagery. Terrestrial land surfaces, by contrast, reflect the energy in many directions, including back towards the sensor, and therefore appear as noisy high backscatter zones. These differences allow flood extent to be mapped using a variety of techniques to an accuracy of ~1 pixel. Pixel sizes range from ~1 m in airborne imagery (e.g. Bates *et al.*, 2006) to between 3 and 75 m in spaceborne imagery (e.g. Baldassarre *et al.* 2009; Horritt, 2000), and can potentially be excellent for flood extent determination. Misclassification errors do occur however, with flattened and wet vegetation behaving, in certain situations, in the same way as open water, and emergent vegetation disrupting the specular reflections in shallow open water to appear more like dry land. Moreover, very few airborne SAR systems exist, and none are routinely used for flood mapping. Satellite SAR systems avoid these problems, but here orbit repeat times are low (7–10 days for RADARSAT, 35 days for ERS-1 and ERS-2) such that there is only a low probability of an overpass occurring simultaneous with a flood and the sensors are designed to be all-purpose instruments and may not be optimal for flood mapping. As a consequence, perhaps only 15–20 consistent inundation extent data sets are currently available worldwide, and even in these errors can be relatively high. The few studies to have obtained simultaneous aerial photo and satellite SAR data have shown that the accuracy of satellite radars in classifying flood extent to only be of the order 80–85% (Biggin and Blyth, 1996; Imhoff *et al.*, 1987). Schumann *et al.* (2008b, see Figure 10.3) have demonstrated how the uncertainty in this classification can be used to improve flood inundation models and design an optimal field campaign to improve future predictions and calibration exercises. Despite the small number of consistent inundation data sets available, such information is critical for validating flood models, and it is only by using flood extent data that one is able to truly validate the two-dimensional performance of flood models. Without such data it is very easy for modellers to continue to use one-dimensional codes for flood inundation modelling, despite the fact that we know that these are physically inappropriate as flooding is predominately a two-dimensional process.

A new data type which may, in the future, be used to validate inundation models are the fine spatial resolution maps of water surface elevation change presented by Alsdorf *et al.* (2007). These data are derived

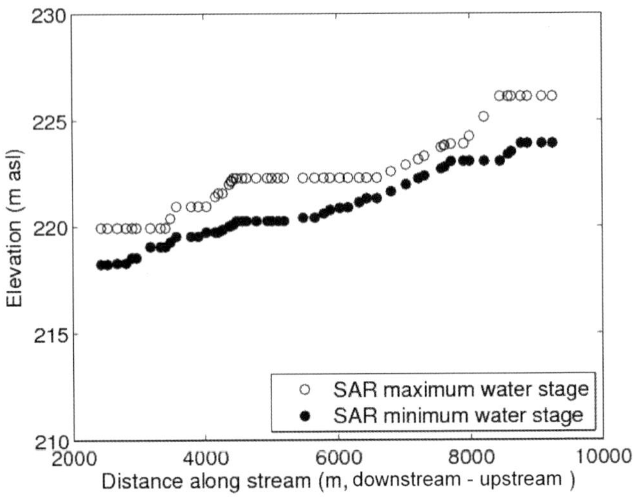

Figure 10.3. Minimum and maximum water stage derived from a SAR image along the Alzette River (taken from Schumann *et al.* 2008b).

from repeat pass satellite radar interferometry, and in the proof-of-concept study by Alsdorf *et al.* are derived from the C band radar onboard the JERS-1 satellite. The technique exploits the fact that at C band flooded vegetation scatters some specularly reflected radar energy back to the antenna and therefore the method can only currently be employed in areas of flooded vegetation. Accordingly, Alsdorf *et al.* demonstrated its utility for an area of the central Amazon mainstem in Brazil where they were able to measure water surface elevation changes to centimetric accuracy and with a spatial resolution of $\sim$200 m across a $\sim$50 km wide floodplain. In the case of JERS-1 the repeat time is 44 days which can be considered as adequate for the Amazon due to the annual timescale of the flood pulse through this reach. Future planned missions, such as the NASA Surface Water Ocean Topography instrument (SWOT, http://swot.jpl.nasa.gov/), will overcome these limitations by using a Ka band radar to remove the need for vegetation to be present to perform the interferometric calculations and decrease the orbit revisit time to $\sim$10 days at the equator (as a worst case), and much less at higher latitudes ($\sim$2–3 days). As well as providing unique measurements of water surface elevation that will revolutionise our ability to test inundation models, SWOT will also provide fine spatial resolution images of flood extent with pixel sizes ranging from $\sim 5 \times 60$ m near nadir to $\sim 70 \times 60$ m at the cross-track range limit. Given the low orbit revisit time this will lead to a significant ($\sim$O(1–2)) increase in the number of sites for which flood extent data are available and potentially make two-dimensional schemes a routine choice for inundation studies.

Finally, whilst flood modellers most often deal with quantitative data, for very many events much so called "soft" data may be available that may also be used in the validation process. Such data include eyewitness accounts, ground and air photos, newspaper reports or any other anecdotal evidence. Whilst it is difficult to be precise about the nature of any errors in these data, they may nevertheless be invaluable in situations where normal methods of data collection have failed. For example, in very extreme events gauges may be bypassed or fail, and resources may rightly be directed to emergency operations rather than meeting the future data needs of flood modellers. An example of the use of such data is provided by Werner (2004) who showed that anecdotal observations of areas that did and did not get inundated during a flood event could be very effective in reducing the uncertainty of an ensemble of simulations.

10.4. Trading-off Speed and Accuracy in Flood Inundation Modelling

In all flood modelling studies there is an inevitable trade-off between computational cost, the physical realism of the inundation model (i.e. the structural errors and conceptual model uncertainty) and the number of simulations that can be used to sample the space of possible model behaviours. Definition of the optimum point within this spectrum is a subjective decision that likely reflects the biases of the particular modeller and his or her experience of the relative strengths of the different types of validation data available for the application at hand. As modelling applications are most often limited by the volume of observed validation data and because these data may have complex, yet poorly known errors, it may be possible for many different classes of model to work equally well (albeit with perhaps very different effective roughness parameters). As accuracy is so difficult to define in a robust way given the limitations of the available data, the modeller may have considerable scope for trading off computational cost and model complexity. Indeed, if we are limited to the use of bulk flow data (stage, discharge) from widely spaced gauging stations, it may be impossible to justify on the basis of objective evidence the use of anything but the simplest wave routing model. Whilst recourse can be made to the argument that physically more realistic models must intrinsically be "better" than simpler ones, in practice this is often very difficult to prove. Moreover, scientific logic based on the principle of Occam's razor, where the simplest explanation that fits the available data is the best one, provides a powerful counter-argument here. For example, Horrit and Bates (2002) show for a reach of the River Severn in the UK that one-dimensional, coupled one-dimensional-two-dimensional and full two-dimensional models were all capable of predicting the inundation extent recorded in satellite SAR images equally well, but with order of magnitude differences in computational cost.

When considering speed and accuracy trade-offs, modellers should also take into account the "whole-life" cost of the modelling process and not just the simulation time. For example, despite being very cheap to run, one-dimensional models applied to complex floodplains require substantial set-up time and skilled operator decision making over the number and shape of floodplain storage areas in order to provide reasonable results. While a structured grid *two-dimensional* model may require the same channel information as a *one-dimensional* model, the set-up complexity for the floodplain may be significantly reduced because floodplain storage areas are defined simply by the model grid scale. Moreover, the reliance on operator decision making in the development of one-dimensional models in order to define floodplain flow paths *a priori* may lead to a degree of circular reasoning whereby results reflect, to too great an extent, the modeller's preconceived ideas of how the system works.

10.5. Error Models for Flood Inundation Data and Uncertainty Estimation

As described in Section 10.1, the uncertainties related to floodplain inundation modelling have their source in input data errors (including the data used for the model calibration), model structural errors and conceptual model uncertainty. In order to include these errors in the estimation of model predictive uncertainty, some assumptions have to be made regarding the error structure. In this section we focus on input and output errors. Structural errors and errors resulting from conceptual model uncertainty can be dealt with only through output error analysis in the case of flood inundation problems.

It is important to note that in typical modelling practice both the availability of calibration/validation data and the goal of the modelling play important roles in the model formulation and the choice of the error model structure. The available data dictate the model choice, whilst the goal of the modelling dictates the choice of the model output and the methods of data assimilation.

The conditioning of model predictions is performed through the assessment of model errors, i.e. comparison of simulated output with the observations, under the assumption that errors are additive, independent, both in time and space, and stationary. When these assumptions are met, computationally efficient, statistical tools for data assimilation may be used. Unfortunately, those assumptions are often violated in flood inundation modelling. The errors related to hydrological variables are usually heteroscedastic, i.e. their variance depends on the magnitude of the predicted variable (Sorooshian and Dracup, 1980) and correlated in time and space (Neal *et al.*, 2007; Sorooshian *et al.*, 1983). One of the methods of dealing with these problems is to introduce the appropriate transformation of error to stabilise the variance and take into account the error correlation/covariance structure (Aronica *et al.*, 1998). These methods require introducing additional parameters to describe the error model structure, which might be unidentifiable when data are scarce. The other option is using less statistically efficient but computationally simpler methods of uncertainty estimation, including fuzzy set approaches and other less formal methods of data assimilation (Beven, 2006).

The assessment of uncertainty in predictions is done based on weights describing the measure of model performance. In a Bayesian set-up the weights are obtained from the posterior distribution of parameters obtained as a result of the application of Bayes theorem (Box and Tiao, 1992). When a fuzzy set approach is used, the weights are obtained from the fuzzy performance measures. The prediction error includes parameter related uncertainty and the uncertainty related to the model and observational errors. In a statistical set-up, the form of the posterior distribution of parameters depends on the assumed distribution of the prediction errors and their correlation structure. When Gaussian, or generalised Gaussian distributions of prediction errors can be assumed (possibly after some required transformation), the logarithm of the likelihood function used to evaluate the parameter posterior distribution has the form of a sum of error measures (square, absolute or minimax) over time and/or space (Romanowicz *et al.* 1996). In that case there is a correspondence between the goodness of fit criteria used in an inverse problem of model calibration and the likelihood function. However, as we note above, it may be difficult to find any form of transformation that would allow a formal statistical approach to be applied and we are left with less efficient but simpler non-formal approaches. In that case the choice of criteria for uncertainty assessment should follow from the goal of the modelling.

The main goals of flood inundation modelling are flood risk estimation (Pappenberger *et al.*, 2007a), flood hazard forecasting and warning (Neal *et al.*, 2007; Pappenberger *et al.* 2005b) and maintaining ecologically required inundation extent on floodplains during specified periods (Romanowicz *et al.* 2008) by means of controlled reservoir discharges. A modeller might aim to build a model for general use and in that case, all the available data may be included in the calibration stage. However, the model should be fit for purpose and the choice of an application range that is too wide might jeopardise the quality of predictions aimed at a specific model use.

In the case of flood hazard estimation, which is the most common goal of flood inundation modelling, the model should give reliable predictions of velocity fields and water level heights over the entire inundation extent as well as the inundation outlines. The length of time when the water levels are over a specified threshold may also be valuable to determine the potential losses to property. However, these requirements are difficult to fulfil for a large inundation area due to computer time limitations, limited observations available to calibrate the model over large regions and errors in the inundation models themselves. Moreover, velocity fields are notoriously difficult to observe *in situ* and therefore often cannot be validated.

When distributed data from satellite images or aerial photographs are available, the spatial pattern of simulations can be assessed. This is usually done by the comparison of binary maps of inundation extent. The error has a discrete, binary form and its assessment is based on the flooded/non-flooded cell count. This approach was applied by Aronica *et al.* (2002), and Bates *et al.* (2006) among others. The application of a discrete approach reduces the amount of information available for model conditioning or data assimilation, but also simplifies the calibration process as shown by Pappenberger *et al.* (2006).

For the purpose of flood forecasting and warning, the predictions of time-to-peak and maximum wave height at specific locations are required, as well as an online data assimilation technique, which can be done by state updating (Neal *et al.*, 2007), error updating (Madsen and Skotner, 2005; Refsgaard, 1997) or prediction weight updating (Romanowicz and Beven, 1997). When an updating procedure is used, the error model may include an estimate of the unknown error model parameters, both for output as well as input variables. In practice though, such parameters may be difficult to estimate.

Inundation predictions for the purpose of water management require estimates of inundation area and time of inundation over the entire region of ecological concern. From the point of view of possible error structures, these requirements are not so different from those for flood hazard estimation, though there is no necessity to predict velocity fields. However, combined reservoir management and flow routing may impose stronger constraints on computation times as the optimisation methods used to derive estimates of control variables require multiple calls of the flow routing model at each step of the optimisation routine.

The map of minimum inundation outlines required by the ecosystem along the river reach between Suraż and Izbiszcze, Poland, (49th–57th cross-section) covers the area of the National Narew Park and it is shown in Figures 10.4 and 10.5 with the dashed line (Romanowicz *et al.*, 2008). The low and high water results indicate that large areas of the wetlands are never flooded, even during high water conditions. Further study is required to define the best management scenario that would allow for inundation probability to be extended over the whole wetland area — in particular during high water conditions.

10.5.1. *Input error modelling*

The input errors in flood inundation modelling focus mainly on the hydrometric data or model errors, related to upstream and downstream boundary conditions, topography and friction models as discussed in Section 10.2. Assimilation of uncertain boundary conditions observations can be done when online updating of the model predictions is applied using, for example, Ensemble Kalman Filter (EnKF) techniques (Neal *et al.*, 2007). In this latter application Neal *et al.* (2007) found that the state space formulation of the flood inundation prediction problem enabled a model of the error in the downstream boundary condition to be included. This error model had the form of a first-order autoregressive model AR(1). The harmonic behaviour of the error forecast model caused by the errors in downstream conditions was modelled by Madsen and Skotner (2005). The other method of introducing upstream boundary uncertainty into a flood inundation model consists of propagating the uncertainties through the system (Pappenberger *et al.*, 2005b) using ensemble input forecasts or uncertainty in the rating curve (Pappenberger *et al.*, 2006a).

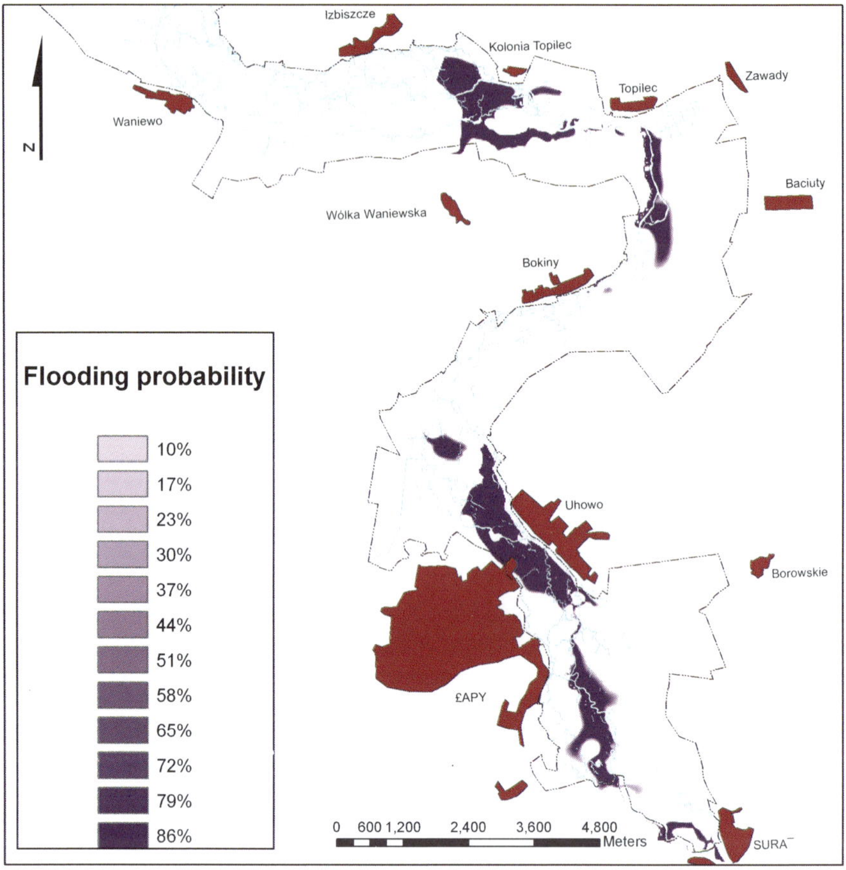

Figure 10.4. Low water level probability map at the National Narew Park area between Suraż and Izbiszcze.

These approaches require assumptions regarding the possible structure of input errors, which in turn follows from the model used for the input predictions. The errors related to model topography are assumed to be additive Gaussian or uniform and are introduced into the model evaluation using multiple model simulations (Pappenberger *et al.*, 2006a).

10.5.2. *Output error models*

The output error is defined as the difference between the simulated and observed output variable. Therefore its form depends on the type of output variable considered and the specific objective function used for

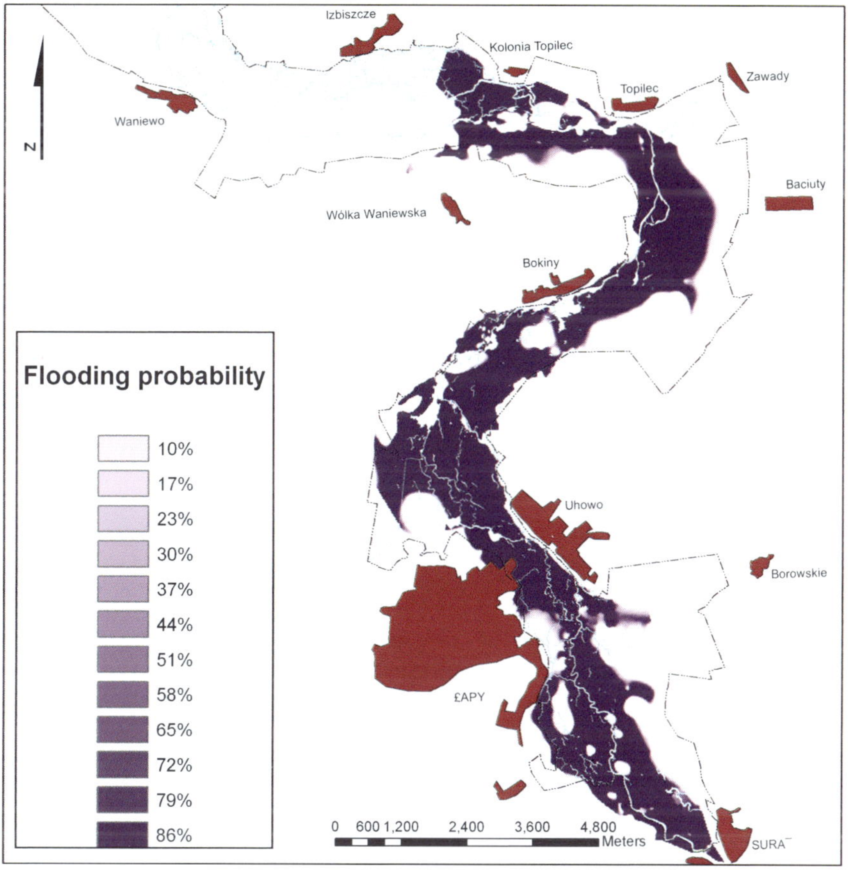

Figure 10.5. High water level probability map at the National Narew Park area between Suraż and Izbiszcze.

the comparison. Typical data consist of hydrographs at a point scale (for example, at the observed cross-sections or at the points where water level sensors are placed), inundation outlines at a vector scale, or spatial inundation patterns on a pixel by pixel basis, or globally (see Section 10.3).

The case of a hydrograph comparison is equivalent to time series analysis from the point of view of the error structure. The errors are usually correlated both in time and space, non-Gaussian, and non-additive; therefore, the assumed error model is unlikely to describe correctly the true error structure. Among the approaches applied in flood inundation modelling are the statistical treatment of the error and fuzzy modelling. When a statistical approach is applied, the error is assumed additive

Gaussian, but it could be also generalised Gaussian, or shaped according to the specific characteristics of the available data (Romanowicz *et al.*, 1996). A multiplicative error can be easily transformed into the additive form through the logarithmic transformation. One method to achieve this is the Generalised Likelihood Uncertainty Estimation (GLUE) approach which allows for a non-additive error structure (Romanowicz and Beven, 2006), but at the expense of statistical efficiency of the estimators. An additive, correlated error structure was introduced in flood inundation modelling by Romanowicz *et al.* (1996), allowing also for the unknown mean and variance of the error of model predictions. A one-dimensional model was applied to the problem of estimating maps of probability of inundation conditioned on the simulations of a two-dimensional flood inundation model treated as observed data. The predictions of water levels at several cross-sections along the river reach during the flood event were used as the model output and the errors had the form of a difference between simulated and observed water levels. Aronica *et al.* (1998) applied two different approaches to estimate the uncertainty of predictions in a flood inundation model with scarce data. The first, statistical approach was based on a heteroscedastic, normal, independent error model and the second applied fuzzy rules to define the modelling errors. In the fuzzy set approach the errors were specified according to classes defined by fuzzy rules. Both approaches gave similar results, but the fuzzy rule approach seemed to be better adjusted to the combination of uncertain information from different sources.

Fuzzy-like model errors were applied by Romanowicz and Beven (2003), where flood inundation maps were estimated conditioned on the historical maps of maximum inundation outlines. In this case, maximum water levels obtained from the simulation of a one-dimensional flood inundation model were compared with water levels at the model cross-sections derived from the historical maximum inundation outlines for the same events. The errors were weighted by a function that took into account the accuracy of the inundation outlines. Pappenberger *et al.* (2007b) also used a fuzzy type measure to calibrate flood inundation models (more details see Section 10.5.3).

Procedures for non-Gaussian error transformation, used in hydrological modelling have not been applied in flood inundation models; however, a procedure similar to the one described in Romanowicz and Beven (2006), accounting for non-additive errors, was applied by Werner (2004) to absolute differences of simulated and observed maximum water levels at a given location. It should be noted that the procedures of error transformation are closely related to the methods of error assessment and the choice of the likelihood function, as described in Section 10.5.3.

Schumann *et al.* (2007) presented the estimation of flood inundation predictions using satellite radar data, flood marks and aerial photography-derived flood extent to condition distributed roughness parameters. Local estimates of flood inundation extent were compared using absolute error between water levels derived from different information sources at model cross-sections. The reach-scale estimates were derived applying clustering of cdfs representing similar model error characteristics. That clustering leads to spatially distributed roughness parameterisations giving acceptable model performance both locally and at the reach scale.

Neal *et al.* (2007) in their EnKF application to the online updating of a one-dimensional hydraulic flow model assumed uncorrelated observation errors and state estimates. As that assumption is not fulfilled in practice it limited the performance of the model. This formulation would also allow one to build in the error model for the input variables and introduce a coloured noise model, but these were not applied in that paper.

The next most explored method of comparing inundation model predictions with observations applies discrete binary maps. In the case of this binary approach, the error model has the form of binary data and is assessed using methods of binary pattern comparison. This method can be used when inundation extent maps are available from SAR images, or inundation outlines. In this case spatial predictions of a flood inundation model are transformed into a pattern of wet and dry cells and are compared with analogous observed binary transformed maps (Aronica *et al.*, 2002; Pappenberger *et al.*, 2006b, 2007a,b; Werner *et al.*, 2005). The comparison can be made on global, sub-domain or local scales, as described by Pappenberger *et al.* (2006b) and in the next section (a representation of the effects of an increase in spatial scale in model evaluation on modelling results has been shown by Schumann *et al.*, 2008b). Fuzzy set approaches with binary maps were also applied by Pappenberger *et al.* (2007a,b) to evaluate uncertainties of predictions of a two-dimensional flood inundation model. The predictions were conditioned on remotely sensed satellite SAR data of inundation extent. The whole observed SAR inundation map was classified according to four inundation categories, i.e. each cell was represented by a four-dimensional vector specifying the degree of belonging. Additionally, the model predictions were fuzzyfied in a similar manner, but taking into account differences in spatial resolution of observed and modelled inundation area. Comparison of two fuzzy category cells was performed using the standard similarity function of Hagen (2003). The authors compared the obtained fuzzy performance measure with traditional discrete binary performance measures that did not take into account the observation errors.

Numerical algorithms used for the solution of distributed flood inundation models involve errors coming from numerical approximations and computer techniques. For example, Pappenberger *et al.* (2005a) has shown that a parameter which controls the accuracy of the numerical scheme can be responsible for larger uncertainties than the so-called physical parameters. Some errors can be estimated approximately and taken into account in the presentation of the results. There might be additional errors following from the inappropriate use of programming techniques, as well as presentation errors and errors connected with inappropriate use of the model by the user (e.g. by using the model for cases beyond its specified application range). Such errors are not usually treated as uncertainties of the modelling and can be reduced by testing and comparison with results obtained using other techniques as well as user support procedures.

10.5.3. *Estimating uncertainties and sensitivities in flood inundation models*

Methods of model error assessment can be classified into local measures and those that are cumulated over the whole simulated floodplain. The example of locally assessed errors was given in Romanowicz and Beven (2003), globally assessed errors were presented by Aronica *et al.* (2002), and comparisons of cumulated and local model performance were given by Pappenberger *et al.* (2006c) and Schumann *et al.* (2007). It is often very difficult to find a model that provides simulation results that are globally and locally consistent with the observations. There are then five possible responses: investigate those regions of the flow domain where there are consistent anomalies between model predictions and the range of observations; avoid using data that we don't believe or that are doubtful; introduce local parameters if there are particular local anomalies; make error bounds wider in some way where data are doubtful; and if none of the above can be done (because, for example, there is no reason to doubt anomalous data) then resort to local evaluations in assessing local uncertainties; or introduce targeted model evaluation (Pappenberger *et al.*, 2007a). The latter consists, for example, of weighting the model

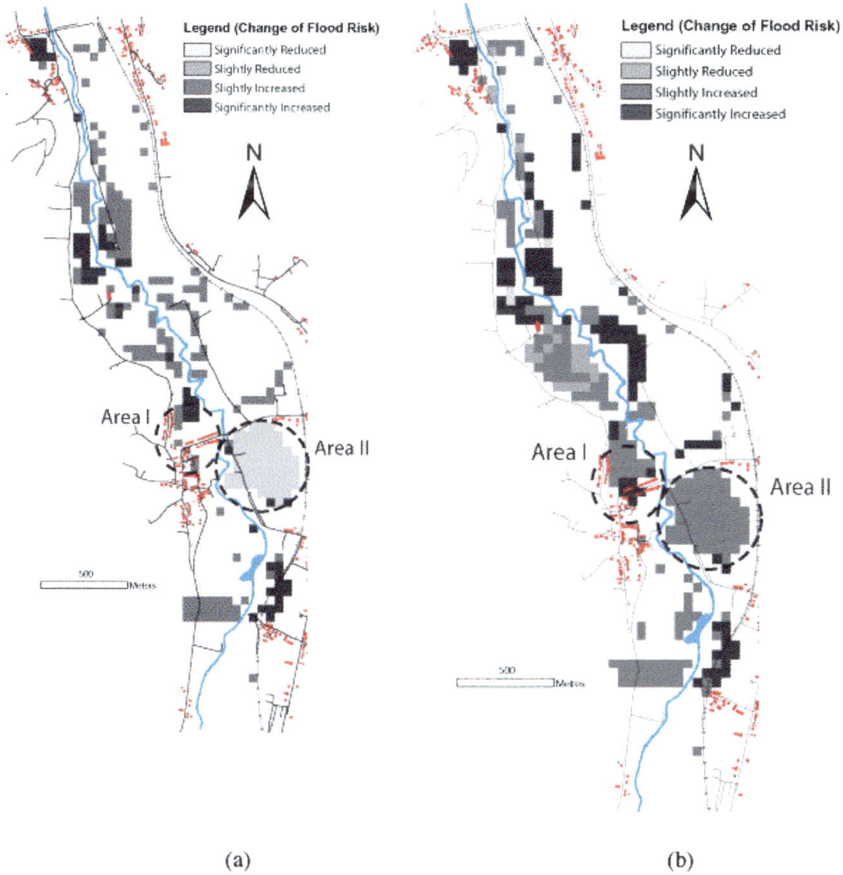

Figure 10.6. Comparison of flood hazard derived from a global performance measure to maps compared to one weighted by road km (left, a) and buildings (right, b). There is no change in flood risk in all white areas. It can be shown that weighting the model calibration process with different types of vulnerability can lead to significantly different flood risk maps (taken from Pappenberger *et al.*, 2007a).

performance with respect to additional information such as vulnerability (see Figure 10.6).

A deterministic model simulation calibrated with a remotely sensed flood inundation image will lead to a deterministic flood inundation map, which can be directly used for planning, model evaluation or other purposes. However, if uncertainty is included into this process and one deals with multiple simulations then an uncertain output map has to be derived. Aronica *et al.* (2002) proposed a method to derive a "probability" map using a discrete binary model output on a pixel by pixel basis by:

$$RCM_j = \frac{\sum_i L_i w_{ij}}{\sum_i L_i}, \tag{10.1}$$

in which L_i is the weight for each simulation i (and which can be any relevant skill score such as the F measure used by Aronica *et al.* (2002)) and the simulation results for the jth model element (e.g. computational cell or node) is $w_{ij} = 1$ for wet and $w_{ij} = 0$ for dry. The weight can be based on the normalised performance measures quoted above which are derived from maps conditioned on remotely sensed flood extent information. RCM_j is the relative confidence measure for each cell j, which expresses the belief that an uncertain prediction is a consistent representation of the system behaviour (for a discussion see Hunter *et al.*, 2008). Probability maps of inundation may also be derived from local water level predictions by interpolation of the quantiles of predicted water levels obtained at cross-sections along the river reach and overlaying them on the digital terrain model of river reach (Romanowicz *et al.* 1996, 2003).

Uncertainties in hydrodynamic model calibration and boundary conditions can have a significant influence on flood inundation predictions. Uncertainty analysis involves quantification of these uncertainties and their propagation through to inundation predictions. The inverse of this is sensitivity analysis, which tries to diagnose the influence that model parameters have on the uncertainty (Cloke *et al.*, 2008; Pappenberger *et al.*, 2006b; Pappenberger and Ratto, this volume). Hall *et al.* (2005) and Pappenberger *et al.* (2008) used a variance-based global sensitivity analysis, the so called “Sobol” method, to quantify the significant influence of variance in the Manning channel roughness coefficient in raster-based and *one-dimensional* flood inundation model predictions of flood outline and flood depth. Pappenberger *et al.* (2008) extended such a study to multiple other factors such as bridge representations and boundary conditions. The methodology allowed the sub-reaches of channel that have the most influence to be identified, demonstrating how far boundary effects propagate into the model and indicating where further data acquisition and nested higher-resolution model studies should be targeted.

10.6. Conclusions

Whilst flood inundation models are relatively parsimonious in terms of their input data requirements and number of unknown parameters, these values may often be subject to complex and poorly understood errors. Moreover, observed flow information available to constrain uncertainty in flood inundation predictions may be extremely limited in terms of either spatial or temporal coverage, as well as itself being subject to significant measurement uncertainty. Whilst the development of better remotely sensed data capture techniques continues to allow progress to be made in providing a more data-rich environment for flood inundation modelling, key uncertainties, particularly in terms of our ability to obtain distributed flow measurements during floods, will undoubtedly remain. The question of whether the main source of uncertainty is either the model parameterisation or errors related to the data cannot be resolved easily as these sources are inter-related. Therefore, the assumption that the researcher can neglect model uncertainty, and take into account only data uncertainty, is not justified. The other problem that arises is when the choice between statistical formal and informal approaches has to be made. As discussed in this chapter, the scarcity of observations and complexity of flood inundation modelling makes the use of formal approaches that would take into account both spatial and temporal correlation of errors computationally prohibitive. Despite these intrinsic limitations, sufficient progress in data capture has been made to allow uncertainty analysis techniques for flood inundation modelling to mature rapidly over the last decade. As this chapter has demonstrated, we now have a number of promising methods and data sets at our disposal which will allow us over the medium term to better understand how these complex non-linear models respond when confronted with non-error-free data, and how this can be used to develop better spatially distributed measures of flood risk that include our degree of belief in these predictions.

References

Ali, S. and Goodwin, P. (2002). "The Predictive Ability of 1D Models to Simulate Floodplain Processes" in: Falconer, R.A., Lin, B., Harris, E.L. *et al.* (eds), *Hydroinformatics 2002: Proceedings of the Fifth International Conference on Hydroinformatics. Volume One: Model Development and Data Management*, IWA Publishing, London, pp. 247–252.

Alsdorf, D.E, Bates, P., Melack, J.M. *et al.* (2007). The spatial and temporal complexity of the Amazon flood measured from space, *Geophys. Res. Lett.*, **34**, paper no. L08402.

Alsdorf, D.E., Melack, J.M., Dunne, T. *et al.* (2000). Interferometric radar measurements of water level changes on the Amazon floodplain, *Nature*, **404**, 174–177.

Aronica, G., Bates, P.D. and Horritt, M.S. (2002). Assessing the uncertainty in distributed model predictions using observed binary pattern information within GLUE, *Hydrol. Process.*, **16**, 2001–2016.

Aronica, G., Hankin, B.G. and Beven, K.J. (1998). Uncertainty and equifinality in calibrating distributed roughness coefficients in a flood propagation model with limited data, *Adv. Water Resour.*, **22**, 349–365.

Babaeyan-Koopaei, K., Ervine, D.A., Carling, P.A. *et al.* (2002). Field measurements for two overbank flow events in the River Severn, *J. Hydraul. Eng.-ASCE,* **128**, 891–900.

Baltsavias, E.P. (1999). A comparison between photogrammetry and laser scanning, *ISPRS J. Photogramm.*, **54**, 83–94.

Bates, P.D., Horritt, M.S., Aronica, G. *et al.* (2004). Bayesian updating of flood inundation likelihoods conditioned on flood extent data, *Hydrol. Process.*, **18**, 3347–3370.

Bates, P.D., Horritt, M.S., Smith, C. *et al.* (1997). Integrating remote sensing observations of flood hydrology and hydraulic modelling, *Hydrol. Process.*, **11**, 1777–1795.

Bates, P.D., Stewart, M.D., Siggers, G.B. *et al.* (1998). Internal and external validation of a two-dimensional finite element model for river flood simulation, *P. I. Civil. Eng.-Water.*, **130**, 127–141.

Bates, P.D., Wilson, M.D., Horritt, M.S. *et al.* (2006). Reach scale floodplain inundation dynamics observed using airborne synthetic aperture radar imagery: data analysis and modelling, *J. Hydrol.*, **328**, 306–318.

Berry, P.A.M., Garlick, J.D., Freeman, J.A. *et al.* (2005). Global inland water monitoring from multi-mission altimetry, *Geophys. Res. Lett.*, **32**, L16401.

Beven, K.J. (1989). Changing ideas in hydrology — the case of physically-based models, *J. Hydrol.*, **105**, 157–172.

Beven, K.J. (2002). Towards a coherent philosophy for environmental modeling, *P. R. Soc. Lond., A*, **458**, 2465–2484.

Beven, K.J. (2006). A manifesto for the equifinality thesis, *J. Hydrol.*, **320**, 18–36.

Beven, K.J. and Binley, A. (1992). The future of distributed models: model calibration and uncertainty prediction, *Hydrol. Process.*, **6**, 279–298.

Biggin, D.S. and Blyth, K. (1996). A comparison of ERS-1 satellite radar and aerial photography for river flood mapping, *J. Chart. Inst. Water E.*, **10**, 59–64.

Blazkova, S. and Beven, K. (1997). Flood frequency prediction for data limited catchments in the Czech Republic using a stochastic rainfall model and TOPMODEL, *J. Hydrol.*, **195**, 256–278.

Box, G.E.P. and Tiao, G.C. (1992). *Bayesian Inference in Statistical Analysis*, John Wiley and Sons, Chichester, p. 608.

Butler, J.B., Lane, S.N. and Chandler, J.H. (2001). Automated extraction of grain-size data from gravel surfaces using digital image processing, *J. Hydraul. Res.*, **39**, 519–529.

Cameron, D.S., Beven, K.J., Tawn, J. *et al.* (1999). Flood frequency estimation by continuous simulation for a gauged upland catchment (with uncertainty), *J. Hydrol.*, **219**, 169–187.

Cloke, H.L., Pappenberger, F. and Renaud, J.P. (2008). Multi-method global sensitivity analysis (MMGSA) for modelling floodplain hydrological processes, *Hydrol. Process.*, **22**, 1660–1674.

Cobby, D.M., Mason, D.C. and Davenport, I.J. (2001). Image processing of airborne scanning laser altimetry for improved river flood modelling, *ISPRS J. Photogramm.*, **56**, 121–138.

Cobby, D.M., Mason, D.C, Horritt, M.S. *et al.* (2003). Two-dimensional hydraulic flood modelling using a finite element mesh decomposed according to vegetation and topographic features derived from airborne scanning laser altimetry, *Hydrol. Process.*, **17**, 1979–2000.

Coles, S. (2001). *An Introduction to Statistical Modelling of Extreme Values*, Springer-Verlag, London.

Costa, J.E., Spicer, K.R., Cheng, R.T. *et al.* (2000). Measuring stream discharge by non-contact methods: a proof-of-concept experiment, *Geophys. Res. Lett.*, **27**, 553–556.

Cunge, J.A. (2003). Of data and models, *J. Hydroinform.*, **5**, 75–98.

Cunge, J.A., Holly, F.M. and Verwey, A. (1980). *Practical Aspects of Computational River Hydraulics*, Pitman, London.

Dawson, R.J., Hall, J.W., Sayers, P. *et al.* (2005). Sampling-based flood risk analysis for fluvial dike systems, *Stoch. Env. Res. Risk A.*, **19**, 388–402.

Di Baldassarre, G., Schumann, G. and Bates, P.D. (2009). Near real time satellite imagery to support and verify timely flood modelling, *Hydrol. Process.*, **23**, 799–803.

Fread, D.L. (1993). "Flood Routing" in: Maidment, D.R (ed.), *Handbook of Applied Hydrology*, McGraw Hill, New York, 10.1–10.36.

Gomes-Pereira, L.M. and Wicherson, R.J. (1999). Suitability of laser data for deriving geographical data: a case study in the context of management of fluvial zones, *ISPRS J. Photogramm*, **54**, 105–114.

Grayson, R.B., Moore, I.D. and McMahon, T.A. (1992). Physically-based hydrologic modelling: II. Is the concept realistic? *Water Resour. Res.*, **28**, 2659–2666.

Hagen, A. (2003). Fuzzy set approach to assessing similarity of categorical maps, *Int. J. Geogr. Inf. Sci.*, **17**, 235–249.

Hall, J.W., Dawson, R.J., Sayers, P.B. *et al.* (2003). A methodology for national-scale flood risk assessment, *P. I. Civil Eng.-Water*, **156**, 235–247.

Hall, J., Tarantola, S., Bates, P.D. *et al.* (2005). Distributed sensitivity analysis of flood inundation model calibration, *J. Hydraul. Eng.-ASCE*, **131**, 117–126.

Hankin, B.G., Hardy, R., Kettle, H. *et al.* (2001). Using CFD in a GLUE framework to model the flow and dispersion characteristics of a natural fluvial dead zone, *Earth Surf. Proc. Land.*, **26**, 667–687.

Hey, R.D. (1979). Flow resistance in gravel bed rivers, *J. Hydr. Eng. Div.-ACSE*, **105**, 365–379.

Hodgson, M.E., Jensen, J.R., Schmidt, L. *et al.* (2003). An evaluation of LiDAR- and IFSAR-derived digital elevation models in leaf-on conditions with USGS Level 1 and Level 2 DEMs, *Remote Sens. Environ.*, **84**, 295–308.

Horritt, M.S. (1999). A statistical active contour model for SAR image segmentation, *Image Vision Comput.*, **17**, 213–224.

Horritt, M.S. (2000). Calibration of a two-dimensional finite element flood flow model using satellite radar imagery, *Water Resour. Res.*, **36**, 3279–3291.

Horritt, M.S. and Bates, P.D. (2001). Predicting floodplain inundation: raster-based modelling versus the finite element approach, *Hydrol. Process.*, **15**, 825–842.

Horritt, M.S. and Bates, P.D. (2002). Evaluation of 1-D and 2-D numerical models for predicting river flood inundation, *J. Hydrol.*, **268**, 87–99.

Hunter, N.M., Bates, P.D., Horritt, M.S. *et al.* (2005). Utility of different data types for calibrating flood inundation models within a GLUE framework, *Hydrol. Earth Syst. Sci.*, **9**, 412–430.

Hunter, N.M., Bates, P.D., Horritt, M.S. *et al.* (2006). Improved simulation of flood flows using storage cell models, *P. I. Civil Eng.-Wat. M.*, **159**, 9–18.

Hunter, N.M., Bates, P.D., Horritt, M.S. *et al.* (2007). Simple spatially-distributed models for predicting flood inundation: a review, *Geomorphology*, **90**, 208–225.

Hunter, N.M., Bates, P.D., Neelz, S. *et al.* (2008). Benchmarking 2D hydraulic models for urban flood simulations, *P. I. Civil Eng.-Wat. M*, **161**, 13–30.

Institute of Hydrology (1999). *Flood Estimation Handbook: Procedures for Flood Frequency Estimation*, Institute of Hydrology: Wallingford, UK.

Kim, Y., Muste, M., Hauet, M. *et al.* (2008). Stream discharge using mobile large-scale particle image velocimetry: a proof of concept, *Water Resour. Res.*, **44**, W09502.

Knight D.W. and Shiono K. (1996). "River Channel and Floodplain Hydraulics" in: Anderson, M.G., Walling D.E and Bates, P.D. (eds), *Floodplain Processes*, John Wiley and Sons, Chichester, pp. 139–182.

Kouwen, N. (2000). Effect of riparian vegetation on flow resistance and flood potential, *J. Hydraul., Eng.-ACSE*, **126**, 954.

Krabill, W.B., Collins, J.G., Link, L.E. *et al.* (1984). Airborne laser topographic mapping results, *Photogramm. Eng. Rem. S.*, **50**, 685–694.

Lane, S.N. (2000). The measurement of river channel morphology using digital photogrammetry, *Photogramm. Rec.*, **16**, 937–957.

Lane, S.N., Bradbrook, K.F., Richards, K.S. *et al.* (1999). The application of computational fluid dynamics to natural river channels: three-dimensional versus two-dimensional approaches, *Geomorphology*, **29**, 1–20.

Madsen, H. and Skotner, C. (2005). Adaptive state updating in real-time river flow forecasting — a combined filtering and error forecasting procedure, *J. Hydrol*, **308**, 302–312.

Marks, K. and Bates, P.D. (2000). Integration of high resolution topographic data with floodplain flow models, *Hydrol. Process*, **14**, 2109–2122.

Mason, D.C, Cobby, D.M., Horritt, M.S. *et al.* (2003). Floodplain friction parameterization in two-dimensional river flood models using vegetation heights derived from airborne scanning laser altimetry, *Hydrol. Process.*, **17**, 1711–1732.

Mason, D.C., Horritt, M.S., Hunter, N.M. *et al.* (2007). Use of fused airborne scanning laser altimetry and digital map data for urban flood modelling,

Hydrol. Process., **21**, 1436–1477.
Matgen, P., Schumann, G., Henry, J.-B. *et al.* (2007). Integration of SAR-derived inundation areas, high precision topographic data and a river flow model toward real-time flood management, *Int. J. Appl. Earth Obs.*, **9**, 247–263.
Neal, J.C., Atkinson, P.M. and Hutton, C.W. (2007). Flood inundation model updating using an ensemble Kalman filter and spatially distributed measurements, *J. Hydrol.*, **336**, 401–415.
Nicholas, A.P. and Mitchell, C.A. (2003). Numerical simulation of overbank processes in topographically complex floodplain environments, *Hydrol. Process.*, **17**, 727–746.
Pappenberger, F., Beven, K.J., Frodsham, K. *et al.* (2007a). Grasping the unavoidable subjectivity in calibration of flood inundation models: a vulnerability weighted approach, *J. Hydrol.*, **333**, 275–287.
Pappenberger, F., Beven, K., Horritt, M. *et al.* (2005b). Uncertainty in the calibration of effective roughness parameters in HEC-RAS using inundation and downstream level observations, *J. Hydrol.*, **302**, 46–69.
Pappenberger, F., Beven, K.J., Hunter N. *et al.* (2005a). Cascading model uncertainty from medium range weather forecasts (10 days) through a rainfall-runoff model to flood inundation predictions within the European Flood Forecasting System (EFFS), *Hydrol. Earth Sys. Sci.*, **9**, 381–393.
Pappenberger, F., Beven, K.J., Ratto, M. *et al.* (2008). Multi-method global sensitivity analysis of flood inundation models, *Adv. Water Res.*, **31**, 1–14.
Pappenberger, F., Frodsham, K., Beven, K.J. *et al.* (2007b). Fuzzy set approach to calibrating distributed flood inundation models using remote sensing observations, *Hydrol. Earth Syst. Sci.*, **11**, 739–752.
Pappenberger, F., Iorgulescu, I. and Beven, K.J., (2006b). Sensitivity analysis based on regional splits (SARS-RT), *Environ. Modell. Softw.*, **21**, 976–990.
Pappenberger, F., Matgen, P., Beven, K. *et al.* (2006a). Influence of uncertain boundary conditions and model structure on flood inundation predictions, *Adv. Water Res.*, **29**, 1430–1449.
Pappenberger F. and Ratto M. (this volume).
Pavelsky, T.M. and Smith, L.C. (2008). Remote sensing of hydrologic recharge in the Peace-Athabasca Delta, Canada, *Geophys. Res. Lett.*, **35**, L08403.
Puech, C. and Raclot, D. (2002). Using geographical information systems and aerial photographs to determine water levels during floods, *Hydrol. Process.*, **16**, 1593–1602.
Rabus, B., Eineder, M., Roth, A. *et al.* (2003). The shuttle radar topography mission — a new class of digital elevation models acquired by spaceborne radar, *ISPRS J. Photogramm.*, **57**, 241–262.
Romanowicz, R.J. and Beven, K.J. (1997). Dynamic real-time prediction of flood inundation probabilities, *Hydrol. Sci. J.*, **43**, 181–196.
Romanowicz, R.J. and Beven, K.J. (2003). Estimation of flood inundation probabilities as conditioned on event inundation maps, *Water Resour. Res.*, **39**, 1073.

Romanowicz, R.J. and Beven, K.J. (2006). Comments on generalised likelihood uncertainty estimation, *Reliab. Eng. Syst. Safety*, **91**, 1315–1321.

Romanowicz, R., Beven, K.J. and Tawn, J. (1996). "Bayesian Calibration of Flood Inundation Models" in: Anderson, M.G., Walling, D.E. and Bates, P.D. (eds), *Floodplain Processes*, John Wiley and Sons, Chichester, pp. 333–360.

Romanowicz, R.J., Kiczko, A. and Napiorkowski, J.J. (2008). Stochastic transfer function simulator of a 1-D flow routing, *Publis. Inst. Geophys., Pol. Acad. Sc.*, **E-10**, 151–160.

Romeiser, R., Runge H., Suchandt, S. *et al.* (2007). Current measurements in rivers by spaceborne along-track InSAR, *IEEE T. Geosci. Remote*, **45**, 4019–4031.

Samuels, P.G. (1990). "Cross Section Location in One-Dimensional Models" in: White, W.R. (ed.), *International Conference on River Flood Hydraulics*, John Wiley and Sons, Chichester, pp. 339–350.

Sanders, B.F. (2007). Evaluation of on-line DEMs for flood inundation modelling, *Adv. Water Res.*, **30**, 1831–1843.

Schubert, J.E., Sanders, B.F., Smith, M.J. *et al.* (2008). Unstructured mesh generation and landcover-based resistance for hydrodynamic modeling of urban flooding, *Adv. Water Res.*, **31**, 1603–1621.

Schumann, G., Matgen, P., Pappenberger, F. *et al.* (2007). Deriving distributed roughness values from satellite radar data for flood inundation modelling, *J. Hydrol.*, **344**, 96–111.

Schumann, G., Matgen, P. and Pappenberger, F. (2008a). Conditioning water stages from satellite imagery on uncertain data points, *IEEE Geosci. Remote S. Lett.*, **5**, 810–813.

Schumann, G., Matgen, P., Pappenberger, F. *et al.* (2008b). Evaluating uncertain flood inundation predictions with uncertain remotely sensed water stages, *Int. J. River Basin Manage.*, **6**, 187–199.

Schumann, G., Pappenberger F. and Matgen, P. (2008c). Estimating uncertainty associated with water stages from a single SAR image, *Adv. Water Resour.*, **31**, 1038–1047.

Sellin, R.H.J. and Willets, B.B. (1996). "Three-Dimensional Structures, Memory and Energy Dissipation in Meandering Compound Channel Flow" in: Anderson, M.G., Walling, D.E. and Bates, P.D. (eds), *Floodplain Processes*, John Wiley and Sons, Chichester, pp. 255–298.

Sorooshian S. and Dracup, J.A. (1980). Stochastic parameter estimation procedures for hydrological rainfall-runoff models: correlated and heteroscedastic cases, *Water Resour. Res.*, **16**, 430–442.

Sorooshian, S., Gupta, V.K. and Fulton, J.L. (1983). Evaluation of maximum likelihood parameter estimation techniques for conceptual rainfall-runoff models: influence of calibration data variability and length on model credibility, *Water Resour. Res.*, **19**, 251–259.

Stewart, M.D., Bates, P.D. Anderson, M.G. *et al.* (1999). Modelling floods in hydrologically complex lowland river reaches, *J. Hydrol.*, **223**, 85–106.

Stoesser, T., Wilson, C.A.M.E., Bates, P.D. *et al.* (2003). Application of a 3D numerical model to a river with vegetated floodplains, *J. Hydroinform.*, **5**, 99–112.

Straatsma, M.W. and Baptist, M. (2008). Floodplain roughness parameterization using airborne laser scanning and spectral remote sensing, *Remote Sens. Environ.*, **112**, 1062–1080.

Vorosmarty, C.J. (2002). Global water assessment and potential contributions from Earth Systems Science, *Aquat. Sci.*, **64**, 328–351.

Werner, M.G.F. (2004). A comparison of predictive reliability in flood extent modeling, *Hydrol. Earth Syst. Sci.*, **8**, 1141–1152

Werner, M.G.F., Hunter, N. and Bates, P.D. (2005). Identifiability of distributed floodplain roughness values in flood extent estimation, *J. Hydrol.*, **314**, 139–157.

Westaway, R.M., Lane, S.N. and Hicks, D.M. (2003). Remote survey of large-scale braided, gravel-bed rivers using digital photogrammetry and image analysis, *Int. J. Remote Sens.*, **24**, 795–815.

Wilson, M.D., Bates, P.D., Alsdorf, D. *et al.* (2007). Modeling large-scale inundation of Amazonian seasonally flooded wetlands, *Geophys. Res. Lett.*, **34**, paper no. L15404.

CHAPTER 11

Flood Defence Reliability Analysis

Pieter van Gelder and Han Vrijling
Delft University of Technology, Netherlands

11.1. Introduction

In this chapter the probabilistic approach of the design and the risk analysis in flood defence engineering is outlined. The application of the probabilistic design methods offers the designer a way to unify the design of engineering structures, processes and management systems. For this reason there is a growing interest in the use of these methods. This chapter is structured as follows. First, an introduction is given to probabilistic design, uncertainties are discussed, and a reflection on the deterministic approach versus the probabilistic approach is presented. The chapter continues by addressing the tools for a probabilistic systems analysis and its calculation methods. Failure probability calculation for an element is reviewed and evaluation methods of a calculated risk level are addressed. The chapter ends with a case study on the reliability analysis and optimal design of a caisson flood defence.

The basis of the deterministic approach is the so-called design values for the loads and the strength parameters. Loads could be, for instance, the design water level and the design significant wave height. Using design rules according to codes and standards it is possible to determine the shape and the height of the cross-section of the flood defence. These design rules are based on limit states of the flood defence system's elements, such as overtopping, erosion, instability, piping and settlement. It is assumed that the structure is safe when the margin between the design value of the load and the characteristic value of the strength is large enough for all limit states of all elements.

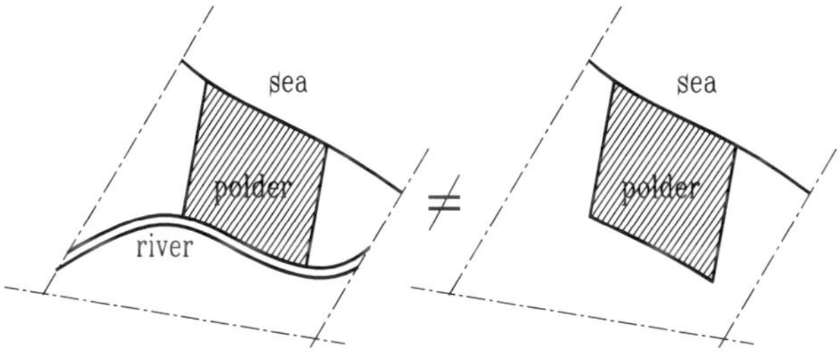

Figure 11.1. Different safety levels for the same design.

The safety level of the protected area is not explicitly known when the flood defence is designed according to the deterministic approach. The most important shortcomings of the deterministic approach are:

- The fact that the failure probability of the system is unknown.
- The complete system is not considered as an integrated entity. An example is the design of the flood defences of the protected area of Figure 11.1. With the deterministic approach, the design of the sea dike is, in both cases, exactly the same. In reality, the left area is threatened by flood from two independent causes: the sea and by the river. Therefore, the safety level of the left area is less than the safety level of the right one.
- Another shortcoming of the deterministic approach is that the length of flood defence does not affect the design. In the deterministic approach the design rules are the same for all the sections independently of the number of sections. It is, however, intuitively clear that the probability of flooding increases with the length of the flood defence.
- With the deterministic design methods it is impossible to compare the strength of different types of cross-sections such as dikes, dunes and structures like sluices and pumping stations.
- And last, but not least, the deterministic design approach is incompatible with other policy fields like, for instance, the safety of industrial processes and the safety of transport of dangerous substances.

A fundamental difference with the deterministic approach is that the probabilistic design methods are based on an acceptable frequency or probability of flooding of the protected area.

The probabilistic approach results in a probability of failure of the whole flood defence system taking account of each individual cross-section and each structure. As a result, the probabilistic approach is an integral design method for the whole system.

11.2. Uncertainties

Uncertainties in decision and risk analysis can primarily be divided into two categories: uncertainties that stem from variability in known (or observable) populations and, therefore, represent randomness in samples (inherent or aleatory uncertainty); and uncertainties that come from a basic lack of knowledge of fundamental phenomena (epistemic uncertainty).

Data can be gathered by taking measurements or by keeping record of a process in time. Research can, for instance, be undertaken with respect to the physical model of a phenomenon or into the better use of existing data. By using expert opinions, it is possible to acquire the probability distributions of variables that are too expensive or practically impossible to measure.

The goal of all this research obviously is to reduce the uncertainty in the model. Nevertheless, it is also thinkable that uncertainty will increase. Research might show that an originally flawless model actually contains a lot of uncertainties, or, after taking some measurements, that the variations of the dike height can be a lot larger. It is also possible that the average value of the variable will change because of the research that has been done.

The consequence is that the calculated probability of failure will be influenced by future research. In order to guarantee a stable and convincing flood defence policy after the transition, it is important to understand the extent of this effect.

11.3. Probabilistic Approach of the Design

The accepted probability of flooding is not the same for every polder or floodplain. This depends on the nature of the protected area, the expected loss in case of failure and the safety standards of the country. For instance, for a protected area with a dense population or an important industrial development, a smaller probability of flooding is allowed than for an area of lesser importance.

For this reason accepted risk is a better measure than an accepted failure probability, because risk is a function of the probability and the consequences of flooding.

The most general definition of risk is the product of the probability and a power of consequences:

$$risk = (probability) \cdot (consequence)^n.$$

In many cases, such as economical analyses, the power n is equal to one.

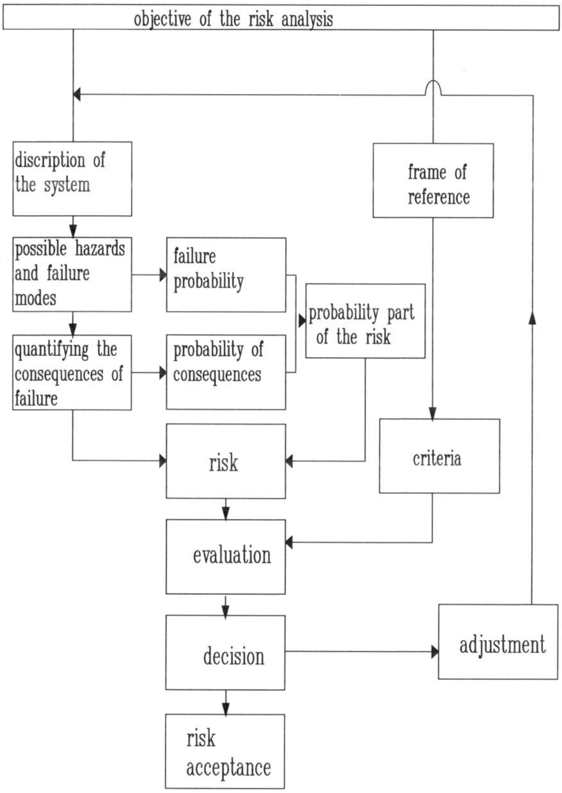

Figure 11.2. Probabilistic approach of the design.

Figure 11.2 shows the elements of the probabilistic approach. First of all the flood defence system has to be described as a configuration of elements such as dike sections, sluices and other structures. Then an inventory of all the possible hazards and failure modes must be made. This step is one of the most important of the analysis because missing a failure mode can seriously influence the safety of the design.

The next step can be the quantifying of the consequences of failure. Hereby it is necessary to analyse the consequence of the failure for all possible ways of failure. Sometimes the consequences of failure of an element of the system are different for each element.

The failure probability and the probability of the consequences form the probability part of the risk. When the risk is calculated the design can be evaluated. For this to happen, criteria must be available, such as a maximum acceptable probability of a number of casualties or the demand of minimising the total costs including the risk. For determining the acceptable risk, we need a frame of reference. This frame of reference can be the national safety level aggregating all the activities in the country. After the evaluation of the risk, one can decide to adjust the design or to accept it with the remaining risk.

11.4. System Analysis

Every risk analysis, which is the core of the probabilistic design, starts with a system analysis. There are several techniques to analyse a system but in this case we will restrict ourselves to the fault tree analysis, which arranges all the events in such a way that their occurrence leads to failure of the system. In Figure 11.3 there is an example given of a fault tree. A fault tree consists of basic events (E1 ... E9), combined events (E10 ... E12), a top event (failure) and gates (and, or). A gate is the relation of the events underneath the gate that lead to the event above the gate.

11.4.1. *Simple systems*

The simplest systems are the parallel system and the series system (Figure 11.4). A parallel system that consists of two elements functions as long as one of the elements functions. When a system fails in the event of only one element failing it is called a "series system".

When the elements and their failure modes are analysed it is possible to make a fault tree. The fault tree gives the logical sequence of all the possible events that lead to failure of the system.

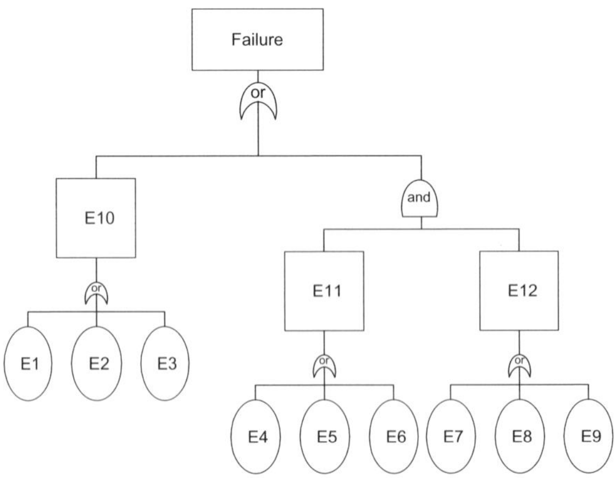

Figure 11.3. Fault tree.

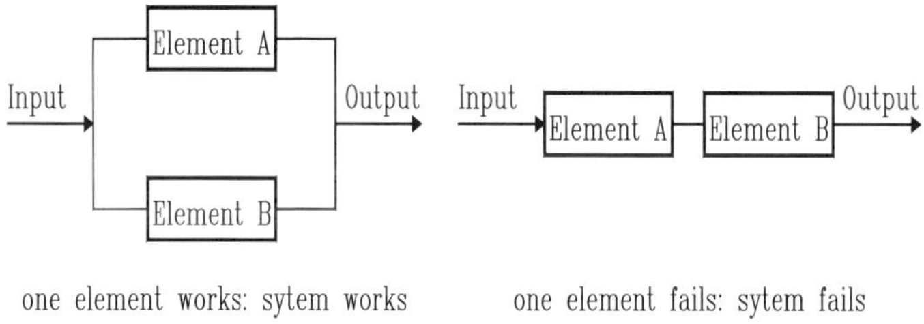

Figure 11.4. Parallel and series system.

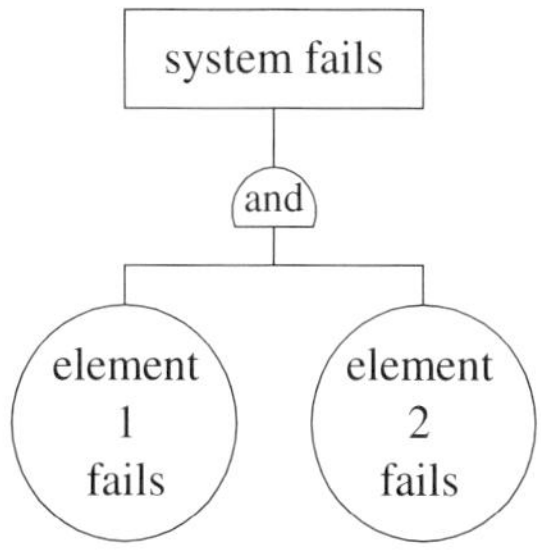

Figure 11.5. Fault tree of a parallel system.

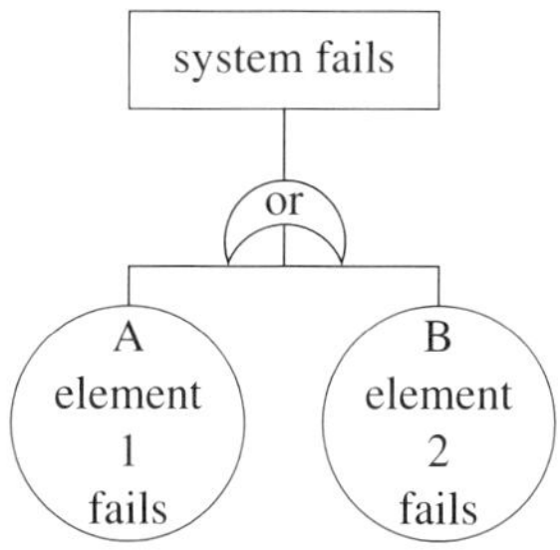

Figure 11.6. Fault tree of a series system.

Take, for instance, the fault tree of a simple parallel system. The basic events in this system are the failure events of the single elements. The failure of the system is called the top event. The system fails only when all single elements fail; the gate between the basic events and the top event is a so-called AND gate (see Figure 11.5).

A series system of two elements fails if only one of the elements fails as depicted by the so-called "or" gate between the basic events and the top event (see Figure 11.6).

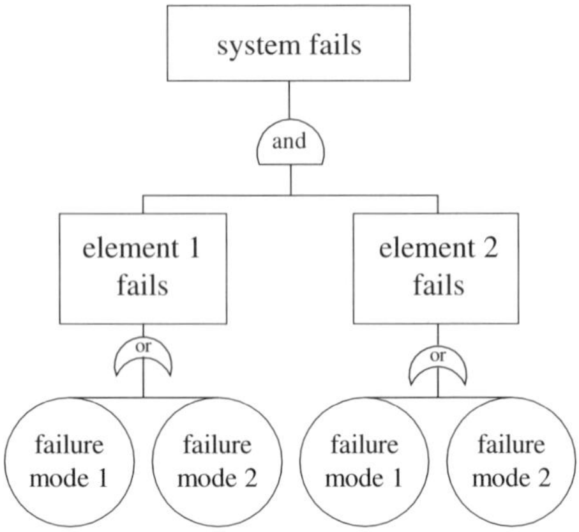

Figure 11.7. Elements of a parallel system as series systems of failure modes.

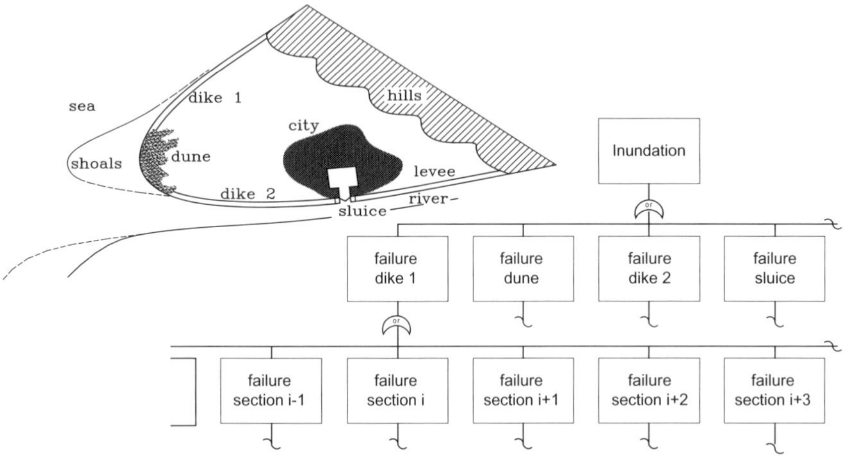

Figure 11.8. Flood defence system and its elements presented in a fault tree.

When there are more failure modes possible for the failure of an element then the failure modes are the basic events and the failure of an element is a so-called composite event.

In Figure 11.7 an example is given of a parallel system of elements in which the system elements are on their turn series systems of the possible failure modes.

For all the elements of the flood defence, all the possible failure modes can be the cause of failure (Figure 11.8). A failure mode is a mechanism

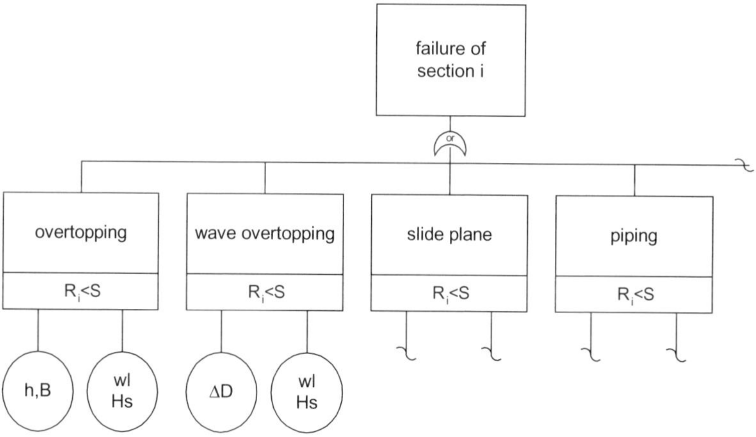

Figure 11.9. A dike section as a series system of failure modes.

that leads to failure. A review of failure modes is provided by Vrijling (1986) and Allsop *et al.* (2007).

The place of the failure modes in the system is demonstrated by a fault tree analysis in Figures 11.9 and 11.10.

An advantage of the probabilistic approach over the deterministic approach is illustrated in Figure 11.10, where human failure to close the sluice is included in the analysis.

The conclusion of this analysis is that any failure mechanism of any element of any subsystem of the flood defence system can lead to inundation of the polder. The system is therefore a series system.

11.5. Failure Probability of a System

This section gives an introduction of the determination of the failure probability of a system for which the failure probabilities of the elements are known.

In Figure 11.11 there are two fault trees given, one for a parallel system and one for a series system, both consisting of two elements.

Event A is the event that element 1 fails and event B is the event that element 2 fails. The parallel system fails if both the elements fail. The failure probability is the probability of A and B. The series system fails if at least one of the elements fail, so the failure probability is the probability of A or B.

The probability of A and B is equal to the product of the probability of A and the probability of B given A. The probability of A or B is equal to the

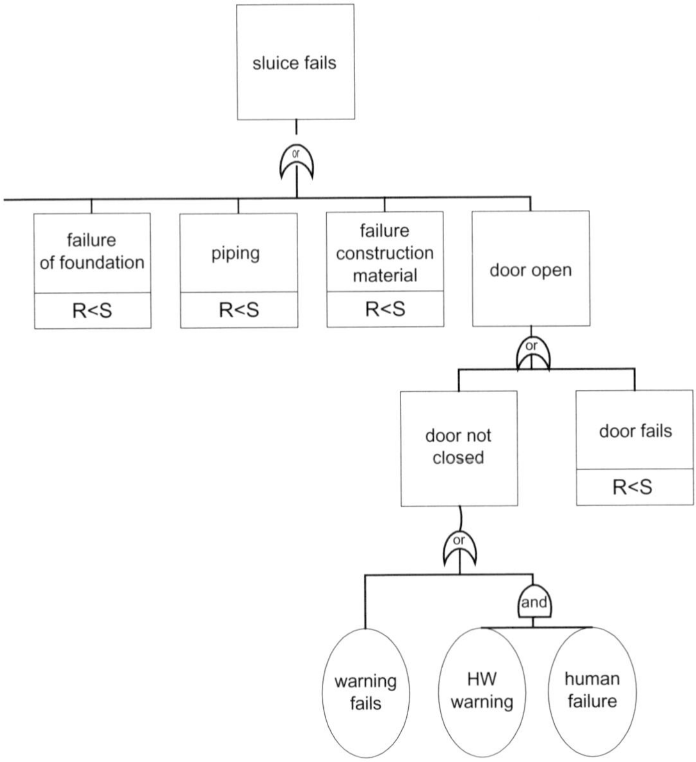

Figure 11.10. The sluice as a series system of failure modes.

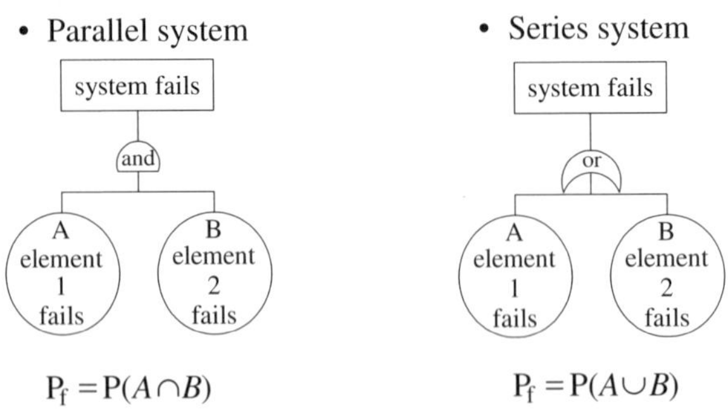

Figure 11.11. Fault trees for parallel and series system.

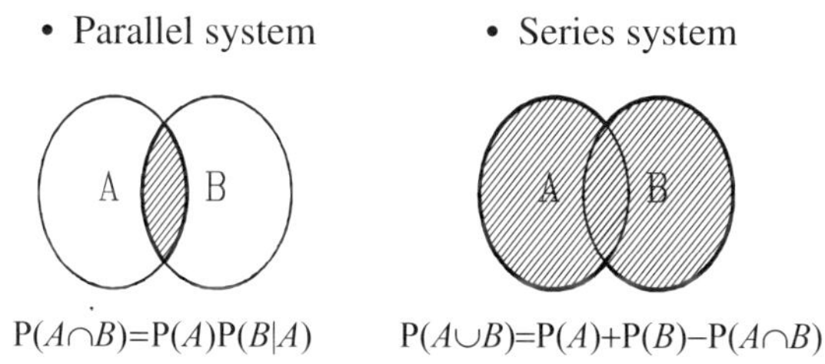

Figure 11.12. Combined events.

sum of the probability of A and the probability of B minus the probability of A and B.

In practice the evaluation of the probability of B given A is rather difficult because the relation between A and B is not always clear. If A and B are independent of each other, then the probability of B given A is equal to the probability of B without A. In this case the probability of A and B is equal to the product of the probability of A and the probability of B:

$$\mathrm{P}(B|A) = \mathrm{P}(B) \Rightarrow \mathrm{P}(A \cap B) = \mathrm{P}(A)\mathrm{P}(B). \tag{11.1}$$

If event A excludes B then the probability B and A is zero:

$$\mathrm{P}(B|A) = 0 \Rightarrow \mathrm{P}(A \cap B) = 0. \tag{11.2}$$

If A includes B then the probability of B given A is 1 and so the probability of A and B is equal to the probability of A:

$$\mathrm{P}(B|A) = 1 \Rightarrow \mathrm{P}(A \cap B) = \mathrm{P}(A). \tag{11.3}$$

In the same way it is possible to determine the probability of A or B. If A and B are independent of each other the probability of A or B is:

$$\mathrm{P}(A \cup B) = \mathrm{P}(A) + \mathrm{P}(B) - \mathrm{P}(A)\mathrm{P}(B). \tag{11.4}$$

If event A excludes B then the probability B and A is zero so the probability of A or B is:

$$\mathrm{P}(A \cup B) = \mathrm{P}(A) + \mathrm{P}(B). \tag{11.5}$$

If A includes B then the probability of B given A is 1 and so the probability of A and B is equal to the probability of A and the probability of A or B is:

$$\mathrm{P}(A \cup B) = \mathrm{P}(A) + \mathrm{P}(B) - \mathrm{P}(A). \tag{11.6}$$

In many cases the events A and B are each described by a stochastic variable respectively Z_1 and Z_2. Event A will occur when $Z_1 < 0$ and event B will occur when $Z_2 < 0$. In Section 11.6 this will be further explained. In many cases, for instance when there is a linear relation between Z_1 and Z_2, the dependency of the events A and B can be described by the correlation coefficient. This is defined by:

$$\rho = \frac{\mathrm{Cov}(Z_1 Z_2)}{\sigma_{Z_2}\sigma_{Z_2}}, \tag{11.7}$$

in which σ_{Z_1} is the standard deviation of Z_1; σ_{Z_2} is the standard deviation of $Z_{2;}$ $\mathrm{Cov}(Z_1 Z_2)$ is the covariance of Z_1 and $Z_{2;}$ $= E((Z_1 - \mu_{Z_1})(Z_2 - \mu_{Z_2}))$ = expected value of $(Z_1 - \mu_{Z_1})(Z_2 - \mu_{Z_2})$; $f_{Z_1}(\xi_1)$ is the probability density function of Z_1.

In the graph of Figure 11.13, the probability of A or B is plotted against the correlation coefficient. The probability of B is the lower limit of the failure probability and the sum of the probability of A and the probability of B is the upper limit of the failure probability. It can be seen that as long as the correlation coefficient is smaller than 90%, the failure probability is close to the upper limit.

In the case of a series system with a large number of elements, the lower and upper bounds are:

- the maximum probability of the failure of a single element;
- and, the sum of the failure probabilities of all the elements.

$$\max(\mathrm{P}(i)) \le \mathrm{P_f} \le \sum_{i=1}^{n} \mathrm{P}(i). \tag{11.8}$$

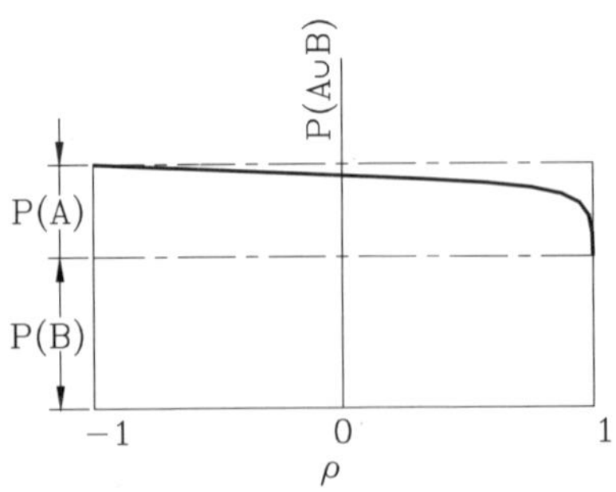

Figure 11.13. Probability of A or B given ρ.

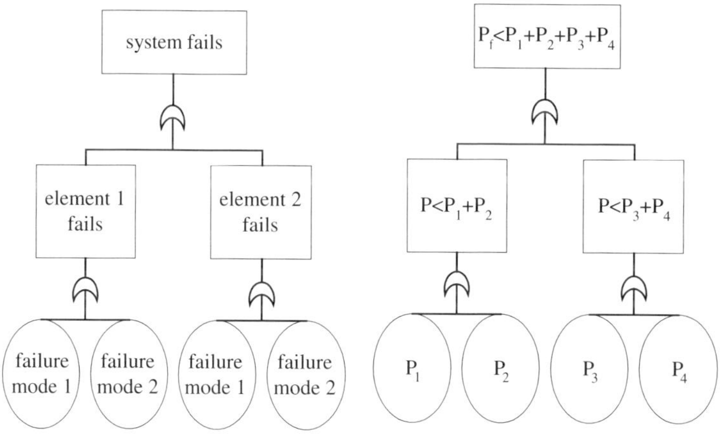

Figure 11.14. Probability of failure of a series system.

Ditlevsen (1979) has narrowed these boundaries to get a better estimation of the failure probability.

$$
\begin{gathered}
\mathrm{P}(1)+\sum_{i=2}^{n} \max \left(\mathrm{P}(i)-\sum_{j=1}^{i-1} \mathrm{P}(i \cap j)\right) \leq \mathrm{P}_{\mathrm{f}} \\
\mathrm{P}_{\mathrm{f}} \leq \sum_{i=1}^{n} \mathrm{P}(i)-\sum_{i=2}^{n} \max_{j<i}(\mathrm{P}(i \cap j)).
\end{gathered}
\tag{11.9}
$$

Let us now look at a series system that consists of two elements each having two failure modes (Figure 11.14). If the probabilities of the potential failure modes are known, it is possible to determine the upper limit of the failure probabilities of the element as the sum of the probabilities of the two different failure modes. The upper limit of the probability of failure of the system can be determined as the sum of the upper limits of the failure probability of the two elements. So the upper limit of the failure probability is the sum of the probability of all the failure modes.

11.6. Estimation of the Probability of a Failure Mode of an Element

After analysing the failure probability of the system as a function of the probabilities of the failure modes we need to know the probabilities of failure modes to estimate the failure probability. These probabilities

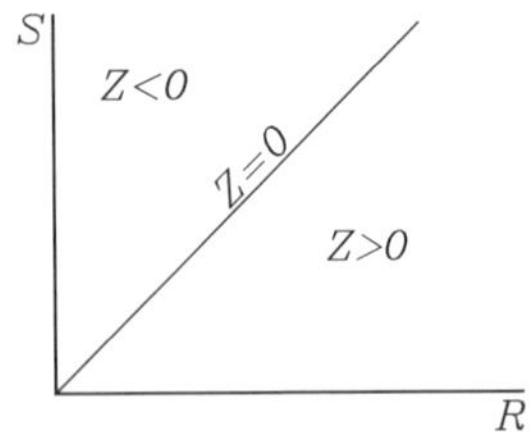

Figure 11.15. Reliability function.

can be determined by analysing historical failure data or by probabilistic calculation of the limit states.

For most cases there is not enough specific failure data available so we have to determine the failure probabilities by computation. The probabilistic computation uses the reliability function and the probability density function of the variables as the basis for the determination of the failure probability. A reliability function is a function of the strength and the load for a particular failure mode. In general, the formulation of the reliability function is: Z = R-S, in which R is the strength and S is the load. The failure mode will not occur as long as the reliability function is positive. The graph of Figure 11.15 shows the reliability function. The line Z = 0 is a limit state. This line represents all the combinations of values of the strength and the loading for which the failure mode will just not occur. So it is a boundary between functioning and failure. In the reliability function the strength and load variables are assumed to be stochastic variables.

If the distribution and the density of all the strength and load variables are known it is possible to estimate the probability that the load has a value x and that the strength has a value less than x:

$$\left.\begin{array}{l}\mathrm{P}(S=x)=\mathrm{f}_S(x)\mathrm{d}x\\ \mathrm{P}(R\leq x)=\mathrm{F}_R(x)\end{array}\right\}\Rightarrow \mathrm{P}(S=x\cap R\leq x)=\mathrm{f}_S(x)\mathrm{F}_R(x)\mathrm{d}x. \qquad (11.10)$$

The failure probability is the probability that $S = x$ and $R < x$ for every value of x. So we have to compute the sum of the probabilities for all possible values of x:

$$\mathrm{P_f} = \int_{-\infty}^{\infty} \mathrm{f}_S(x)\mathrm{F}_R(x)\mathrm{d}x. \qquad (11.11)$$

This method can be applied when the strength and the load are independent of each other.

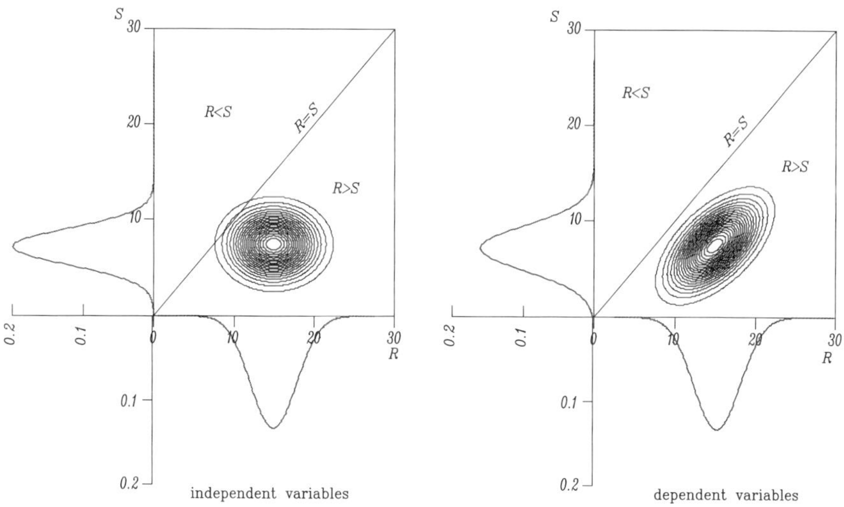

Figure 11.16. Joint probability density function.

Figure 11.16 gives the joint probability density of the strength and the load for a certain failure mode in which the strength and the load are not independent. The strength is plotted on the horizontal axis and the load is plotted on the vertical axis. The contours give the combinations of the strength and the load with the same probability density. In the area (Z < 0) the value of the reliability function is less than zero and the element will fail.

In a real case the strength and the load in the reliability function are nearly always functions of multiple variables. For instance, the load can consist of the water level and the significant wave height. In this case the failure probability is less simple to evaluate. Nevertheless, with numerical methods like numerical integration and Monte Carlo simulation it is possible to solve the integral

$$P_f = \iiint_{Z<0} \mathrm{f}_{\mathrm{r}_1,\mathrm{r}_2,\ldots,\mathrm{r}_n,\mathrm{s}_1,\mathrm{s}_2,\ldots,\mathrm{s}_m}(r_1, r_2, \ldots, r_n, s_1, s_2, \ldots, s_m) \times dr_1 dr_2 \ldots r_n ds_1 ds_2 \ldots ds_m. \quad (11.12)$$

These methods which take into account the real distribution of the variables are called level III probabilistic methods (EN, 1991). In the Monte Carlo simulation method, a large sample of values of the basic variables is generated and the number of failures is counted. The number of failures equals:

$$N_f = \sum_{j=1}^{N} 1(g(\mathbf{x}_j)), \tag{11.13}$$

in which N is the total number of simulations. The probability of failure can be estimated by:

$$P_f \approx \frac{N_f}{N}. \tag{11.14}$$

The coefficient of variation of the failure probability can be estimated by:

$$V_{P_f} \approx \frac{1}{\sqrt{P_f N}}, \tag{11.15}$$

in which P_f denotes the estimated failure probability.

The accuracy of the method depends on the number of simulations. The relative error made in the simulation can be written as:

$$\varepsilon = \frac{\frac{N_f}{N} - P_f}{P_f}. \tag{11.16}$$

The expected value of the error is zero. The standard deviation is given as:

$$\sigma_\varepsilon = \sqrt{\frac{1-P_f}{NP_f}}. \tag{11.17}$$

For a large number of simulations, the error is normal distributed. Therefore, the probability that the relative error is smaller than a certain value E can be written as:

$$P(\varepsilon < E) = \Phi\left(\frac{E}{\sigma_\varepsilon}\right), \tag{11.18}$$

$$N > \frac{k^2}{E^2}\left(\frac{1}{P_f} - 1\right). \tag{11.19}$$

The probability of the relative error E being smaller than $k\sigma_\varepsilon$ now equals $\Phi(k)$. Requiring a relative error of $E = 0.1$ lying within the 95% confidence interval ($k = 1.96$) results in:

$$N > 400\left(\frac{1}{P_f} - 1\right). \tag{11.20}$$

The equation shows that the required number of simulations and thus the calculation time depend on the probability of failure to be calculated.

Most structures in coastal and river engineering possess a relatively high probability of failure (i.e. a relatively low reliability) compared to structural elements/systems, resulting in reasonable calculation times for Monte Carlo simulation. The calculation time is independent of the number of basic variables and, therefore, the Monte Carlo simulation should be favoured over the Riemann method in case of a large number of basic variables (typically more than five). Furthermore, the Monte Carlo method is very robust, meaning that it is able to handle discontinuous failure spaces and reliability calculations in which more than one design point is involved.

The problem of long calculation times can be partly overcome by applying importance sampling. This is not elaborated upon here. Reference is made to Bucher, 1987.

If the reliability function (Z) is a sum of a number of normal distributed variables, then Z is also a normal distributed variable. The mean value and the standard deviation can easily be computed with these equations:

$$Z = \sum_{i=1}^{n} a_i x_i, \quad \mu_Z = \sum_{i=1}^{n} a_i \mu_{x_i}, \quad \sigma_Z = \sqrt{\sum_{i=1}^{n} (a_i \sigma_{x_i})^2}. \tag{11.21}$$

This is the base of the level II probabilistic calculation. The level II methods approximate the distributions of the variables with normal distributions and they estimate the reliability function with a linear first-order Taylor polynomial, so that the Z-function is normal distributed.

If the distribution of the Z-function is normal, and the mean value and the standard deviation are known, it is rather easy to determine the failure probability:

$$\begin{aligned} \mathrm{P_f} &= \mathrm{P}(Z < 0) = \Phi(-\beta) \\ \text{with}: \quad \beta &= \frac{\mu_Z}{\sigma_Z}. \end{aligned} \tag{11.22}$$

11.6.1. *Non-linear Z-function*

In case of a non-linear Z-function, it will usually be approximated by a Taylor polynomial. The function will then depend on the point around which it is linearised. The mean value and the standard deviation of the linear Z-function can be calculated analytically using the low-order terms of the Taylor expansion. If the reliability function is estimated by a linearised Z-function at the point where all of the variables have their mean value ($x_i^* = \mu_{x_i}$), then we speak of a mean value approach. The so-called design

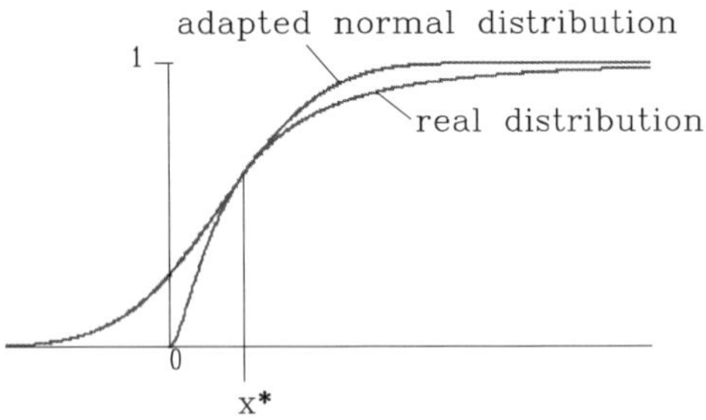

Figure 11.17. Adapted normal distribution.

point approach estimates the reliability function by a linear function at the point on Z = 0 where the value β has its minimum. Finding the design point is a minimisation problem.

11.6.2. *Non-normally distributed basic variables*

If the basic variables of the Z-function are not normally distributed the Z-function will be unknown and probably non-normally distributed. To cope with this problem the non-normally distributed basic variables in the Z-function can be replaced by a normally distributed variable. In the design point the adapted normal distribution must have the same value as the real distribution. As the normal distribution has two parameters (μ and σ) one condition is not enough to find the right normal distribution. Therefore, the value of the adapted normal probability density function must also have the same value as the real probability density function (see Figure 11.17).

The two conditions give a set of two equations with two unknown which can be solved:

$$\left.\begin{aligned} \mathrm{F_N}(x^*) &= \mathrm{F_x}(x^*) \\ \mathrm{f_N}(x^*) &= \mathrm{f_x}(x^*) \end{aligned}\right\} \Rightarrow \mu_N, \sigma_N. \tag{11.22}$$

This method is known as the Approximate Full Distribution Approach (AFDA).

11.7. Choice of Safety Level

To construct a flood defence that always performs its function and is perfectly safe from collapse is at the very least an uneconomical pursuit, and most likely an impossible task too. Although expertly designed and well constructed, there will always be a small possibility that the structure fails under severe circumstances (ultimate limit state). The acceptable probability of failure is a question of socio-economic reasoning.

In a design procedure one has to determine the preferred level of safety (i.e. the acceptable failure probability). For most civil engineering structures the acceptable failure probability will be based on considerations of the probability of loss of life due to failure of the structure. In general, two points of view for the acceptable safety level can be defined (Vrijling *et al.*, 1995):

The individual accepted risk. The probability accepted by an individual person to die in case of failure of the structure; in western countries this probability is of the order 10^{-4} per year or smaller.

The societal accepted risk. Two approaches are presented, depending on the relative importance of the total number of lives lost in case of failure on the one hand and the total economic damage on the other. If the number of potential casualties is large, the likelihood of failure should be limited accordingly. The accepted probability of occurrence of a certain number of casualties in case of failure of a structure is then restricted proportional to the inverse of the square of this number (Vrijling *et al.*, 1995). If the economic damage is large, an economic optimisation equating the marginal investment in the structure with the marginal reduction in risk should be carried out to find the optimal dimensions of the structure.

The two boundary conditions based on the loss of human lives form the upper limits for the acceptable probability of failure of any structure. In case of a flood defence without amenities the probability of loss of life in case of failure is very small. In that case the acceptable probability of failure can be determined by economical optimisation, weighing the expected value of the capitalised damage in the life of the structure (risk) against the investment in the flood defence. The next section provides more background on this concept.

If, for a specific flood defence structure, failure would include a number of casualties, the economic optimisation should be performed under the constraint of the maximum allowable probability of failure as defined by the two criteria related to loss of life.

The explicit assessment of the acceptable probability of failure as sketched above is only warranted in case of large projects with sufficient means. For smaller projects a second approach is generally advised. This second approach to the acceptable safety level is based on the evaluation of the safety of existing structures supplemented by considerations of the extent of the losses involved in case of failure. Consequently, the assumption is made that the new structure should meet the safety requirements that seem to be reasonable in practice. This approach is found in many codes where a classification of the losses in case of failure leads to an acceptable probability of failure. Most structural codes provide safety classes for structures (Eurocode 1, EN, 1991). The structure to be designed might fit into one of these safety classes providing an acceptable probability of failure. It should be noted, however, that for most structural systems loss of life is involved. The following classification and table with acceptable probabilities of failure was developed in the EU PROVERBS project, especially for vertical breakwaters (Oumeraci, 2001):

- *Very low safety class*, where failure implies no risk to human injury and very small environmental and economic consequences.
- *Low safety class*, where failure implies no risk to human injury and some environmental and economic consequences.
- *Normal safety class*, where failure implies risk to human injury and significant environmental pollution or high economic or political consequences.
- *High safety class*, where failure implies risk to human injury and extensive environmental pollution or very high economic or political consequences.

Limit state design requires the structure to satisfy two principal criteria: the Ultimate Limit State (ULS) and the Serviceability Limit State (SLS). To satisfy the ULS, the structure must not collapse when subjected to the peak design load for which it was designed. To satisfy the SLS criteria, a structure must remain functional for its intended use subject to routine (everyday) loading, and as such the structure must not cause occupant discomfort under routine conditions.

11.8. Reliability-Based Design Procedures

11.8.1. *General formulation of reliability-based optimal design*

Generally, in a design process one pursues the cheapest design that fulfils the demands defined for the structure. The demands can be expressed in two fundamentally different ways:

- The total expected lifetime costs of the structure consisting of the investment and the expected value of the damage costs are minimised as a function of the design variables;

- if a partial safety factor system is available, one can optimise the design by minimising the construction costs as a function of the design variables under the constraint that the design equations related to the limit state equations for all the failure modes are positive.

The minimisation of the lifetime costs can be formalised as follows:

$$
\begin{aligned}
&\min_{\mathbf{z}} C_T(\mathbf{z}) = C_I(\mathbf{z}) + C_{F;ULS}P_{F;ULS}(\mathbf{z}) + C_{F;SLS}P_{F;SLS}(\mathbf{z}) \\
&\text{s.t} \quad z_i^L \leq z_i \leq z_i^U \qquad\qquad i = 1, \ldots, m \\
&\qquad P_{F;ULS}(\mathbf{z}) \leq P_{F;ULS}^U \\
&\qquad P_{F;SLS}(\mathbf{z}) \leq P_{F;SLS}^U,
\end{aligned}
\tag{11.23}
$$

in which: $\mathbf{z} = (z_1, z_2, \ldots, z_m)$ is the vector of design variables; $C_T(\mathbf{z})$ is the total lifetime costs of the structure; $C_I(\mathbf{z})$ is the investment in the structure as a function of the design variables $\mathbf{z}$; $C_{F;ULS}$ is the damage in monetary terms in case of ULS failure; $C_{F;SLS}$ is the damage in monetary terms in case of SLS failure; $P_{F;ULS}(\mathbf{z})$ is the probability of ULS failure as a function of the design variables; $P_{F;SLS}(\mathbf{z})$ is the probability of SLS failure as a function of the design variables; z_i^L, z_i^U is the lower and upper bound of design variable i; $P_{F;ULS}^U, P_{F;SLS}^U$ is the upper bound of the failure probability for ULS failure and SLS failure respectively.

Generally the design variables will be subjected to constraints. For instance, all geometrical quantities should be greater than zero. Furthermore, the failure probabilities can be subject to constraints, especially for structures where human lives are involved. In that case, the maximum failure probabilities are enforced by regulations. In cases that loss of human lives is not involved in case of failure of the structure, formally the constraint on the failure probabilities can be set to 1 and the acceptable failure probability as well as the optimal design are completely decided by the lifetime costs only. If relevant, maintenance costs and inspection costs can be added to the total expected lifetime costs.

Obtaining accurate assessments of the damage in case of failure is not always practically possible. In that case, the optimal design can be found by minimising a cost function which only comprises of the investment and imposing a constraint on the failure probability which expresses a qualitative idea of the economic optimal failure probability.

If the design is performed using a code based on a partial safety factor system, the following optimisation problem is applicable:

$$\min_{\mathbf{z}} C_T(\mathbf{z}) = C_I(\mathbf{z})$$
$$\text{s.t} \quad z_i^L \leq z_i \leq z_i^U,$$
$$G_i(\mathbf{z}, \mathbf{x}^{\mathbf{C}}, \gamma) > 0 \qquad (11.24)$$

in which: $C_I(\mathbf{z})$ is the investment in the structure as a function of the design variables $\mathbf{z}$; $G_I(\mathbf{Z}, \mathbf{X}^{\mathbf{C}}, \gamma)$ is the limit state function for failure mode i as a function of the design variables $\mathbf{z}$, the characteristic values of the random variables as defined in the partial safety factor system $\mathbf{x}^{\mathbf{c}}$ and the vector of partial safety factors $\boldsymbol{\gamma}$.

Generally, partial safety factors are available for several target probabilities of failure or safety classes (see Table 11.1). Since the choice of the safety factors involves implicitly the choice of a target probability of failure and expected costs of failure, the same optimal design should be obtained from (11.23) and (11.24).

Table 11.1. Overview of safety classes.

	Safety class			
Limit state type	Low	Normal	High	Very high
SLS	0.4	0.2	0.1	0.05
ULS	0.2	0.1	0.05	0.01

11.8.2. *Cost optimisation*

If loss of life in case of failure of the structure is not an issue for the structure under consideration, no constraint is set on the failure probability and the acceptable probability of failure equals the economic optimal probability of failure. A procedure for probabilistic optimisation of vertical flood defences has been developed in the PROVERBS project (Oumeraci, 2001).

The optimisation can be written as:

$$\min_{\mathbf{z}} C(\mathbf{z}) = C_{I;0} + C_I(\mathbf{z})$$
$$+ \sum_{n=1}^{N} \left(\frac{365 C_{F;SLS} P_{F;SLS}(\mathbf{z}) + C_{F;ULS} P_{F;ULS}(\mathbf{z})}{(1 + r' - g)^n} + \frac{C_{maint}}{(1 + r')^n} \right)$$
$$\text{s.t} \quad 0 \leq z_i, \qquad (11.25)$$

in which: $\mathbf{z}$ is the vector of design variables; $C_{I;}$: is initial costs, not depending on the design variables; $C_I(\mathbf{z})$ is construction costs as a function of the design variables; $C_{F;SLS}$ is costs-per-day in case of serviceability failure; $P_{F;SLS}(\mathbf{z})$ is the probability of serviceability failure per day; $C_{F;ULS}$ is costs-per-event in case of ultimate limit state failure; $P_{F;ULS}(\mathbf{z})$ is the probability of ultimate limit state failure per year; C_{maint} is maintenance costs for the flood defence per year; r' is the net interest rate per year; g: is the yearly rate of economical growth, expressing growth and development of the harbour; N: is the lifetime of the structure in years.

An inspection of Equation (11.25) shows that the total lifetime costs consist of investment costs and the expected value of the damage costs. In principle, for every year of the structure's lifetime, the expected damage has to be taken into account, and not all the costs are made at the same time. Therefore, the influence of interest, inflation and economical growth has to be taken into account in order to make a fair comparison of the different costs.

The expected value of the damage costs is a function of the failure probability. The failure probability is a function of the design variables. Therefore, minimisation of Equation (11.25) results in the optimal geometry and at the same time the optimal failure probability. Ready-at-hand minimisation algorithms can be applied to find the optimal set of design variables.

When implementing the cost function in any programming language, the failure probability as a function of the design variables has to be included. Due to the specific character of the optimisation process, the choice of the probabilistic procedure is not an arbitrary one. One should be aware of the following points:

- The minimisation process comprises a large number of evaluations of the cost function, each evaluation involving a probability calculation. Therefore, time-consuming methods should be avoided;
- The values of the cost function for any given point should be stable. The Monte Carlo procedure in particular provides probability estimates that contain an error, which is inherent to the procedure. This (small) error generally presents no problem, but in this case it causes variations of the cost function that disturb the optimisation process (see Figure 11.18).

The points of attention mentioned above leads to the conclusion that level II methods are suitable for application in an optimisation process.

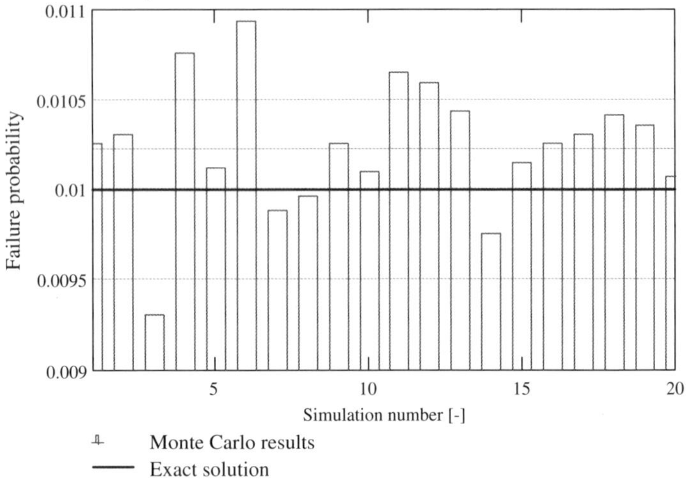

Figure 11.18. Result of 20 calculations of failure probability by Monte Carlo.

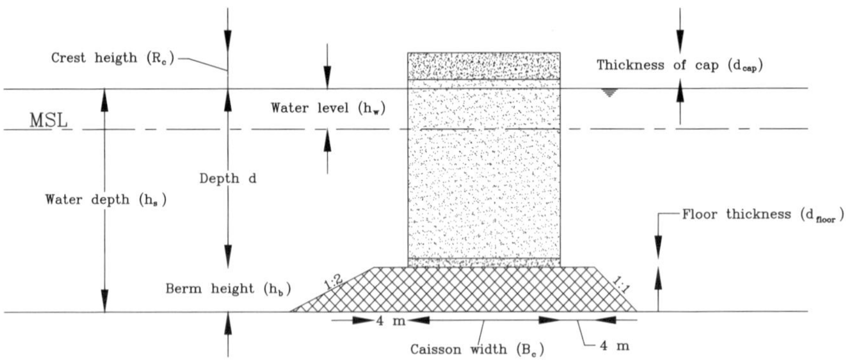

Figure 11.19. Overview of conceptual breakwater design for economic optimisation.

Level III methods will generally lead to too much computational effort or will disturb the optimisation process.

The procedure described above has been applied to a fictitious design case of a vertical breakwater in a water depth of 25 m. Three design variables are considered: the height and width of the caisson and the height of the rubble berm (see Figure 11.19).

As a first step, a deterministic optimisation for chosen wave heights was performed. This step is meant to show the connection between the deterministic optimisation for a given safety level and the full probabilistic approach. As a result of this, the choice of the input values comparable

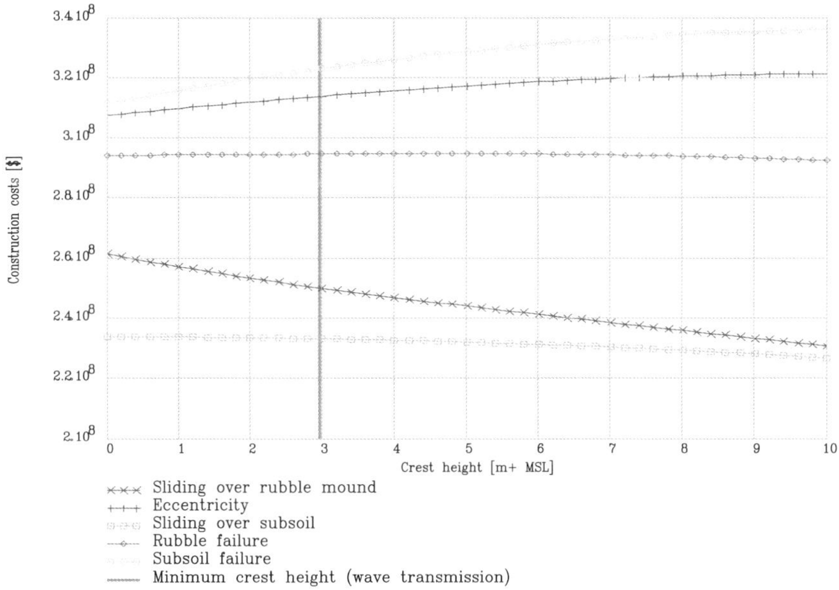

Figure 11.20. Construction costs of the breakwater as a function of the crest height (berm height 6 m).

to the characteristic values in Equation (11.25) does not correspond to the choice made for the partial safety factor system. Furthermore, all safety factors have been set to 1 and the berm height is fixed at a value of 6 m. For this situation it is possible to find a minimum caisson width as a function of the crest height for every single failure mode. Once the crest height and the caisson width are known, the construction costs of the caisson structure can be calculated (see Figure 11.20).

Generally speaking, bearing capacity failure of the subsoil controls the largest minimum caisson width. Furthermore, the results show that in general a lower crest height leads to a narrower caisson and, thus, to lower construction costs. However, the minimum crest height required is determined by wave transmission. In the deterministic approach the minimum crest height related to wave transmission imposes a constraint on the crest height. Thus, the optimal geometry is decided by wave transmission and by the bearing capacity failure of the subsoil.

While at first sight it seems reasonable to have an equal probability of failure for all the failure modes in the system, probabilistic optimisation shows that, as with the deterministic approach, bearing capacity failure of

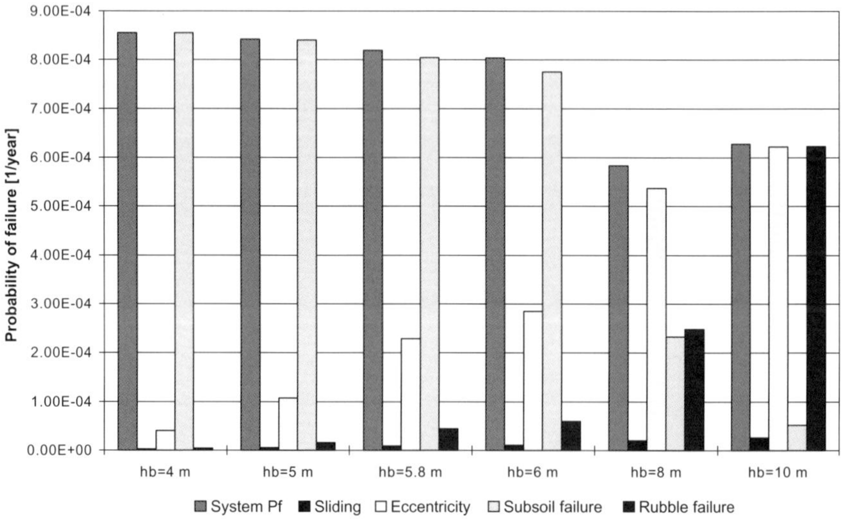

Figure 11.21. Overview of ULS failures probabilities for several berm heights.

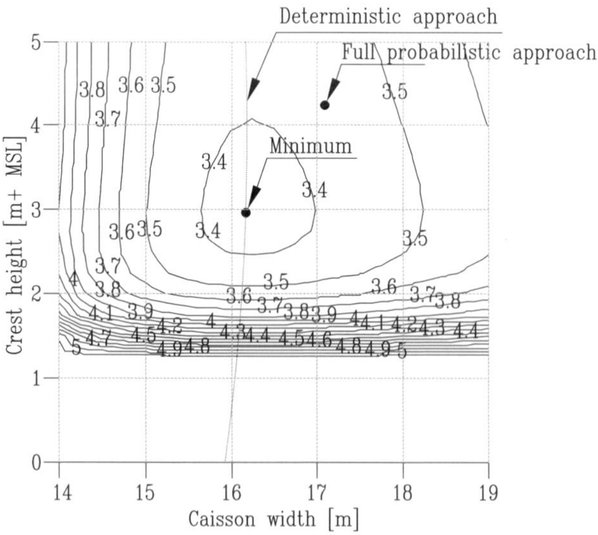

Figure 11.22. Contour plot of total lifetime costs in 10^8 US \$ (random wave height only) and optimal geometries for different levels of modelling. From Voortman *et al.* (1999).

the subsoil largely determines the probability of ultimate limit state failure (see Figure 11.21).

Inspection of the lifetime costs as a function of crest height and caisson width indicates that also in the probabilistic approach the crest height is limited by wave transmission (see Figure 11.22).

The optimal probability of system failure is quite low in comparison to existing structures ($8 \cdot 10^{-4}$). This could be caused by the choice of the cost figures or by a limited spreading of the random variables.

11.9. Concluding Remarks

This chapter has presented the probabilistic approach of the design and the risk analysis of flood defences, in which uncertainties in load, resistance variables and models can be taken into account. A systems point of view has been adopted. The application of the probabilistic design methods offers the designer a way to unify the design of engineering structures, processes and management systems. A case study on the reliability analysis and optimal design of a caisson breakwater has illustrated the presented methods.

References

Bucher, G.C. (1987). Adaptive sampling, an iterative fast Monte Carlo procedure, internal report, University of Innsbruck, Innsbruck, Austria.

Ditlevsen, O. (1979). Narrow reliability bounds for structural systems, *J. Struct. Mech.*, **7**, 453–472.

EN (1991). *Eurocode 1: Actions on structures.* Institution of Structural Engineers: London. Available at http://www.eurocode1.com/en/eurocode1.html/ (Accessed on 20/04/2012).

Oumeraci, H., Kortenhaus, A., Allsop, N.W.H. *et al.* (2001). *Probabilistic Design Tools for Vertical Breakwaters*, Balkerna, Rotterdam.

Thoft-Christensen, P. and Baker, M.J. (1982). *Structural Reliability Theory and its Applications*, Springer, New York.

van Gelder, P.H.A.J.M. and Vrijling, J.K. (1997). Risk-averse reliability-based optimization of sea-defences, *Risk Based Decis. Making Water Resour.*, **8**, 61–76.

Voortman, H.G., van Gelder, P.H.A.J.M. and Vrijling, J.K. (1999). "Reliability Analysis of the Dynamic Behaviour of Vertical Flood Defences" in: Schueller, G.I. and Kafka, P. (eds), *Safety and Reliability: Proceedings of ESREL '99, 10th European Conference*, Munich-Garching, September 13–17, 1999.

Vrijling, J.K. (1986). "Probabilistic Design of Water Retaining Structures" in: Duckstein, L. and Plate, E.J. (eds), *Engineering Reliability and Risk in Water Resources*, Springer, Dordrecht.

Vrijling, J.K. and van Gelder, P.H.A.J.M. (1998). "The Effect of Inherent Uncertainty in Time and Space on the Reliability of Flood Protection" in: Lyderson, S., Hansen, G.K. and Sandtorv, H.A. (eds), *Safety and Reliability: Proceedings of ESREL '98: European Safety and Reliability Conference 1998*, Trondheim, Norway, 16–19 June, 1998. pp. 451–456.

Vrijling, J.K., van Hengel, W. and Houben, R.J. (1995). A framework for risk evaluation, *J. Hazard. Mater.*, **43**, 245–261.

CHAPTER 12

Uncertainties in Flood Modelling in Urban Areas

Slobodan Djordjević
Centre for Water Systems, University of Exeter, UK

Zoran Vojinović
UNESCO-IHE, Netherlands

Richard Dawson
School of Civil Engineering and Geosciences, Newcastle University, UK

Dragan A. Savić
Centre for Water Systems, University of Exeter, UK

12.1. Challenges of Uncertainty Analysis in Urban Flood Models

In order to predict consequences of *pluvial* flooding, urban flood models need to simulate flow in two different systems — the buried pipe network and the urban surface, including complex interactions between these two flow domains. Thus, coupled sub-surface/surface urban flood models are sometimes referred to as *dual drainage* models and the two sub-systems are often called the *minor system* (constructed pipes and open channels) and the *major system* (natural channels and surface flow pathways). The latter can be modelled either as a network of 1D (one-dimensional) open channels and ponds or as a 2D computational flow domain. Hence, we talk about 1D/1D or 1D/2D urban flood models (Mark *et al.*, 2004).

Moreover, when combined *pluvial/fluvial* or *pluvial/fluvial/coastal* flooding is the problem at hand, urban flood models need to take river flow or the sea water level variations into account either as a downstream boundary condition or as a third sub-system that should be simulated simultaneously with the two dual drainage sub-systems. Whilst the modelling of flows in storm sewer networks and in rivers has been well developed over the past few decades and, consequently, the modelling tools have almost become standardised, advancements are still being made in modelling of flow on urban surfaces and, in particular, how the flows below and above ground are coupled and how they interact with each other. Uncertainties related to these interactions have not yet been studied in detail.

On the other hand, the availability of exceptionally detailed and accurate terrain and land use data in urban surfaces has risen dramatically in the past few years, with horizontal resolution and vertical accuracy becoming very high indeed and related costs going down. In addition, the quality of rainfall data has also increased significantly following more widespread application of weather radars, online rain gauges and advances in rainfall forecasting. These developments have influenced not only the approaches to hydraulic modelling of urban floods, but, going one step back, they have opened new possibilities for more reliable prediction of surface run-off hydrographs in urban areas. Ultimately, the split between the hydrological and hydraulic phases of urban flood modelling — whereby the former only produces input to the latter — will be lost and fully integrated rainfall-runoff-urban drainage-river channel-flooding models will become operational.

With this background in mind, the potential for applying uncertainty analysis in urban flood modelling depends on the level of complexity of the simulation engine used and on the size and data resolution of an urban area. These dictate how computationally demanding the simulation may become and consequently if a particular type of uncertainty analysis is feasible at the current level of technology or not.

In that sense, one approach is to conduct uncertainty analysis based on a large number of runs using a series of storms, with or without varying catchment and drainage system parameters. This can be efficient with state-of-the-art coupled models only by assuming a rather coarse spatial resolution, i.e. with a simplified pipe network and/or very low resolution of the surface flow domain *or* with approximate urban flood models in which the computational treatment of flow either in the minor or in the major system is significantly simplified or not even considered.

The alternative approach is to run simulations for only a limited number of rainfall events using high resolution, higher dimensionality full-dynamic coupled models. This implies limitations in terms of what kind of analysis can be done because formal probability theory is out of the question here and the application of formal statistical methods is rather limited. However, with this approach even some simple forms of sensitivity analysis may enable issues specific for urban flood risk assessment to be highlighted.

In effect, the biggest challenge in uncertainty analysis in urban flood models is the application of any formal method of analysis given how computationally demanding the coupled full-dynamic models are. It might be expected that the computational expense will gradually become less of an obstacle with the ever increasing computer speed, particularly with progress in the domain of parallel computing.

This chapter focuses on uncertainties and flood risk assessment and management methods specific for *urban flood* modelling from the perspective of an urban drainage engineer. The emphasis is intentionally placed on uncertainty analysis related to the minor system and its interactions with the major system, which are either not present or not as relevant in other types of flood models whose uncertainty issues are extensively analysed elsewhere in this book. For the same reasons, attention is paid to uncertainties linked to 1D modelling, whilst some 2D modelling issues are considered. Detailed analysis of problems of 2D approaches in urban areas is beyond the scope of this chapter.

Following this introduction, Section 12.2 describes sources of uncertainty in urban flood models by identifying them in data, mathematical models, numerical methods and calibration procedures. Section 12.3 looks at flood risk analysis through different mapping methods. A method for calculating urban flood risk and attributing risk between different organisations responsible for urban flood risk management is illustrated in Section 12.4. Selected results and key lessons from case studies the authors have been involved in are given in Section 12.5, which points to references where readers may find more substantial descriptions of the studies.

12.2. Sources of Uncertainty of Urban Flood Models

General issues related to sources of uncertainty in flood models have been reviewed in Chapter 3. Sources of uncertainty of *urban* flood models can be classified as follows:

1. Different issues related to various domains of data.
2. Equations for conservation laws and other relationships describing water flow in dual drainage systems.

3. Numerical solutions and specifics of their implementation.
4. Calibration procedures.

12.2.1. *Terrain, land use and sewer system data*

Urban terrain data is nowadays routinely available in 1 m resolution by Light Detection and Ranging (LiDAR) and increasingly so in higher resolution, even up to 0.25 m. More recently mobile mapping technologies are enabling an even higher resolution and offering the potential to measure road camber and kerb heights. The first step in processing raw LiDAR data is distinguishing between buildings and trees, hence filtering out of trees and other surface features (e.g. cars) that do not represent urban terrain is the first potential source of uncertainty. Whilst a very high resolution of LiDAR data is good news when it comes to reduction of related uncertainties, it adds further computational overheads.

In 1D/1D urban flood modelling — as well as in urban drainage modelling without consideration of flooding — using of Digital Elevation Model (DEM) data for automatic sub-catchment delineation is essential because it reduces uncertainties caused by manual or distance-based delineation that do not take terrain slopes and surface features into account. Furthermore, remotely sensed data, including photogrammetry, can be used to automatically parameterise key model variables such as roofed and other impervious areas, their connection type, population equivalent, surface roughness, etc. Runoff can be estimated by merging information from imaging and DEM information. The uncertainty associated with land use classification is a function of the quality of the imaging data, the accuracy of the algorithms used to interpret this data and the coarseness of the classification. This can lead to a range of classifications resolutions:

1. A limited classification of a handful of land use types, where "urban" may encompass a range of building densities and surface types.
2. Distinguishing key features such as streets, car parks, housing, green areas, etc. as *surface types* and assigning assumed imperviousness and roughness accordingly.
3. A high level of detail in which all surface features including individual buildings are described by corresponding imperviousness.

Even the highest level of detail in land use description may be a source of uncertainty if connections between impervious areas and the drainage system are not known. There will be differences in flood *volumes* depending on if the runoff from roofs is spilled to a pervious area — in which case part of it would be infiltrated — or not. In addition, there are differences in urban flood *dynamics* depending on if the runoff from roofs is introduced to sewer pipes directly or if it is spilled onto an impervious area (e.g. pavement) before reaching the sewer system.

Another potential source of uncertainty related to land use data is urban creep. Whilst *external* urban growth is usually noticeable and — at least in principle — easy to record, paving of small gardens is difficult to monitor but it can cumulatively have huge impact on surface run-off. Therefore urban flood models are in a constant need of *re-calibration* in order to keep land use data up to date.

The uncertainty about pipe diameters and elevations is always present, either due to incomplete or inaccurate database or because the effective pipe diameters are reduced due to sedimentation. This uncertainty can be kept low if CCTV inspections are conducted regularly and any siltation is taken into account.

The structural condition of sewer pipes gradually deteriorates in time. However, it is subject to improvement following interventions that affect pipe roughness and, consequently, the capacity of the drainage system. Related uncertainty may be reduced if calibrated models are updated regularly based on maintenance records and using built-in pipe deterioration models that gradually slide pipe roughness values in time.

Flap-valves at the outfalls may influence the flood dynamics. Although the structure of flap-valve data is very simple — existent or non-existent — if this data is unknown it may be an additional source of uncertainty.

Where pumping stations are part of flood management strategy, operational rules and the effectiveness of their application can introduce uncertainties. Like pipes, they degrade with time and may lose some of their capacity. This can be reduced by updating pump curves based on the data from literature.

Specific to 1D/1D modelling are the uncertainties that arise from the creation of surface network of open channels and ponds. *Manual* creation of the 1D surface network (usually based on street profiles) is subjective. Uncertainties related to *automatic* creation of the surface flow network (Maksimović *et al.*, 2009) are present in every step of this process:

1. Definition of flood pathways in the plan view by the bouncing ball technique is sensitive to the choice of snap distance.
2. Automatic identification of bridges as passable features (Evans, 2008) — as opposed to obstructions in the LiDAR DEM data — is possible;

however, the uncertainty in the connectivity of surface flow paths may be introduced by routes not easily detectable from DEM data such as covered pathways between buildings.
3. Computation of paths' cross-sections by intersecting DEM with vertical planes locally perpendicular to a path involves uncertainty related to chosen density of cross-sections and the length of path over which cross-section geometry is averaged.
4. The threshold defining minimum size of a pond — which can be set either as minimum area or minimum volume — is a matter of choice, subject to a desired resolution of the surface network.

In part due to differences in measurement methods and because a Digital Terrain Model (DTM) elevation is an average representative value for a grid cell, the elevations of manhole covers and local terrain may differ (Figure 12.1). The former is used to identify the elevation at which floodwater surcharges from the urban drainage system whilst the latter is used in surface flow modelling. The manner in which these datasets are integrated into the model may bring in uncertainty in the simulation of sub-surface/surface interactions.

Two-dimensional flood models that work on a *regular grid* take DTM data directly from LiDAR. This data may be a source of uncertainty, especially when a large area needs to be modelled, in which case the original set of data cannot be used without the reduction of resolution and

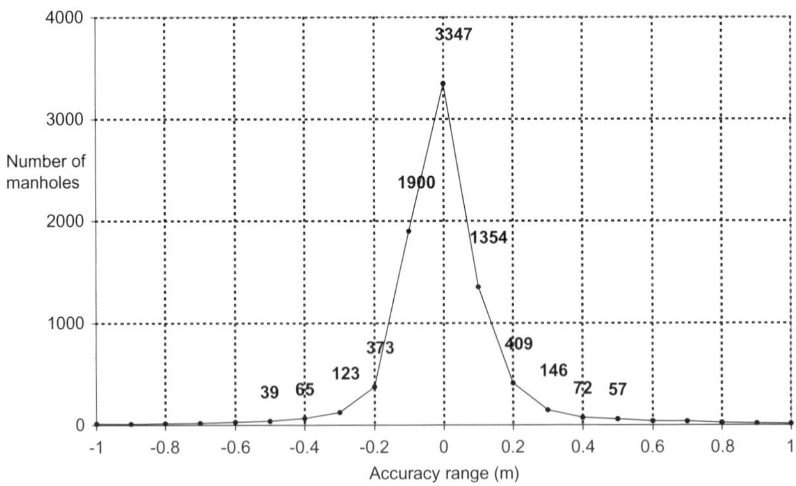

Figure 12.1. Distribution of manhole cover levels compared with DTM (after Adams and Allitt, 2006).

subsequent averaging (generalising), because not only the computational speed but also computer memory may become a constraint. The uncertainty in 2D urban flood modelling with *low grid resolution* can be reduced in different ways, e.g. by introducing a "multi-layered" flow simulation based on information about buildings' coverage, location and direction within each low resolution grid cell (Chen *et al.*, 2012).

Two-dimensional flood models that work on an *unstructured mesh* are advantageous in that they are generally more computationally efficient and, in principle, they have a greater potential to capture all surface features more precisely and at a lower cost. The way in which an irregular surface computational mesh is generated may be a source of uncertainty if it is not done carefully. As outlined by Ettrich (2007), an urban surface is best triangulated using irregularly distributed data points and contours of constant heights in conjunction with data about polygons representing surface features such as curbs, house borders, etc. Ettrich used 150,000 triangles of varying size to cover a total area of 1.5 km^2, whilst a much larger number of cells would be required in order to provide a similar level of detail on this area with a regular grid, though this problem may be reduced if an *adaptive* grid is implemented.

Data about sub-surface/surface links (inlets, gullies and manholes, s-s/s links thereafter) are potentially a great source of uncertainty because this data is commonly not readily available. Even though not every single gully is to be modelled individually, detailed data is required in order to generate parameters of *equivalent elements* that represent groups of s-s/s links in the model (Leandro *et al.*, 2007). Uncertainty in manhole data is associated with the information about manhole size and shape (which both influence local energy losses), but also with the data about manhole covers — if it is sealed, loosely attached or resting. These three states indicate if the cover removal is possible (partially or completely) or not, and consequently if, during surcharging, the s-s/s link cross-section area can increase (gradually or suddenly) or not.

12.2.2. *Flow equations in dual drainage models and their parameters*

Flow in drainage pipes and in the 1D surface channels is described by the mass and momentum conservation equations (Saint Venant equations):

$$\frac{\partial h}{\partial t} + \frac{1}{B}\frac{\partial Q}{\partial x} = 0, \tag{12.1}$$

$$\frac{\partial Q}{\partial t} + \frac{\partial}{\partial x}\left(\frac{Q^2}{A}\right) + gA\left(\frac{\partial h}{\partial x} + S_f - S_0\right) = 0, \tag{12.2}$$

where h = water depth, t = time, B = water surface width, Q = discharge, x = space coordinate, A = cross-sectional area, g = gravitational constant, S_f = friction slope and S_0 = bed slope. When written in this (complete) form, these equations are also referred to as full dynamic wave model. In addition to already mentioned uncertainties in pipe/channel roughness and geometry, Saint Venant equations may be the source of modelling uncertainty if the momentum equation is reduced by dropping one or more terms. In practice this is done either to speed up the simulation or to enable stable simulation of transcritical flows. Whilst dropping the first two terms would still only introduce minor uncertainty in a number of situations (unless a rapidly varying unsteady flow is modelled), neglecting also the third term and hence reducing the Saint Venant equations to the kinematic-wave form is unacceptable. This is because sub-surface and surface systems make a multiple — horizontally and vertically — looped dual drainage network in which backwater effects must be taken into account, particularly in surcharging pipe networks. The ramifications of using simplifications of the 2D flow equations are elaborated in Chapter 10.

Flow through s-s/s links can be described by a free or submerged weir formula, orifice equation or using an experimentally determined relationship between flow rate and hydraulic head of flows below and above ground. Coefficients in these relationships are in principle different for the two flow directions — from surface to sub-surface and vice versa. In addition, the transitions between different regimes and levels of submergence are dependant on a number of factors. Finally, whilst empirical values of the capacity of some types of inlet structures are available, experimental and Computational Fluid Dynamics (CFD) investigations of outflows thorough s-s/s links are still limited. Therefore, parameters of these links may introduce great uncertainty into urban flood modelling, especially in pluvial and combined pluvial/fluvial flood events.

12.2.3. *Numerical solutions and their implementation*

Numerical methods commonly implemented in 1D pipe/channel models have been well described in the literature (e.g. Abbott and Minns, 1998). Finite difference schemes — Abbott–Ionescu method, Preissmann four-point method and Delft method being the most popular ones — may all suffer from inaccuracies resulting from numerical diffusion and numerical instabilities. The experience gained in numerous applications over the past few decades enables a choice of computational parameters that can help keep the resulting uncertainties at minimum.

Sources of uncertainties hidden in numerical solutions in coupled 1D/1D and 1D/2D urban flood models worth attention here are primarily related to simulation of pressurised flow in pipe networks. The vast majority of sewer models deal with this using the well-known Preissmann open-slot concept, in which pressurised pipe flow is simulated by solving free-surface flow equations following imaginary "opening" of the pipe by adding a vertical narrow open slot on the top of it. Although this is very practical and one of the most elegant ideas in computational hydraulics, it involves several problems.

Firstly, theoretical widths of open slots are so small that they would tend to introduce numerical instabilities, particularly during the transition from pressurised flow to free-surface flow and vice versa, when the wave celerity swiftly changes over two orders of magnitude. Therefore, slot widths used in practice are always a trade-off between accuracy and robustness. Moreover, there is an uncertainty surrounding the theoretical wave celerity corresponding to pressurised flow because its effective value is reduced by lateral connections (Watanabe and Kurihara, 1993).

Secondly, pipe area fictitiously extended by the slot is incorrect. This problem can be minimised by different tricks, some of which are discussed, e.g. by Ukon *et al.* (2008).

Thirdly, models based on the open-slot approach cannot simulate movement of trapped air.

Finally, the classical Preissmann slot cannot handle negative pressure in a pressurised pipe because it assumes one-to-one relationship between the pressure in the surcharged pipe and the corresponding water level in the slot. The solution to this problem has been suggested by Vasconcelos *et al.* (2006) who developed so called *two-component pressure approach.* By the time of the writing of this book this method has not been implemented in any of the mainstream sewer models.

Trade-off between accuracy, on the one hand, and model robustness and efficiency, on the other hand, is also present in the simulation of supercritical and transcritical flows in 1D models. The most common approach is to implement two approximations (Havnø *et al.*, 1985):

1. Gradual reduction of the inertia term in the momentum Equation (10.2) as the flow approaches the supercritical regime.
2. The use of an algorithmic structure for the boundary conditions inherent to subcritical flow to all regimes.

The former may result in an unrealistically diffused hydraulic jump (when applied at a relatively small Froude number), or it may lead to non-amplifying but persistent water level oscillations downstream from the hydraulic jump (Djordjević *et al.*, 2004). The latter approximation enables the application of the efficient general solution algorithm for finite difference problems (Friazinov, 1970) and its variations, but an inadequate representation of internal boundary conditions introduces some uncertainties.

12.2.4. *Calibration procedures in urban flood modelling*

Since urban drainage systems are designed to handle two types of flows for combined systems (waste water and storm water) and one type of flow for separate systems, the models used to replicate their functioning must also contain either two (dry and wet weather) or one type of flow. Clearly, storm water quantities are more relevant for predictions of flood extent, depths and velocities, but waste water quantities are equally (if not more) relevant for assessing *health impacts* of urban floods. The process of adjusting model parameters to replicate Dry Weather Flows (DWF) and/or Wet Weather Flows (WWF) can be defined as a DWF or WWF calibration-verification process.

Calibration of both dry weather and wet weather flows requires the catchment to be divided into monitored sub-catchments (i.e. the sub-catchments upstream of each individual flow monitor). Typically in this process, the parameters of each monitored sub-catchment are adapted to produce the best fit between observed and predicted data. At all times it is important to ensure that the parameters remain within sensible limits and do not vary dramatically within a small area without justification.

The degree of accuracy of the monitoring equipment and differences between the system geometry in the model and in reality are the factors that influence the success of model calibration efforts with respect to different criteria. Therefore, two key issues for urban drainage model calibration are *monitoring* and *model instantiation activities.*

In general it is normal that monitoring devices are used to measure the depth and velocity of flow in urban drainage networks at critical points of interest along with the rain gauges and/or weather radar. In order to better understand model calibration process and its limits, it is necessary to understand more thoroughly the monitoring (or measurement) process. This process involves a number of steps and at each step there is a possibility to incur errors, which eventually have an impact on the overall results and their uncertainty. These errors are a fundamental limit of the accuracy to which the subsequent numerical models can be calibrated.

Measurement uncertainty is usually expressed as a parameter, which is associated with the result of a measurement that defines the range of the values that could reasonably be attributed to the measured quantity, e.g. depth or velocity of flow. It is caused by the process of measurement not replicating the true value the model is attempting to fit. Similar to the urban drainage models, measurement devices also require significant calibration efforts. Therefore, the measurement uncertainty is the sum of the following uncertainties:

1. Sampling errors: an example of this is sampling at a location in the network that does not give relevant values of flow.
2. Gauge errors: gauge measurements are often biased; they give measured values consistently below or above the true value.
3. Measurement errors of model input: gauge rainfall not being measured accurately or the gauge is far from the centroid of the local catchment where flow rates are measured.
4. Measurement density: the number of monitoring points.

In terms of the model instantiation process, there are several steps involved in that, and at each step there is a possibility to introduce uncertainties. For example, precise knowledge about sub-catchment characteristics and their boundaries is hard to define. The physical characteristics of pipe network or open channel systems (effective diameters, cross-sections, slopes, ancillaries, connectivity, structural condition, etc.) are only partially known. Furthermore, instantiation of urban drainage models often require simplification of system details in order to allow numerical computations to be undertaken within practical timeframes. Any form of model simplification will cause unaccounted head losses due to omitted structures, and missing storage capacity attributed to omitted pipes will create an impediment to achieve more reliable model calibration. Errors generated in this process provide a fundamental limit to how well the model can explain the variability in the measured data.

To adequately calibrate a model for both wet and dry conditions introduces further challenges. In addition to the errors associated with the input to the model, there are many other forms that modelling uncertainty can take. Some of them are: the model physics is too simple to describe the phenomena; suboptimal values of model parameters; the calibration data does not cover the full range of response values the model needs to predict and, therefore, model fit is poor in the range of values not represented in the calibration, etc.

All mathematical models, including those of complex urban catchments, are inevitably abstractions of reality. Consequently, some parameters of urban drainage models can be experimentally determined, while some others have little or no physical meaning and their values must be estimated through the comparison with field measurements. This problem, referred to as model calibration, can also be viewed from an optimisation perspective as: finding the set of parameter values that maximise/minimise some criteria — calibration criteria here — that express the degree of agreement between simulated model outputs and measured data sets (di Pierro *et al.*, 2005). Calibration in this sense is considered predominantly as the process by which the parametric model uncertainty is reduced.

The first step in model calibration is to perform sensitivity analysis and identify those parameters whose variation does not impact significantly on the model behaviour. Those parameters can be excluded from the calibration process. Formal approaches to sensitivity analysis and model identifiability analysis are described by Freni *et al.* (2009).

The next step is to identify what criteria are to be taken into account in the model calibration process. Several aspects of model behaviour must be considered when comparing measured data to the model output and they could be approached in a *certain order*, namely: first minimise volume error, then bias (i.e. a constant displacement), and finally, timing error. The question arises as to which features of the computed and measured outputs should be emphasised in determining the efficacy of the model. This problem is unfortunately even more difficult when the model calibration is done on series of rainfall events as the dimensionality of the calibration problem increases. A parameter to which the simulation is not sensitive for extreme events may become very influential at lower flow regimes, and vice versa. In doing such work, there is no single calibration criterion that is of universal relevance. Indeed, the criteria should be selected according to the purpose for which the model is to be applied. For example, a flood model should emphasise more extreme rainfall events, but a Combined Sewer Overflow (CSO) model should be orientated towards more frequent events. Hence, the preferred solution would be to develop a series of criteria that focus upon the more important aspects of model behaviour rather than to rely on a single index.

The process of adjusting model parameters has to be selected as it may range from a simple trial-and-error procedure to the most sophisticated mathematical optimisation approaches under conditions of uncertainty. Since the manual calibration process is essentially a trial-and-error process of minimising the differences between the computed and measured values, a multi-criteria calibration procedure based upon a global optimisation algorithm by which different objective functions are satisfied simultaneously may be preferred (Khu *et al.*, 2006). Furthermore, additional information needed for the proper use of the calibrated model may be provided by optimisation approaches that obtain not only the best parameters set, but also a probability distribution of parameters — e.g. Kanso *et al.* (2003) used the Metropolis algorithm to calibrate an urban storm water pollution model. Using measured data, they generated posterior distribution of accumulation rate D_{accu} for a 42 ha urban catchment and for a 186 m^2 street surface, both for the two different initial mass conditions. However, any automatic procedure still requires specialist input and the results (i.e., "optimal parameter values") must not be blindly accepted. Sometimes, the entire calibration might be deemed meaningless if the geometrical features of the domain were not represented accurately. Therefore, a modeller should first remove the uncertainty associated with input data as much as possible before proceeding with model calibration.

If the model calibration has been accomplished, the model *validation* exercise should be carried out on a set of measurements different from the calibration data set. In this context, the validation would refer to analysing and explaining the difference between model results and measurements, and if necessary, further adjustment of model parameters. Since extreme urban flood events are relatively rare, the amount of high quality field data for model calibration/validation is usually modest. Even when dynamic measurements of water levels and velocities in the sewer system and rainfall intensities are available with satisfactory spatial and temporal resolutions, surface observations are usually sparse, limited to flood extents and maximum depths and rarely linked to timing.

The uncertainty associated with the predictions from urban drainage models should be included in the decision analysis, or, if it is not done, then making decisions on the basis of a single model output should be treated with great caution.

12.2.5. *The modeller*

A big source of uncertainty is the *modeller*. Given the same set of system data, the same simulation model, the same data for model calibration and the same project objective, two people will come up with different solutions. Not only will an inexperienced modeller usually make a number of errors along the way without being aware of most of them, but also any two well-versed users would typically produce somewhat different results. This source of uncertainty may be reduced if guidelines such as codes of practice for hydraulic modelling are followed (e.g. WaPUG, 2002), but it cannot be fully eliminated.

12.3. Urban Flood Risk Mapping

12.3.1. *Geographic information systems*

Mapping of flood risk in urban areas has evolved a lot, following the advances in modelling and supporting technologies. As the early methods for flood risk mapping were based on the results of simulation of flow in sewer system alone, i.e. without considering the surface flow, the extent of surcharging was often mapped by changing colours of pressurised pipes. The frequency of surcharging over a series of events has also been used as a measure of flood risk (Verworn, 2002). A slightly more advanced mapping method — though also almost redundant today — was to show "blobs" around overloaded sewer network nodes on the map, the size of which would be proportional to the maximum volume of water that surcharged from a manhole. The corresponding presentation in a longitudinal sewer profile was the water level in "virtual reservoirs."

The dual drainage approach (Mark *et al.*, 2004) has been a step forward that, supported by Geographic Information Systems (GIS) ultimately enabled more realistic and more comprehensive analysis of flood risk, from the presentation of simulated flood extents, depths and velocities (and their combinations), counting of flooded properties, flooded roads, calculation of flood damage and, ultimately, presenting a range of measures of flood hazards in a highly visual form.

GIS and Remote Sensing (RS) play an important role in the flood management context. GIS tools have been developed to geo-reference time varying results from hydrodynamic models to a spatial framework or to a grid which includes a model of the terrain. For any computational time step, depth, velocity and hazard can be determined for each grid element in the 2D spatial framework. Such maps, which represent the extent of flood hazards, can provide a basis for defining suitable flood management measures including emergency response actions. The flood visualisation component of GIS enables engineers and emergency response planners to become familiar with the potential behaviour of flooding, its rates of rise, evolving flood extents, and areas of high flood hazard with lead times prior to the area concerned being flooded. With the use of GIS technology, model results can be linked to the cadastre and property databases and as such used efficiently in the overall urban planning process. Future developments will aim at more efficient use of virtual reality environments for flood animation in 3D.

GIS mapping techniques are needed in order to adequately show the numerical model results in space and time. In this respect, 2D models are much easier to handle since they have their own spatial framework, whereas mapping of the 1D model results is less straightforward because development of the appropriate framework is necessary.

12.3.2. *Mapping of 1D modelling results*

One way of producing the flood-extent maps for the 1D model is to project the longitudinal profile of maximum water levels onto a 2D map. Assuming that the water level is constant within each cross-section, the cross-sections need to be widened to fit the topography and their points geo-referenced and assigned by extrapolation with the corresponding water level values, as shown in Figure 12.2 (this process is necessary because it is difficult to create series of cross-sections with sufficiently accurate topography). These points are then interpolated using a Triangulated Irregular Network (TIN) method to form the continuous water surface. Thus, the obtained water surface finally needs to be intersected with the DTM and only positive values should be presented on a GIS map. Figure 12.3 illustrates how 1D model results (depths and velocities) can be superimposed with cadastre

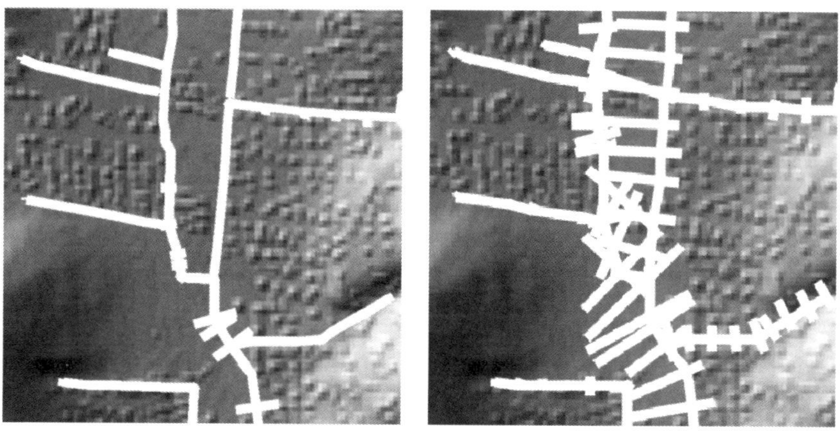

Figure 12.2. Widening of 1D model cross-sections.

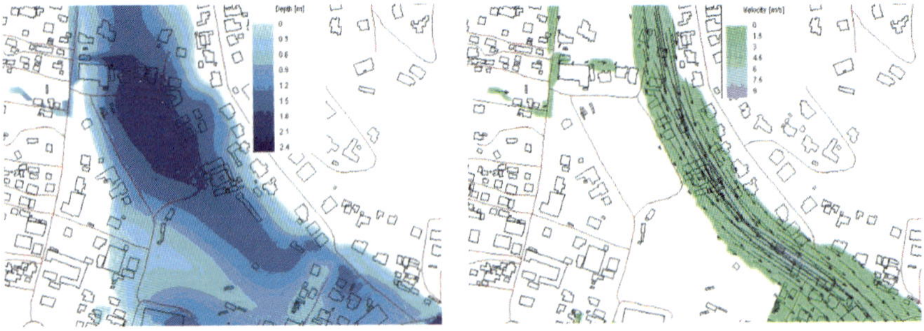

Figure 12.3. A GIS representation of 1D model results — left: water depths; right: velocities (after Vojinovic and van Teeffelen, 2007).

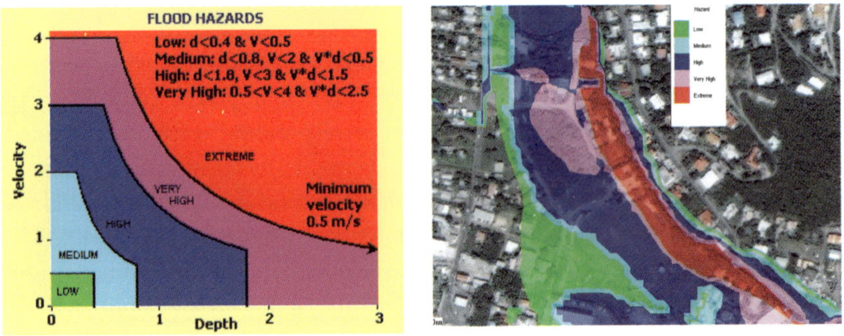

Figure 12.4. Left: flood hazard categories based on calculated velocities and depths (after Price and Vojinovic, 2008); right: a GIS mapping of 1D model results (hazards).

images within a GIS framework. Figure 12.4 shows how these computed depths and velocities can be *combined* to reflect different hazard categories defined as low, medium, high, very high and extreme in accordance with the diagram shown on the same figure. Similar diagrams are part of flood risk management regulations in some countries.

12.3.3. *Remote sensing technologies*

Remote sensing technologies, such as Airborne Laser Scanning (ALS) and LiDAR, are used to provide a comprehensive topographic coverage of entire floodplain areas in an accurate and economic manner. The sufficiently fine resolution of ALS data can provide a good spatial framework to compensate for a coarse resolution of the hydrodynamic model. Such a framework makes it possible to map water level and velocity results onto a terrain model and to visualise the quantification of flood hazards across the floodplain.

In addition to the use of numerical models for developing flood hazard maps, RS images can also be used to produce similar information. In this respect, the use of Advanced Synthetic Aperture Radar (ASAR) satellite images plays an important role. The ASAR technology utilises radar instruments which can work in the night time and which can have much higher resolution than other sources of satellite observations. Similar to the processing of numerical model results, a framework for processing ASAR data is also needed.

Typically, the backscatter coefficient of the targets on the ground needs to be differentiated in terms of the wet (flooded) and dry (non-flooded) areas. Once this is done, it is then necessary to derive the flood depths along the floodplain. Normally, a DTM of the study area is used in combination with ASAR data to complete this work. Furthermore, an interpolation algorithm is then needed to produce the water surface with elevation information in each cell. Although there are several interpolation methods which are embedded in commercial GIS tools, the studies to date have demonstrated that they are not necessarily very efficient and, hence, better interpolation methods are needed (e.g., Pengxiao, 2008). Once the flood depths are derived, the analysis of flood frequencies for each cell needs to be carried out. Finally, by combining flood depth and flood frequency information, a form of a hazard map can be produced, as illustrated in Figure 12.5.

12.4. The Risk Attribution Problem and Potential Solutions

12.4.1. *Integrated urban flood risk management*

Whilst fluvial and coastal flood risk analysis is routine (e.g. Dawson and Hall, 2006) quantitative flood risk assessment in the urban area represents a genuine challenge due to the complex interaction of natural and engineered processes, some of which operate at very local scales.

Integrated Urban Flood Risk Management (IUFRM) explicitly recognises the interrelationships between all sources of flooding and the effectiveness and cost of flood risk management measures, within changing social, economic and environmental contexts. The main sources of flooding include intense pluvial runoff that leads to sewers surcharging and surface flows, coastal storm surges, fluvial flooding caused by high river flows, and perhaps

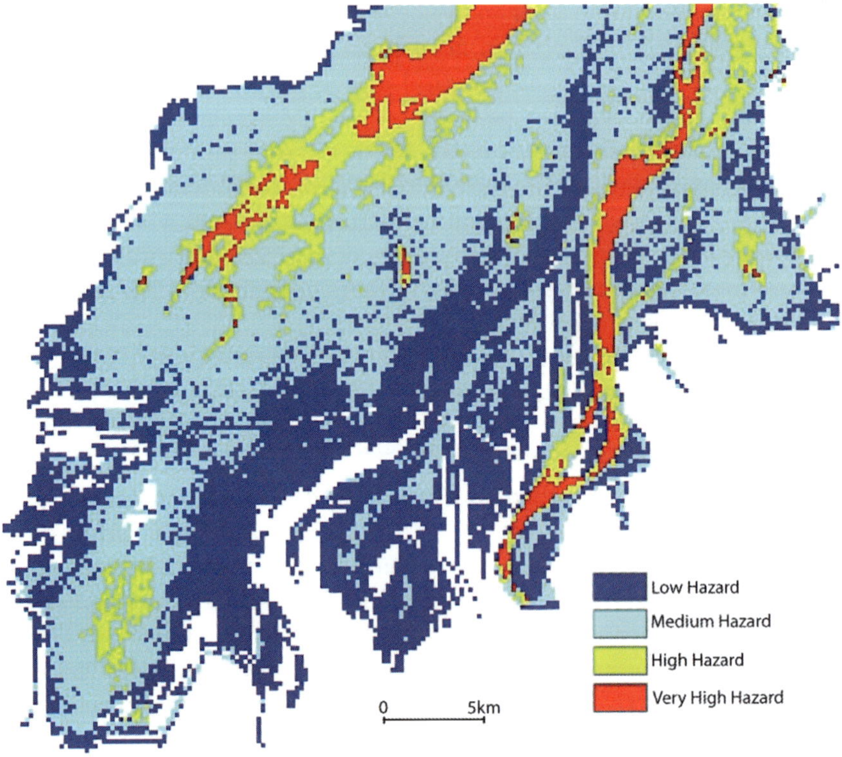

Figure 12.5. Example of a flood hazard map derived form remote sensing images (after Pengxiao, 2008).

also groundwater floods. A given flood event could be caused by a single source, or several sources acting in combination.

Severe flooding in urban areas in the UK in summer 2007 acted as a stimulus to the development of more integrated approaches to urban flood risk management. Currently in the UK, as in many other countries, urban flood risk management is fragmented. Ownership and responsibility for urban infrastructure and flood management is in the hands of a variety of public and private actors. Defra, the government department with lead responsibility for flooding, is promoting a more integrated approach to urban flood risk management (Defra *et al.*, 2005) in which the various organisations with a role in urban flooding work together to understand the processes of flooding and develop integrated solutions that tackle flooding in an efficient way.

Integrated solutions may involve a number of measures, for example, infrastructure investments and spatial planning regulations, which are designed together to achieve the desired level of risk reduction. Although the organisational context differs in many countries, the challenge of addressing integrated urban flood risk analysis has been identified in the USA (Rangarajan, 2005) and elsewhere (Andjelkovic, 2001). There is potential to support these institutional initiatives with a new generation of flood modelling tools that can simulate the effects of sewer and surface flows (Mark and Djordjević, 2006). Flood simulations can act as a vehicle for collective learning about system performance by various stakeholders in FRM. However, for this to be achieved a transformation of the standard approach to urban drainage modelling is necessary.

In the past modelling systems were designed and used with the prime objective of designing urban drainage systems to a certain standard with little consideration of situations that exceeded that standard. A risk-based approach, by contrast, involves consideration of a wide range of loading conditions, including conditions that exceed the design standard and lead to extensive surface flooding (Hall *et al.*, 2003). A precondition for this transformation is the development of core concepts for a framework for unified systems-based flood risk analysis. In this framework risk can be used as a "common currency" to compare risks from different sources on a common basis and in a situation where there are several organisations responsible for risk management. We wish to be able to *disaggregate* the total risk and *attribute* it to different components in the system and/or agents with responsibility for risk reduction.

12.4.2. *Risk attribution*

Risk attribution has been introduced (Dawson *et al.*, 2008) as a process for calculating the relative contribution towards risk from different flooding sources and components of flooding pathways, including infrastructure components. Risk attribution provides essential information for a number of urban flood risk management purposes:

1. *Risk ownership.* There are frequently several organisations with a role in flood risk management. We wish to know, in broad terms, what proportion of the risk each is responsible for.
2. *Estimation of capacity to reduce risk.* Ideally, risk should be owned by organisations with the greatest capacity to manage it. Capacity to reduce flood risk is related to the potential to change the characteristics of the

flooding system, e.g. by replacing the infrastructure or modifying surface flow paths. We wish to identify those organisations with the capacity to reduce risk.

3. *Asset management.* Given limited resources, an organisation with responsibility for management of flood defence or drainage infrastructure should rationally invest those resources so that they maximise benefits in terms of risk reduction. Within a specified set of system components we therefore wish to identify those components that contribute most to risk, and to compare the measures to reduce risk with the cost of implementing those measures in order to develop an optimum intervention strategy.
4. *Data acquisition strategies.* Monitoring strategies should be targeted so that resources are invested in data acquisition that makes the greatest contribution to reducing uncertainty.

12.4.3. *The risk calculation*

Consider a system that is described in terms of X basic variables, where X comprises a vector of loading variables S, which may be spatially and temporally variable, and a vector of variables that describe the flood management infrastructure system R that might include the height or other dimensions of dykes, the dimensions of surface water courses or the dimensions of the sewer system. Their variation might be continuous (e.g. a height variable) or discrete (e.g. a "blocked" or "not blocked" descriptor of a pipe). Capital notation (e.g. X) denotes a random variable and lower case (e.g. x) denotes a fixed value of that variable.

The variability in the loading and resistance is described by a joint probability distribution $\rho(x)\colon x \geq 0$. There is a damage function $d(x)$, where the units of d are £ (British pounds) or some suitable currency, which gives the flood damage in the systems for a given vector x that completely describes the system state. For many states of the system $d(x) = 0$. Indeed we only expect $d(x) > 0$ when S is large or when there are some inadequacies in system design or some failure, for example due to deterioration or blockage. The risk r associated with the system is:

$$r = \int_0^\infty \rho(x)d(x)dx. \tag{12.3}$$

The temporal dimension of this risk estimate is implicit in $\rho(x)$, so when, for example, $\rho(x)$ measures annual probability then r is an Expected

Annual Damage (EAD). However, $d(X)$ is seldom in an explicit form, but is evaluated by a combination of numerical models of flooding and economic damage functions. Moreover, the joint distribution $\rho(x)$ may not be convenient either and may be available in the form of an observed or synthetic time series of joint loading variables. Under these circumstances risk may be approximated by a finite sum, considering discrete samples of vectors of loading variables and simulating the corresponding damage using an appropriate model. The expected annual damage is estimated from:

$$r \approx \frac{N}{n} \sum_{i=1}^{n} d(X_i), \tag{12.4}$$

where n is the total number of events X_i simulated and N is the number of events per year.

12.4.4. *Methods of risk attribution*

There are a number of possible methods for attributing risk between flood management infrastructure and owners.

Source attribution. Urban flood risk managers may be interested in the sources of water that led to a particular flood event. For example, if flooding was caused by a combination of sewer surcharging and overtopping of river flood defences, then flood risk managers will wish to know the proportions of water, at a particular site, that originated from these two sources. Source attribution uses hydrodynamic particle tracking methods (Fischer *et al.*, 1979) that enable the water that ends in a particular location to be tracked back to its source. Balmforth *et al.* (2006) used this type of approach to calculate the total flood volumes conveyed in the sewers, overland and in the urban water courses. However, this analysis was conducted for only one event, whereas, in keeping with the principles outlined earlier in this section, this calculation should be repeated over a range of loading events, to identify the expected value.

Design standards attribution. Standards-based attribution quantifies the performance of different engineering components in the system at their "design standard". For example, an organisation with responsibility for urban drainage (UDO), provides a specified level of service to discharge rainfall events up to return period T_s. If the system floods in any rainfall event with return period $T'_s \leq T_s$, then the flood damage is the responsibility of the UDO as they have not fulfilled the standard to which they are committed. If the system floods only in events for which $T\psi > \psi T_s$ then the damage is not the responsibility of the UDO. However, if the system has capacity $T_s' \leq T_s$, and an event with return period $T\psi > \psi T_s$ occurs, then a proportion of the damages are the responsibility of the UDO. A flood model can be used to estimate the damage $d(l_T)$ given rainfall l_T with return period T. By definition $d(l_T) = 0$ when $T\psi < \psi T_s'$. Therefore the expected damage attributable to the UDO, r_{UDO}, given a probability density $\rho(l_T)$ of rainfall is:

$$r_{UDO} = \int_0^{l_{Ts}} \rho(l)d(l)dl + d(l_{Ts}) \int_{l_{Ts}}^{\infty} \rho(l)dl. \tag{12.5}$$

This may be extended further to consider situations where there is blockage or some other sewer failure (Dawson *et al.*, 2008).

Sensitivity-based risk attribution. This approach apportions risk between the system variables that influence the total flood risk on the basis of estimates of actual or potential variation. In particular it helps identify variables in the system that might be most influential in risk reduction. It can also help to identify uncertain variables that should be the target for data acquisition to improve the accuracy of the flood risk estimate.

If each of the loading variables, S (e.g. fluvial flows, rainfall and surge tides) were the unequivocal responsibility of a particular agent, then risk ownership could be disaggregated on the basis of sensitivity to the relevant loading variable. However, rainfall, for example, is dealt with in sewer and highway drainage systems as well as urban water courses. In that case it is necessary to also consider the variables R that define system performance. There are many variables that may, for practical reasons, be known precisely (to within some tolerance), e.g. pipe diameter. Nonetheless, we may wish to understand the potential for risk reduction by changing the value of such a variable. Under these circumstances we can specify a range of potential variation and corresponding probability distribution.

There are a wide range of potential sensitivity measures, summarised in Table 12.1 and explored in more detail by Hall *et al.* (2009). Here we demonstrate Variance-Based Sensitivity Analysis (VBSA) (Saltelli *et al.*, 2000) which measures the amount by which the variance in r is reduced if one or more of X_i were fixed at some value. The variance V can be decomposed into contributions from each of the input factors acting on their own or in increasingly high order interactions (Saltelli *et al.*, 1999;

Table 12.1. Summary of risk attribution methods.

Attribution method	Risk ownership	Estimation of capacity to reduce risk	Asset management
Standards-based attribution	Provide risk ownership is well defined according to specified standards.	Limited.	With further expert diagnosis of reasons why system is not performing to standard.
Source attribution	Provide risk ownership only where unequivocally related to the source of flooding.	Good where capacity to reduce risk is strongly related to controlling water volume.	With further expert diagnosis to understand flood flow paths through assets.
Linear regression	For (approximately) linear systems. Easiest where responsibility is allocated according to loading variables.	Good for (approximately) linear systems.	Good for (approximately) linear systems.
Partial derivatives	Difficult to interpret globally for non-linear systems. Easiest where responsibility is allocated according to loading variables.	Difficult to interpret globally for non-linear systems.	Difficult to interpret globally for non-linear systems.
Discrete systems analysis	Not applicable.	Good for discrete systems.	Good for discrete systems.
Variance-based sensitivity analysis (VBSA)	Good. Easiest where responsibility is allocated according to loading variables.	Based on *current values* of variables, so requires further analysis to deal with future change.	Provides a rational basis for evaluation of inspection and maintenance/repair strategies.
Partial expected value of perfect information	Good for more general decisions problems than VBSA. Easiest where responsibility is allocated according to loading variables.	Based on *current values* of variables, so requires further analysis to deal with future change.	Provides a rational basis for evaluation of inspection and maintenance/repair strategies.

Sobol, 1993):

$$V = \sum_i V_i + \sum_{\substack{i,j \\ i<j}} V_{ij} + \sum_{\substack{i,j,l \\ i<j<l}} V_{ijl} + \cdots + V_{12\ldots k}, \tag{12.6}$$

where

$$V_i = V[E(r|X_i = x_i^*)], \tag{12.7}$$

$$V_{ij} = V[E(r|X_i = x_i^*, X_j = x_j^*)] - V_i - V_j, \tag{12.8}$$

and so on. $V[E(r|X_i = x_i^*)]$ is referred to as the Variance of the Conditional Expectation (VCE) and is the variance over all values of x_i^* in the expectation of r given that X_i has a fixed value x_i^*. VCE measures the amount by which $E(r|X_i = x_i^*)$ varies with the value of x_i^*, while all the effects of the $X_j, j\psi \neq i$, are averaged. The ratio $S_i = V_i/V$, therefore a measure of the sensitivity of r with respect to X_i.

Total sensitivity indices, S_{Ti}, represent the average variance that would remain as long as X_i stays unknown and provide an indicator of interactions within the model. Factors with small first-order indices but high total sensitivity indices affect the risk, r, mainly through interactions. These are calculated using (Homma and Saltelli, 1996):

$$S_{Ti} = 1 - \frac{V[E(r|X_{\sim i} = x_{\sim i}^*)]}{V(r)}, \tag{12.9}$$

where $X_{\sim i}$ denotes all of the factors other than X_i.

The computational expense of the methodology described here can be considerable, even for a rather small system. In practice, urban flooding systems involve tens of thousands of variables. The only feasible approach to tackling this problem is therefore by hierarchical simplification of the system, with the attribution analysis being applied at several levels, with initial screening to identify the most important variables. Sensitivity-based risk attribution is demonstrated in Sections 12.5.4 and 12.5.5.

12.5. Case Studies in Urban Flooding Uncertainty Analysis

12.5.1. *Uncertainties related to urban drainage models and their calibration*

As stated in Section 12.2.4, validation of a calibrated model should be done using data sets not used in the calibration, because it is possible to derive many different model parameter sets for the same study area that will satisfy given calibration criteria. However, when it comes to the simulation of pipe flow models with such parameters, when carried out on a data set which is different from the calibration data set, they may produce very different results in terms of flooding or overflow emissions.

An informative example of uncertainty analysis in sewer pipe network performance analysis is described by Hansen *et al.* (2005). They applied Monte Carlo methodology by sampling from the probability distribution of the urban drainage model inputs/parameters, simulating the pipe flow and water levels in the manholes with the MOUSE model (DHI Software, 2004), which enabled determination of probability distribution of the maximum water level in the manholes. Among other conclusions, they used uncertainty analysis to illustrate how calibration can improve the quality of the model output.

Figure 12.6 shows the range of maximum water levels in a manhole with non-calibrated and calibrated model (pipe top and ground levels are also shown for reference). The benefit of the calibration in terms of reducing the range of levels varied among manholes and often led to lower maximum water levels, i.e. prediction of flooding was less uncertain with the calibrated model. Not that this conclusion was surprising, but this analysis enabled quantification of the benefits of calibration.

Thorndahl *et al.* (2008) conducted an uncertainty analysis of an application of the urban drainage model, also using MOUSE and applying the GLUE methodology (Beven, 2006). They performed 10,000 simulations of several observed events on a small urban drainage system, for three

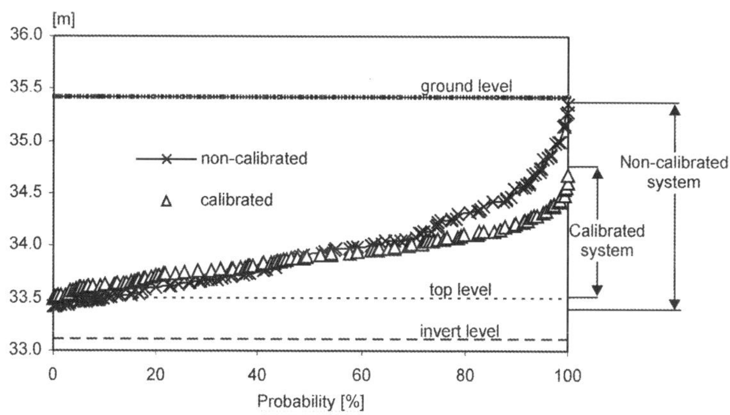

Figure 12.6. Range of maximum water level in a manhole in two situations: non-calibrated system and calibrated system (after Hansen *et al.*, 2005).

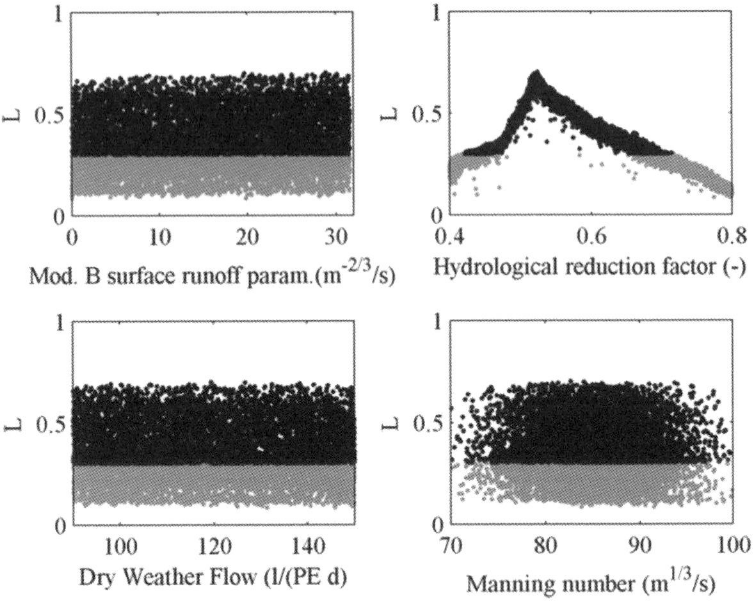

Figure 12.7. Combined likelihoods over all calibration events as a function of parameter values. Accepted simulations with empirical likelihood $L > 0.3$ (6091 of 10,000 runs) are shown in black (after Thorndahl *et al.*, 2008).

different model setups. This was deemed to be a sufficient number because the Kolmogorov–Smirnov d between overflow volume cdfs with 10,000 and 20,000 runs was smaller than 0.01. Based on the analysis the methodology has been shown to be very applicable — it was possible to provide useful prediction bounds and identify limitations of the model, and some results are shown here.

Figure 12.7 shows "dotty plots" that illustrate the likelihood of every single simulation (represented by one dot) as a function of values of four (out of six) model parameters. The most conspicuous is the narrow peak of hydrological reduction factor (part of the impervious area contributing to the runoff) that corresponds to the mean of the event specific optimum reduction factors. Interestingly, maximum likelihood of the reduction factor was higher for the model setup with area weighted rainfall input from two rain gauges than for the model setup with one single rain gauge. Also, applying the same likelihood threshold, the greater number of simulations was accepted in the model setup with spatially variable rainfall input (44%) than in the single rain gauge case (38%). This is an obvious result of the use of reduction factor to compensate in uncertainties in the rainfall input. Furthermore, 61% of simulations were accepted in the model setup with two rain gauges and the more complex surface runoff model (kinematic-wave surface runoff model), which is the case shown in Figure 12.7.

The other four MOUSE model parameters (initial loss, dry weather flow, pipe roughness and local energy loss coefficient) showed a clear indication of equifinality, i.e. that it is possible to have the same maximum likelihood regardless of the parameter value. This result either indicates prediction insensitivity to these parameters or that some parameters interact closely in producing behavioural models. Sensitivity analysis confirmed that the impact of these four parameters appeared to be minimal, whilst the interaction between their values in producing behavioural parameter sets is negligible.

In urban drainage modelling, the most relevant results are the combined sewer overflow volumes and water levels in surcharging manholes (that lead to flooding). Figure 12.8 shows cdf-plot of the maximum water level in the manhole that surcharged most frequently in the study by Thorndahl *et al.* (2008). Compared to cdf-plots of overflow volume (not shown here) the prediction interval is quite narrow — varying only 10 cm between 95% and 5% prediction interval — which indicates a rather small uncertainty on the estimation of maximum water level in that manhole.

12.5.2. *Sensitivity of results to the capacity of sub-surface/surface links*

Figure 12.9 shows flow rate in one sub-surface/surface (s-s/s) link and the sum of flow rates through *all* s-s/s links during the flood event following

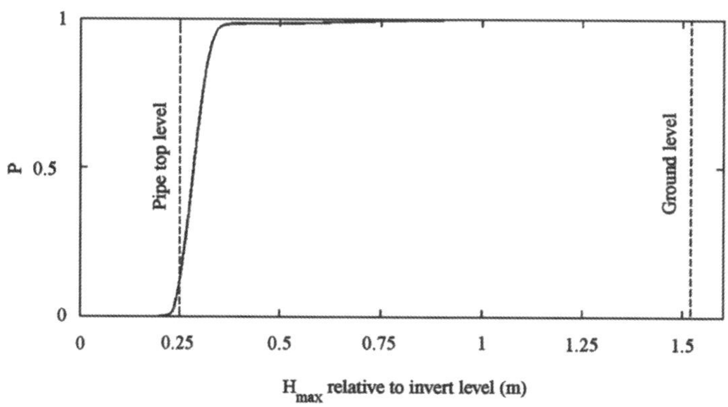

Figure 12.8. Cdf of water level in a critical manhole (after Thorndahl *et al.*, 2008).

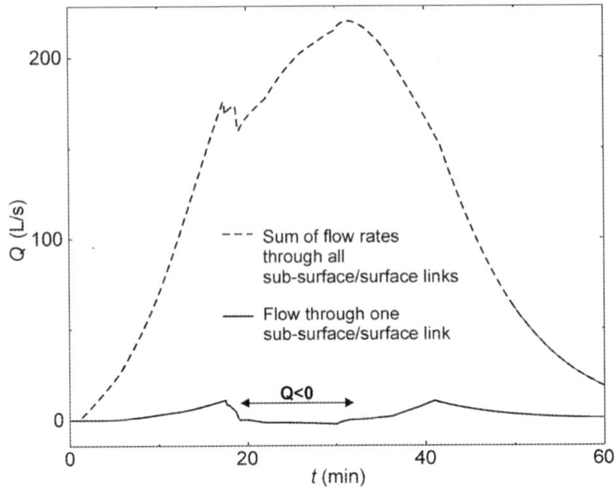

Figure 12.9. Flow rates in one s-s/s link and in all s-s/s links combined (negative flow means surcharging from the manhole to the surface).

a 30-minute block rainfall of high intensity on a small urban catchment described by Djordjević *et al.* (2005). Positive values in this diagram represent water from the surface entering the minor drainage system (i.e. s-s/s links working as inlets), whereas negative values correspond to the outflow from the surcharging sewer system to the surface system (i.e., s-s/s links working as storm sewer overflows). Surface runoff hydrographs were introduced to the 1D surface network nodes in this example. Figure 12.10 illustrates how the uncertainty in capacity of s-s/s links may affect urban flood dynamics simulated by the 1D/1D Simulation of Interaction between Pipe flow and Surface Overland flow in Networks (SIPSON) model (Djordjević *et al.*, 2004):

1. Figure 12.9 shows flow rates in: one of the outfall sewer pipes (dashed line), main surface open channel outfall (dotted line) and the sum of flow rates through all s-s/s links (solid line), simulated with an assumed capacity of s-s/s links.
2. Figure 12.10 shows corresponding results simulated with the s-s/s link capacities (i.e. weir crest length) equal to 50% of the value assumed in the first case. Consequently, the sewer system is filled more slowly, surcharging and the outflow from the sewer system to the surface are delayed and less extensive, and, overall, surface flood flow rates are much higher.

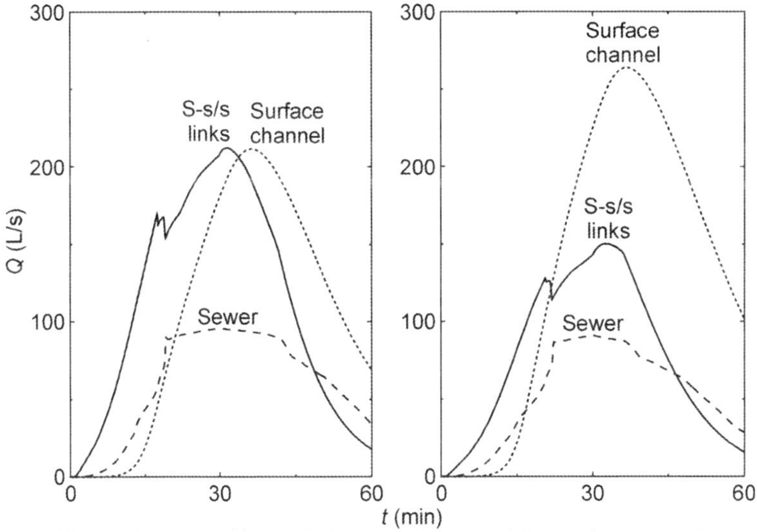

Figure 12.10. Sensitivity of flows to s-s/s links parameters; case shown on the right-hand side is calculated with link capacity 50% of the capacity assumed in the simulation on the left-hand side.

Another example of the sensitivity of results to the capacity of sub-surface/surface links is described in Leandro *et al.* (2009). They calibrated 1D/1D urban flood model of a small urban catchment via results of 1D/2D model, with the parameters of s-s/s links as calibration parameters and the absolute difference in cumulative volumes of flow through s-s/s links in the two models as the objective function. Figure 12.11 shows flow rates in selected manholes *before* and *after* calibration. Contrary to the diagrams shown in Figure 12.9, positive values shown in Figure 12.11 represent the outflow from the surcharging sewer system to the surface, and surface runoff hydrographs were introduced to the sewer network nodes in this example. The latter assumption is incorrect for those manholes that surcharge while runoff hydrographs still have significant values; however, it is purposely kept in order to enable a consistent comparison between results of the two modelling approaches (i.e. 1D/1D and 1D/2D model).

12.5.3. *Sensitivity of results with respect to DTM resolution*

This case study, details of which are given in Mwalwaka (2008), investigated effects of DTM resolution on urban flood risk. Four DTM sets, 5 m, 10 m, 15 m and 20 m grid size were produced. Figure 12.12 displays terrain

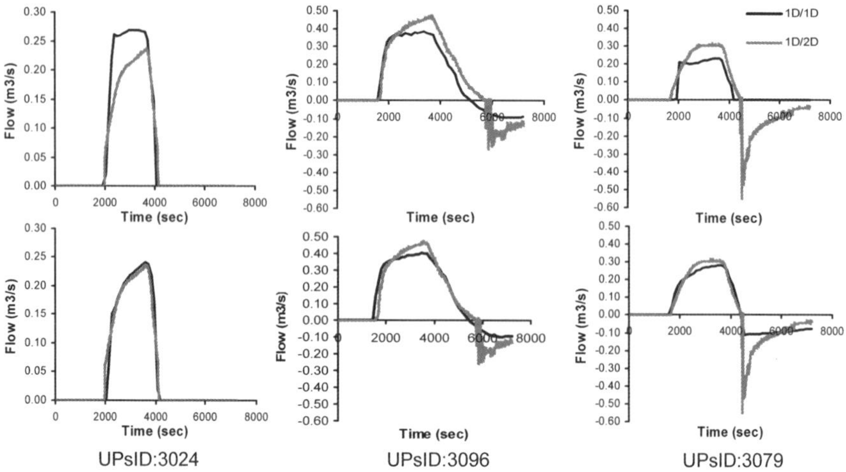

Figure 12.11. Discharges at three selected s-s/s links, before calibration (top row) and after calibration (bottom row); positive flow means surcharging from the sewers to the surface (after Leandro *et al.*, 2009).

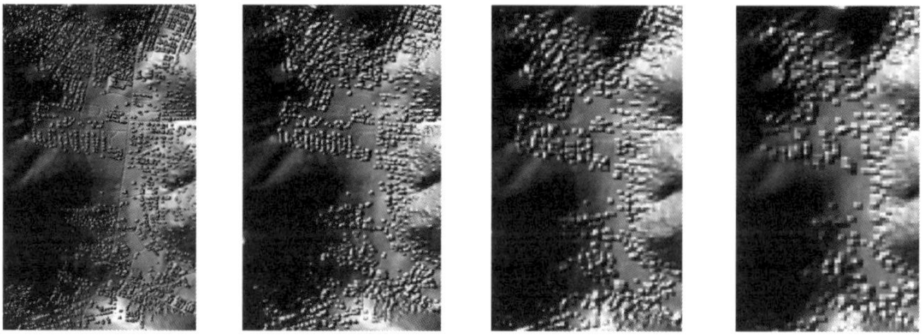

Figure 12.12. DTM in four resolutions: 5 m, 10 m, 15 m and 20 m (from left to right).

characteristics with four different resolutions that were used as a basis for MIKE FLOOD model (DHI Software, 2008).

The model was run with four terrain resolutions for the rainfall event of 100 mm in 1 hour, which corresponds to 100-year Average Recurrence Interval (ARI) for that area. Three points of interest (A, B, C) where the computed variables were compared among different model results are all located along the floodplain, i.e. in low-lying areas which are known to be the most affected areas. The difference in model results, as shown in Figure 12.13, could be mainly attributed to the averaging or upscaling

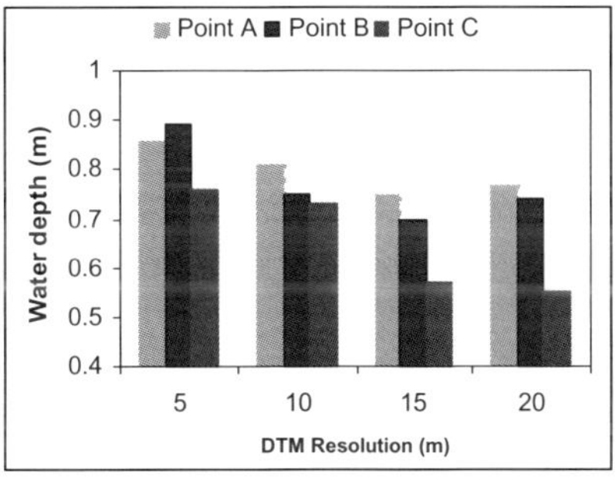

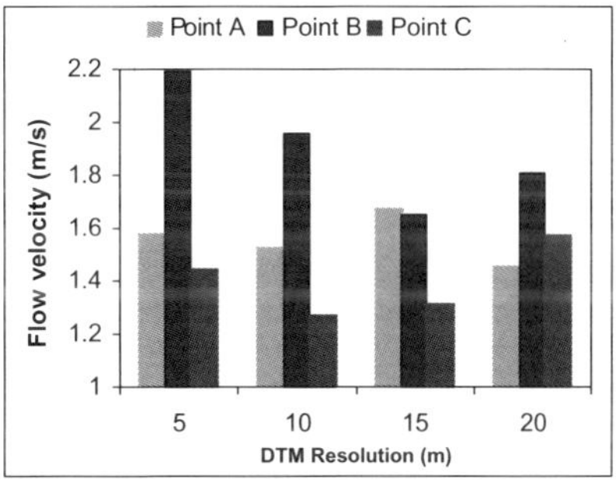

Figure 12.13. Maximum flood depths (left) and flow velocities (right) at three selected points, simulated using four different DTM resolutions (after Mwalwaka, 2008).

effect of the high resolution to the low resolution DTM. Such effects are represented by the loss of small-scale topographic information which is present in higher DTM resolutions. As a result, 2D models with lower DTM resolution are likely to generate floods over more wide spread areas and with smaller depths than high DTM resolution 2D models. Furthermore, the difference in results can be also explained by fact that the level and scale of physics calculated within 2D models with higher resolution are more complex as the morphology of the domain is more irregular when compared to a 2D model with low resolution.

In addition to this, head losses due to flow over or around such structures are also represented differently with different models. With higher resolution models, small geometric "discontinuities" such as road or pavement curbs can play a significant role in diverting the shallow flows that are generated along roads and around buildings, and in several instances they were found contrary to the flows calculated by the low resolution models.

From this case study it can be concluded that the DTM resolution has a significant effect on simulation results and therefore it carries a degree of uncertainty. It can be observed from the obtained results that the flood simulation characteristics affected by different DTM resolution are: inundation extent, flow depth, flow velocity and flow patterns across the model domain. Whilst not allowing for the deduction of universal rules from the limited number of modelling experiments, nor for defining a sufficiently high DTM resolution in order to keep its influence acceptably small, these results indicate the importance of evaluating the uncertainty related to DTM resolution in urban flood risk assessment. Consequently, model parameters calibrated on a DTM of certain resolution are not directly transferable to a lower resolution model.

12.5.4. *Integration of fluvial and urban flood risk analysis*

Flood risk attribution methodology described in Section 12.4.4 is here demonstrated on a synthetic case study. More detail is provided by Dawson *et al.* (2008) with the salient points repeated here.

Consistent meteorological boundary conditions using methods described by Burton *et al.* (2004) drive a hydrological model (Todini, 1996) of an upstream catchment and provide direct rainfall inputs to the urban catchment. The upstream hydrology model provides boundary conditions of river flow next to the urban area in which the flooding is simulated by the 1D/1D coupled surface and sewer flow model SIPSON (Djordjević *et al.*, 2005).

Maximum flood depths during an event are subsequently extracted from the model and integrated with depth-damage curves (Penning–Rowsell *et al.*, 2003) assuming one property at every 20 m of surface flood flowpaths. That way damages for a given flood event are estimated, and subsequently risks (Figure 12.14). Multiple samples of the model variables are generated and used to attribute risk to infrastructure and other system components. It is important to note that the risk attribution methodology is not tied to the specific model components used in this study and will be suited to any system of models and methodologies that calculates flood damage according to any metric(s) of interest. The risk for the system was calculated to be an expected annual damage of £576 k.

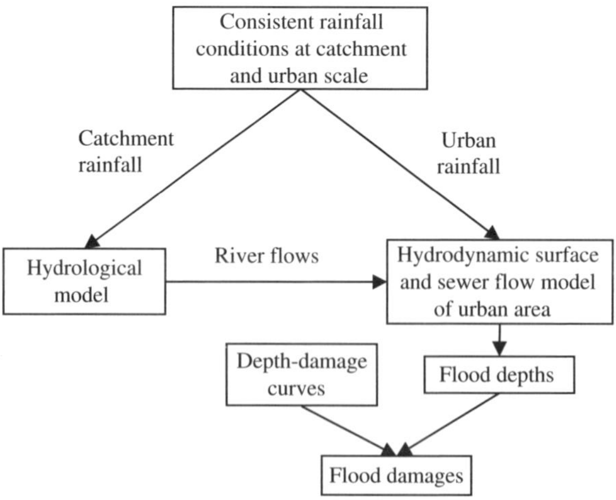

Figure 12.14. Overview of urban flood risk analysis modelling process.

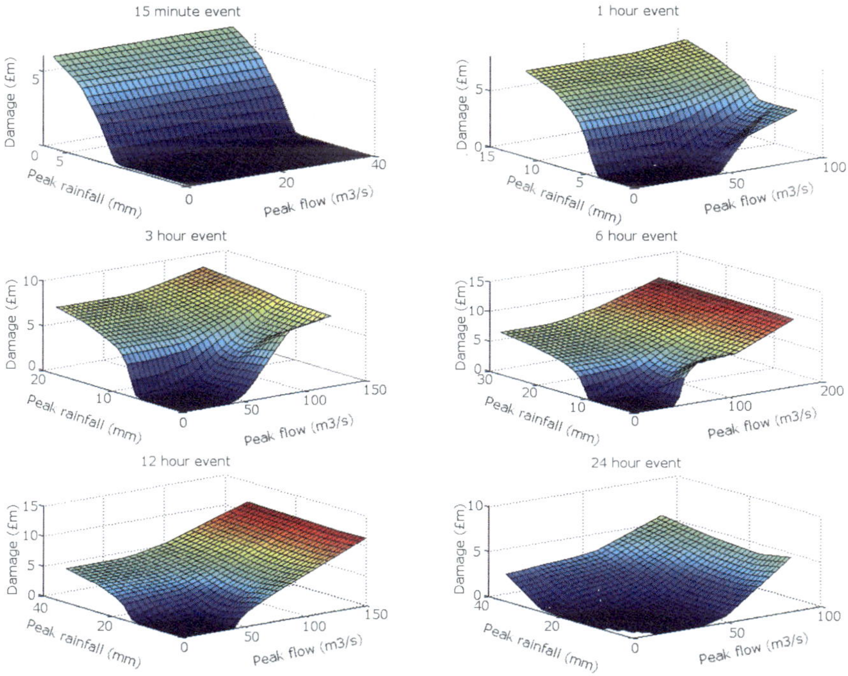

Figure 12.15. Relationship between damage (z-axis), peak rainfall, peak flow rate in the river for different event durations.

Figure 12.15 demonstrates the importance of analysing risk over a wide range of loading conditions by plotting the relationship between damage, peak rainfall and river flow for different rainfall event durations. This shows that damage is dominated by rainfall under short intense rainfall events, but river flow becomes increasingly important as the duration of the rainfall event increases.

The sensitivity analysis for the model are shown in Table 12.2. They indicate that the highest proportion of the risk is from the event duration, peak rainfall and pipe diameter. The total indices illustrate that the same variables that dominate the first-order indices (event duration, peak rainfall and pipe diameter) are, in this case, also most actively involved in interaction.

Figure 12.16 illustrates the influence of pipe diameter, river capacity and permeability of urban surface upon the resultant flood damages. Varying pipe diameter over the range of values analysed here leads to the largest changes to flood damage. Damage increases linearly with the proportion of impervious surfaces in the urban area, but the difference

Table 12.2. Sensitivity indices for key variables.

Variable	Sensitivity indices	
	First order	Total
Duration	0.19	0.65
Peak flow rate	0.02	0.10
Peak rainfall	0.15	0.48
Pipe diameter	0.04	0.30
River width	0.00	0.00
Impermeable area	0.01	0.03

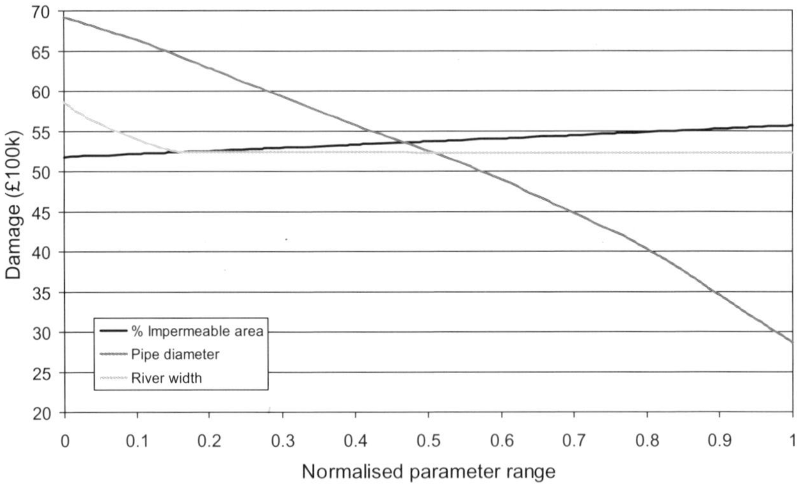

Figure 12.16. Influence of pipe diameter, river capacity and permeability of urban surface upon flood damage. Independent variable normalised by range (after Dawson *et al.*, 2008).

between 30% and 90% impervious surface alters the damage by only ~£400 k. Whilst river capacity shows a non-linear interaction with damage, the maximum change in damage is ~£600 k compared to £4 m for the pipe diameter.

12.5.5. *Pipe blockage*

The catchment and the system used in the previous section (described in detail by Dawson *et al.*, 2008) is used here to illustrate the methodology for identification of the most critical pipes in the sewer network with regard to flood risk resulting from pipe blockage. However, even for this small

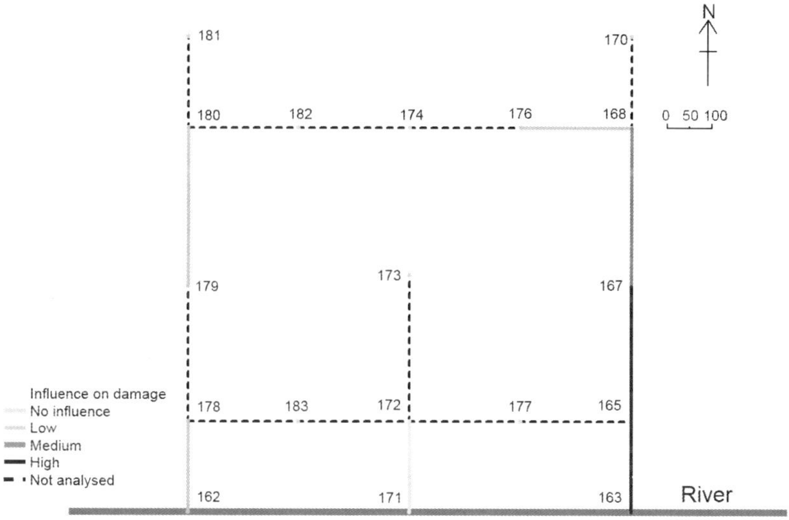

Figure 12.17. Pipes that, when blocked, have the greatest influence on flood damages (after Dawson *et al.*, 2008).

system, it is impractical to simulate 2^n pipe blockage combinations (n is the number pipes, in this case $n = 18$, giving over 250 k combinations). A tiered approach was developed that:

1. Analysed the damage from single blockages in each pipe.
2. Identified the pipes that, when blocked, lead to the greatest increase in flood damages for the design standard (1 in 10-year event) of the sewer system.
3. Calculated the sensitivity indices for these pipes (Figure 12.17).

As might be expected intuitively, important components in the urban drainage system are the lowest pipes (167–165 and 165–163), which also exhibit the strongest interactions with other pipe blockages. Of the three outfalls, the middle one (172–171) is the least critical because its blockage can be handled with the other two, whereas the most critical outfall is the most downstream one (165–163). This implies that the most successful flood risk reduction strategy would, in this case, be to increase the capacity of these critical pipes, whilst monitoring activities should be targeted to ensuring these pipes do not block. The obvious advantage of this methodology is in its ability to identify most critical pipes in complex systems where that is less obvious than in this simple example.

References

Abbott, M.B. and Minns, A.W. (1998). *Computational Hydraulics, Second Edition*, Ashgate, Aldershot.

Adams, A. and Allitt, R. (2006). Using DTMs — The Technical and Financial Case, *WaPUG Spring Meeting 2006*, Coventry.

Andjelkovic, I. (2001). *Guidelines on Non-Structural Measures in Urban Flood Management*, IHP-V Technical Documents in Hydrology, 50, UNESCO, Paris.

Balmforth, D., Digman, C., Kellagher, R. *et al.* (2006). *Designing for Exceedance in Urban Drainage — Good Practice*, CIRIA Report C635, London.

Beven, K.J. (2006). A manifesto for the equifinality thesis, *J. Hydrol.*, **320**, 18–36.

Burton, A., Kilsby, C., Moaven-Hashemi, A. *et al.* (2004). *Built Environment: Weather Scenario for Investigation of Impacts and Extremes*, Technical Briefing Note 2. Available at http://www.cru.uea.ac.uk/cru/projects/betwixt/ (Accessed on 02/05/2012).

Chen, A.S., Evans, B., Djordjević, S. *et al.* (2012). Multi-layered coarse grid modelling in 2D urban flood simulations, *J. Hydrol.*, **470–471**, 1–11.

Dawson, R.J. and Hall, J.W. (2006). Adaptive importance sampling for risk analysis of complex infrastructure systems, *P. R. Soc. A*, **462**, 3343–3362.

Dawson, R.J., Speight, L., Hall, J.W. *et al.* (2008). Attribution of flood risk in urban areas, *J. Hydroinform.*, **10**, 275–288.

Defra, Department of Transport, Office of the DPM and HM Treasury (2005). *Making Space for Water: First Government Response to the Autumn 2004 Consultation Exercise*, Defra. Available at http://archive.defra.gov.uk/environment/flooding/documents/policy/strategy/strategy-response1.pdf/ (Accessed on 02/05/2012).

DHI Software (2004). *MOUSE — Reference Manual*, Hørsholm.

DHI Software (2008). *MIKE FLOOD — Reference Manual*, Hørsholm.

di Pierro, F., Djordjević, S., Kapelan, Z. *et al.* (2005). Automatic calibration of urban drainage model using a novel multi-objective genetic algorithm, *Water Sci. Technol.*, **52**, 43–52.

Djordjević, S., Prodanović, D., Maksimović, Č. *et al.* (2005). SIPSON — simulation of interaction between pipe flow and surface overland flow in networks, *Water Sci. Technol.*, **52**, 275–283.

Djordjević, S., Prodanović, D. and Walters, G.A. (2004). Simulation of transcritical flow in pipe/channel networks, *J. Hydraul. Eng.-ASCE*, **130**, 1167–1178.

Ettrich, N. (2007). Surface-sewer coupling and detailed elevation models for accurate urban drainage modeling, *Special Aspects of Urban Flood Management,Proceedings Cost Session Aquaterra Conference*, Amsterdam, 2007, pp. 183–196.

Evans, B. (2008). Automated bridge detection in DEMs via LiDAR data sources for urban flood modeling, *11th International Conference on Urban Drainage*, Edinburgh, CD-ROM.

Fischer, H.B., List, E.J., Koh, R.C.Y. *et al.* (1979). *Mixing in Inland and Coastal Waters*, Academic Press, New York.

Freni, G., Mannina, G. and Viviani, G. (2009). Identifiability analysis for receiving water body quality modeling, *Environ. Modell. Softw.*, **24**, 54–62.

Friazinov, I.V. (1970). Solution algorithm for finite difference problems on directed graphs (trans), *J. Comput. Math. Mathematic. Phys.*, **10**, 474–477.

Hall, J.W., Boyce, S., Dawson, R.J. *et al.* (2009). Sensitivity analysis for hydraulic models, *J. Hydraul. Eng.*, **135**, 959–969.

Hall, J.W., Meadowcroft, I.C., Sayers, P.B. *et al.* (2003). Integrated flood risk management in England and Wales, *Nat. Haz. Rev.*, **4**, 126–135.

Hansen, A.C., Liu, L., Linde, J.J. *et al.* (2005). Accounting for uncertainty in urban drainage system performance assessment using safety factors applied to runoff, *10th Int. Conf. on Urban Drainage*, Copenhagen, CD-ROM.

Havnø, K., Brorsen, M. and Refsgaard, J.C. (1985). Generalized mathematical modeling system for flood analysis and flood control design, *2nd International Conference on the Hydraulics of Floods & Flood Control*, Cambridge, pp. 301–312.

Homma, T. and Saltelli, A. (1996). Importance measures in global sensitivity analysis of nonlinear models, *Reliab. Eng. Syst. Safe.*, **52**, 1–17.

Kanso, A., Gromaire, M.-C., Gaume, E. *et al.* (2003). Bayesian approach for the calibration of models: application to an urban stormwater pollution model, *Water Sci. Technol.*, **47**, 77–84.

Khu, S.–T., di-Pierro, F., Savić, D. *et al.* (2006). Incorporating spatial and temporal information for urban drainage model calibration: an approach using preference ordering genetic algorithm, *Adv. Water Resour.*, **29**, 1168–1181.

Leandro, J., Chen, A.S, Djordjević, S. *et al.* (2009). A comparison of 1D/1D and 1D/2D coupled (sewer/surface) hydraulic models for urban flood simulation, *J. Hydraul. Eng.-ASCE*, **135**, 495–504.

Leandro, J., Djordjević, S., Chen, A.S. *et al.* (2007). The use of multiple-linking-element for connecting sewer and surface drainage networks, *32nd Congress IAHR*, Venice.

Maksimović, Č., Prodanović, D., Boonya-aroonnet, S. *et al.* (2009). Overland flow and pathway analysis for modelling of urban pluvial flooding, *J. Hydraul. Res.*, **47**, 512–523.

Mark, O. and Djordjević, S. (2006). While waiting for the next flood in your city... *7th International Conference on Hydroinformatics*, I, Nice, pp. 252–265.

Mark, O., Weesakul, S., Apirumanekul, C. *et al.* (2004). Potential and limitations of 1D modelling of urban flooding, *J. Hydrol.*, **299**, 284–299.

Mwalwaka, M. (2008). *Effects of Model Schematisation, Geometry and Parameter Values on Flood Wave Approximation in Urban Areas*, MSc thesis, UNESCO-IHE, Delft, the Netherlands.

Pengxiao, C. (2008). *Flood Hazard Analysis Using GIS and Remote Sensing Data*, MSc Thesis, UNESCO-IHE, Delft, the Netherlands.

Penning-Rowsell, E.C., Johnson, C., Tunstall, S.M. *et al.* (2003). *The Benefits of Flood and Coastal Defence: Techniques and Data for 2003*, Middlesex University Flood Hazard Research Centre, Middlesex.

Price, R.K. and Vojinovic, Z. (2008). Urban flood disaster management, *Urban Water J.*, **5**, 259–276.

Rangaragan, S. (2005). Challenges and solutions in urban drainage planning and management, *Geophys. Res. Abstr.*, **7**, 1288.

Saltelli, A., Chan, K. and Scott, M. (2000). *Sensitivity Analysis*, Wiley, New York.

Saltelli, A., Tarantola, S. and Chan, K. (1999). A quantitative model-independent method for global sensitivity analysis, *Technometrics*, **41**, 39–56.

Sobol, I, (1993). Sensitivity analysis for non-linear mathematical models, *Math. Mode. Comput. Exp.*, **1**, 407–414.

Thorndahl, S., Beven, K.J., Jensen, J.B. *et al.* (2008). Event based uncertainty assessment in urban drainage modeling, applying the GLUE methodology, *J. Hydrol.*, **357**, 421–437.

Todini, E. (1996). The ARNO rainfall-runoff model, *J. Hydrol.*, **175**, 339–382.

Ukon, T., Shigeta, N., Watanabe, M. *et al.* (2008). Correction methods for dropping of simulated water level utilizing Preissmann and MOUSE slot models, *11th International Conference on Urban Drainage*, Edinburgh, CD-ROM.

Vasconcelos, J.G., Wright, S.J. and Roe, P.L. (2006). Improved simulation of flow regime transition in sewers: two-component pressure approach, *J. Hydraul. Eng.*, **132**, 553–562.

Verworn, H.-R. (2002). Advances in urban-drainage management and flood protection, *Philos. Transact. A. Math. Phys. Eng. Sci.*, **360**, 1451–1460.

Vojinovic, Z. and van Teeffelen, J. (2007). An integrated stormwater management approach for small islands in tropical climates, *Urban Water J.*, **4**, 211–231.

WaPUG (2002). *Code of Practice for the Hydraulic Modelling of Sewer Systems*, Wastewater Planning Users Group. Available at http://www.wapug.org.uk.

Watanabe, K. and Kurihara, T. (1993). "Practical Simulation Method of Surcharged Flow Using Pressure-Relaxation Effect in Manhole" in: Marsalek, J. and Torno, H.C. (eds), *6th International Conference on Urban Storm Drainage*, I, Niagara Falls, pp. 128–133.

CHAPTER 13

The Many Uncertainties in Flood Loss Assessments

John Chatterton, Edmund Penning-Rowsell and Sally Priest
Flood Hazard Research Centre,
Middlesex University, UK

It is often assumed in the assessment of flood risk — gauged as probability times consequences — that all the main uncertainties lie in the hydrological and hydraulic fields, and that the human domain is inherently more predictable. This assumption arises more from ignorance than from a careful analysis of the evidence. The reality is that the human dimension in floods and flood risk assessment is also full of uncertainty, and this is the issue addressed in this chapter.

13.1. Flood Damage Assessments

13.1.1. *Methods of flood damage assessment*

The publication of "Making Space for Water" in 2004 by UK's Department of the Environment Food and Rural Affairs (Defra, 2004; 2005a,b), heralded a new policy approach in England, characterised by a move away from flood defence and towards Flood and Coastal Erosion Risk Management (FCERM). As a result, the Environment Agency, which manages both strategic initiatives and day-to-day activities with respect to FCERM, adopted the S-P-R-C (Source, Pathway, Receptor and Consequence) model to represent the continuum or throughput from storm to damages caused by that storm to people's property and the environment. In future, risk would be managed by seeking to influence all four elements, rather than simply managing flood pathways and hence flood probability.

This policy move therefore shifts attention to the whole S-P-R-C continuum. The uncertainties with regard to source and pathway parameters — affecting the frequency and depth of flooding — are widely discussed, including in this volume. However, data and models to represent the receptor and consequence parameters have generally been unquestioned, and regarded in some way as "absolute", and somehow based on robust and easily understood science. This characterisation is examined in this chapter with the view to setting out a more balanced situation. In this respect, receptor variables are represented by land use or built property, infrastructure, utilities, etc., and consequences relate to the resultant flood damage, traditionally categorised as direct or indirect, and tangible or intangible (Table 13.1), including potential loss of life in floods.

The approach to evaluating flood damages and the benefits of flood alleviation in UK and throughout the world has been largely led by Middlesex University's Flood Hazard Research Centre's investigations since 1970 and published in its series of monographs known as the "Blue", "Red", "Yellow" and "Multi-Coloured" Manuals (MCM) (e.g. Penning–Rowsell *et al.*, 2005). In 1977, the "Blue Manual" first provided guidance on appraising flood hazards in the UK, involving both damage to urban properties and the benefits of protecting agricultural land. This was followed by the "Red Manual" in 1987, which investigated in greater detail the indirect or secondary effects of floods, as well as updating material on industrial, commercial and retail flood damages. In 1992 the "Yellow Manual" focused on coastal erosion and flooding problems, and systematised the assessment of the "intangible" impacts of coastal erosion on beach recreation and other use values at the coast previously left as unquantifiable. Finally, in 2005 the "Multi-Coloured Manual" brought

Table 13.1. Direct, indirect, tangible and intangible flood impacts, with examples (from Penning-Rowsell *et al.*, 2005, Table 3.1).

		Measurement	
		Tangible	Intangible
Form of loss	Direct	Damage to building and contents	Loss of life; loss of an archaeological site
	Indirect	Loss of industrial production	Inconvenience of post-flood recovery

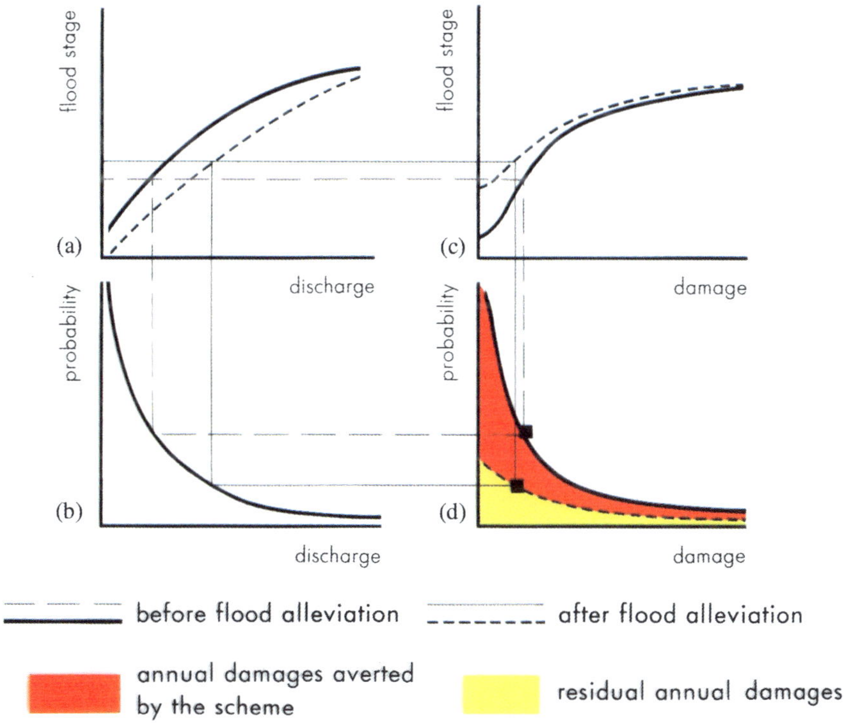

Figure 13.1. The classic four-part risk assessment diagram (Penning-Rowsell *et al.*, 2005).

together and updated in a single volume the range of techniques presented previously, based on Defra-funded research between 2002 and 2005.

Figure 13.1 illustrates the classic four-part diagram summarising the calculation of annual average flood losses. This approach has been adopted as the basis for Estimating Expected (or Average) Annual Damages (EAD or AAD) to inform national policy (NaFRA, 2006), strategic initiatives at catchment or coastal cell level (e.g. Catchment Flood Management Planning or Shoreline Management Planning). It is also used at a project level to evaluate alternative Flood and Coastal Erosion Risk Management (FCERM) options. The EAD parameter summarises the hazard at a place, in a region or nationally, by summing the impacts of floods of all probabilities of occurrence, rather than focusing on a single flood (say the "100-year" or 1% probability flood).

When calculating EAD, we need to know about potential flood damages. To this end, depth/damage curves for a suite of residential and

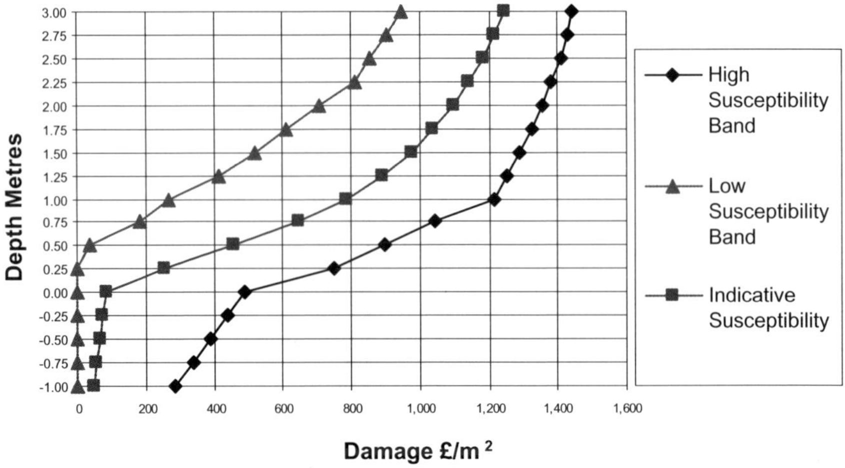

Figure 13.2. Depth/damage curves for a high street retail shop, showing high, low and indicative (i.e. typical) susceptibility (from Penning-Rowsell *et al.*, 2005).

Table 13.2. Results from the 2006 NaFRA national flood risk assessment (£ millions).

Country	EAD residential	EAD Non-residential	EAD total
England	747.4	401.3	1,148.7
Wales	173.8	88.5	262.3
Total	**921.2**	**489.8**	**1,411.0**

non-residential properties have been systematically developed by the Flood Hazard Research Centre (FHRC) since 1973 and these are applied in this process to the appropriate property at risk from flooding at a site or on the nation's floodplains. Using this type of data (e.g. as in Figure 13.2), linked to the RASP (Risk Assessment for Strategic Planning) model (Sayers *et al.*, 2003) to evaluate the probability of overtopping or breaching of flood defences, has provided a year-on-year evaluation of national flood exposure in England and Wales, expressed as national EAD. These results have been used to inform the government's expenditure plans for FCERM within its comprehensive spending reviews.

The 2006 EAD was estimated at £1.4 billion, affecting 2.3 million properties (at a mean EAD of £608 each) and up to 4 million people (National Flood Risk Assessment: NaFRA, 2006) (Table 13.2). However, EAD figures such as these are far from stable through time. Table 13.3 shows

Table 13.3. A comparison of annual flood risk assessments (NaFRA, 2006).

Flood risk assessment outcomes					
Assessment	NAAR 2000	NAAR 2001	RASP 2002	NaFRA 2004	NaFRA 2005
Damages (£bn/yr)	3.0	0.801	1.060	N/A	2.332
Properties affected (000s)	1,830	1,909	1,741	N/A	2,218

Notes: 1. NAAR = National Assessment of Assets at Risk
2. NaFRA = National Flood Risk Assessment
3. The NAAR result for 2000 is the "Do nothing" AAD. The NAAR 2001 result assumes the maintenance of the current level of protection. The NAAR 2001 "Do nothing" value is £3.5bn.

the significant fluctuation in national EAD calculated in this way over a five-year period, largely reflecting inconsistencies in data inputs and modelling techniques; the inputs are too uncertain to track with any confidence the real annual movement in the risk that Defra and the Environment Agency are seeking to manage. Data on sources and pathways of flooding, and modelling these processes, of course contributes to this uncertainty, but headline values of EAD are also very much influenced by the quality of receptor and consequence data.

As a result, the way in which we count properties at risk, measure their floor areas, estimate their thresholds of flooding, select depth/damage data appropriate to the land use present will all affect the headline EAD. Taking the last item first — the depth/damage data — each such dataset (Figures 13.2 and 13.3) has an "indicative", "high" and "low" susceptibility "curve", reflecting the range of damages surveyed in the sample of properties the dataset represents. Where the range is wide, as with industrial premises (Figure 13.3), which could vary from chemical production to a scrap-metal yard, the uncertainty is substantial when applying the curve to all properties at risk coded as "factory" in land use inventories. In NaFRA this sensitivity was regarded as "very high" when applying the range of susceptibility data to the national property database.

13.1.2. *Indirect flood losses*

The types of flood damage or disruptions that are included will also influence the level of EAD. Thus, in England and Wales, the calculation of damages has concentrated largely on impacts on residential and non-residential receptors (NaFRA, 2006). Although evaluation techniques for

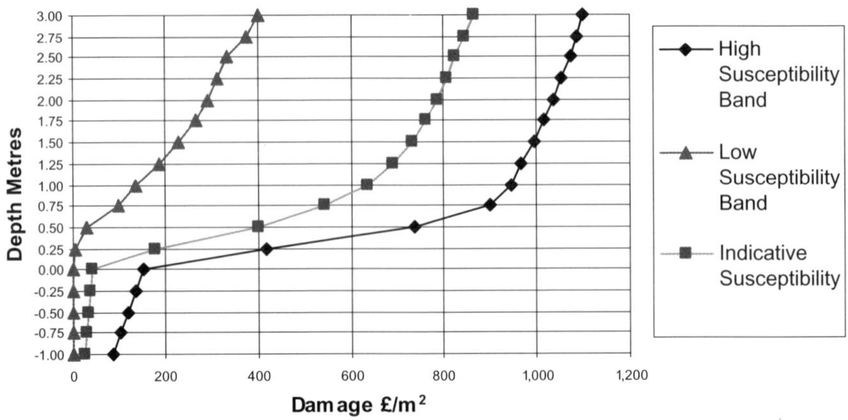

Figure 13.3. Depth/damage curves for industrial premises (from Penning-Rowsell *et al.*, 2005). Systematic and measurement errors need to be eliminated and data collection methods stabilised year-on-year to give confidence in the data and therefore confidence in the assessment of changing risk. The NaFRA project has set perturbation limits based on expert judgment (rather than statistical methods) for the expected variations in measurements. As an example, simply by taking the low and high susceptibility depth/damage extremes as against the indicative depth/damage curve decreases or increases EAD by 42% and 27% respectively.

all direct and indirect impacts are relatively well understood (Penning-Rowsell *et al.*, 2005), analysts have tended to shy away from quantification (applying money as a numeraire) to indirect flood losses (Table 13.1). This is partly because of time and resource constraints but largely because of the uncertainty as to the scale of the impacts and hence damages to the many facilities concerned; for example, to an electricity sub-station, a water treatment plant or to a flood disrupted road. None of these impacts is included in current national flood risk assessments.

However, the floods in England and Wales in summer 2007 (affecting around 65,000 properties through direct flooding) have highlighted the fact that it may be these hitherto ignored impacts (consequences) that need to be recorded to reflect the real cost of flooding (Pitt, 2008). Thus the Mythe Water Treatment Works in Tewkesbury was inundated and cut off the water supply to some 180,000 people for several weeks. Electricity sub-stations, also serving communities in Gloucestershire and beyond, with a hinterland population of over 500,000 people, narrowly escaped flooding. The social impacts of flooding (at worst, often ignored, or radically underestimated at best) could in these circumstances be as significant as the financial or economic impacts of flooding.

Table 13.4. Risk matrix.

IMPACT	Low (Probability of flooding)	Moderate (Probability of flooding)	Significant (Probability of flooding)
Significant	Medium Risk	**High Risk**	**Very High Risk**
Moderate	Low Risk	Medium Risk	High Risk
Low	Negligible Risk	Low Risk	Medium Risk

PROBABILITY OF FLOODING

A filter approach is being developed to determine the scale of impact of these wider consequences. The Environment Agency's flood zone mapping (www.environment-agency.gov.uk/homeandleisure/floods/) divides the nation's floodplains into having "significant", "moderate" and "low" probability of flooding, based on 100 m grids. Developing metrics to define the consequences or impacts of this flooding to individual receptors as "significant", "moderate" or "low" and combining with flood likelihood will enable high risk receptors to be distinguished from low risk receptors (Table 13.4). Concentrating on the economic and social consequences to receptors in the "high" and "very high" risk boxes of this matrix will help to develop a consistent approach to gaining an understanding of the overall scale of potential damage and disruption, either nationally or at a local level.

The necessary metrics (for utilities/infrastructure) will need to describe:

- The *susceptibility* of a utility/infrastructure and supporting networks, etc. to physical damage through contact with flood water;
- The *dependency* of properties served by utility plants and networks both within the flood zone and beyond;
- The ease of *transferability* of production or supply from a non-flooded site (the degree of redundancy built into the system);
- The *size* of the utility/infrastructure (e.g. population served; output/throughput; daily road users; etc.).

Efficient development of this system of metrics will lead to a better understanding of national and regional EAD which currently lacks this key indirect loss dimension.

13.1.3. *Uncertainties in data and classification for property at risk*

The underestimation of EAD as a result of a poor understanding of the impacts of floods on particular receptors is only one aspect of uncertainty in flood loss assessments. Quality of data attributes (size, flood threshold, etc.) and the classification of property to match appropriate depth/damage characteristics are key to appropriate damage estimates.

A comparison of NaFRA 2006 economic damages for an area of Carlisle also covered by a detailed Project Appraisal Report (PAR) showed the contrasts evident in Table 13.5. The major differences are that 277 more residential properties were represented in the PAR than in NaFRA, because of different methods of data assembly and spatial aggregation. As a result, direct economic damages per residential property in the PAR were nearly four-times those in the NaFRA and the direct economic damages per commercial property in the PAR were over six-times those in NaFRA. The split of potential damages between residential and non-residential (commercial) sectors is broadly similar between the PAR (44%:56%) and NaFRA (39%:61%), but the PAR shows substantially greater potential damages.

There are several main reasons for this: differences in assigning property types; determining average floor areas, particularly non-residential properties; the treatment of basements and application of basement depth/damage data; the assessment of flood depths from modelling (water surface and terrain); the assessment of property thresholds; property valuations; and the assessment of the number of properties at risk.

There are therefore significant uncertainties that surround the estimation of economic damages from both NaFRA and the PAR. These reside both in the data used and methods applied. When some ground truthing of the NaFRA outputs at this site was undertaken, for comparison to the PAR, this showed that the national assessment provides an underestimate of total

Table 13.5. Project appraisal and NAFRA data for Carlisle, UK.

Residential	**PAR 2007**	**NaFRA**
Number of properties	1,688	1,411
EAD Economic damages: do minimum	£966,148	£204,480
Damage per property (EAD)	£572	£145
Commercial (capped)	**PAR 2007**	**NaFRA**
Number of properties	390	434
EAD Economic damages: do minimum	£1,483,986	£260,031
Damage per property (EAD)	£3,805	£599

risk, although a decision as to which is the most accurate assessment cannot be taken without further detailed study. It is equally as likely that there are gross discrepancies and systematic measurement errors in the PAR process as in NaFRA.

13.1.4. *A method to improve data quality: the data quality score system*

A method described by Penning-Rowsell *et al.* (2005) in the MCM has been developed to improve data quality in flood risk assessments. This method is applied to the depth-damage data[a] used, the property ground floor area (since damage, especially to non-residential properties is presented as pound-per square metre), and the property type code.

The method recognises that the Environment Agency has developed a National Property Database[a] (NPD), combining Ordnance Survey address point data (locating property spatially) with Government Valuation Office Agency (VOA) property type lists which are in turn linked to MCM depth/damage codes and, amongst other attributes, floor space. As with any large hybrid database, gaps in records lead to algorithms being created to plug the missing data gaps, leading to potential error and inconsistency. Data on basement and upper storey property records can also be suspect. To combat potential error a data quality score system has been developed to identify and flag where the input data need review and improvement (Table 13.6).

The data quality scores are applied as in Table 13.7. If a score total across fields "A" to "C" exceeds a pre-set threshold (e.g. a score of 7) then

Table 13.6. Data Quality Scores (DQS).

Score	Description	Explanation of data
1	"Best of Breed"	No better available: unlikely to be improved in the near future
2	Data with known deficiencies	To be replaced as soon as the third parties responsible for the data reissue it with an update, etc.
3	Gross assumptions	Not invented but deduced by the project team from experience or related literature/data sources
4	Heroic assumptions	No data sources available or yet found: data based on no more than educated guesses

[a]The NPD (National Property Dataset) was superseded in 2010 by NRD (National Receptor Database).

Table 13.7. Application of the data quality score system (Penning-Rowsell *et al.*, 2005).

A ***Depth/damage (D/D) data applied***

Score and its associated data attributes:

1. D/D data from MCM is available for the MCM land use in question and cannot be improved without further research
2. D/D data from MCM is available for the MCM land use in question but sample size is known to be small and improvements are definitely possible with further research
3. D/D data is not available and an equivalent D/D set is suggested or an aggregate weighted mean suggested e.g. Bulk classes, all shops or stores (21), or the non residential property weighted mean
4. No reasonable data is available (e.g. golf courses, football stadia, cemeteries, playing fields etc.) or "miscellaneous" land uses are recorded

B ***Ground floor area***

Score and its associated data attributes:

1. If flagged as 1 in NPD and validated in Ordnance Survey's MasterMap (OSM)
2. Unchanged from NPD having been through the 2006 property area screening method
3. If screening shows discrepancy between RV (Rateable Value) and floor area take MCM Mean floor area for MCM code in question
4. No area in NPD, take MCM mean

C ***Land Use Type (MCM code)***

Score and its associated data attributes:

1. MCM code validated from other sources e.g. GOAD town centre surveys or ground truthing
2. VOA matches address point
3. None
4. Random match between VOA and address point to unmatched property

the EAD for each property calculated is deemed unfit for the purpose of assessing risk until the data record attributes are improved. In NaFRA this test is largely applied by impact zones (100 m grids) for all properties contributing more than 1% of total EAD for that catchment. The process tends to eliminate overestimated EAD, but it may equally apply to show where EAD in impact zones is likely to have been underestimated.

13.1.5. *Ground truthing*

To identify types of errors and related issues, some of the strategic properties that were inundated by the flooding events of July 2007 have been examined to inform the fitness for purpose debate (see Figures 13.4 and 13.5). The results indicate the following:

- There are properties on the ground that are not represented at all in the NPD;
- There are properties on the ground that are given incorrect floor areas in the NPD;
- There are cases where the sum of floor areas in a multi-address property (e.g. a block of flats) is less than the size of the property as a whole.

Thus Figure 13.4 shows that the Trebor Bassett confectionery factory adjacent to the River Don in Sheffield, which suffered many millions of pounds of flood damages in June 2007, is not represented in NPD at all. Figure 13.5 shows the floor area for the Mythe water treatment works in Tewkesbury is actually four times that recorded in NPD. Finally, Figure 13.6 shows the high risk impact zone for Worcester, where the EAD of £662,673 is radically overestimated because most of the 47 residential properties are raised above ground level and therefore would not suffer physical damages in a flood, although they would be subject to disruption.

The widespread nature of these property omissions and the associated impact on underestimation of the EAD make it important that a solution is developed in the future to correct for these property omissions and underestimates so that an estimate of the true EAD can be made. The NRD is responding to these needs but its algorithms are still some way from determining wholly accurate land use and property characteristics, essential for damage estimations.

13.1.6. *Combining better ground truthing with the DQS system*

To obtain a better understanding of the uncertainty of flood damage assessments with changes in data inputs has led to the development of iterative procedures where by we add refinement to the input data in response to excessive data quality scores, often via substantial ground truthing of the data.

To this end, Figure 13.7 summarises the Present Value of Damages (PVd) baseline "do nothing" option (against which plausible FCERM options are gauged) for several thousand properties exposed to flood risk in the fluvial Lower Thames, west of London. The first cut assessment of potential flood damages, with initial data and modelling, was in excess of £1 billion but was stabilised at around £200 million by targeting improvement of key data input attributes for properties contributing individually more than 0.1% of the total calculated floodplain PVd.

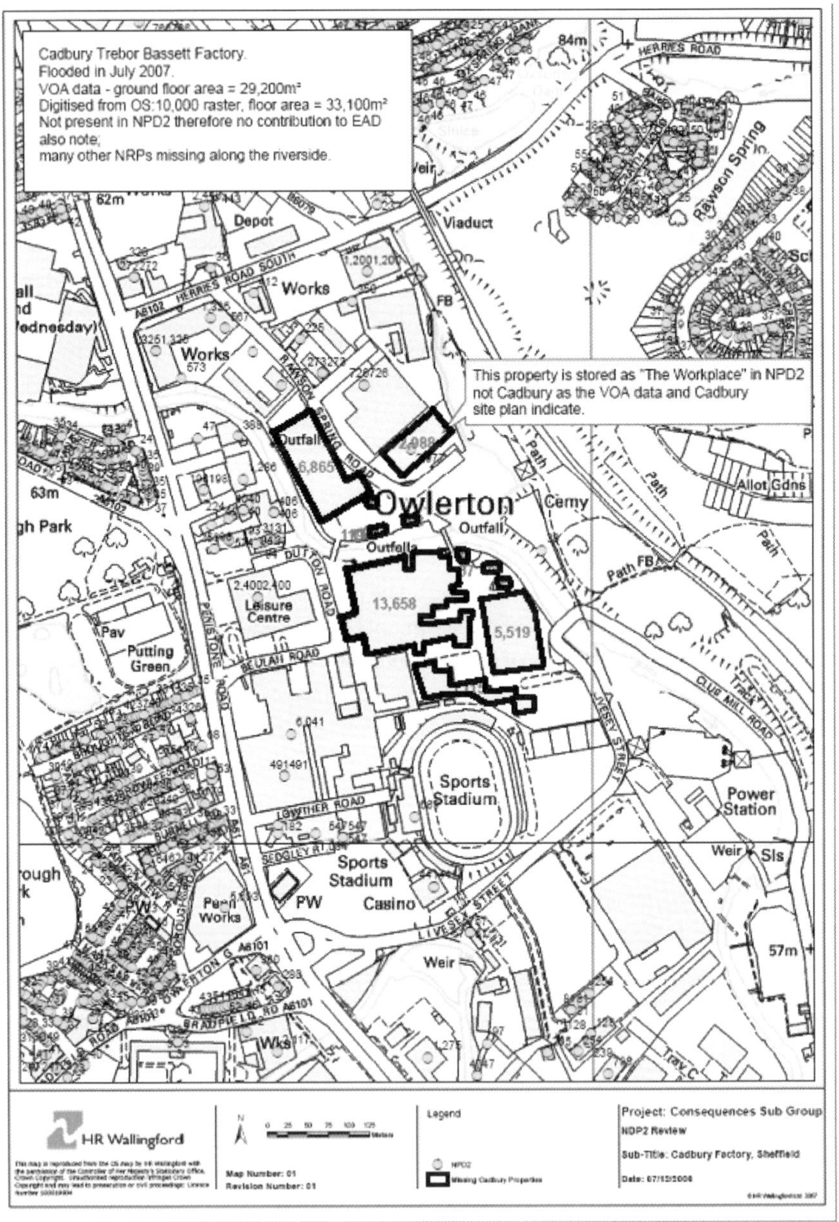

Figure 13.4. The Trebor Bassett site, Sheffield (properties missing in the NPD are shown by the heavy black outline).

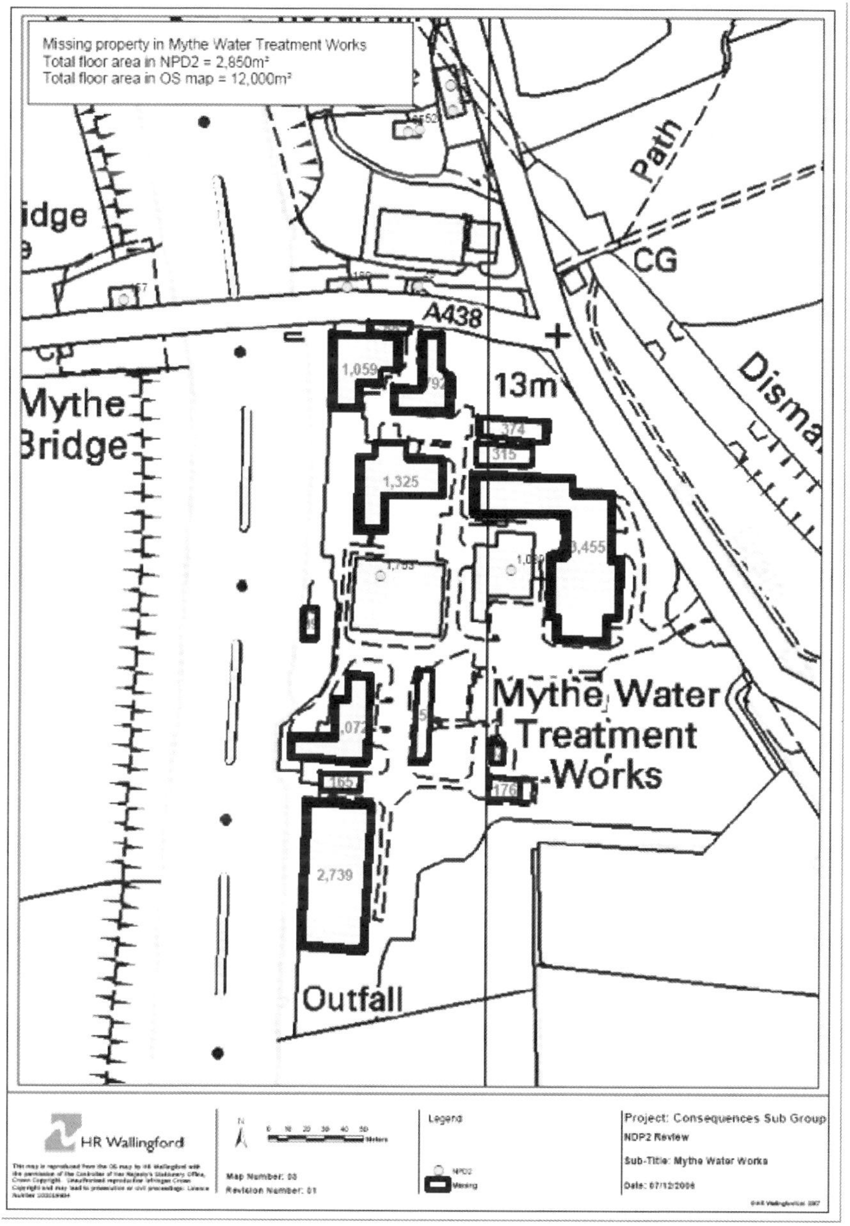

Figure 13.5. The Mythe Water Treatment Works, Tewkesbury: buildings missing in the NPD shown by the heavy black outline.

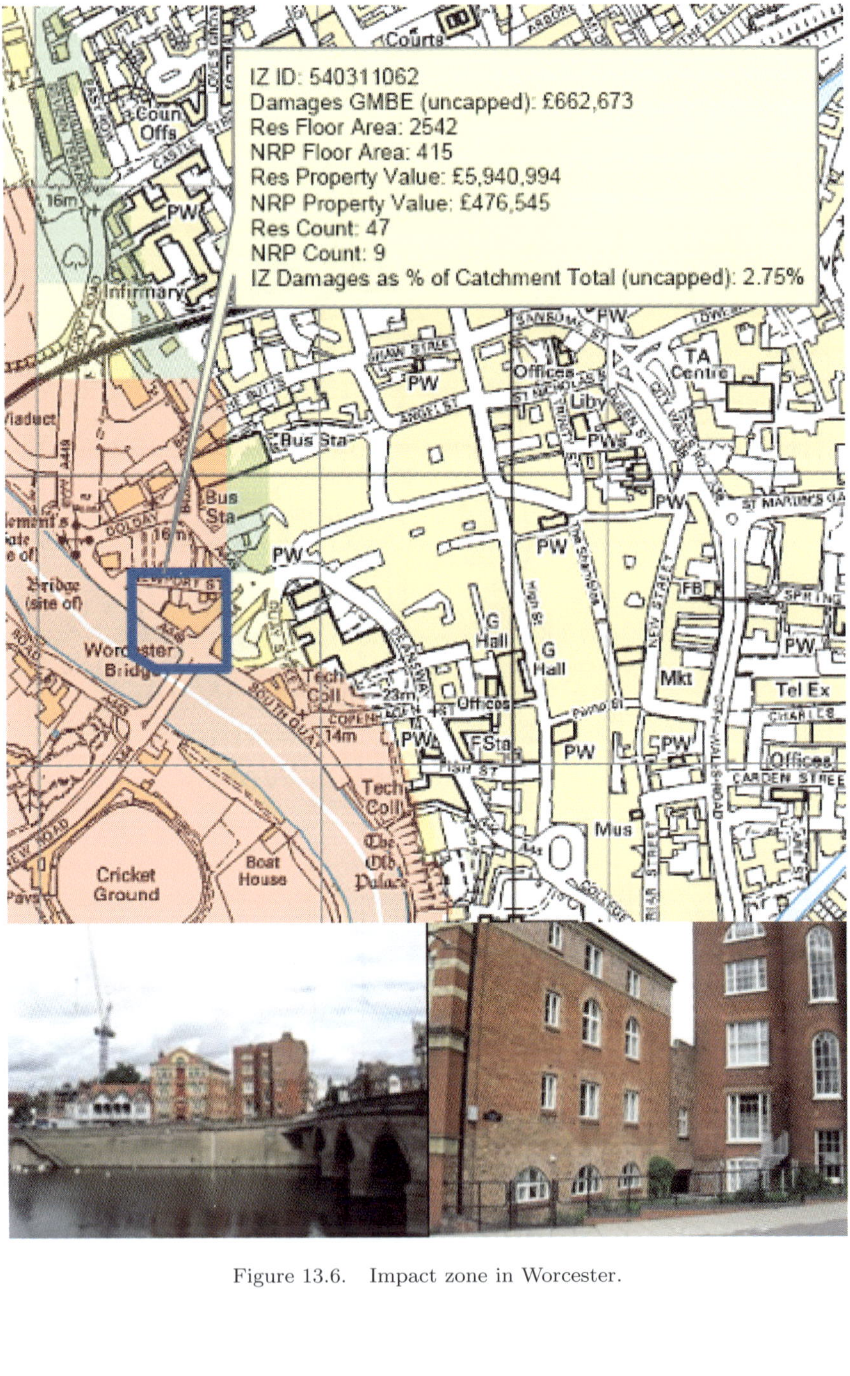

Figure 13.6. Impact zone in Worcester.

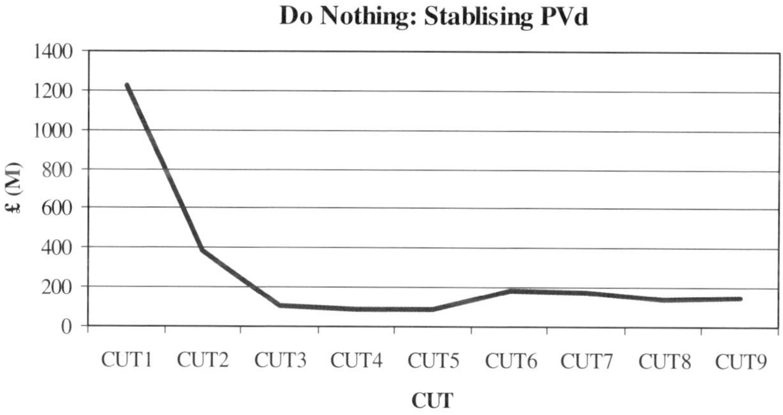

Figure 13.7. Stabilising PVd for the Lower Thames project (the different results represent progressive refinements to the input data sets, based on the data quality score system, designed to remove uncertainty and increase accuracy).

That ground truthing specifically focused on obtaining better data on the thresholds of flooding, property ground floor areas, property type, and the depth/damage data applied. At each stage of the data improvement process the risk assessment was redone until the results were stable. A very substantial error in the final risk assessment was avoided in this way, through targeting areas of maximum data uncertainty.

We need to see this kind of improved risk assessment in its policy context. Although around £600 was allocated in England and Wales for flood defence projects in financial year 2007/8, the Government programme of departmental cuts from 2010 has significantly reduced this allocation from the projected spend of about £1 billion per year. Without careful audit of the accuracy or otherwise of the S-P-R-C parameters that influence EAD calculations, and hence an understanding of the uncertainty of this risk assessment process, this reduced expenditure cannot be effectively prioritised.

13.2. Loss of Life in Floods: Uncertainties and Data Issues

13.2.1. *The research context*

Research carried out by Priest *et al.* (2008) focused on developing a methodology to estimate loss of life in flood events. The context and rationale for this work is that in order to reduce the risk to life in floods it is necessary to understand the causes of that loss of life. To do that it is necessary to pinpoint where, when and how loss of life is more likely to occur in floods, and what type of interventions and flood risk management measures may be effective in eliminating or reducing these serious injuries and fatalities.

The objectives of this research were therefore to develop further a model, or models, to provide insight into, and estimates of, the potential loss of life in floods, based on research already undertaken in the UK and new data collected on flood events in continental Europe. We also aimed to build partly on existing parallel work (e.g. Jonkman and Penning-Rowsell, 2008), through the use of GIS, to develop systems to map the outputs of the Risk to Life models that were developed, thus providing some case example estimates of the potential loss of life in floods. The research took as a starting point the Risk to People model developed in the UK (HR Wallingford, 2003; 2005) and assessed the applicability of this model for flood events in continental Europe, where these events tend to be more severe and life threatening than in the UK. Data on flood events were gathered from 25 locations across six European countries as well as data from an additional case study in the UK.

A number of problems were identified with the current Risk to People model when applied to the flood data collected from continental Europe. In particular the model was not designed for the major rivers and mountainous catchments that are common there compared with the UK and it therefore resulted in dramatic over-predictions of injuries and fatalities. Moreover, the model was found to contain several structural weaknesses, such that predicted fatalities could exceed the population at risk.

Research conducted into the factors surrounding European flood fatalities also highlighted the importance of institutional arrangements and mitigating factors such as evacuation and rescue operations in affecting — i.e. reducing loss of life "on the ground" vis-à-vis model predictions. Finally, the UK model was found to be hugely (and over) sensitive to variables measuring "people vulnerability", which in much of the flooding in the wider European context is arguably of less importance in than it is in the UK. This is because the European floods for which information about flood fatalities were available had water depths and velocities of such a magnitude that would threaten the safety of most people, not only the elderly and infirm highlighted by the UK model. A more detailed critique of this model is provided by Priest *et al.* (2008).

13.2.2. *A new "threshold" model*

As a result of our research a new semi-qualitative "threshold" model which combines hazard and exposure thresholds and mitigating factors has been developed (Table 13.8) to assess risk to life from flooding in the wider European context (herein termed the Risk to Life model). The model has been designed to be flexible enough to be used and applied at a range of scales, from a broad assessment at a regional or national scale, to a more detailed local scale. This flexibility is essential as not all European countries have detailed flood extent, flood velocity or flood depth data — or similar information — that is readily available. However, it is envisaged that the model could be used to allow flood managers to make general and comparative assessments of risk to life in their localities, and to consider using the results to assist in targeting resources before, during and after flooding so that loss of life is minimised.

Regarding uncertainty in the resulting predictions of loss of life in European floods (i.e. using Table 13.8 in practice), we consider the following seven factors to be of particular significance:

1. Calibration issues. There are many data issues here, but good quality data on the numbers of deaths in floods are surprisingly limited and therefore it is necessary to work with what we currently have and try to improve these data-sets and our understanding of the factors involved (Priest *et al.*, 2008, Section 4.3, p. 32).
2. The core variables. The reliance of the Risk to Life model on the need to be able to produce depth/velocity product model (as in the Risk to People model also) means that if this cannot be done, the results will be invalid, not just uncertain. Therefore the inherent uncertainties in that kind of flood inundation modelling mean that risk to life and injury cannot always be predicted. Indeed, the locations where this is now possible are in the minority in England and Wales, and also in continental Europe.
3. Zoning of the risk. Within the original Risk to People method, an area at risk needed to be designated and data provided for that area. It thereby is assumed in modelling risk to life and injury for an area that all of the data are homogenous across the same zone. To achieve consistency with this assumption leads to considerable problems in defining these risk zones, notable amongst which is the different descriptor variables that should be used to do this: depth, velocity, etc. (Priest *et al.*,

Table 13.8. The Risk to Life model (for full model see Priest *et al.*, 2008).

DEPTH x VELOCITY MID-RANGE	OUTDOOR HAZARD	NATURE OF THE AREA	STRUCTURAL DAMAGES	RISK TO LIFE CATEGORIES WITHOUT MITIGATION	MAIN FACTOR LEADING TO FATALITIES	MF	RtoL
$>7 m^2s^{-1}$	Extreme Dangerous for all	**3. High vulnerability** **2. Medium vulnerability** **1. Low vulnerability**	Total collapse may occur. Structural damages probable in particular for properties with poor quality building fabric	Risk to life in this scenario is extreme as not only are those in the open very vulnerable to the effects of the flood waters but those who have also sought shelter are also very vulnerable due to the fact that building collapse is a real possibility (Red)	**Hazard and building collapse dominated**	1 2 3 4	R R O Y
1.1 to 7 m^2s^{-1}		**3. High vulnerability**		All those exposed to the hazard outside will be in direct danger from the floodwaters. Those living in mobile homes will be at risk from the high depths and velocities and those in bungalows will be at risk from not being able to escape to upper floors. Those in very poorly constructed properties will also be vulnerable from structural damages/collapse (Red)		1 2 3 4	R O Y Y
		2. Medium vulnerability	Structural damages Possible	All those exposed to the hazard outside will be in direct danger from the floodwaters. Damages to structures are possible. Those in unanchored wooden frames houses are particularly vulnerable. With very deep waters there is the risk of some not being able to escape. (Orange)	**Hazard dominated**	1 2 3 4	O O Y G
		1. Low vulnerability		All those exposed to the hazard outside will be in direct danger from the floodwaters. In this scenario those residing in these properties have the lowest risk although structural damages are still possible in wooden properties (Yellow)		1 2 3 4	Y Y Y G
0.5.to 1.1 m^2s^{-1}	High Dangerous for most	**3. High vulnerability**	Structural damages and collapse possible for properties with poor quality building fabric	Those outside are vulnerable from the direct effects of from the floodwaters. In addition, those in bungalows will be vulnerable in deeper waters. People will also be afforded little protection in mobile homes and campsites. Those in very poorly constructed properties will also be vulnerable from structural damages and/or building collapse. Vehicles are likely to also stall and lose stability. (Orange)		1 2 3 4	O O Y G
		2. Medium vulnerability	Structural damages – less likely and less severe	Anyone outside in the floodwaters will be in direct danger from the floodwaters. It is here at this point where behaviour becomes significant as structural damages are less likely so those inside should be on the most part protected. Vehicles are likely to stall and lose stability. Are people undertaking inappropriate actions such as going outside where is it not necessary? (O)	**Behaviour dominated**	1 2 3 4	O Y Y G
		1. Low vulnerability		Anyone outside in the floodwaters will be in direct danger from the floodwaters. It is here at this point where behaviour becomes significant as structural damages are less likely so those inside should be on the most part protected. Vehicles are likely to stall and lose stability. Are people undertaking inappropriate actions such as going outside where is it not necessary? (Y)		1 2 3 4	Y G G G
0.25 to 0.5 m^2s^{-1}	Moderate Dangerous for some	**3. High vulnerability**	Structural damages possible for properties with very poor quality building fabric	Only the most vulnerable should be in direct danger from the floodwaters. (e.g. children and the elderly). They are obviously most vulnerable as they are less able to save themselves from the flood waters and in this category the shelter may not protect them. Motor vehicles may become unstable at these depths and velocities. Those in very poorly constructed properties may also be vulnerable from structural damages. (Yellow)	**People vulnerability dominated though some behaviour-related**	1 2 3 4	Y Y G G
		2. Medium vulnerability	Unlikely	Only the most vulnerable should be in direct danger from the floodwaters (e.g. children and the elderly). Motor vehicles may become unstable at these depths and velocities. Those who seek shelter should be safe.(Yellow)		1 2 3 4	Y G G G
		1. Low vulnerability		Only the most vulnerable should be in direct danger from the floodwaters (e.g. children, the elderly). Motor vehicles may become unstable at these depths and velocities. Those who seek shelter should be safe (Green)		1 2 3 4	G G G G
$<0.25 m^2s^{-1}$	Low Caution	**3. High vulnerability** **2. Medium vulnerability** **1. Low vulnerability**	Unlikely	A very low risk to adults either out in the open or who is in a property. There may be a threat to the stability of some vehicles even at these low depth-velocity factors. (Green)	**Low risk**	Mitigation factors here are unlikely	G

2008, Section 4.4. p. 35). Within the new Risk to Life model (Priest *et al.*, 2008, p. 111) this procedure is not strictly necessary as the methodology permits areas of different sizes to be overlapped in a layering approach, as illustrated in the French example for the town of Ales (Priest *et al.*, 2008, Section 10.3, p. 118).

4. The data used in the model. The data that are used in the Risk to Life model causes some degree of uncertainty. This is firstly due to the proxies that are used (e.g. old/young age as a surrogate for vulnerability, as age does not take into consideration of the degree of help that people receive in the flood event). Secondly, uncertainty is generated when deciding how to estimate the number of people who are at risk (e.g. what data is best to use — census data; local counts, etc.) as it will never be possible accurately to estimate the numbers of people who are truly exposed (for instance, how many people will be on the street versus how many people will be able to shelter), time of day of the flood, season, etc. Priest *et al.* (2008, Chapter 6) show that the process of applying the Risk to People model to the Boscastle case study highlights many of the problems/uncertainties with the approach.
5. Banding. The banding of the physical characteristics (e.g. the depth-velocity product) within the new approach (Table 13.8) does allow for some uncertainty with these figures but of course this will be particularly susceptible at the boundaries of the different categories (Priest *et al.*, 2008, Section 9.8, p. 114).
6. The approach of the Risk to Life model. The fact that the new model provides bands and scales (Table 13.8) means that it is perhaps less precise — and arguably therefore more uncertain — than the Risk to People approach (with its deterministic estimation of deaths and injuries) but it does mean that the bands can take many of the uncertainties and complexities into account and provides a guide rather than an absolute answer that may be orders of magnitude wrong in its predictions of fatalities (and as was highlighted by some of the continental European cases). The new model also allows the tailoring of the approach, and therefore possibly a reduction in the uncertainty, by local flood risk managers through the application of more site-specific data (Priest *et al.*, 2008, Section 9.8, p. 114).
7. Human behaviour and "chance" factors. Accounting for variations in human behaviour and chance factors cannot be accommodated in either the Risk to People or the Risk to Life models, yet the treatment of these factors will always lead to uncertainties in the numbers of

people who are predicted as killed or injured in floods. Many authors in this area of research cite this as a problem, and it is mentioned in many places where we have looked at case studies. We have tried to include this in the model where it assesses the bands where healthy adults should be safe unless they do something "foolish" but this is an imperfect solution to a difficult problem (Priest *et al.*, 2008, Chapter 5, Section 5.3.1, p. 43).

13.2.3. *Future directions*

A number of problems remain in further refining the Loss of Life model. Firstly, the results generated from the application of the new approach are, as with other models of this type, hugely sensitive to the data input into the model and the different values attributed to the model components.

Secondly, this factor, along with general limitations in the availability of data, have highlighted the need to establish reliable, systematic and consistent methods for collecting data on injury and loss of life following flood events across Europe, as well as for the need to make available the data that is collected. A key constraint relates to who is responsible for collecting such data; which at present varies from agencies at local, regional and national levels. Therefore we have suggested (Priest *et al.*, 2008) that protocols are needed to address this issue.

Thirdly, several questions can also be raised at this point about the purpose of modelling risk to life. For instance, is it aimed at modelling a worse case scenario? It is unlikely that it will ever be possible to estimate accurately the number of deaths from a flood event. Therefore, should the modelling simply be used as a guide to the identification of those areas which are most likely to suffer fatalities from flooding? It is also possible to question the feasibility of trying to apply one model to assess the risk to life for floods across the whole of Europe, due to the large differences in types of flood hazard, the areas affected, people's vulnerability, and the varying institutional arrangements (particularly regarding emergency response and evacuation).

13.3. Conclusions

The human dimensions of flood risk have received far greater attention in the last decade than previously (e.g. Jonkman, 2007; Tapsell *et al.* 2002). However, the uncertainties in predicting human aspects of floods have often been neglected in the enthusiasm of agencies and institutions for appraising investment in FCERM with techniques such as benefit–cost analysis which suggest that precision in the economic and human aspects of that calculus is far greater than it really is.

This chapter has illustrated this lack of precision in the area of risk assessment with regard to property at risk in UK floodplains and the propensity of floods to lead to loss of life. We see that predictions in this area can be at variance with reality by orders of magnitude, not just small degrees of uncertainty or "error". The causes of these differences are multifarious: data issues, modelling biases, and conceptual problems.

It will seem trite and self-serving for us to suggest that "more research needs to be done in this field". The only excuse we put forward to support this contention is that these topics have received systematic attention only in the last decade or so, whereas other areas of flood risk assessment have a history of many decades or more. What is needed is balance: we should not be content to polish the exponents of equations predicting flood probability if the "consequences" variables suffer from the kinds of errors and uncertainties that this chapter has explored.

References

Defra (2004). *Making Space for Water: Developing a New Government Strategy for Flood and Coastal Erosion Risk Management in England: A Consultation Exercise*, Defra, London.

Defra (2005a). *Making Space for Water: Taking Forward a New Government Strategy for Flood and Coastal Erosion Risk Management in England. First Government Response to the Autumn 2004: Making Space for Water Consultation Exercise*, Defra, London.

Defra (2005b). *Making Space For Water: Taking Forward a New Government Strategy for Flood and Coastal Erosion Risk Management in England, Delivery plan*, Defra, London.

HR Wallingford (2003). *Flood Risks to People Phase 1. Final Report Prepared for Defra/Environment Agency Flood and Coastal Defence R&D Programme*, HR Wallingford, Wallingford, Oxon.

HR Wallingford (2005). *R&D Outputs: Flood Risks to People, Phase 2. FD2321/TR1 The Flood Risk to People Methodology*, Defra/Environment Agency Flood and Coastal Defence R&D Programme, HR Wallingford, Wallingford, Oxon.

Jonkman, S.N. (2007). *Loss of Life Estimation in Flood Risk Assessment: Theory and applications*, PhD dissertation, Delft Cluster, Delft, the Netherlands.

Jonkman, S.N. and Penning-Rowsell, E.C. (2008). Human instability in flood flows, *J. Am. Water Resour. As.*, **44**, 1–11.

Met Office (2002). Severe Weather Impact Model. Available at http://www.metoffice.gov.uk/weather/uk/links.html (Accessed 01/05/12).

NaFRA (2006). *National Flood Risk Assessment 2006*, Environment Agency, London.

Penning-Rowsell, E.C., Johnson, C., Tunstall, S. *et al.* (2005). *The Benefits of Flood and Coastal Risk Management: A Manual of Assessment Techniques*, Middlesex University Press, London.

Pitt, M. (2008). *Learning lessons from the 2007 floods: An Independent Review by Sir Michael Pitt*, UK Government, Cabinet Office, London.

Priest, S., Wilson, T., Tapsell, S. *et al.* (2008). Building a model to estimate risk to life for European flood events, Floodsite Report Number T10-07-10.

Sayers, P.B., Hall, J.W. Dawson, R.J. *et al.* (2003). "Risk Assessment for Flood and Coastal Defence Systems for Strategic Planning (RASP) — A National Scale Application and a Look Forward to More Detailed Methods" in: *Proceedings of the 38th Defra Flood and Coastal Management Conference, Keele, UK, July 16–18 2003*, pp. 5.2.1–5.2.13.

Tapsell, S.M., Penning-Rowsell, E.C., Tunstall, S.M. *et al.* (2002). Vulnerability to flooding: health and social dimensions, *Philos. T. R. Soc. A*, **360**, 1511–1525.

CHAPTER 14

Uncertainty and Sensitivity Analysis of Current and Future Flood Risk in the Thames Estuary

Jim W. Hall
Environmental Change Institute, University of Oxford, UK

Hamish Harvey
Bill Harvey Associates Ltd., Exeter, UK

Owen Tarrant
Environment Agency, UK

14.1. Introduction to Flood Risk Management in the Thames Estuary

London and the Thames Estuary are threatened by flooding from storm surges in the North Sea and high fluvial flows in the River Thames and its tributaries. The area at risk is low-lying, densely populated and of critical importance to the economic wellbeing of the UK. In 1953 a storm surge in the North Sea breached flood defences at several locations in East Anglia and the Thames Estuary, resulting in disastrous flooding in which 309 people died. Following the 1953 floods, plans were put in place for construction of a comprehensive system of flood defences in the Thames Estuary, the centrepiece of which was the Thames Barrier, completed in 1982 (Gilbert and Horner, 1984). As well as this moveable barrier, the flood defence system incorporated 337 km of raised defences together with eight other major barriers across tributaries to the Thames. It has been estimated that the current flood risk management system will provide a flood defence standard of 1:1000 years in the year 2030 for most of the tidal Thames floodplain, and at present the standard is higher.

With more than 20 years passed since the barrier was constructed, the Environment Agency embarked upon the Thames Estuary 2100 (TE2100) project to review strategic options for flood risk management in the Thames Estuary over the course of the 21st century. Over this time frame sea level rise and other climatic changes can be expected to increase flood risk. Continued development in the natural floodplain to promote economic growth could potentially further increase the consequences of future flooding.

The TE2100 project has involved comprehensive analysis of flood risk in the Thames Estuary and of options for future flood risk management. Summaries of the project are available at the Environment Agency's website http://www.environment-agency.gov.uk/te2100. In keeping with the theme of this volume, in this chapter our emphasis is upon the analysis of uncertainties in flood risk analysis in the Thames Estuary. We do not deal with the analysis of options or decisions in TE2100, though the analysis of uncertainties described here is intended to inform the process of deciding between flood risk management strategies.

The principles and practice of uncertainty analysis are described elsewhere in this volume. The aim of this chapter is to provide a practical example of uncertainty analysis in a complex system in which:

- the consequences of flooding are potentially very serious, in societal and economic terms;
- the probabilities of flooding are, on the whole, low so require careful analysis to quantify and cannot be estimated from observed flood frequencies;
- the potential investment of government resources is high;
- some aspects of the present flood risk, for example, the reliability of flood defences, are quite uncertain.

This combination of circumstances means that special attention has been paid to the analysis of uncertainty in TE2100, which means that it is an illustrative example for similarly high profile applications worldwide.

14.2. Flood Risk Analysis of the Thames Estuary System

The Thames Estuary flood defence system is illustrated in Figure 14.1. Flooding within the Thames Estuary is a consequence of extreme water

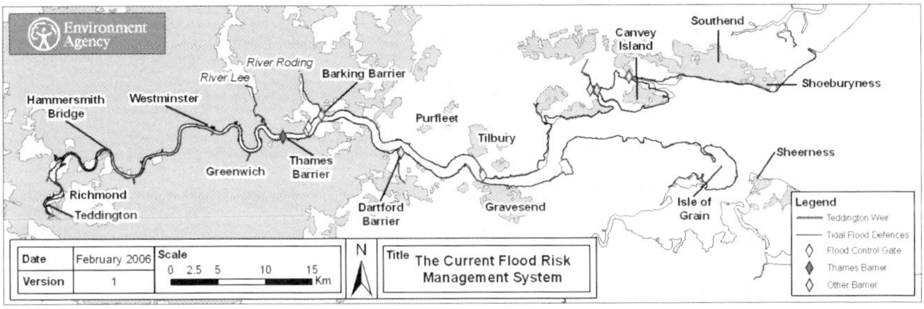

Figure 14.1. Main components of the Thames Estuary flood defence system.

levels in the southern North Sea (at the mouth of the estuary in the east) and/or extreme flows in the River Thames, which has its tidal limit at Teddington in the west (the nearby gauging station is at Kingston, which is referred to in the description below). Practically the entire length of the river/estuary east of Teddington is protected by raised defences, as well as by tide-excluding barriers, most notably the moveable Thames Barrier, which is designed to prevent extreme tides from penetrating up the river into London. Flooding in this system may therefore occur by the following mechanisms:

- extreme water levels in the estuary exceeding the crest level of the flood defences;
- breaching of the flood defences;
- failure to close the Thames Barrier or other barriers.

Flooding may also occur due to extreme flows in the tributaries that flow into the estuary (e.g. River Lee and River Roding) or intense local rainfall, but these mechanisms are not addressed in this chapter.

Analysis of flood risk in the estuary requires a system model that integrates:

- the joint probability of water levels in the estuary and flows in the River Thames (together referred to as "boundary conditions");
- a hydrodynamic model, which is used to compute water levels throughout the estuary for given combinations of boundary conditions;
- the probability of barrier failure to close;
- reliability analysis of the flood defence structures;
- two-dimensional models of inundation of the floodplain in the event of flood defence overtopping or breaching;

- damage functions (e.g. depth-damage functions) that allow computation of the consequences of flooding.

This process may be summarised in the following equation for Expected Annual Damage (EAD):

$$EAD = \int p(\mathbf{x})D(\mathbf{x})d\mathbf{x}, \tag{14.1}$$

where $D(\mathbf{x})$ denotes the damage associated with event-defining vector $\mathbf{x}$ and $p(\mathbf{x})$ denotes the associated probability density. In other words, EAD is an average over all *aleatory* uncertainty (that to do with natural variability — Hall and Solomatine, 2008), encoded in $p(\mathbf{x})$. The purpose of uncertainty analysis is to explore the influence of *epistemic* uncertainty (that to do with a deficit of knowledge — Hall and Solomatine, 2008), whether with respect to parameter values or model structure.

The aleatory uncertainties integrated out in Equation (14.1) are:

- The tidal variation of sea level at the mouth of the estuary (taken as the tide gauge at Southend) and discharge entering the estuary at Kingston. It is assumed that an event can be adequately described by a pair of scalars — peak water level at Southend and the corresponding discharge at Kingston — where probability is captured in the form of a joint probability density function over these boundary conditions.
- The reliability of passive structures (defences) and active structures, such as frontager gates, embedded in the defence line. These reliabilities are expressed using "fragility curves" (Dawson and Hall, 2002), which express the conditional probability of defence failure (in some specified mode) given water level in the estuary against the defence ("load").
- The reliability of active structures (barriers or barrages), in terms of their "failure to operate". If an active structure fails it is assumed that this failure results in the structure being completely open throughout the event. The probability of failure to operate is assumed to be unconditional.

Estimation of EAD involves two distinct elements of calculation. One involves estimating the behaviour of the physical system for given "events", where an event is defined by the boundary conditions, the configuration of the barrier system and the state of the defence system. The other involves estimating the probability of occurrence of those situations. These two are then brought together in a probability-weighted sum (integration of continuous distributions is approximated numerically by sampling and summation).

In general, clarity and flexibility are maximised if modelling of the physical system and consideration of the probable occurrence of particular events are kept separate. On the other hand, it is sometimes necessary to introduce optimisations which violate this separation. Any such optimisation should be undertaken with care. In particular, the intended modes of use of the model must be taken into account, as otherwise limitations may be introduced which inhibit these uses. Of particular note is the possibility of rendering it difficult or impossible to examine uncertainty in particular variables.

Figure 14.2 shows a high level data flow diagram for an idealised version of the expected damage estimation. The two paths through the calculation are marked "S" (statistical) and "P" (physical). In the same way as other complex applied risk analysis studies (Helton, 1993; Storlie and Helton, 2006), the Thames analysis relied upon Monte Carlo simulation to compute the risk estimate. If events are sampled according to probability density, all of the weights are equal to $1/n$, where n is the number of samples. The possibility of weighting the final sum, rather than using brute force Monte Carlo integration, is included for two reasons. Firstly, in the case of discrete random variables, such as that of barrier system state, it may be appropriate to simulate each possibility and take a probability weighted sum of the results to find the expectation. Secondly, it enables a potentially efficient class of optimisations to be accommodated. This class includes importance

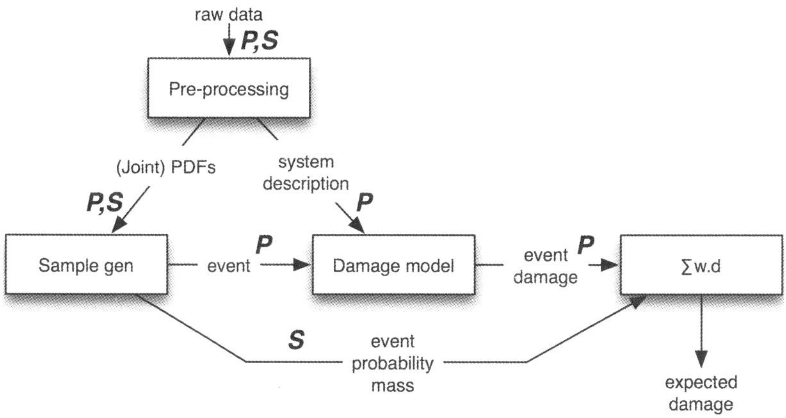

Figure 14.2. Data flow schematic of the risk (expected damage) estimation process, showing the separate data flows associated with physical system modelling (P) and probabilities of occurrence (S).

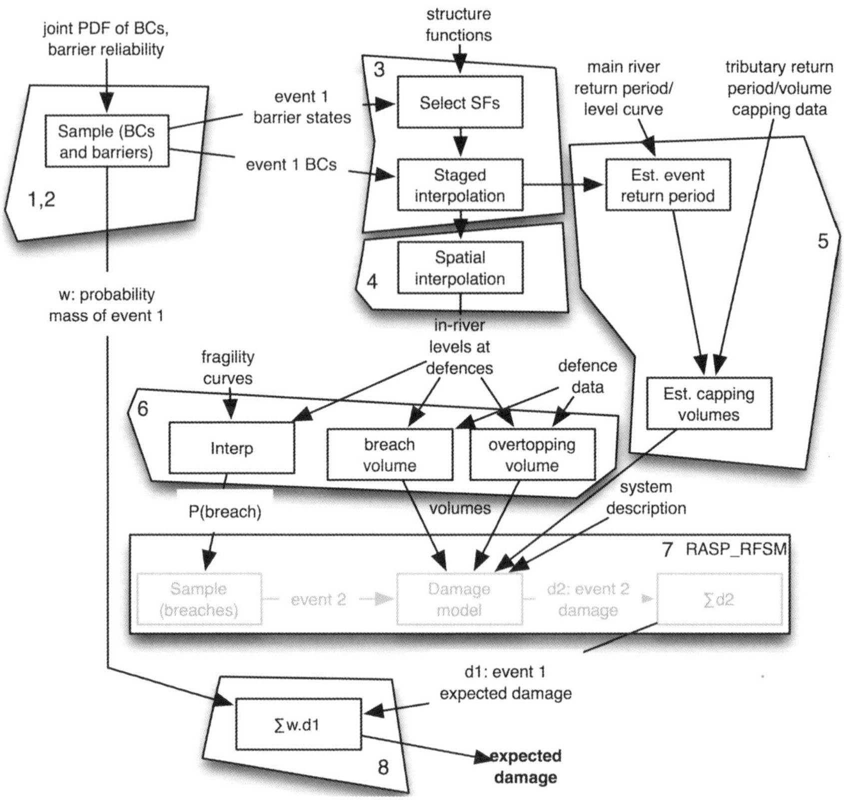

Figure 14.3. Data flow diagram of the expected damage calculation as implemented.

sampling, which is not used here, and a pre-processing approach, which is. Figure 14.3 summarises the procedure by means of which EAD is estimated.

1. Choose a sample of main river Boundary Conditions (BCs) and associated weights appropriate to the integration method.
2. Evaluate the barrier control rule. If the barrier should be closed, propagate using both open barrier and closed barrier structure functions, partitioning weight from Step 1 according to probability of failure to close the barrier. If the barrier should be open, propagate once only, using open barrier structure functions and weight from Step 1.
3. Interpolate into structure functions selected at Step 2 using main river boundary conditions to find the main river water levels.
4. Interpolate in-river water levels spatially from hydraulic model nodes onto defences.

5. Estimate return period of the water level at the mouth of each tributary, and select tributary levels to match by interpolation into return period/level curve from database.
6. Given a set of in-river water levels, estimate the probability of breaching of each defence, and volumes given breach and overtopping.
7. Estimate expected damage conditional on these probabilities and volumes.
8. Accumulate the sum of product of weights from Step 2 and the event conditional expected damages from Step 7.

The final result of the accumulation in Step 8 is then an estimate of expected annual damage.

14.3. Sources of Uncertainty in the Flood Risk Calculation

Inevitably there are uncertainties in each step of the analysis described above. These originate from the scarcity of data with which to estimate the relevant input distributions and uncertainties in the various models that are used to compute system states for given inputs. These sources of uncertainty are summarised in Table 14.1. They relate only to uncertainties in the present day risk estimate. Looking over the 21st century risk is expected to change for a variety of reasons, including climate change, development in the floodplain and deterioration of and/or improvements to the flood defences. Uncertainties in these future changes are dealt with in Section 14.7.

Table 14.1 also summarises the evidence from which probability distributions representing these epistemic uncertainties have been estimated. In Table 14.2 the range of different qualities of evidence for uncertainty analysis are summarised. Whilst for some variables there are ample observations from which to estimate uncertainties, in other cases the evidence is from analogous cases or expert judgment. This classification can be compared with the results of sensitivity analysis reported later in this chapter to identify those variables which are a priority for improving the quality of available evidence.

14.4. Process of Uncertainty Analysis

The evidence summarised in Table 14.1 was used to estimate probability distributions to represent each of the epistemic uncertainties. A large Monte Carlo sample was taken from each of these distributions and the EAD calculation was run with each member of this sample.

Table 14.1. Summary of sources of uncertainty in the flood risk analysis calculation.

Variable or function	Sources of present day uncertainty	Sources of evidence about uncertainty
Distribution of water level at Southend	Statistical uncertainties in estimating extreme sea levels	Standard error estimates from extreme value analysis of Southend water level (Dixon and Tawn, 1994; Dixon and Tawn, 1995; Dixon and Tawn, 1997). Spread of various fitted distributions from alternative extreme value distribution functions (see Figure 14.4)
Distribution of discharge at Kingston	Statistical uncertainties in extreme discharge	Confidence intervals on extreme value distribution fitted to daily flow data. Inter-comparison of alternative extreme value distributions
Probability of failure to close barriers	Limited empirical evidence of failure probabilities Uncertain reliability of mechanical and human systems	Reliability analysis of barrier systems
Water level at defences	Hydraulic modelling of water levels in the river, given boundary conditions. Includes model parameterisation (channel/ defence/floodplain geometry, roughness parameterisation) and numerical errors	Comparison with observations with model predictions, particularly during extreme events
Defence crest level	Scarcity/accuracy of measurements	Generic information on typical errors in survey methods e.g. ground survey error $\approx \pm 10$ mm, low level LiDAR error $\approx \pm 270$ mm etc.
Probability of breaching of flood defences (conditional probability distribution)	Scarcity of information about defence composition and condition. Limitations of quantified knowledge of failure modes	Lower and upper bounds on defence fragility functions, which have been computed from uncertainty analysis of reliability calculations

(*Continued*)

Table 14.1. (*Continued*)

Variable or function	Sources of present day uncertainty	Sources of evidence about uncertainty
Probability of failure to close frontage gates and other in-line structures	Limited empirical evidence of failure probabilities Uncertain reliability of mechanical and human systems	Reliability analysis of gate systems
Condition grade of flood defence structures	Inaccuracies in condition assessments of structures Limited knowledge about the relationship between condition grade and probability of failure.	Benchmarking of condition grade assessments against examples of defences of known condition (a total of 17860 the condition grades assigned by delegates on Environment Agency condition assessment courses)
Inflow volume to floodplain in the event of defence failure.	Assumptions about breach dimensions and inflow volume (parameterised as: weir constant, breach width multiplier, breach duration, hydrograph multiplier)	Uncertainty analysis of influential variables in for breach dimensions and inflow volume
Ground elevation in the floodplain	Errors in LiDAR survey of the floodplain	Ground truth at thirteen points in the survey area (root-mean-square error (RMSE) ranged from 0.021 m to 0.061 m)
Floodplain water levels	Model uncertainties in numerical models of floodplain inundation	Benchmarking of the approximate flood spreading model used versus simulations using a full shallow water equation solver
Location and type of properties and people in the floodplain	Classification and aggregation of properties	Benchmarking of property database
Property threshold levels	Scarcity of precise threshold measurements	Survey of 127 properties yielding a mean threshold height of 0.303 m and a standard deviation of 0.135 m
Depth-damage functions	Sampling and systematic errors in data on flood damages (Penning-Rowsell, 2003) Regional and local variations Omission of indirect and intangible damages	Published depth-damage functions have three curves: lower, indicative and upper damage susceptibility (Penning-Rowsell, 2005). Indirect and intangible damages were omitted

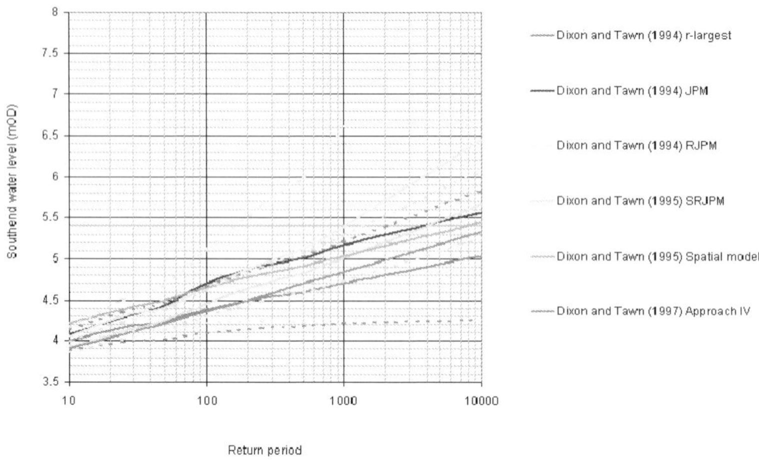

Figure 14.4. Comparison of published extreme value distributions of water levels at Southend (with 90% confidence intervals shown dotted where available).

Table 14.2. Summary of types of evidence available upon which to base uncertainty estimates.

Quality of information available for uncertainty analysis	Examples of variables
Direct observations from which to estimate uncertainties	Distribution of water level at Southend
	Distribution of discharge at Kingston
	Water level at defences
	Defence crest level
	Ground elevation in the floodplain
	Location and type of properties and people in the floodplain
	Property threshold levels
Observations at locations that may not be directly applicable to the Thames Estuary	Depth-damage functions
	Condition grade of flood defence structures
No, or very limited, relevant observations. Model inter-comparisons available	Floodplain water levels
No relevant observations. Model-based analysis	Probability of failure to close barriers
	Probability of breaching of flood defences
	Probability of failure to close frontage gates and other in-line structures
	Inflow volume calculation

A distribution from the resulting sample of EAD was then constructed to obtain a picture of how uncertainty in input parameters and model performance translate into uncertainty in the value of EAD.

Imperfect representation of physical reality by the model components is expressed as a probability distribution over the value of an intermediate variable conditional on the model-predicted value of that variable and, in some cases, the values of other variables, such as spatial location. As an example, hydraulic model error is expressed as a normal distribution with mean equal to the model prediction and standard deviation a function of the location in the estuary (the error model is based upon previous benchmark studies of the Thames hydraulic models).

14.4.1. *Dependence*

Unless otherwise stated, variables were treated as being independent and functions as being independent, conditional upon their inputs. Thus the reliability of difference sections of flood defence was treated as being independent conditional upon the input water level.

The tidal and fluvial boundary conditions (z, q) of the system do show statistical dependence (Svensson and Jones, 2002). Joint probability analysis of tidal elevations and flows has been conducted in several recent studies. Write the joint density function on $z \times q$ as $f_{Z,Q}(z, q)$. A large (2,400,000 member) Monte Carlo sample from $f_{Z,Q}(z, q)$ was available from previous studies, developed using the method described in Hawkes *et al.* (2002). We write this large sample as $\mathcal{S} = \{v_i = (z_i, q_i)\colon i = 1, \ldots, 2,400,000\}$. The marginal cumulative distributions on z and q are written $F_Z(z)$ and $F_Q(q)$ respectively.

Whilst the uncertainty in the dependence function is acknowledged, the approach adopted here is to incorporate only the uncertainty in the two marginal distributions (flow at Kingston and Southend water level). The dependence between these two variables is assumed to be known and constant. The uncertainties surrounding flow at Kingston and Southend water level were presented in the form of distributions of errors, incorporating both sampling errors and model uncertainties. For any given realisation of this error distribution there is some corresponding distribution $F'_Z(z)$ at Southend or $F'_Q(q)$ at Kingston (we take these errors to be independent). The joint sample $\mathcal{S}$ can be transformed so that its marginals correspond to $F'_Z(z)$ and $F'_Q(q)$ by the transformation $z'_i = F'^{-1}_Z(F_Z(z_i))$ and similarly for q'_i, for all i, yielding a transformed sample $\mathcal{S}'$. This process can be repeated as many times as necessary to generate a stable Monte Carlo estimate of the implication of the uncertainties in marginal distributions.

14.4.2. *Response surfaces*

A naïve risk analysis implementation using the method set out in Figure 14.3 would propagate the full Monte Carlo sample $\mathcal{S}$, calculating an event expected damage for each member of the sample and setting all weights w to the reciprocal of the sample size $1/|S|$. The computational cost of this approach would be prohibitive, however.

If event expected damage varies smoothly in each boundary condition, it allows an approach based on interpolation. Event expected damage was calculated at a grid of points covering the boundary condition space. Event expected damage for any event can then be estimated using bi-linear interpolation. The resulting response surface is illustrated in Figure 14.5, which shows event expected damage for the barrier open and barrier closed situations at a location a few kilometres upstream from the Thames Barrier. The boundary condition space is represented by the horizontal plane. Damage is represented by position on the vertical axis and colouration. The response surfaces illustrate how tides dominate when the barrier is open, while with the barrier closing both tide (by controlling water level at time of closure) and flow (through its accumulation during the period of closure) show some influence.

14.5. Uncertainties in Future Changes

Appraisal of options for flood risk management in the Thames Estuary involves quantification of future as well as present day risks. Looking into the future, flood risk will change, due to a variety of processes of change, and the range of uncertainties also increases. In Table 14.3 the potential drivers of change in flood risk are summarised. Whilst evidence upon which to base estimates of these future changes and associated uncertainties is rather scarce, for TE2100 we have been able to quantify a number of the key uncertainties.

Future changes in relative sea level in the Thames Estuary are attributable to both eustatic and isostatic effects. Bingley *et al.* (2007) studied vertical land movement in the Thames Estuary, which they estimated to be in the range 1.09+/−0.64 mm/yr. Table 14.4 reports projections of relative sea level rise in 2095 along with the associated uncertainties, which

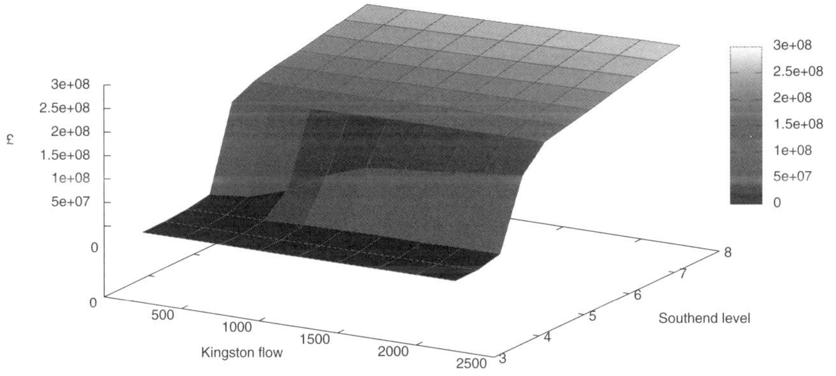

(a)

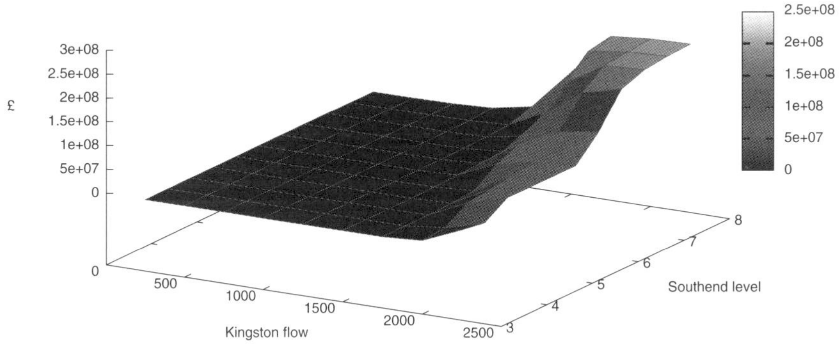

(b)

Figure 14.5. Response surfaces of expected damage (integrated over defence state) for a flood area upriver of and near Thames Barrier, (a) barrier open and (b) barrier closed.

are reported as 90% confidence interval. There remains the question of how the emissions scenarios should be combined. Some discussion of this problem is provided by Hall *et al.* (2007) and references therein.

Howard *et al.* (2008) employed an ensemble of Hadley Centre Regional Climate Model (RCM) runs, coupled with the POL CS3 surge model to analyse potential future changes in storm surge frequency and height. The analysis showed no significant trend in storm heights in an SRES A1B scenario.

Analysis of the impacts of climate change on flows in the River Thames has been conducted by Reynard *et al.* (2004), Kay *et al.* (2008) and

Table 14.3. Drivers of future change in variables or functions that determine flood risk in the Thames Estuary.

Variable or function	Driver of change in the future
Variable: Extreme water level at Southend	Mean sea level rise Changing storm surge frequency
Variable: Discharge at Kingston	Changing rainfall Changing land use
Variable: Discharge in tributaries	Changing rainfall Changing land use
Variable: Probability of failure to close barriers (that influence river water levels)	Deterioration of mechanical/control systems Technological changes may enable improved reliability in future
Function: Water level at defences	Changes in channel morphology
Variable: Defence crest level	Settlement
Variable: Probability of breaching of flood defences (conditional probability distribution)	Deterioration
Variable: Probability of failure to close frontage gates and other in-line structures	Deterioration of mechanical/control systems Changes in frontage usage
Variable: Condition grade	Deterioration
Variable: Weir equation constant	None
Variable: Breach width	None
Variable: Event duration	None
Variable: Inflow volume calculation	None
Function: Water level at points in the floodplain	None
Variable: Location and type of properties and people in the floodplain	Land use and demographic changes
Variable: Property threshold levels	None
Function: Depth-damage relationships	Changing wealth and household contents

Table 14.4. Thames Region relative time mean sea level change (metres) from present day to 2095 under 3 different scenarios with 90% confidence intervals (after Howard *et al.*, 2008).

	A1Fi	A1B	B1
Mean	0.56	0.47	0.40
Min	0.24	0.21	0.19
Max	0.88	0.73	0.61

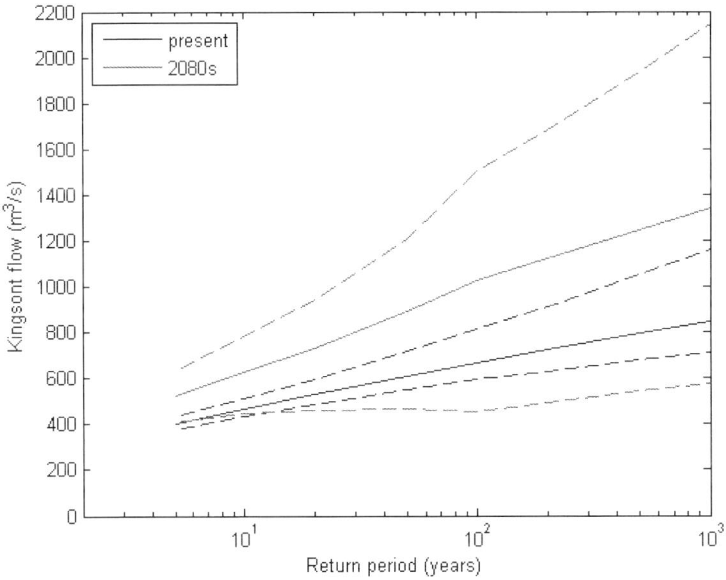

Figure 14.6. Estimated future Kingston flow estimates (with 95% confidence intervals) obtained by combining present day uncertainties with G2G ensemble estimates of potential increases.

Bell *et al.* (2008), using a range of downscaling methods, including a grid-based hydrological model (G2G) coupled with Hadley Centre RCM ensemble outputs mentioned above. The spread of these various model studies provides some evidence for quantification of the range of potential uncertainties in future flow estimates. Each of these model predictions, along with the associated uncertainties, have been combined with the (uncertain) estimated present day flows to give the future flows illustrated in Figure 14.6. This has been done by sampling the ensemble members (with equal probability) and applying the sampled increase factors to samples from the (skew) distribution of flow at given return period. The projected changes associated with the 1:100-year flow has been applied to flows above this return period, because to extrapolate the growth in increase to high return periods would result in implausibly high increases (e.g. approximately 130% at 10,000 years). These estimates should be treated with considerable caution because they are derived from a relatively small sample of outputs from only one combination of climate and hydrological models. The extension of the results to high return periods is not warranted by the analysis and is conducted here out of necessity to provide uncertainty estimates of Kingston flows at all return periods of relevance to TE2100. Of particular concern is the extent to which the highest postulated increases in flows at very high return periods are physically realisable given the floodplain storage that will be utilised at very high flows.

The analysis of future changes in flow and surge has been carried out separately (though driven by the same Hadley regional climate model output). These separate analyses will not provide evidence upon which to assess the potential change in dependency between flow and surge. Svensson and Jones (Svensson and Jones, 2004, 2005a,b) used regional climate model and shelf-seas model outputs to make a preliminary assessment of changes in the dependence between sea surge and river flow, using precipitation as a proxy for river flow. Several locations on the south- and west-coast of Britain show significant increases in the dependence between sea surge and precipitation in the period 2071–2100, compared to the control run 1961–1990. Svensson and Jones did not study the Thames Estuary, so conclusions for TE2100 are hard to draw.

With regard to uncertainties in future defence condition, Buijs *et al.* (2009) report on analysis of embankment consolidation as well as providing quantified analysis of deterioration for a selected number of defences on the Dartford to Gravesend embayment. The analysis indicates the embankments constructed in the 1970s now have a consolidation rate of 4 mm/year with a range of between 1 mm/year and 6 mm/year.

Location and type of properties in the floodplain has been explored by means of a combination of demographic, economic and land use modelling. The results of the analysis are reported in Hall *et al.* (2009) and Hall *et al.* (2010), who adopted a scenario-based approach to explore potential land use and socio-economic futures for London and the Thames Estuary (Figure 14.7).

14.6. Results of Uncertainty Analysis

Figure 14.8 presents the cumulative distribution function for Expected Annual Damage (EAD) for the Thames Estuary, i.e. it represents the uncertainty in the EAD taking into account all of the uncertainties listed in Table 14.1. Risk estimates for the present and for 2100 are illustrated. The contributions to uncertainty in 2100 are discussed in Section 14.7. Mean estuary-wide EAD is estimated as £560 million with 10th and 90th percentiles of £270 and £970 respectively.

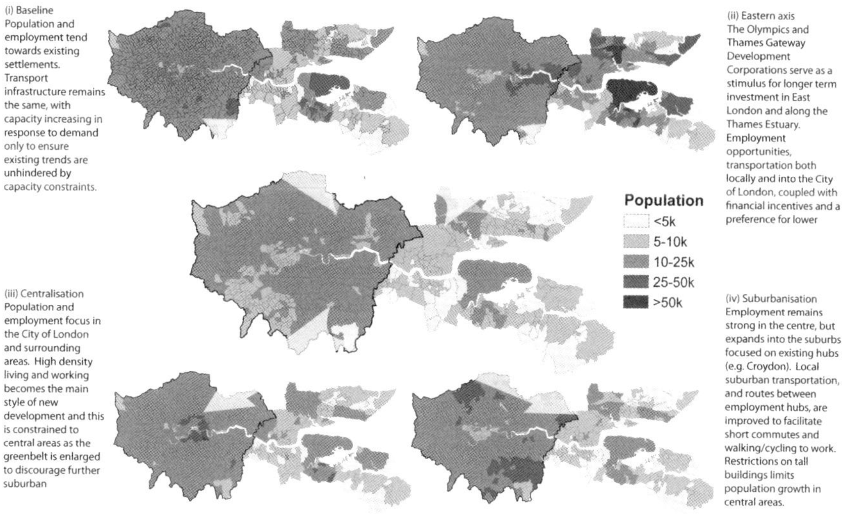

Figure 14.7. Potential socio-economic change in London and the Thames Estuary: maps of population by ward for present day (centre) and 2100 where clockwise from top left four maps show different development scenarios (from Hall *et al.*, 2009).

The implication of uncertainties in future changes (with exception of the socio-economic uncertainty which was dealt with as scenarios rather than probabilistically) is illustrated in the cumulative distribution function in Figure 14.8. The mean estuary-wide expected annual damage is estimated to increase to £1890 million in 2100, with 10th and 90th percentiles of £540 and £2790 respectively. It is clear that future changes add considerably to the total risk in the Thames Estuary (even without considering changing vulnerability due to socio-economic change) and also to the uncertainty in the risk estimate.

The panel on the right-hand side of Figure 14.8 presents the evolution of the values of the deciles of the distribution of EAD as the number of samples is increased. These give an indication of the robustness of the results displayed in the main figure. Examination of these plots indicates that at the final sample sizes (240 for present day, 220 for 2100) the general form of the cumulative distribution functions is stable, although local adjustments continue. These adjustments relate predominantly to two situations.

1. Quantiles at high cumulative probabilities show lower stability than those at low cumulative probabilities. Much larger sample sizes are

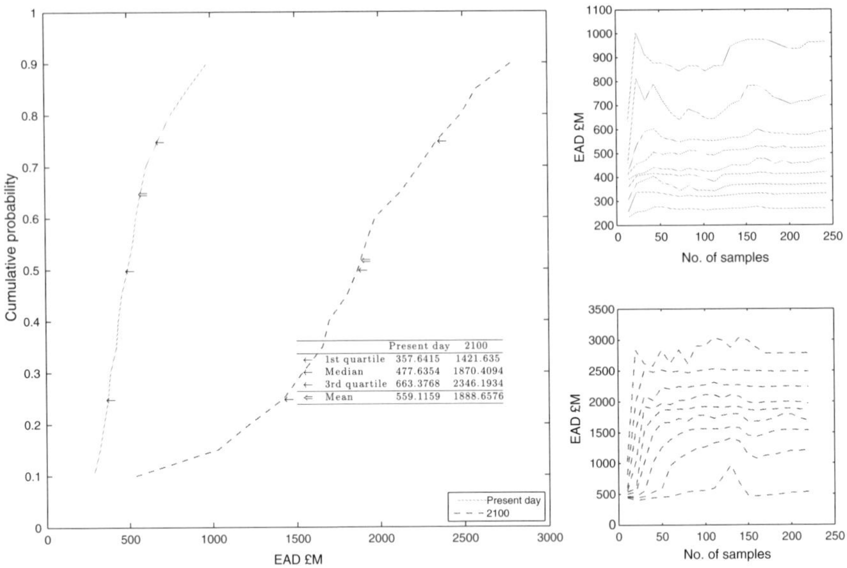

Figure 14.8. Distribution of EAD for the whole Thames Estuary (solid line: present day; dashed line: 2100). Right-hand panels show convergence of the risk estimates during the Monte Carlo sampling procedure.

needed to estimate the upper tail of the distribution precisely, as the probability density is low and only a very small proportion of the overall sample provides information in this region.

2. Stability is also low around regions of low gradient in the cumulative density plots, where EAD increases rapidly with a small increase in cumulative probability. Here, the variation relates to the sample density needed to precisely locate such steps in the function.

Analysis was halted at the point shown in these diagrams as considerable additional computational expense would have been required for relatively minor improvements in precision.

Figure 14.9 illustrates how uncertainty in the flood risk estimate varies through the estuary, by presenting the coefficient of variation of the risk estimate for a number of distinct flood areas. This EAD is located spatially largely upriver of the Thames Barrier, with lesser contribution from elsewhere in the estuary. This pattern holds in 2005 and 2100. The particular forms of the curves of cumulative probability of EAD (not illustrated here) vary considerably depending on location, reflecting the nature of the topography and distribution of building stock.

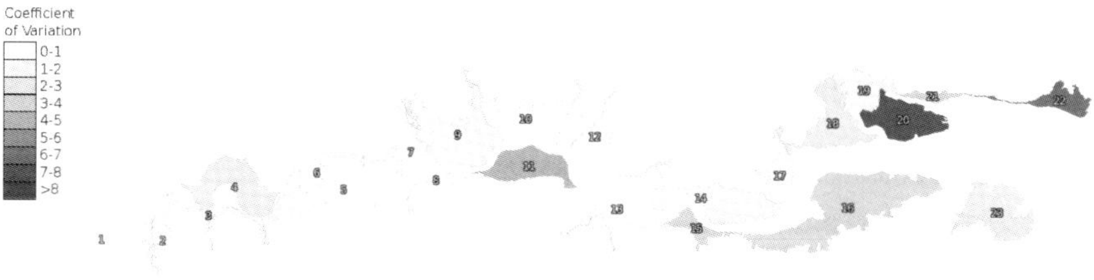

Figure 14.9. Map of coefficient of variation of EAD.

14.7. Sensitivity Analysis

Variance-based sensitivity analysis (Saltelli *et al.*, 2000) provides a means of identifying variables that make the greatest contribution to the variation in EAD. It was noted above that some of the uncertainty estimates for input variables to the risk calculation are more reliable than others. Sensitivity analysis provides a means of identifying the extent to which the EAD estimate is sensitive to these assumptions. It also helps to identify variables that should be targets for future data collection.

Variance-based sensitivity analysis requires a large number of model runs, with the number of runs increasing with the number of input variables. The run time of the system model described above is sufficient that it was necessary to reduce the problem order. First, the input variables were grouped by type. Thus, for example, it is assumed that crest level error is fully dependent at the scale of the flood areas shown in Figure 14.9, such that, for a given sample member, all defences in a flood area will have the crest levels corresponding with the same cumulative probability. The result of this grouping was a set of 12 variables. Next, a screening method was applied to identify and remove variables from the analysis that make no noticeable contribution to the uncertainty in the risk estimate.

14.7.1. *Screening using the Method of Morris*

The Method of Morris (Morris, 1991; Saltelli *et al.*, 2000) was applied to the 12 variable groups. This method provides an indication of which variables influence the output of the model most strongly and, more importantly, which have little or no influence. The results of the Morris analysis are, for each input variable, a pair of measures, μ and σ, which are measures, respectively, of the *total effect* of a given variable and the extent to which it interacts with other variables. The Method of Morris does not provide a robust ranking of variables' contribution to variance in output (EAD), but

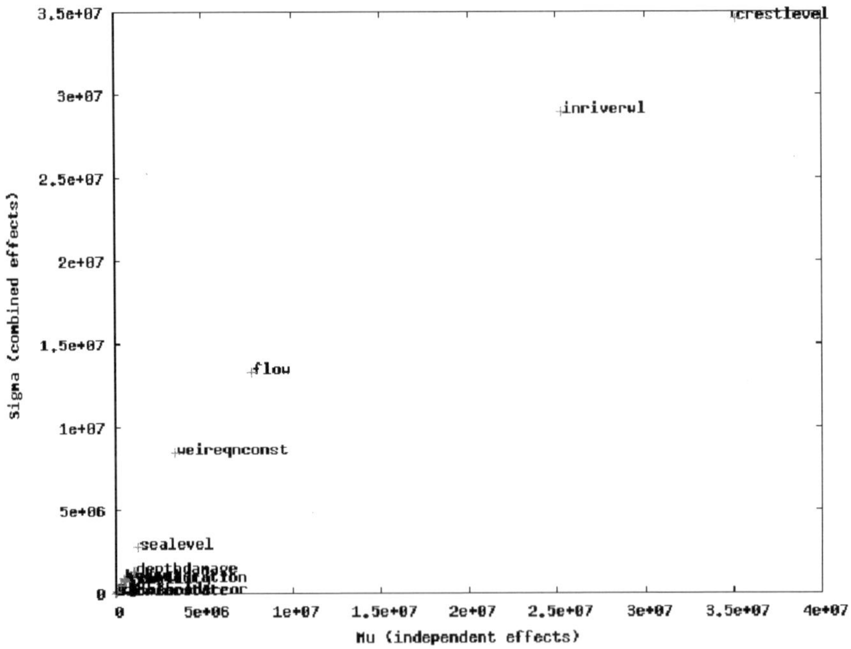

Figure 14.10. Output of Method of Morris for a typical flood area.

a more qualitative indication. Its purpose is to separate those variable with a significant effect from those without.

Figure 14.10 shows example results for a location a few kilometres upstream of the Thames Barrier. We see that defence crest level and hydraulic model error dominate, with the flow boundary condition and the choice of value of the weir equation constant following. However, note that crest level will tend to be over-emphasised because of the assumption of full dependence in variations in level of all relevant defence sections. Sea level uncertainty is also clear of the main group of less influential variables.

The results of the Method of Morris were used to filter variables with no noticeable influence on the estimate of EAD so that the subsequent variance-based analysis could focus upon the most influential variables, which were found to be: Southend water level; Kingston flow; water level at defences; crest level; inflow volume calculation; probability of breaching; condition grade; floodplain water levels; property threshold levels; and depth damage functions.

14.7.2. *Variance-based sensitivity analysis*

Variance-based sensitivity analysis decomposes the variance V in the EAD estimate (labelled here as Y) into the contributions from the various input factors, which can be labelled $X_1, \ldots, X_t$. The sensitivity index S_i represents the fractional contribution of a given factor X_i to the variance in EAD. In order to calculate the sensitivity indices the total variance V in the model output Y is apportioned to all the input factors X_i as (Sobol, 1993):

$$V = \sum_{i<j} V_i + \sum_{i<j} V_{ij} + \sum_{i<j<l} V_{ijl} + \cdots + V_{1,2,\ldots,k},$$

where

$$V_i = V[E(Y|X_i = x_i^*)],$$
$$V_{ij} = V[E(Y|X_i = x_i^*, X_j = x_j^*)] - V_i - V_j,$$

$V[E(Y|X_i = x_i^*)]$ is referred to as the Variance of the Conditional Expectation (VCE) and is the variance over all values of x_i^* in the expectation of Y given that X_i has a fixed value x_i^*. This is an intuitive measure of the sensitivity of Y to a factor X_i, as it measures the amount by which $E(Y|X_i = x_i^*)$ varies with the value of x_i^*, while all the effects of the X_js, $j \neq i$, are averaged. The first order (or "main effect") sensitivity index S_i for factor X_i is therefore defined as:

$$S_i = V_i/V.$$

Also of interest is the influence of factor X_i when acting in combination with other factors. There are $2^k - 1$ of such interactions, so it is usually impractical to estimate the effect of all of them. A more practical approach is to estimate the k total sensitivity indices, S_{Ti} (Homma and Saltelli, 1996):

$$S_{Ti} = 1 - \frac{E(Y|X_{\sim i} = x_{\sim i}^*)}{V(Y)},$$

where $X_{\sim i}$ denotes all of the factors other than X_i. The total sensitivity index therefore represents the average variance that would remain as long as X_i stays unknown. The total sensitivity indices provide an indicator of interactions within the model. For example, factors with small first order indices but high total sensitivity indices affect the model output Y mainly through interactions — the presence of such factors is indicative of redundancy in the model parameterisation.

For more in-depth description of variance based sensitivity analysis, the reader is referred to Saltelli *et al.* (2000) and Saltelli *et al.* (2004). For the purposes of the sensitivity indices in this study, the software SimLab[a] was used.

Variance based sensitivity analysis was applied to the ten variable identified as being influential by the Morris analysis described above, setting the variables not included in these shortlists to median values. In total, 768 samples were run, and results for 384 and 768 samples compared. Few changes in rank order of importance were found in the dominant variables, and additional runs were deemed unlikely to offer further insights.

Results are presented in Figures 14.11 and 14.12 for first order (S_i) and total order (S_{Ti}) indices respectively. The legend order is consistent between these plots to aid comparison. Variables with an index value of less than 5% of the maximum are excluded. Variables that have a much higher total sensitivity index than the first-order sensitivity index indicate redundancies in the model. This is to be expected in the present model as,

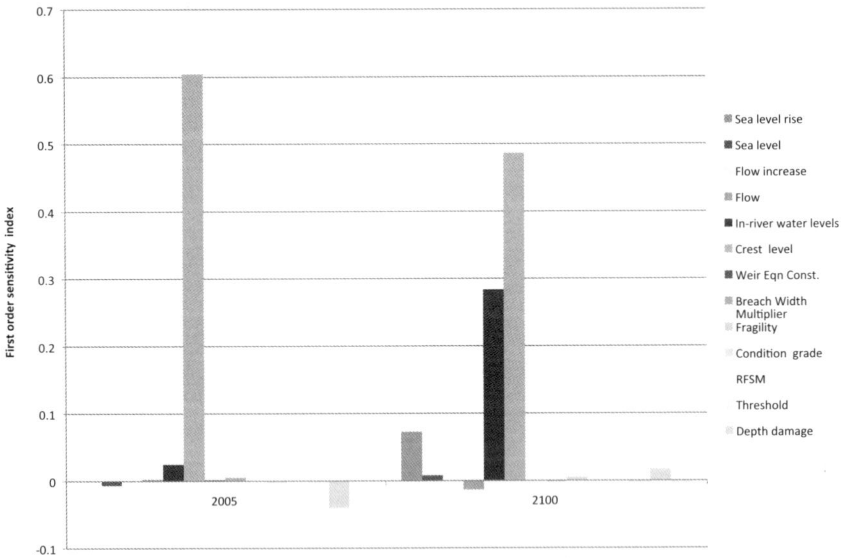

Figure 14.11. Estuary-wide first-order sensitivity indices.

[a]SimLab was developed by the EU Joint Research Centre and is freely available for non-profit use from: http://www.jrc.cec.eu.int/uasa/prj-sa-soft.asp.

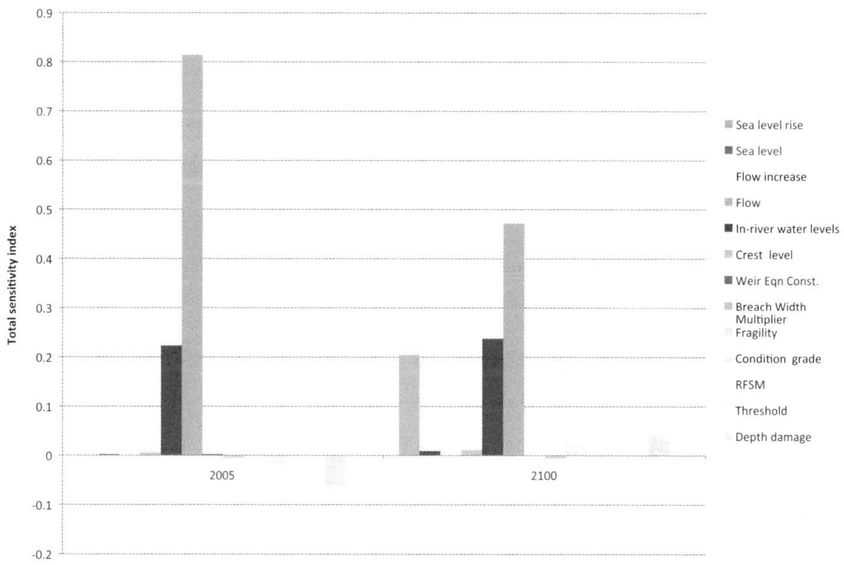

Figure 14.12. Estuary-wide total sensitivity indices.

for example, a systematic reduction in defence crest levels is very similar in effect to a systematic increase in water levels against those defences.

Variance-based sensitivity analysis indicates that the input variables which contribute most to uncertainty in estimates of EAD are defence crest level and water level at the defences. Crest level and river water level are closely related. That they both appear ahead of others suggests that the total error in the maximum head achieved during an event is a dominant effect. Other variables which show influence in certain areas are: river flow at the western boundary (Kingston), inflow volume, water level at the eastern boundary (Southend), and floodplain water levels. Uncertainty on Southend water level is highest on low probability events, the influence of which is correspondingly limited. In general there is high consistency between first order and total order rank orders.

The dominance of crest level uncertainty in the central and upper reaches of the estuary (which contribute most to flood risk) is consistent with the fact that almost all defences in this area have crest level standard deviation set at 0.34 m, based on the known quality of survey data available for this study. Better quality survey data would reduce the contribution of crest level to the total uncertainty in the risk estimate.

The sensitivity results for 2100 show that sea level rise also contributes significantly, and depth-damage curve error shows some influence. The uncertainty in sea levels (see Figure 14.4) might be expected to be more influential (both at present and in 2100), but this is highest on low probability events, the influence of which is correspondingly limited. Sea level rise, meanwhile, influences events of all probabilities of occurrence equally, so is actually more influential. The appearance of depth-damage curves in 2100 when it does not show for the present day is because more flooding is predicted to occur in 2100 (assuming no change to the flood defence system).

As with the Method of Morris, samples of vector valued variables (e.g. defence crest level) were generated with equal cumulative probability. This is in contrast with the uncertainty analysis, where each element of some such variables (thus, for example, the crest level of each individual defence) is sampled independently. This limitation will tend to over-emphasise the influence on total variance of vector-valued variables where errors are largely independent, and the greater the multiplicity of the variable (that is, the size of the vector), the greater the extent of this bias. The dominating presence of crest level error may be to some extent an artefact of this computationally necessary simplification. The variables associated with the boundary conditions, Southend water level and Kingston flow, are singletons. Their influence may, as a corollary of the above, be underestimated relative to vector-valued variables. The strong showing of some these variables is therefore particularly notable: boundary condition uncertainty contributes significantly to output uncertainty. An intermediate case is that of river model error (referred to as "water level at defences" error). In uncertainty analysis this is assumed to be fully dependent for a given sub-system of defences, and fully independent between. The possible bias resulting from an assuming estuary-wide dependence in sensitivity analysis is thus considerably less than for defence crest level. A further corollary is that those variables with high multiplicity but low influence can be presumed to have very low contribution to output uncertainty.

14.8. Conclusions

Decisions upon investment in major capital works of flood defence, such as in the Thames Estuary, are based upon quantified estimates of flood risk. Given the significance of these decisions it is important to scrutinise the uncertainties in risk estimates. These originate from incomplete information about input variables in the risk calculation and from uncertainties in the process-based models used to compute water depths in the floodplain and consequential damage. In this chapter we have described in outline the system of models used to compute economic flood risk in the Thames Estuary, and an associated uncertainty and sensitivity analysis.

The principles and practice of Monte Carlo uncertainty analysis are well known. Methods of variance-based sensitivity analysis are also becoming more widely used (Hall *et al.*, 2005; Hall *et al.*, 2009). However, complex flooding systems such as the one described in this chapter still represent a considerable challenge for uncertainty and sensitivity analysis.

In order to implement the uncertainty analysis, the system model has been reconfigured so that the process of sampling from random variables is separated from the process of calculating flood depths and damage. In this way it is possible to access and manipulate each of the sampling procedures and analyse their influence.

The theoretical structure for flood risk analysis with systems of flood defences is now well known (Dawson *et al.*, 2005). In the case of the Thames Estuary it has been necessary to pay additional attention to the treatment of barrier states and barrier failure. The approach that is proposed here is flexible and scalable so that it could be readily applied to adapted system configurations with new barriers.

The boundary conditions in the analysis (flows and tide water levels at opposite ends of the system) are known to be correlated. A procedure for resampling the joint distribution of these two variables was implemented, based upon rescaling the marginal distributions.

The computational expense of a single realisation of the risk analysis to compute EAD is considerable. In order to conduct uncertainty analysis involving repeated propagation of large Monte Carlo samples, a family of response surfaces was constructed. Cloud computing facilities were used to distribute the analysis on multiple compute nodes (Harvey and Hall, 2010). Reasonably stable quantiles for the distribution of EAD were obtained based upon 220 samples of the input vectors.

Variance-based sensitivity analysis for systems with large numbers of input variables is computationally more intensive. Non-influential variables were screened out using the Method of Morris before applying Sobol's method for calculating the variance-based sensitivity indices. The computational expense has been reduced by the grouping of variables, though at the cost of having to make a conservative assumption about their dependence, which is then reflected in an over-estimate of their influence on the EAD estimate. It was not necessary to make this assumption in the uncertainty analysis of EAD.

Acknowledgements

The study described in this chapter was conducted as part of the Environment Agency's Thames Estuary 2100 project. It was also supported in part by the EPSRC Flood Risk Management Research Consortium (Grant EP/F020511). The advice and review of colleagues in the Environment Agency, HR Wallingford and Halcrow is gratefully acknowledged. Analysis of flood risk in the Thames Estuary has advanced rapidly in the TE2100 project since the work described in this chapter, so the presented results do not necessarily represent best current knowledge of the state of flood risk, and associated uncertainties, in the Thames Estuary.

References

Bell, V.A., Moore, R.J., Cole, S.J. *et al.* (2008). *Area-Wide River Flow Modelling for the Thames Estuary 2100 Project Model Formulation and Assessment*, Centre for Ecology and Hydrology, Wallingford.

Bingley, R.M., Teferle, F.N., Orliac, E.J. *et al.* (2007). *Absolute Fixing of Tide Gauge Benchmarks and Land Levels: Measuring Changes in Land and Sea Levels around the coast of Great Britain and along the Thames Estuary and River Thames using GPS, Absolute Gravimetry, Persistent Scatterer Interferometry and Tide Gauges*, Defra/Environment Agency, London.

Buijs, F.A., Hall, J.W., Sayers, P.B. *et al.* (2009). Time-dependent reliability analysis of flood defences, *Reliab. Eng. Syst. Safe.*, **94**, 1942–1953.

Dawson, R.J. and Hall, J.W. (2002). "Improved condition characterisation of coastal defence infrastructure? in: Allsop, N.W.H. (ed.), *Coastlines, Structures and Breakwaters*, Thomas Telford, London, pp. 123–134.

Dawson, R.J., Hall, J.W., Sayers, P.B. *et al.* (2005). Sampling-based flood risk analysis for fluvial dike systems, *Stoch. Env. Res. Risk Analysis.*, **19**, 388–402.

Dixon, M.J. and Tawn, J.A. (1994). Extreme Sea-Levels at the UK A-Class Sites: Site-By-Site Analysis, POL Internal Document Number 65, Proudman Oceanographic Laboratory, Liverpool.

Dixon, M.J. and Tawn, J.A. (1995). Extreme Sea-Levels at UK A-Class Sites: Optimal Site-By-Site Analysis and Spatial Analysis for the East Coast, POL Internal Document No.72, Proudman Oceanographic Laboratory, Liverpool.

Dixon, M.J. and Tawn, J.A. (1997). Estimates of Extreme Sea Conditions: Spatial Analysis for the UK Coast, POL Internal Document No.112, Proudman Oceanographic Laboratory, Liverpool.

Gilbert, S. and Horner, R. (1984). *The Thames Barrier*, Thomas Telford, London.

Hall, J.W., Boyce, S.A., Wang, Y. *et al.* (2009). Sensitivity analysis for hydraulic models, *J. Hydraul. Eng.*, **135**, 959–969.

Hall, J.W., Dawson, R.J., Batty, M. *et al.* (2010). "City-Scale Integrated Assessment of Climate Impacts, Adaptation and Mitigation? in: Bose, R.K. (ed.), *Energy Efficient Cities: Assessment Tools and Benchmarking Practices*, World Bank, New York, pp. 43–64.

Hall, J.W., Dawson, R.J., Walsh, C.L. *et al.* (2009). *Engineering Cities: How Can Cities Grow Whilst Reducing Emissions And Vulnerability?*, Tyndall Centre for Climate Change Research, Newcastle University, http://www.tyndall.ac.uk/sites/default/files/engineeringcities.pdf.

Hall, J.W., Fu, G. and Lawry, J. (2007). Imprecise probabilities of climate change: aggregation of fuzzy scenarios and model uncertainties, *Climatic Change*, **81**, 265–281.

Hall, J.W. and Solomatine, D. (2008). A framework for uncertainty analysis in flood risk management, *J. River Basin Manage.* **6**, 85–98.

Hall, J.W., Tarantola, S., Bates, P.D. *et al.* (2005). Distributed sensitivity analysis of flood inundation model calibration, *J. Hydraul. Eng.*, **131**, 117–126.

Harvey, H. and Hall, J.W. (2010). *Uncertainty Analysis Using Amazon web Services*, Hydroinformatics 2010. Proceedings of the 9th International Conference on Hydroinformatics, Tianjin, China.

Hawkes, P.J., Gouldby, B.P., Tawn, J.A. *et al.* (2002). The joint probability of waves and water levels in coastal engineering, *J. Hydraul. Res.*, **40**, 241–251.

Helton, J. (1993). Uncertainty and sensitivity analysis techniques for use in performance assessment for radioactive waste disposal, *Reliab. Eng. Syst. Safe.*, **42**, 327–367.

Homma, T. and Saltelli, A. (1996). Importance measures in global sensitivity analysis of nonlinear models, *Reliab. Eng. Syst. Safe.*, **52**, 1–17.

Howard, T., Lowe, J.A., Pardaens, A. *et al.* (2008). Met Office Hadley Centre Projections of 21st Century Extreme Sea Levels for TE2100, Met Office, Hadley Centre.

Kay, A.L., Bell, V.A. and Lowe, J.A. (2008). *Area-Wide River Flow Modelling for the Thames Estuary 2100 Project: Climate Change Impacts on Flood Frequency*, Centre for Ecology and Hydrology, Wallingford.

Morris, M. (1991). *Factorial sampling plans for preliminary computational experiments*, *Technometrics*, **33**, 161–174.

Penning-Rowsell, E.C. (ed.) (2003). *The Benefits of Flood and Coastal Defence: Techniques and Data for 2003*, Middlesex University, Enfield.

Penning-Rowsell, E.C. (ed.) (2005). *The Benefits of Flood and Coastal Risk Management; A Manual of Assessment Techniques*, Middlesex University Press, London.

Reynard, N.S., Crooks, S.M. and Kay, A.L. (2004). Impact of climate change on flood flows in river catchments, Environment Agency Report, CEH Wallingford.

Saltelli, A., Chan, K. and Scott, M. (2000). *Sensitivity Analysis*, Wiley, New York.

Saltelli, A., Tarantola, S., Campolongo, F. *et al.* (2004). *Sensitivity Analysis in Practice: A Guide to Assessing Scientific Models*, Wiley, New York.

Sobol, I. (1993). Sensitivity analysis for non-linear mathematical models, *Math. Modelling Comp. Exp.*, **1**, 407–414.

Storlie, C.B. and Helton, J.C. (2006). Multiple Predictor Smoothing Methods for Sensitivity Analysis, Sandia National Laboratory, Albuquerque, NM.

Svensson, C. and Jones, D.A. (2002). Dependence between extreme sea surge, river flow and precipitation in eastern Britain, *Int. J. Climatol.*, **22,** 1149–1168.

Svensson, C. and Jones, D.A. (2004). Dependence between sea surge, river flow and precipitation in south and west Britain, *Hydrol. Earth Syst. Sci.*, **8**, 973–992.

Svensson, C. and Jones, D.A. (2005a). Climate change impacts on the dependence between sea surge, precipitation and river flow around Britain, *Proc. 40th Defra Flood and Coastal Management Conference*, University of York, UK.

Svensson, C. and Jones, D.A. (2005b). Dependence between extreme sea surge, river flow and precipitation: a study in south and west Britain, *Hydrol. Earth Syst. Sci.*, **8**, 973–992.

SECTION IV

UNCERTAINTIES IN REAL-TIME FLOOD FORECASTING

CHAPTER 15

Operational Hydrologic Ensemble Forecasting

Albrecht H. Weerts
Deltares, Delft, Netherlands

Dong-Jun Seo
The University of Texas at Arlington, TX, USA

Micha Werner
Deltares, Delft, Netherlands

John Schaake
Consultant to National Weather Service, Silver Spring, MD, USA

15.1. Introduction

Ensemble forecast techniques are beginning to be used for hydrological prediction by operational hydrological services throughout the world. Ensemble weather and climate prediction systems have been established, with the first of these systems becoming operational in the early 1990s, predicting the evolution of the atmosphere from weather to climate scales. Examples are the EPS (Ensemble Prediction System) operated by the European Centre for Medium Range Weather Forecasts (ECMWF) (Molteni *et al.*, 1996) and the GEFS (Global Ensemble Forecast System) operated by the National Centers for Environmental Prediction in the US (Tracton and Kalnay, 1993).

Atmospheric ensemble forecasts are generated by perturbing the initial conditions and introducing different model physics, assumed *a priori* to be equally likely, and computing the evolution of the atmosphere due to these perturbed initial conditions and different model physics. With the availability of predicted ensemble forcings, such as precipitation and temperature, producing hydrological ensemble forecasts from them is a logical next step (Bartholmes and Todini, 2005). Such ensembles have found application in several operational fluvial forecasting systems used for the short and medium ranges. The European Community Sponsored EFAS system (De Roo *et al.*, 2003; Gouweleeuw *et al.*, 2005, Pappenberger *et al.*, 2005) and the Hydrological Ensemble Forecast System (HEFS) in the US National Weather Service (NWS, 2007a) are prime examples.

For water resources planning and management, ensemble forecasts with lead times as long as several months are necessary. Such long-range hydrologic forecasts may be obtained through the use of long-range atmospheric ensemble forecasts from, for example, historical observations, the ECMWF seasonal ensemble forecasting system, the NCEP Climate Forecast System (CFS) (Saha *et al.*, 2006) or downscaled General Circulation Models (GCM) predictions (Luo *et al.*, 2007). These ensembles can then be processed through a hydrological-hydraulic model cascade to provide a long-range ensemble water resources forecasts. An example of such long-range forecasting based on historical observations is the Ensemble Streamflow Prediction (ESP) procedure (Day, 1985) used by the US NWS. In this procedure an empirical ensemble of precipitation and temperature inputs is sampled from the long-term historical time series of catchment-average temperature and precipitation. Ensemble forecasts are attractive because they not only provide an estimate of the most probable future state of the system, but also an estimate of the range of possible outcomes. Indeed, many users are risk-averse, and are often more concerned with having a quantitative estimate of the probability that catastrophic outcome may occur than with having a single estimate of the most probable future outcome.

Not only does ensemble prediction offer a general approach to probabilistic hydrologic prediction, it also offers an approach to improve the absolute accuracy of hydrologic forecasts by helping identify and address all major sources of uncertainty. To produce reliable (i.e. probabilistically unbiased) and skilful hydrologic ensemble forecasts, the forecast system must be able to reduce and account for uncertainties from a wide range of sources. These include uncertainty in precipitation and other atmospheric forcing inputs, uncertainty in initial hydrological conditions, uncertainty in structures and parameters of hydrologic models, and uncertainty in human regulation and control of hydrologic variables. Figure 15.1 identifies major sources of uncertainty that need to be addressed in a comprehensive hydrologic ensemble forecast system (NWS, 2007b).

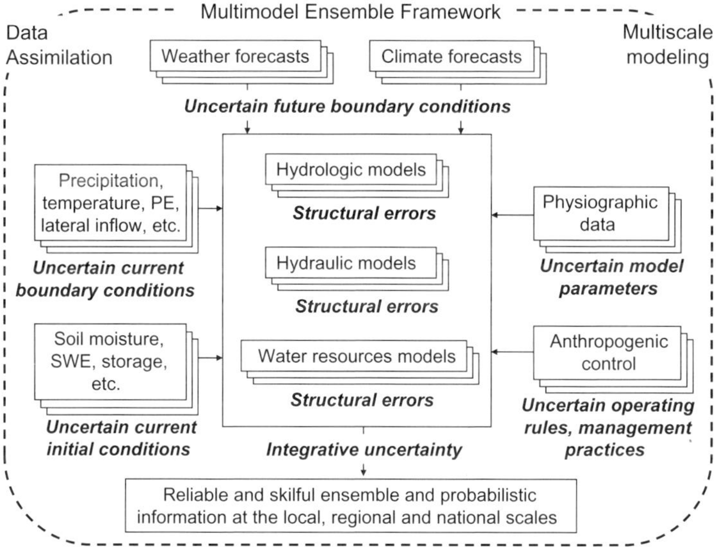

Figure 15.1. Major sources of uncertainty in a hydrological ensemble forecasting system (NWS, 2007b). PE stands for Potential Evaporation, SWE stands for Snow Water Equivalent.

The existing atmospheric and hydrological ensemble forecast techniques do not fully account for important uncertainties (Pappenberger *et al.*, 2005). As these techniques rely on imperfect models of the processes they represent, model forecasts (both atmospheric and hydrologic) contain a complex set of biases that must be removed to meet the user requirements. In this chapter, we describe all important aspects of hydrological ensemble forecasting from the viewpoint of developing an integrated, end-to-end hydrologic ensemble forecast system that leverages recent scientific advances and, if available, human forecasters: atmospheric aspects, hydrological aspects, data assimilation (DA), verification and post-processing as shown in Figure 15.2.

The term ensemble preprocessing in Figure 15.2 refers to the process to create an unbiased and skilful ensemble NWP forecast from either a raw deterministic NWP forecast or a raw ensemble NWP forecast (see Section 15.2). The data assimilator in Figure 15.2 uses available measurements to determine an ensemble of hydrological states (see Section 15.3.1). The ensemble post-processor in Figure 15.2 creates a bias-corrected ensemble from a deterministic hydrological forecast or an ensemble hydrological forecast (see Section 15.3.2 and 15.3.3). The preprocessor, data assimilator

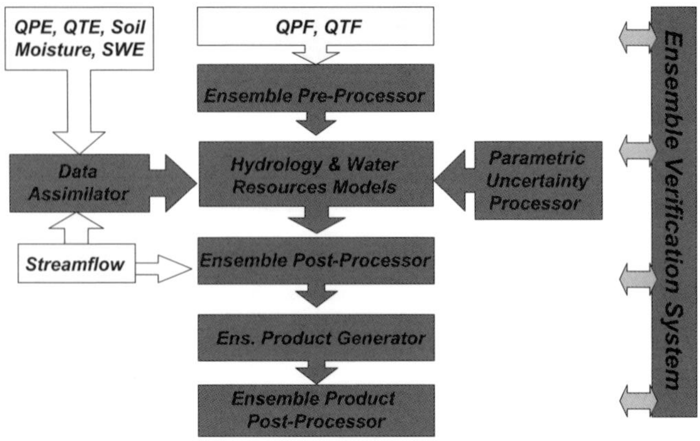

Figure 15.2. Major components in an operational hydrological ensemble forecasting system (NWS, 2007b). QPF stands for Quantative Precipitation Forecast and QTF stands for Quantative Temperature Forecast, SWE stands for Snow Water Equivalent.

and post-processor can be used and are often researched independently from each other. Ideally, they should be used in combination with one another. However, research into the combined use of all components envisioned is at an early stage and a number of significant challenges remain (see Section 15.5). Combined use of the preprocessor and the post-processor has been realised in the HEPS system, which is currently under testing for deployment by the NWS.

15.2. Atmospheric Uncertainty Aspects

The ideal input into a hydrological ensemble prediction system would be an ensemble of weather and climate forecasts that has three essential properties (WMO, 2007; Ebert, 2007):

- *reliability*: the agreement between the forecast probability and the observed frequency over many cases;
- *sharpness*: the tendency to forecast with a concentration of large probabilities around some value, as opposed to small probabilities spread over a wide range of values;
- *resolution/discrimination*: the ability of the forecast to produce different probabilities of exceedance for different events/discriminate between true events and true non-events.

An important atmospheric aspect of hydrologic ensemble prediction is whether current atmospheric forecasts account for all of the important meteorological and climatological uncertainties. Existing raw ensemble weather and climate forecasts meet the above properties to only a limited degree. This is not only due to the number of ensemble members being limited by computing resources (and hence subject to sampling uncertainty), but also because the EPS does not currently account for all significant sources of uncertainty. The result is that the forecasts are not necessarily reliable; they can be biased in the mean and typically do not display enough variability, leading to an underestimation of the uncertainty (Buizza *et al.* 2005).

Different approaches have been proposed to derive reliable probabilistic forecasts from raw model ensembles, a process that involves a combination of bias correction and downscaling in some form. Most of these methods are based on the idea of correcting the current forecast using past forecasts and corresponding observations, as has been done for deterministic forecasts. These include the Bayesian Processor of Output (BPO) (Krzysztofowicz, 2004), Bayesian Model Averaging (BMA) (Raftery *et al.*, 2005), ensemble dressing/best-member method (Roulston and Smith, 2003), non-homogeneous Gaussian regression (Gneiting *et al.*, 2005), logistic regression (Hamill *et al.*, 2004; Hamill and Whitaker, 2006), analogue techniques (Hamill and Whitaker, 2006), forecast assimilation (Stephenson *et al.*, 2005), and constrained bivariate modelling of forecast and observed variables that accounts for temporal scale-dependent relationships in both forecast errors and precipitation and temperature variability over the entire forecast period (Schaake *et al.* 2007).

Reggiani and Weerts (2008a) present an application of the BPO for a deterministic precipitation forecast. Slougther *et al.* (2007) applied BMA to a multi-model ensemble of spatially-distributed rainfall forecasts for the coastal region of the north-western US (Oregon). Improvements of the best-member methods have been proposed by Wang and Bishop (2005) and Fortin *et al.* (2006). Wilks and Hamill (2007) compared logistic regression, non-homogenous Gaussian regression and ensemble dressing using a 15 member reforecast dataset with a lead time of 15 days from 1979 to present (Hamill and Mullen, 2006). An important conclusion by Wilks and Hamill (2007) is that there appears to be no single best forecast method for all applications, and that extensive work is necessary on the ensemble model output statistics methods in the future.

As the development and application of bias correction and downscaling techniques are usually location-specific, it is difficult to compare objectively their performance over a range of space-time scales, and to assess systematically their strengths and weaknesses. Toward that end, significant community-wide effort is necessary. From the point of view of the operational centres, data requirements for bias correction and downscaling are a large issue as they require some form of systematic hindcasting over a long period (Hamill and Mullen, 2006). Also, it is expected that, due to the highly variable nature of precipitation, data requirements for precipitation ensemble hindcasts, particularly for significant to extreme precipitation events, may differ greatly from those for other atmospheric variables.

15.3. Hydrologic Uncertainty Aspects

In addition to atmospheric uncertainty as described above, a complete hydrologic ensemble forecast system must be able to reduce, to the greatest extent possible, and account for all major sources of hydrologic uncertainty. These include model initial conditions (e.g. soil moisture and snow water equivalent), model parameters, model structures, and human control of hydrologic variables (e.g. reservoir operations, irrigation) (see Figure 15.1). It is expected that even with timely implementation of "best available" science and technology, there will always exist significant "residual" hydrologic uncertainty, which will have to be accounted for via statistical modelling.

Given the above, a hydrologic ensemble forecast system may include several components to deal with various hydrologic uncertainties (see Figure 15.2). In Section 15.3.1 we describe DA post-processing and multi-model ensembles in some detail.

15.3.1. *Data assimilation*

The role of DA in hydrological ensemble prediction is to produce the best possible estimates of initial hydrological conditions and as such to constrain uncertainty at the start of the forecast. Besides a realistic representation of the system state (soil moisture, snow water equivalent, groundwater, etc.) by the ensemble mean, the ensemble members must provide realistic estimation of the uncertainty in the system state.

Observed precipitation and temperature data will typically be used in establishing (hydrological) model boundary conditions. Available observations can be effectively used in quantifying and reducing the error in the modelled water levels and discharges from the process models. These observations may include *in situ* measurements of water levels, discharge, snow water equivalent, soil moisture and groundwater levels. They may include remotely sensed observations from radar or satellites.

Sequential DA techniques provide the general framework in real-time forecasting for explicitly taking into account input uncertainty, model uncertainty and output uncertainty. The Extended Kalman Filter (EKF), an extension of the Kalman filter (Kalman, 1960) for linear systems, became popular in hydrologic forecasting in the late 1970s and early 1980s (Chiu, 1978; Georgakakos, 1986a,b; Kitanidis and Bras, 1980). However, the linearisation in the EKF is notoriously inaccurate if the non-linearities are strong. A possible way to circumvent these problems is by letting the errors evolve with the non-linear model equations by performing an ensemble of model runs. This has led to the development of the well-known Ensemble Kalman Filter (EnKF) (Burgers *et al.*, 1998; Evensen, 1994) and particle filter techniques such as sequential importance resampling filter and residual resampling filter (Arulampalam *et al.*, 2002; Doucet *et al.*, 2001; Gordon *et al.*, 1993).

Dual parameter and state estimation was investigated using EnKF techniques (Moradkhani *et al.*, 2005b; Vrugt, 2004). Weerts and El Serafy (2006) investigated the use of particle filters and EnKF using a conceptual hydrological rainfall runoff model for improving flood forecasts. Slater and Clark (2006) investigated the use of EnKF for assimilating snow water equivalent data into a streamflow forecasting model SNOW-17 for possible implementation in the US NWS operational forecasting system. Francois *et al.* (2003) showed the potential to improve streamflow simulations by assimilating ERS-1 SAR data into a conceptual hydrological model. El Serafy and Mynett (2004) showed the potential of EnKF by comparing it with EKF using a 1D hydrodynamic model of the Rhine to improve flood forecasting. Other applications of using EnKF to update hydraulic flood forecasting models are given by Neal *et al.* (2007), Madsen *et al.* (2003) and Shiiba *et al.* (2000).

15.3.2. *Post-processing*

Despite a long history in the hydrologic literature, automatic DA is relatively new in operational hydrology for various reasons. Similarly, explicit accounting of parametric uncertainty via a parametric uncertainty processor and accounting of structural uncertainty via, e.g. multi-model ensemble techniques are only in their infancy in the operational arena. As implementing these advances requires rather significant upgrades to current forecast systems, it is likely that in the early days of operational hydrologic ensemble forecasting, a purely statistical "catch-all" ensemble post-processor is necessary to reduce and account for the integrated hydrologic uncertainty (Seo *et al.*, 2006).

Post-processing of deterministic hydrological forecasts is often done by applying simple Autoregressive- (AR) and/or Moving Average (ARMA) type models (Broersen and Weerts, 2005; Madsen *et al.* 2000). This type of post-processing is used in many operational flood forecasting systems around the world. Recently, new initiatives on post-processing of hydrological forecasts (including ensemble forecasts) from a probabilistic viewpoint have been described in the literature. Reggiani and Weerts (2008b) applied the BPO method (Krzysztofowicz, 2004) to post-process deterministic hydrological forecasts of the River Rhine. Reggiani *et al.* (2008) applied the BPO to ensemble hydrological forecasts, again for the River Rhine.

In the US NWS, an ensemble post-processor of an ARX type has been developed and implemented (Seo *et al.* 2006). A combination of linear regression and probability matching, it attempts to correct biases in model streamflow simulation in the mean, and to account for all hydrologic uncertainties in a lump-sum manner. While shown to produce reliable streamflow ensembles in dependent validation, independent validation indicates that the procedure is subject to potentially large uncertainties due to sampling and/or non-stationarities (see Figure 15.3).

The final observation is that post-processing of hydrological forecasts follows the trend in the atmospheric community on post-processing of deterministic and ensemble forecasts. Methods proposed in the literature are promising but yet need to prove themselves in an operational real-time forecast setting.

15.3.3. *Multi-model ensemble*

Estimation theory states (see, e.g. Schweppe, 1973) that combining informative forecasts from different models reduces forecast uncertainty (Georgakakos *et al.*, 2004). It is also well known that structural errors in hydrologic models are very difficult to correct. As these errors tend to be strongly correlated in time (for lumped models) or in space and time (for distributed models), addressing them through post-processing requires complex and often heavily parameterised statistical modelling. Multi-model ensembles offer potential for accounting for structural uncertainty without such data- and parameter-intensive statistical modelling.

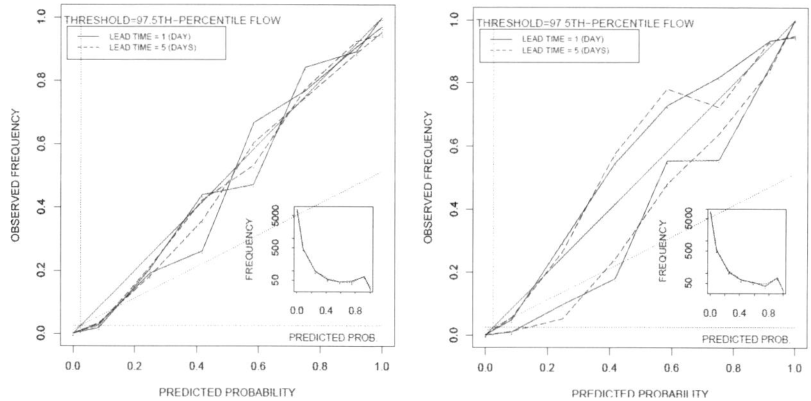

Figure 15.3. Reliability diagrams for probabilistic prediction at Newport, Pennsylvania, US, based on post-processed ensemble traces in the dependent (left-hand side) and independent (right-hand side) validation periods. The diagrams for 1- and 5-day-ahead predictions are in solid and dashed lines, respectively. The two lines for each lead time correspond to the two dependent (left-hand side) and independent (right-hand side) validation periods. The threshold is the 97.5th-percentile flow. The solid and dotted lines in the lower-right corner are the histograms of the predicted probability for lead times of 1 and 5 days, respectively (from Seo *et al.* 2006).

Vrugt and Robinson (2007) compared BMA and EnKF for probabilistic streamflow forecasting. The results from that study suggest that for the watershed under consideration, BMA cannot achieve a performance matching that of the EnKF method. Beckers *et al.* (2008) applied the BMA method to a multi-model ensemble, made up of several meteorological forecasts combined with a hydrological model and a model cascade of a hydrological-hydraulic model for the forecasting point at Lobith on the Rhine. Table 15.1 shows the Root Mean Squared Error (RMSE) of the different forecasts used in the BMA method. Figure 15.4 shows an example of the outcome of the BMA.

15.4. Verification

The methods discussed above are used to improve the reliability, skill and resolution of probabilistic hydrological forecasts. Forecast verification techniques may be applied to assess these attributes. As with the ensemble forecasting approach, these techniques have been developed primarily within the atmospheric sciences, but are often equally applicable to other disciplines, such as the hydrological sciences (Wilks, 2006).

Table 15.1. RMSE of the individual forecast models and the BMA mean forecast for different lead times, with the lowest RMSE's highlighted in grey. All calculations used a training period of 28 days (from Beckers, Sprokkereef and Roscoe, 2008).

Forecast	Meteorological input	Hydrological/ hydraulic model	RMSE (24–48 hrs)	RMSE (48–72 hrs)	RMSE (72–96 hrs)
1	HIRLAM	HBV	0.252	0.329	0.428
2	ECMWF	HBV	0.249	0.313	0.379
3	DWD-LM	HBV	0.249	0.302	0.347
4	DWD-GME	HBV	0.249	0.306	0.345
5	HIRLAM	HBV/SOBEK	0.196	0.258	0.381
6	ECMWF	HBV/SOBEK	0.196	0.250	0.340
7	DWD-LM	HBV/SOBEK	0.195	0.238	0.314
8	DWD-GME	HBV/SOBEK	0.195	0.239	0.303
9	LobithW (statistical model)		0.176	0.250	0.366
BMA mean forecast			0.179	0.235	0.307

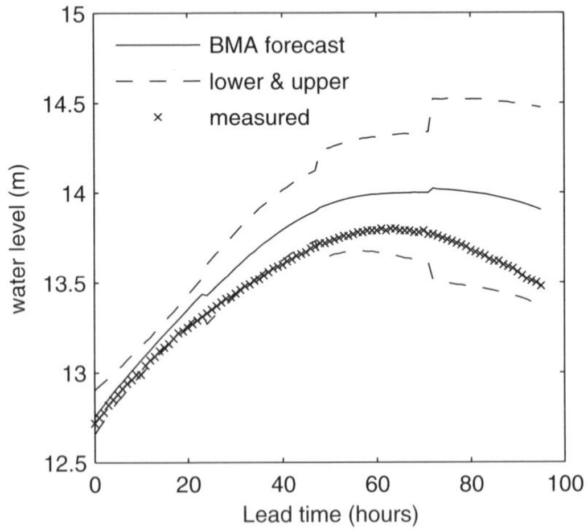

Figure 15.4. BMA water level (m) forecast (solid line) with the lower and upper confidence bounds (dashed lines) versus measured water level (m) (marker x) at Lobith for a water level peak in March 2007 (from Beckers, Sprokkereef and Roscoe, 2008).

From the viewpoint of operational hydrologic forecasting, there are at least three types of verification of interest: (i) diagnostic; (ii) trend; and (iii) prognostic. Diagnostic verification is concerned with assessing different attributes of ensemble forecasts, such as reliability, skill, resolution, discrimination, etc., to diagnose the performance of the forecast system and process so that cost-effective improvements may be made. Trend analysis is concerned with being able to discern and assess improvement in forecast quality over time. Prognostic verification is concerned with being able to provide the users of the forecast, such as the forecasters and the emergency managers, with verification information that may be directly used for decision making. Such verification information would come from translating and casting all available verification information into the context of the forecasting and decision-making problem at hand.

Methods for verification of forecasts are well established (Wilks, 2006), and such verification provides clear insight into value and skill of the ensemble predictions at different lead times, giving valuable information to the forecaster in interpreting the forecast products.

Roulin and Vannitsem (2005) evaluated the skill of a hydrological ensemble prediction system, integrating a water balance model with ensemble precipitation forecasts from ECMWF — EPS, for two Belgian catchments. Recently, verification of ensemble forecasting in the Rhine basin was carried out by the German Bundesanstalt für Gewässerkunde (BfG or Federal Institute of Hydrology) in cooperation with Dutch Waterdienst (Renner and Werner, 2007). Results clearly show that bias in the ensemble hydrological forecasts is not only influenced by the quality of the ensemble meteorological forecasts, but also by the catchment size and the variability of the hydrological regime.

To ascertain the quality of probabilistic hydrological forecasts, several verification techniques can be applied. Reliability can be assessed through, for example, reliability diagrams or attribute diagrams. An example is given in Figure 15.5. These diagrams measure the agreement between predicted probabilities and observed frequencies. If the forecast is reliable, then whenever the forecast probability of an event occurring is P, that event should occur a fraction of P of the time (Ebert, 2007).

Measures for assessing the overall quality of ensemble forecasts include the Brier Score (BS), which measures the mean squared error in the probability space for a specific threshold. The Brier Skill Score (BSS) measures skill relative to a reference forecast (usually climatology or naïve forecast). The Ranked Probability Score (RPS) is another way of assessing the overall quality of the probabilistic forecast. RPS measures the squared difference in probability space when there are multiple categories (when there are only two categories RPS is equal to the BS). As with the BSS, the ranked probability skill score measures skill relative to a reference forecast.

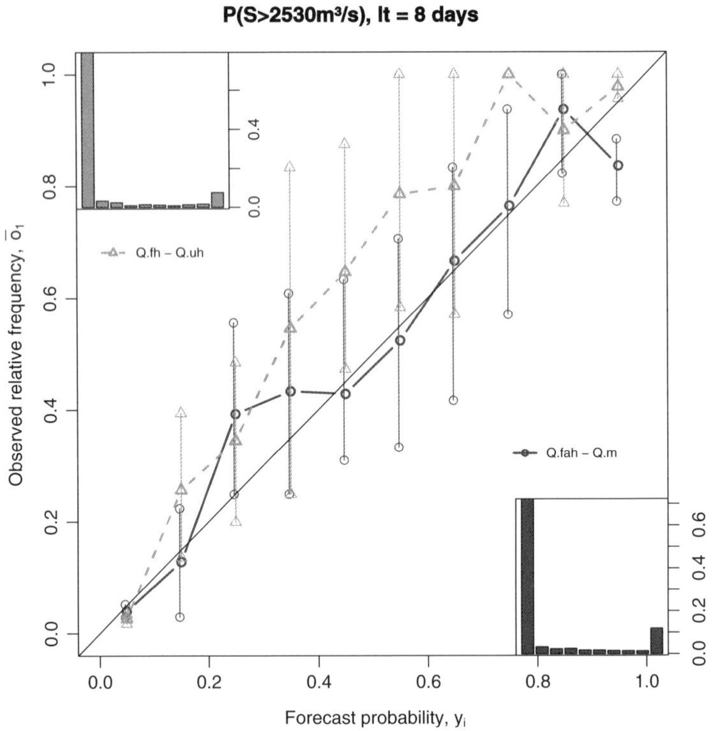

Figure 15.5. Example of a reliability diagram for the gauging station at Andernach on the Rhine. The dark-grey line with circles shows the reliability of the error corrected flow forecasts (Q.fah) at Andernach, verified against the observations (Q.m). The light-grey line with triangles shows the reliability of the uncorrected flow forecast (Q.fh) verified against the baseline simulation (Q.uh). Confidence intervals are given for $\alpha = 0.05$. Histograms of forecast probabilities are given in the upper-left and lower-right corners (from Renner and Werner, 2008).

RPS applies when there are a discrete number of categories, but it can be extended to continuous categories as the Continuous Ranked Probability Score (CRPS). CRPS is particularly attractive in that it does not depend on the particular choice of thresholds, and that it allows comparative verification with single-value forecasts, for which CRPS reduces to absolute mean error.

The Relative Operation Characteristic (ROC) is a measure to assess the ability of the forecast to discriminate between events and non-events. The ROC curve plots the hit rate (Probability of Detection (POD)) against the false alarm rate (Probability of False Detection (POFD)). The curve is created using increasing probability thresholds to make the yes/no

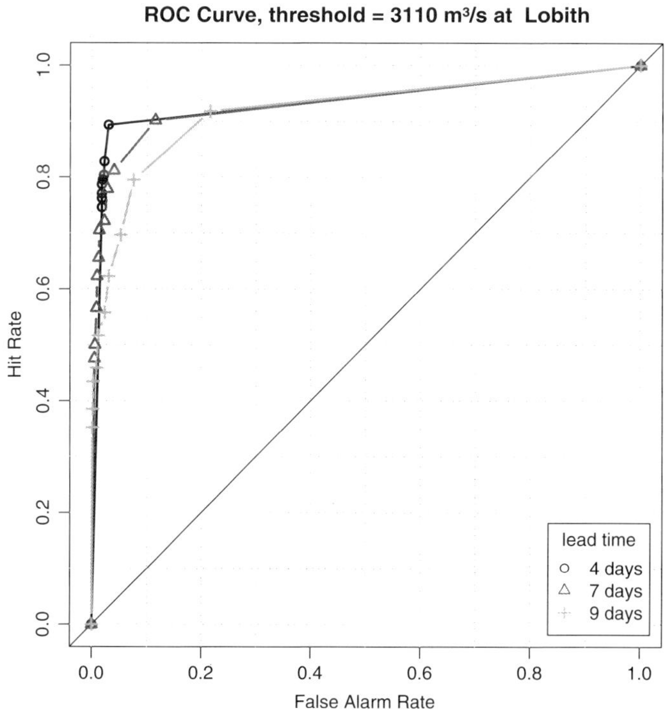

Figure 15.6. Example of a ROC diagram for the gauging station at Lobith on the River Rhine. In this case the ensemble flow forecast is verified against the observed flow at a threshold of 3110 m^3/s, which is the 90% exceedance frequency. The curves show the ROC for lead times of 4, 7 and 9 days (from Renner and Werner, 2008).

decision (WMO, 2007). Figure 15.6 shows an example of the ROC for the ECMWF–EPS discharge forecast for Lobith (River Rhine) for 90% exceedance frequency at three different lead times (4, 7 and 9 days) determined over three years of ensemble forecasts (20052007). As expected, the skill measured by the ROC curves decreases with increasing lead times.

15.5. Promises and Challenges

It is clear that the use of ensembles of atmospheric inputs, combined with advanced methods of reducing and accounting for hydrologic uncertainties have significant benefits in providing a more complete, and statistically sound quantification of uncertainty in the hydrological forecast. The utility of ensemble forecasts in increasing the skill and lead time of hydrological predictions has been shown by several authors (Regimbeau *et al.*, 2007; Roulin; 2007), and Roulin (2007) showed that using the ensemble forecast, or even the ensemble mean, to underpin flood warning has more relative economic value than using the deterministic forecast alone. Such translation of the probabilistic forecast into an economical value that balances the expected damage due to a flood event against the expected loss due to a false warning (and possible ensuing evacuation) can thus provide stakeholders with an instrument for rational decision making in the presence of uncertainty (Raiffa and Schlaifer, 1961).

Development of hydrologic ensemble forecasting techniques is at an early stage. Developing improved techniques will require major international interdisciplinary collaborations involving forecast producers and forecast users as well as the science community. Accordingly, an international effort, the Hydrologic Ensemble Prediction Experiment (HEPEX), was launched in 2004 (Schaake *et al.*, 2007). HEPEX has organised several workshops and a number of test-bed projects to address specific science issues and to demonstrate the strengths and limitations of existing methods. More information about HEPEX can be found at http://www.hepex.org.

15.5.1. *Ensemble DA*

In the US NWS, forecasters play an integral part in DA. They routinely practice manual DA, referred to as "Run-Time Modifications" (MODS). To be operationally viable, automatic DA must be able to perform comparably to manual DA, but also allow forecaster intervention and control. Maintaining complementarities between automatic DA and manual control is essential in an operational forecast system as in the real world there will be critical times when automatic DA may not work for various reasons. Accordingly, it is necessary to objectively evaluate the performance of automatic DA (Seo *et al.*, 2008) and ascertain the optimal balance between automatic and manual DA that best capitalises on scientific and computational advances and human forecasters.

Most DA techniques used in or proposed for hydrologic forecasting are adopted and/or adapted from electrical engineering, weather forecasting and oceanography. Many hydrologic DA problems are, however, very nonlinear not only in the physical processes but also in the observational processes. Also, almost all hydrologic systems modelled for operational forecasting are not well observed, and hence subject to very large degrees of freedom. Such systems are likely to pose very underdetermined inverse problems. Lastly, operational hydrologic forecasting is very often concerned with extreme events, such as floods and droughts, whereas, with the exception of particle filtering, all DA techniques are "optimal" only in the second-order moment sense.

For the above reasons, general and flexible techniques that combine strengths of variational assimilation and ensemble Kalman filter, and that allow easy transition from single-value to ensemble DA, such as the maximum likelihood ensemble filter (MLEF, Zupanski, 2005), are extremely attractive. For example, the US NWS is developing a variant of MLEF that is essentially an ensemble extension of 2D Variational Assimilation (2DVAR) for lumped hydrologic models (Seo *et al.*, 2003a). Note that, in this approach, operational experience gained from the single-value DA is directly transferable to ensemble DA. This is an extremely important consideration for operational forecasting, given that single-value forecasting is expected to continue to be practiced for the foreseeable future, even after introduction of ensemble forecasting capabilities. Such a phased approach (from single-value to ensemble DA) is pursued in NWS also for DA for distributed hydrologic models (Seo *et al.*, 2003b) and for hydrologic routing using a variable three-parameter Muskingum model (O'Donnell, 1985).

15.5.2. *Post-processing*

Statistical post-processing is an amalgamation of prediction, bias correction and scaling, for each of which different solutions may be possible. As the ensemble forecast system improves over time, it may be expected that statistical post-processing will become largely a problem of bias correction and scaling. For example, for streamflow each post-processed ensemble member should be probabilistically unbiased at all temporal scales of aggregation. Such a requirement arises from the fact that the user must be able to generate reliable probabilistic products from post-processed ensembles for any chosen forecast time window. To ascertain strengths of weaknesses of various competing techniques for these different aspects of post-processing, community-wide efforts are needed.

15.5.3. *Ensemble verification*

Ensemble verification as it is applied in operational hydrology today borrows heavily from the atmospheric science community. One of the distinguishing aspects of streamflow or precipitation ensembles is that they are multi-scale in nature, and hence should be verified over a wide range of spatiotemporal scales of aggregation. Unlike verification measures for single-value forecasts, most of the measures for ensemble forecasts are not expressed in physically meaningful units. While this poses little problem for diagnostic verification, it makes the use of verification information for real-time forecasting and decision making very difficult. This is an extremely important aspect of hydrologic ensemble forecasting; its promise can be realised only if the user is able to use the probabilistic information with ease and clarity in real-time decision making. The US NWS is developing such verification measures and the results will be reported in the near future.

Acknowledgements

We would like to acknowledge Maik Renner for providing the figures on the verification of the ensemble forecasts of the Rhine. We would also like to acknowledge Joost Beckers and Kathryn Roscoe for providing the data on the BMA forecasts of the Rhine. We also would like to thank the forecasting office of the Dutch Ministry of Transport and Waterways and the German Federal Office of Hydrology for granting permission for the use of the operational Rhine Forecasting system.

References

Arulampalam, M.S., Maskell, S., Gordon, N. *et al.* (2002). A tutorial on particle filters for online nonlinear/non-Gaussian Bayesian tracking, *IEEE T. Signal Proces.*, **50**, 174–188.

Bartholmes, J. and Todini, E. (2005). Coupling meteorological and hydrological models for flood forecasting, *Hydrol. Earth Syst. Sci.*, **9**, 333–346.

Beckers, J.V.L., Sprokkereef, E. and Roscoe, K. (2008). "Use of Bayesian model averaging to determine uncertainties in river discharges and water level" in: *4th International Symposium on Flood Defence: Managing Flood Risk, Reliability and Vulnerability*, Toronto, Ontario, Canada, May 6–8 2008.

Broersen, P.M.T. and Weerts, A.H. (2005). "Automatic error correction of rainfall-runoff models in flood forecasting systems" in: *Proceedings of the IEEE/IMTC Conference*, Ottawa, Canada, 1531–1536.

Buizza, R., Houtekamer, P.L., Toth, Z. *et al.* (2005). A compariosn of ECWMF, MSC, and NCEP global ensembles prediction systems, *Mon. Weather Rev.*, **133**, 1076–1097.

Burgers, G., van Leeuwen, P. and Evensen, G. (1988). Analysis scheme in the ensemble Kalman filter, *Mon. Weather Rev.*, **126**, 1719–1724.

Chiu, C.-L. (1978). *Applications of Kalman Filter to Hydrology, Hydraulics, and Water Resources*, paper presented at the Proceedings of AGU Chapman Conference, May 22–24 1978, Pittsburgh, Pennsylvania, USA.

Day, G.N. (1985). Extended streamflow forecasting using NWSRFS, *J. Water Res. Pl.-ASCE*, **111**, 157–170.

De Roo, A., Gouweleeuw, B., Thielen, J. *et al.* (2003). Development of a European Flood Forecasting System, *Int. J. River Basin Manage.*, **1**, 49–59.

Doucet, A., de Freitas, N. and Gordon, N. (2001). "Sequential Monte Carlo Methods in Practice" in: Jordan, M., Lauritzen, S.L., Lawless, J.F. *et al.* (eds), *Statistics for Engineering and Information Science*, Springer, New York, pp. 1–581.

Ebert, E. (2007). Verification of probability forecasts, 3rd International Verification Methods Workshop, ECMWF, Reading, UK. Available at http:/ www.ecmwf.int/newsevents/meetings/workshops/2007/jwgv/ (Accessed on 20/04/2012).

El-Serafy, G.Y. and Mynett, A.E. (2004). "Comparison of EKF and EnKF in SOBEK-River: Case study Maxau-Ijssel" in: Liong, S.-Y., Phoon, K.K. and Babovic, V. (eds), *Proceedings of the 6th International Conference on Hydroinformatics*, World Scientific Publishing, Singapore, pp. 513–520.

Evensen, G. (1994). Sequential data assimilation with a non-linear quasi-geostrophic model using Monte Carlo methods to forecast error statistics, *J. Geophy. Res.*, **97**, 905–924.

Fortin, V., Favre, A.-C. and Saïd, M. (2006). Probabilistic forecasting from ensemble prediction systems: improving upon the best member method by using a different weight and dressing kernel for each member, *Q. J. Roy. Meteor. Soc.*, **132**, 1349–1369.

Georgakakos, K.P. (1986a). A generalized stochastic hydrometeorological model for flood and flash flood forecasting. Part 1 Formulation, *Water Resour. Res.*, **22**, 2083–2095.

Georgakakos, K.P. (1986b). A generalized stochastic hydrometeorological model for flood and flash flood forecasting. Part 2 Case studies, *Water Resour. Res.*, **22**, 2096–2106.

Georgakakos, K.P., Seo, D.-J., Gupta, H. *et al.* (2004). Towards the characterization of streamflow simulation uncertainty through multimodel ensembles, *J. Hydrol.*, **298**, 222–241.

Gneiting, T., Raftery, A.E., Westerveld III, A.H. *et al.* (2005). Calibrated probabilistic forecasting using ensemble model output statistics and minimum CRPS estimation, *Mon. Weather Rev.*, **133**, 1098–1118.

Gordon, N.J., Salmond, D.J. and Smith, A.F.M. (1993). Novel approach to nonlinear/non-Gaussian Bayesian state estimation, *IEE Proc.-F*, **140**, 107–113.

Gouweleeuw, B.T., Thielen, J., Franchello G. *et al.* (2005). Flood forecasting using medium-range probabilistic weather prediction, *Hydrol. Earth Syst. Sci.*, **9**, 365–380.

Hamil, T. and Mullen, S.L. (2006). Reforecasts: an important new dataset for improving weather predictions, *B. Am. Meteorol. Soc.*, **87**, 33–46.

Hamill, T. and Whitaker J.S. (2006). Probabilistic quantitative precipitation forecasts based on reforecasts analogs: theory and application, *Mon. Weather Rev.*, **134**, 3209–3229.

Hamill, T., Whitaker, J.S. and Wei, X. (2004). Ensemble forecasting: improving medium-range forecast skill using retrospective forecasts, *Mon. Weather Rev.*, **132**, 1434–1447.

Kitanidis, P.K. and Bras, R.L. (1980). Real time forecasting with a conceptual rainfall runoff model. 1. Analysis of uncertainty, *Water Resour. Res.*, **16**, 1025–1033.

Krzysztofowicz, R. (2004). "Bayesian Processor of Output: A new technique for probabilistic weather forecasting" in: 17th Conference on Probability and Statistics in the Atmospheric Sciences, 84th AMS Annual meeting, Seattle, 11–15 January 2004, AMS. Available at, http://ams.confex.com/ams/84Annual/17PROBSTA/abstracts/69608.html (Accessed on 14/04/2012).

Luo, L., Wood, E.F. and Pan, M. (2007), Bayesian merging of multiple climate model forecasts for seasonal hydrological predictions, *J. Geophys. Res.*, **112**, D10102.

Madsen, H., Butts, M., Khu, S. *et al.* (2000). "Data assimilation in rainfall runoff forecasting" in: *Proceedings of the 4th Hydroinformatics Conference, Iowa, USA*, International Association of Hydro-Environment Engineering: Cedar Rapids, Iowa, USA.

Madsen, H., Rosbjerg, D., Damgard, J. *et al.* (2003). "Data assimilation into MIKE 11 flood forecasting system using Kalman filtering, water resources systems — hydrological risk, management and development" in: *Proceedings of symposium HS02b held during IUGG2003, Sapporo*, IAHS publication 281, pp. 75–81.

Molteni, F., Buizza, R., Palmer, T.N. *et al.* (1996). The ECMWF ensemble prediction system: methodology and validation, *Q. J. Roy. Meteor. Soc.*, **122**, 73–119.

Moradkhani, H., Sorooshian, S., Gupta, H.V. *et al.* (2005). Dual state-parameter estimation of hydrological models using ensemble Kalman filter, *J. Adv. Water Res.*, **28**, 135–147.

Neal, J.C., Atkinson, P.M. and Hutton, C.W. (2007). Flood inundation model updating using an ensemble Kalman filter and spatially distributed measurements, *J. Hydrol.*, **336**, 401–415.

NWS (2007a). *The Experimental Ensemble Forecast System (XEFS) Design and Gap Analysis*, report of the XEFS Design and Gap Analysis Team, NOAA/NWS, Silver Spring, MD, p. 50.

NWS (2007b). *Office of Hydrologic Development Hydrology Laboratory Strategic science plan*, NOAA/NWS, Silver Spring, p. 149.

O'Donnell, T. (1985). A direct three-parameter Muskingum procedure incorporating lateral inflow, *Hydrol. Sci. J.*, **30**, 479–496.

Pappenberger, F., Beven, K.J., Hunter, N. *et al.* (2005). Cascading model uncertainty from medium range weather forecasts (10 days) through a rainfall-runoff model to flood inundation predictions within the European Flood Forecasting System (EFFS), *Hydrol. Earth Syst. Sci.*, **9**, 381–393.

Raftery, A.E., Gneiting, T., Balabdaoui, F. *et al.* (2005). Using Baysesian model averaging to calibrate forecast ensembles, *Month. Weather Rev.*, **133**, 1155–1174.

Raiffa, H. and Schlaifer, R. (1961). *Applied Statistical Decision Theory*, MIT Press, Cambridge, Massachusetts.

Reggiani, P.M., Renner, M., Weerts, A.H. *et al.* (2008). Uncertainty assessment via Bayesian revision of ensemble stream flow predictions in the operational River Rhine forecasting system, *Water Resour. Res.*, **45**, W02428–W02442.

Reggiani, P.M. and Weerts, A.H. (2008a). Probabilistic quantitative precipitation forecast for flood prediction: an application, *J. Hydrometeorol.*, **9**, 76–95.

Reggiani, P.M. and Weerts A.H. (2008b). A Bayesian approach to decision-making under uncertainty: an application to real-time forecasting in the River Rhine, *J. Hydrol.*, **356**, 56–89.

Regimbeau, F., Habets, F., Martin, E. *et al.* (2007). Ensemble streamflow forecasts Over France, *ECMWF Newsletter*, **111**, 21–27.

Renner, M. and Werner, M. (2007). *Verification of Ensemble Forecasting in the Rhine Basin*, Delft Hydraulics Research Report, Q4378.

Renner, M. and Werner, M. (2008). Verification of ensemble flow forecasts for the River Rhine, *J. Flood Risk Manage.*, **376**, 463–475.

Roulin, E. (2007). Skill and relative economic value of medium-range hydrological ensemble predictions, *Hydrol. Earth Syst. Sci.*, **11**, 725–737.

Roulin, E. and Vannitsem, S. (2005). Skill of medium-range hydrological ensemble predictions, *J. Hydrometeorol.*, **6**, 729–744.

Roulston, M.S. and Smith, L.A. (2003). Combining dynamical and statistical ensembles, *Tellus*, 55A, 16–30.

Saha, S., Nadiga, S., Thiaw, C. *et al.* (2006). The NCEP climate forecast system, *J. Climate*, **19**, 3483–3517.

Schaake, J., Demargne, J., Hartman, R. *et al.* (2007). Precipitation and temperature ensemble forecasts from single-value forecasts, *Hydrol. Earth Syst. Sci. Discuss.*, **4**, 655–717.

Schaake, J.C., Hamill, T.M., Buizza, R. *et al.* (2007). HEPEX: the hydrological ensemble prediction experiment, *B. Am. Meteorol. Soc.*, **88**, 1541–1547.

Schweppe, F.C. (1973). *Uncertain Dynamic Systems*, Prentice-Hall, New Jersey, p. 563.

Seo, D.-J., Cajina, L., Corby, R. *et al.* (2008). Automatic state updating for operational streamflow forecasting via variational data assimilation, *J. Hydrol.*, **367**, 255–275.

Seo, D.-J., Herr, H.D. and Schaake, J.C. (2006). A statistical post-processor for accounting of hydrologic uncertainty in short-range ensemble streamflow prediction, *Hydrol. Earth Syst. Sci. Discuss.*, **3**, 1987–2035.

Seo, D.-J., Koren, V. and Cajina, L. (2003a). Real-time variational assimilation of hydrologic and hydrometeorological data into operational hydrologic forecasting, **4**, 627–641.

Seo, D.-J., Koren, V. and Cajina, N. (2003b). Real-time assimilation of radar-based precipitation data and streamflow observations into a distributed hydrological model, *Int. Assoc. Hydrol. Sci. Pub.*, **282**, 138–142.

Shiiba, M., Laurenson, X. and Tachikawa, Y. (2000). Real-time stage and discharge estimation by a stochastic-dynamic flood routing model, *Hydrol. Process.*, **14**, 481–495.

Slater, A.G. and Clark, M.P. (2006). Snow data assimilation via an ensemble Kalman filter, *J. Hydrometeorol.*, **7**, 478–493.

Sloughter, J.M., Raftery, A.E. and Gneiting, T. (2007). Probabilistic quantitative precipitation forecasting using Bayesian model averaging, *Mon. Weather Rev.*, **135**, 3209–3220.

Stephenson, D.B., Coelho, C.A.S., Balmasedam, M. *et al.* (2005). Forecast assimilation: a unified framework for the combination of multi-model weather and climate predictions. *Tellus*, **57A**, 253–264.

Tracton, M.S. and Kalnay, E. (1993). Operational ensemble prediction at the National Meteorological Center, *Weather Forecast.*, **8**, 379–398.

Vrugt, J.A. (2004). Towards improved treatment of parameter uncertainty in hydrologic modelling, Ph.D, University of Amsterdam.

Vrugt, J.A. and Robinson, B.A. (2007). Treatment of uncertainty using ensemble methods: comparison of sequential data assimilation and Bayesian model averaging, *Water Resour. Res.*, **43**, doi 10.1029/2005WR004838.

Wang, X. and Bishop, C.H. (2003). A comparison of breeding and ensemble transform Kalman filter ensemble forecast schemes, *Q. J. Roy. Meteor. Soc.*, **131**, 965–986.

Weerts, A. and El Serafy, G.Y. (2006). Particle filtering and ensemble Kalman filtering for state updating with hydrological conceptual rainfall-runoff models, *Water Resour. Res.*, **42**, W09403, doi:10.1029/2005WR004093.

Wilks, D.S. (2006). *Statistical Methods in the Atmospheric Sciences 2nd Edition, International Geophysics Series, Volume. 91*, Academic Press, San Diego.

Wilks, D.S. and Hamill, T. (2007). Comparison of Ensemble-MOS methods using GFS reforecasts, *Mon. Weather Rev.*, **135**, 2379–2390.

WMO (2007). Forecast verification — issues, methods and faq, WWRP/WGNE joint working group on verification. Available at http://www.cawcr.gov.au/projects/verification/ (Accessed on 05/05/2012).

Zupanski, M. (2005). Maximum likelihood ensemble filter: theoretical aspects, *Mon. Weather Rev.*, **133**, 1710–1726.

CHAPTER 16

A Data-Based Mechanistic Modelling Approach to Real-Time Flood Forecasting

Peter C. Young*, Renata J. Romanowicz[†] and Keith Beven
Lancaster Environment Centre,
Lancaster University, UK
[†]*Institute of Geophysics, Polish Academy of Sciences,*
Warsaw, Poland

16.1. Introduction

Recent extensive flood events in Britain have focussed the interest of the hydrological community on improved methods of flood forecasting that could help to decrease flood losses under the increasing probability of flood occurrence. Disasters such as floods, which might arise from human-induced environmental changes, are not only dangerous for the natural environment but also for human beings. Consequently, new methodologies are required to mitigate the effects of such flooding both before and during events. In this chapter, we consider various aspects of model-based, real-time flood forecasting, concentrating on the research being carried out at Lancaster under the aegis of the Flood Risk Management Research Consortium (FRMRC). In particular, we discuss the development of models and methods that allow for the real-time updating of forecasts and model parameters in flood forecasting systems.

Refsgaard (1997) gives a review of different updating techniques used in real-time flood forecasting systems, as well as the comparison of two

---

*Peter Young is also Adjunct Professor in the Integrated Catchment Assessment and Management (iCAM) Centre, Fenner School of Environment and Society, Australian National University, Canberra.

different updating procedures applied to a conceptual hydrological model of a catchment, including rainfall-flow and the flow routing models. In relation to his classification, the methods developed in the present chapter utilise both parameter and state updating. However, unlike approaches such as the Extended Kalman Filter (EKF) algorithm used by Refsgaard, where the updating is carried out within a single, nonlinear, state space setting, with the parameters considered as adjoined state variables, our state and parameter estimation procedures are carried out separately but concurrently, employing co-ordinated recursive estimation algorithms. This avoids the well-known deficiencies of the EKF (such as problems with covariance estimation and convergence for multi-dimensional, nonlinear models) and yields more statistically efficient estimates of the model parameters. (i.e. it reduces the uncertainty on these estimates). Based on the results of previous research (Romanowicz *et al.*, 2004), the present study also considers an implementation based on the modelling and forecasting of water level (stage) rather than flow. This approach avoids the errors introduced by the conversion of levels to flow; it can be applied in the situation where flow observations are not available; and it directly yields the forecasts of water levels, with their associated uncertainty estimates, that are normally required for flood forecasting and warning.

The real-time adaptive updating procedures applied in this project follow the methodology for the online forecasting of rainfall-flow processes described originally by Young (2001a, 2002). This exploits the top-down, Data-Based Mechanistic (DBM) approach to the modelling of environmental processes (e.g. Young, 2001b, 2003, 2009, 2011 and the prior references therein), concentrating on the identification and estimation of those "dominant modes" of dynamic behaviour (Young, 1999a) that are most important for the objective of high quality flood prediction. This DBM approach involves inductive modelling, where the dominant mode model structure is identified from the available data; the parameters that characterise this model structure are estimated from the same data; and the model is validated against other, separate data sets. In the resulting model, which can be interpreted directly in hydrological terms, the most important hydrological processes active in the catchment are modelled using the State Dependent Parameter (SDP) method (Young, 2000, 2001c; Young *et al.*, 2001) of estimating the effective rainfall nonlinearity and nonlinear water level routing, together with a Stochastic Transfer Function (STF) method (Young, 2005) for characterising both the linear effective rainfall-water level behaviour and linear water level routing processes.

The complete model consists of these linear and nonlinear stochastic, dynamic, hydrological elements connected in a manner that represents the physical structure of the catchment, accounting for both surface and groundwater effects on river water level or flow. Bearing in mind that the objective of flood forecasting is to predict the water level of the river at future times, the Lancaster system has been developed specifically for water level (stage) forecasting, thus avoiding the uncertainties associated with the stage–discharge calibration relationship. It is important to note, however, that there have been numerous previous studies where DBM models have been concerned with flow modelling and forecasting, and that the procedures developed here for water level forecasting could be used equally well for flow forecasting, if this is required.

The adaptive forecasting system utilises a state space form of the complete catchment model, which provides the basis for Data Assimilation (DA) and forecasting of water levels at specified geographical locations using a modified Kalman Filter (KF) (Kalman, 1960) forecasting engine. Here, the predicted model states (water levels) and important adaptive gain parameters are updated recursively, in real-time, in response to input data received in real-time from remote sensors in the catchment. In this way, the extraction of information content in the data is maximised, leading to improved stochastic forecasts of water level or flow.

16.2. Research Background

The main objective of the research described in this chapter is the development of a methodology for operational real-time flood forecasting systems at extended lead times and a strategy for updating both the states (here river water levels or flows) and associated model parameters in real time. The research reflects the developments underway in the National Flood Forecasting System (NFFS), a dedicated user interface that provides for the incorporation of user-specified hydrological and hydraulic models (see Chapter 18); as well as the efficient import and processing of numerical weather predictions, radar data and online meteorological and hydrological data. Therefore, the tools and routines that are described here must fulfil the requirements of the NFFS. In this regard, the team at Lancaster have co-operated with the developers of the NFFS, Delft Hydraulics, in the initial incorporation of the Lancaster River Severn forecasting system into the NFFS, thus demonstrating that the system is compatible with the NFFS.

The DBM approach allows for online, real-time updating of parameters and states in either DBM models or the related hybrid-metric-conceptual (HMC) forecasting models (Wheater *et al.*, 1993). The same basic approach can be applied to any model of this general, quasi-distributed type: i.e. lumped parameter models for rainfall-level and level routing (or flow equivalents), connected in a manner that represents the geographical nature of the catchment and the location of rainfall and flow gauges within the catchment. However, some difficulties in parameter updating may be experienced in large models whose parameters are not identifiable from the data because the model is over-parameterised. Although the basic tools are also appropriate for fully distributed (partial differential equation) hydrological and hydraulic models, they have not been developed nor tested in this framework and further research would be necessary in this connection.

We have also considered the use of other procedures, such as the topical numerical Bayesian approach to forecasting, as in the ensemble Kalman filter (EnKF: see e.g. Moradkhani *et al.*, 2005b) and other more complex methods that exploit Monte Carlo Simulation (MCS) analysis. However, since these methods are computationally expensive and complicated to implement because of the need for multi-realisation MCS, they should only be used if the nonlinear stochastic structure is such that much simpler, and computationally inexpensive, analytic Bayesian procedures, such as the KF are inappropriate. In the present context, this does not seem to be the case.

In particular, DBM model identification, echoing many previous conceptual catchment modelling studies, shows that rainfall-flow (or water level) processes can be represented by the serial connection of an "effective rainfall" nonlinearity, that represents the nonlinear catchment storage (soil moisture) dynamics, and a linear process, here in the form of a transfer function, which is simply the discrete-time equivalent of a differential equation model (Young, 2005). The impulse response function of this linear module nicely reflects the underlying unit hydrograph properties of the catchment and it can be decomposed into a parallel pathway form that represents fast (surface and near-surface) processes, as well as slow, groundwater processes that affect the base-flow of the river. Stochastic forecasting using such a "Hammerstein" model can be carried out by a simple modification of the KF that allows for the nonlinear input and the heteroscedasticity (changing variance) of the prediction errors.

It could be argued that the simple distributional assumptions that are required for the Bayesian interpretation of the KF are not entirely valid. However, it should be realised that the KF does not require such assumptions (Kalman developed the algorithm using orthogonal projection theory) and it is robust to their violation. Moreover, from a practical standpoint, recent comparison of the results obtained from the EnKF (Moradkhani *et al.*, 2005) and analytical KF results using a DBM model (Young, 2006, 2010a, 2010b, 2013) for the Leaf River in the USA show that the simpler KF-based approach produces results that are as good, if not better, than the much more computational expensive and complicated EnKF approach.

An online flood forecasting system should be capable of producing forecasts over a whole a range of lead-times. Long lead-time forecasts may be important in flood warning, because of the need to make decisions about demountable defences. Although the accuracy of such long lead-time forecasts is naturally compromised by the inevitable uncertainty propagation that affects the forecasts, reasonable results have been obtained so far (Romanowicz *et al.*, 2006). On the other hand, the short-to-medium (4 to 10-hour-ahead) forecasts in NFFS are critical in decisions about public flood warnings and it is here where an adaptive stochastic system, inherently optimised for the minimisation of forecasting errors and uncertainty at specified lead-times, has particular advantages over more conventional deterministic alternatives.

The development of real-time updating methods for online flood forecasting requires the development of generic tools. All of the results presented in the present chapter were obtained using routines available in the CAPTAIN Toolbox (Taylor *et al.*, 2007) for Matlab^{TM} (see later, Section 16.7).

16.3. The Lancaster Real-Time Flood Forecasting System

In this chapter, flood forecasting is understood in a specific sense: namely the derivation of real-time updated, online forecasts of the flood level at certain strategic locations along the river, over a specified time horizon into the future, based on the information about the rainfall and the behaviour of the flood wave upstream. Depending on the length of the river reach and the slope of the river bed, a realistic forecast lead time, obtained in this manner, may range from hours to days. The information upstream can include the observations of river water levels and/or rainfall measurements. In the situation where meteorological ensemble forecasts are available, they can be used to further extend the forecast lead times (Cluckie *et al.*, 2006), as in the approach presented by Krzysztofowicz (1999, 2002) and Pappenberger *et al.* (2005), but such information was not necessary for the lead times required in the present study.

The flow forecasting procedures described here are incorporated within an online, two-step DA procedure based on DBM models, formulated within a stochastic state space setting for the purposes of recursive state estimation and forecasting. In the first step, available observations of rainfall and river water levels at different locations along the river are sequentially assimilated into the forecasting algorithm, based on the statistically identified and estimated stochastic, dynamic DBM models, in order to derive the multi-step-ahead forecasts. These forecasts are then updated in real-time, using a Kalman filter-based approach, when new data become available. The incorporation of new observations into the dynamic model via DA is performed online for every time step (here every hour) of the forecasting procedure.

DA techniques have found wide application in the fields of meteorology and oceanography and an extensive review of sequential DA techniques, together with examples of their application in oceanography is presented in Bertino *et al.* (2003). The problems described there involve the integration of multi-dimensional, spatiotemporal observations into fully distributed numerical ocean models and thus differ from the flood forecasting systems, which have much smaller spatial dimensionality. However, certain aspects of the problem remain the same and this has led to recent applications of the EnKF (see earlier) in a flow forecasting context.

The EnKF was developed (Evensen, 1994) as an alternative to the EKF approach. The EnKF is an adaptation of the standard, analytic KF algorithm to nonlinear systems using Monte Carlo sampling in the prediction (or propagation) step and linear updating in the correction (or analysis) step. It has been applied to rainfall-flow modelling (Moradkhani *et al.*, 2005b; Vrugt *et al.*, 2005), where sequential, EnKF-based estimation is exploited to update both the hydrological model parameters and the associated state variables for a single rainfall-flow model (in contrast to the present paper which considers a much more complicated, quasi-distributed catchment model, involving a number of inter-connected rainfall-water level and routing models). Moradkhani *et al.* (2005a) also apply a Particle Filter (PF) algorithm to implement sequential hydrologic DA. Yet another approach to DA is presented by Madsen and others (Madsen *et al.*, 2003; Madsen and Skotner, 2005), who apply updating to the modelling error of the distributed Mike-11 flow forecasts, at the observation sites, using a constant-in-time, proportional gain depending on the river "chainage". Essentially, this gain is included to adjust for a hydrological model bias.

One problem with both PF and EnKF is that they are computationally intensive approaches to DA and forecasting, requiring many Monte Carlo realisations at each propagation step. Moreover, due to the high complexity of these approaches, there may be questions about the identifiability of parameters involved in the different aspects of the applied routines (e.g. the estimation of the variance hyperparameters associated with stochastic inputs: see later, Section 16.4.1.2). In order to reduce the computational burden of EnKF and other MCS-based schemes, regularisation may be introduced (Sørensen *et al.*, 2004). However, as we shall see in this chapter, the relatively simple nonlinear nature of the rainfall-flow and level routing processes means that there are simpler and computationally much less intensive alternatives to DA that are able to provide comparable forecasting performance.

There are a number of simplified, conventional approaches to flow routing that have been applied to flow forecasting. Amongst others, for example, these include the Muskingham model with multiple inputs (Khan, 1993) and Multiple Regression (MR) models (Holder, 1985). The Muskingum model is deterministic and does not give the required uncertainty bounds for the forecasts. Moreover, these more conventional models tend to have a completely linear structure and so, naturally, they do not perform as well as nonlinear alternatives within the nonlinear rainfall-flow context.

In contrast, a considerable amount of research has been published recently on the application of nonlinear methods in flood forecasting. Amongst others, Porporato and Ridolfi (2001) present the application of a nonlinear prediction approach to multivariate flow routing and compare it successfully with ARMAX model forecasts. However, this approach does not have a recursive form and requires online automatic optimisation when applied to online forecasting. Another nonlinear approach is the application of neural networks for flood forecasting (Park *et al.*, 2005; Thirumalaiah and Deo, 2000). As discussed in these papers, neural network models can yield better forecasts than conventional linear models and, if designed appropriately, they also allow online DA. However, they normally have an overly complex nonlinear structure and so can provide over-parameterised representations of the fairly simple nonlinearity that characterises the rainfall-flow process (see Section 16.4.1). They are also the epitome of the black-box model and provide very little information on the underlying physical nature of the rainfall-flow process: information that can instill confidence in the model and allows for better implementation of the forecasting algorithm within a recursive estimation context.

In this chapter, we consider another approach to simplified modelling that utilises statistical estimation to identify the special, serially connected nature of the rainfall input nonlinearity in the rainfall-water level process and then exploits this in order to develop a computationally efficient forecasting algorithm. This is based on statistically estimated, stochastic-dynamic DBM models of the rainfall-water level and level routing components of the system. These are then integrated into an adaptive, modified version of the standard recursive KF state estimation algorithm that generates both the state variable forecasts and their 95% confidence bounds. Here, the state variables are defined as the "fast-flow" and "slow-flow" water levels, that together characterise the main inferred variables in the identified DBM rainfall-water level models. The serial input nonlinearity in the rainfall-flow model represents an effective rainfall transformation of the measured rainfall, the nature of which is estimated directly from the data using SDP estimation (Young, 2000, 2001c; Young *et al.*, 2001).

In contrast to the model predictions produced by a fully distributed parameter model, the water level forecasts in this case are made only at the location of measurements, in accordance with the goal of the flood forecasting system under consideration. Of course, the resulting water level forecasts could be used to update the predictions of inundation risk along a river by conditioning the risk predictions of a fully distributed flood forecasting model in real time (Romanowicz and Beven, 1998; Romanowicz *et al.*, 2004).

16.4. Methodology

In this chapter, we develop further an earlier approach to forecasting system design for a single component rainfall-flow model (Young, 2001a, 2002) so that it can be used with a complete, multi-component, quasi-distributed model covering a large part of the River Severn catchment in the UK. In Section 16.5, we describe, in some detail, a considerably enhanced version of an earlier system (Romanowicz *et al.*, 2004), where water levels are forecast instead of flows and there is a modified approach to the online adaption of the variance hyper-parameters. These enhancements were introduced to facilitate the incorporation of the forecasting system into the National Flood Forecasting System (NFFS) developed for the

UK Environment Agency (EA) by Delft Hydraulics. The extension also includes the application of a larger number of rainfall gauging stations; the development of an enhanced method of dealing with the rainfall-water level nonlinearity; and a simpler, more robust method of accounting for the heteroscedastic variance. As a result of these changes, the forecasts have smaller bias, lower uncertainty levels and longer lead times.

The use of water levels, instead of flows, enables much better utilisation of existing water level observations, including those stations for which rating curves do not exist or are not reliable. All of the river gauging stations provide water level measurements which, as usual, could be transformed into flow using the rating curve specific to each station. However, variations in flow velocity and the way that losses of energy due to friction change with water level and gradient mean that this transformation is not very well defined, particularly for high flows. Moreover, it is normally based on historic calibration that may well have become out-of-date or could have been changed during an extreme flood event. Hence, by using water level measurements instead of the flows, we bypass those uncertainties related to water level-flow conversion. Additionally and conveniently, we also tend to decrease the heteroscedasticity of the prediction errors. Since water levels are usually measured relative to a base level (i.e. the historically justified minimum water level at a given gauging station) a correction for this reference datum is introduced for each gauging station.

16.4.1. *The nonlinear rainfall-water level model*

The DBM model used in the present catchment modelling context is derived from the linear STF model. Comprehensive tutorials on transfer function models for linear and nonlinear hydrologic systems are available as Appendix 2 of the report presented to the FRMRC (Young *et al.*, 2006) and an encyclopaedia article by the first author (Young, 2005). The reader who is not familiar with such models is advised to consult these tutorials, which show how such models are simply convenient forms of the differential equations, that are the basis of many hydrological models, or their discrete-time equivalents. These discrete-time STF models are particularly straightforward and convenient when modelling from discrete-time sampled data and have been used in the present study. However, equivalent continuous-time STF models could have been utilised and have certain advantages in the case of rapidly sampled or non-uniformly sampled datas (Young, 2004, 2008, 2011).

In standard discrete-time STF modelling, it is assumed that the rainfall $r(t)$ and water level $y(t)$ measurements are sampled uniformly in time at a sampling interval of Δt time units (hours in this case). These discrete-time, sampled measurements, $\{r(t), t_1 < t < t_N\}$ and $\{y(t), t_1 < t < t_N\}$, are denoted by $r_k = r(t_k)$ and $y_k = y(t_k)$, respectively, where N is the total number of samples. The linear STF model relating r_k to y_k takes the form,

$$y_k = \frac{B(z^{-1})}{A(z^{-1})} r_{k-\delta} + \xi_k, \tag{16.1}$$

where δ is a pure, advective time delay of $\delta\Delta t$ time units, while $A(z^{-1})$ and $B(z^{-1})$ are polynomials of the following form:

$$A(z^{-1}) = 1 + a_1 z^{-1} + a_2 z^{-2} + \cdots + a_n z^{-n}$$
$$B(z^{-1}) = b_0 + b_1 z^{-1} + b_2 z^{-2} + \cdots + b_m z^{-m},$$

in which z^{-r} is the backward shift operator, i.e. $z^{-r} y_k = y_{k-r}$. In Equation (16.1), ξ_k is a noise term that is assumed to account for all the uncertainty associated with the inputs affecting the model, including measurement noise, unmeasured inputs, and uncertainty in the model. Note that by multiplying throughout this model by $A(z^{-1})$ and applying the backward shift operator, it can be written in the alternative discrete-time equation form:

$$y_k = -a_1 y_{k-1} - \cdots - a_n y_{k-n} + b_0 r_{k-\delta} + \cdots + b_m r_{k-\delta-m} + \eta_k, \tag{16.2}$$

where $\eta_k = A(z^{-1})\xi_k$ is the transformed noise input. This shows that the river water level at the kth hour is dependent on water level and rainfall measurements made over previous hours, as well as the uncertainty η_k arising from all sources.

Unfortunately, a linear STF model such as (16.1) is unable to characterise the relationship between rainfall and flow because it cannot describe how the level in the river responds to rainfall under different catchment wetness conditions. However, it has been shown (Romanowicz *et al.*, 2004a,b; Young, 1993, 2002, 2003, 2004, 2006; Young and Beven, 1994; Young and Tomlin, 2000) that the nonlinearities arising from the relationship between measured rainfall r_k and "effective" rainfall, denoted here by u_k, can be approximated using gauged flow or water level as a surrogate measure of the antecedent wetness or soil-water storage in the catchment. In particular, the scalar function describing the nonlinearity between the rainfall r_k and effective rainfall u_k, as a function of the soil moisture surrogate y_k, is initially identified non-parametrically using State Dependent Parameter (SDP) estimation (Appendix 16.2). As in the case of the rainfall-flow modelling and forecasting studies considered previously and cited above, the SDP nonlinearity can be parameterised by a power-law relationship. In the present rainfall-level application, however, the power-law relationship is not quite such a good approximation and it has been modified to take into account the somewhat changed non-parametric SDP estimation results, where there is a flattening of the flood wave due to the over-bank flooding. It should be mentioned that this flattening effect is visible both for the flows and water levels. The modified relation, which allows the gain associated with the power law to change for higher water levels, takes the form:

$$u_k = c_0 \cdot c(y_k) \cdot y_k^{\gamma} \cdot r_k; \quad c(y_k) = \begin{Bmatrix} 1 & \text{for } y_k < y_0 \\ c_p & \text{for } y_k \geq y_0 \end{Bmatrix}, \tag{16.3}$$

where u_k denotes the effective rainfall; c_0 is an arbitrary scaling constant[a]; $c_p(0 < c_p \leq 1.0)$ is a constant describing the degree of flattening of the flood wave; and y_0 is related to the bank-full water level.

It should be noted here that the non-parametric estimation results suggest a more complicated function than (16.3): this minor modification to the standard power law was selected for simplicity at this phase of the study and a superior function, that better matches the non-parametrically estimated shape of the nonlinearity, is being investigated using more flexible radial basis functions (Beven *et al.*, 2011). Indeed, in the case of the nonlinear level routing models, described later in Section 16.5.3, the nonlinear relationships are no longer approximated at all by a power law and so they have been parameterised using such radial basis functions.

The power-law exponent γ and constants $\{c_p$ and $y_0\}$ in (16.3) are estimated by a special optimisation procedure that includes the concurrent estimation of the following linear STF model between the delayed effective rainfall $u_{k-\delta}$ and water level:

$$y_k = \frac{B(z^{-1})}{A(z^{-1})} u_{k-\delta} + \xi_k, \tag{16.4}$$

[a]This is normally selected to ensure that the relationship between u_k and y_k makes physical sense (e.g. if they have similar units, then the sum of the u_k is equal to the sum of the y_k).

where δ is a pure, advective time delay of $\delta\Delta t$ time units and ξ_k is usually a heteroscedastic (changing variance) noise term that is assumed to account for all the uncertainty associated with the inputs affecting the model, including measurement noise, unmeasured inputs, and uncertainty in the model. The orders of the STF denominator and numerator polynomials, n and m respectively, are identified from the data during the estimation process and are usually in the range 1–3. The triad [n m δ] is normally used to describe this model structure. Finally, combining Equations (16.3) and (16.4), the complete rainfall-flow model can be written as:

$$y_k = \frac{B(z^{-1})}{A(z^{-1})} u_{k-\delta} + \xi_k \quad u_k = \{c(y_k) y_k^{\gamma}\} r_k, \tag{16.5}$$

where the scalar constant c_p is incorporated into $c(y_k)$. It can also be written in the equation form of (16.2), with r_k replaced by u_k. Note also that, while the STF relationship between effective rainfall and water level is linear, the noise η_k is dependent on the model parameters, thus precluding linear least squares estimation and requiring more advanced STF model estimation procedures available in the CAPTAIN Toolbox (see Section 16.7). And, of course, the complete model between measured rainfall and water level is quite heavily nonlinear because of the effective rainfall nonlinearity in (16.5).

16.4.1.1. *State space formulation of the rainfall-water level model*

In a typical STF model describing the hourly changes of water levels in response to rainfall inputs, $n = m = 2$, so that the general model (16.5) reduces to:

$$y_k = \frac{b_0 + b_1 z^{-1}}{1 + a_1 z^{-1} + a_2 z^{-2}} u_{k-\delta} + \xi_k \quad u_k = \{c(y_k) y_k^{\gamma}\} r_k. \tag{16.6}$$

The decomposition of the STF component of this DBM model (Young, 2005; Young *et al.*, 2006) into the fast and slow water level components $y_{1,k}$ and $y_{2,k}$ is based on a partial fraction expansion of the second order STF in (16.6) and it has the form:

$$\text{fast component: } y_{1,k} = \frac{\beta_1}{1 + \alpha_1 z^{-1}} u_{k-\delta}, \tag{16.7}$$

$$\text{slow component: } y_{2,k} = \frac{\beta_2}{1 + \alpha_2 z^{-1}} u_{k-\delta}, \tag{16.8}$$

where $\alpha_1, \alpha_2, \beta_1$ and β_2 are parameters derived from the model parameters in (16.6) and the total gauged flow is the sum of these two components and the noise ξ_k, i.e.,

$$y_k = y_{1,k} + y_{2,k} + \xi_k. \tag{16.9}$$

The associated residence times (time constants) $\{T_1,\ T_2\}$, steady state gains $\{G_1,\ G_2\}$ and partition percentages$\{P_1,\ P_2\}$, are given by the following expressions:

$$T_i = \frac{\Delta t}{\log_e(\alpha_i)}; G_i = \frac{\beta_i}{1+\alpha_i}; P_i = \frac{100 G_i}{G_1 + G_2} \quad i = 1, 2.$$

The parameters of this DBM model are derived from statistical model identification and estimation analysis based on the observed rainfall-water level data. The value of the DBM method depends on the amount of information available in these data to statistically estimate the model parameters. Here, the Simplified Refined Instrumental Variable (SRIV) option of the rivbjid algorithm in the CAPTAIN Toolbox, as well as associated DBM statistical modelling concepts, are used to identify the order of the STF model (the values of n, m and δ) and to estimate the associated parameters (see the cited references).

For forecasting purposes, the DBM model (16.6), considered in its decomposed form (16.7) to (16.9), is converted to a stochastic state space form. This then allows for the solution of the resulting state equations within a KF framework. It also facilitates the online, real-time estimation of both water level components, as well as the optimisation of the hyperparameters (parameters associated with the stochastic inputs: see Section 16.4.1.2) of the state space model, based on a multi-step-ahead forecast error criterion. The state equations of the decomposed DBM model have the form:

$$\begin{aligned} \boldsymbol{x}_k &= \mathbf{F}\boldsymbol{x}_{k-1} + \mathbf{G}u_{k-\delta} + \boldsymbol{\zeta}_k \\ y_k &= \mathbf{h}^T \boldsymbol{x}_k + \xi_k, \end{aligned} \tag{16.10}$$

where

$$\mathbf{F} = \begin{bmatrix} -\alpha_1 & 0 \\ 0 & -\alpha_2 \end{bmatrix} \quad \mathbf{G} = \begin{bmatrix} \beta_1 \\ \beta_2 \end{bmatrix} \quad \boldsymbol{\zeta}_k = \begin{bmatrix} \zeta_{1,k} \\ \zeta_{2,k} \end{bmatrix} \quad \mathbf{h}^T = [1 \quad 1].$$

Here, the elements of the state vector $\boldsymbol{x}_k = [x_{1,k}\ x_{2,k}]^T$ are, repectively, the fast and slow components (16.7), (16.8), of the rainfall-water level process; while the system noise variables in $\boldsymbol{\zeta}_k$ are introduced to allow for un-measurable stochastic inputs to the system. For simplicity, it is assumed that these are zero mean, serially uncorrelated and statistically independent random variables, with a purely diagonal covariance matrix.

In hydrological applications, the observation noise ξ_k is often complex in form, being serially correlated in time (coloured noise), heteroscedastic (changing variance: see Sections 16.4.1.2 and 16.4.1.5) and it does not necessarily have rational spectral density. However, in the circumstances where it can be assumed to have rational spectral density, it can be modelled using an Autoregressive (AR) or Autoregressive, Moving Average (ARMA) process (Box and Jenkins, 1970; Young, 2005, 2011) of an order that is identified from the data. This can then be incorporated into the KF DA and forecasting procedure by extending the state vector. This is an alternative to the "error correction" or "error updating" procedures that have been suggested over a number of years (e.g. Goswami *et al.*, 2005). In this connection, note that the present paper utilises STF models for rainfall-water level and flood routing forecasts, taking account of system nonlinearities, but the models used by Goswani *et al.* are linear STF models that are not appropriate to the modelling of nonlinear hydrological processes, except perhaps for single events, because they do not account for catchment storage effects.

16.4.1.2. *The modified Kalman filter forecasting engine*

For water level forecasting purposes, the state space model discussed in the previous section is used as the basis for the implementation of the following special KF state estimation and forecasting algorithm, which is a simply modified version of the standard KF algorithm:

***a priori* prediction:**

$$\hat{\boldsymbol{x}}_{k|k-1} = \mathbf{F}\hat{\boldsymbol{x}}_{k-1} + \mathbf{G}\{c(y_{k-\delta})y_{k-\delta}^{\gamma}\}_{\delta} \qquad (16.11a)$$

$$\mathbf{P}_{k|k-1} = \mathbf{F}\mathbf{P}_{k-1}\mathbf{F}^T + \sigma_k^2\mathbf{Q}_r, \qquad (16.11b)$$

***a posteriori* correction:**

$$\hat{\boldsymbol{x}}_k = \hat{\boldsymbol{x}}_{k|k-1} + \mathbf{P}_{k|k-1}\mathbf{h}[\sigma_k^2 + \mathbf{h}^T\mathbf{P}_{k|k-1}\mathbf{h}]^{-1}\{y_k - \mathbf{h}^T\hat{\boldsymbol{x}}_{k|k-1}\} \qquad (16.11c)$$

$$\mathbf{P}_k = \mathbf{P}_{k|k-1} - \mathbf{P}_{k|k-1}\mathbf{h}[\sigma_k^2 + \mathbf{h}^T\mathbf{P}_{k|k-1}\mathbf{h}]^{-1}\mathbf{h}^T\mathbf{P}_{k|k-1} \qquad (16.11d)$$

$$\sigma_k^2 = \lambda_0 + \lambda_1(y_k^2). \qquad (16.11e)$$

In these equations, $\mathbf{P}_k$ is the error covariance matrix associated with the state estimates; $\mathbf{Q}_r$ is the 2×2 Noise Variance Ratio (NVR) matrix, as discussed in Section 16.4.1.3; and the subscript notation $k|k-1$ denotes the estimate at the kth sampling instant (here hour), based on the estimate at the previous $(k-1)^{\text{th}}$ sampling instant. The differences between this version of the KF and the standard version of the algorithm are: the nonlinear input entering in Equation (16.11a); the state-dependent NVR matrix term $\sigma_k^2\mathbf{Q}_r$, replacing the constant covariance matrix $\mathbf{Q}$; and the state-dependent variance defined in Equation (16.11e) and appearing in Equations (16.11b), (16.11c) and (16.11d).

The f-step-ahead forecasts are obtained by simply repeating the prediction f times, without correction (since no new data over this interval are available). The f-step ahead forecast variance is then given by:

$$\text{var}(\hat{y}_{k+f|k}) = \hat{\sigma}_k^2 + \mathbf{h}^T\mathbf{P}_{k+f|k}\mathbf{h}, \tag{16.11f}$$

where $\mathbf{P}_{k+f|k}$ is the error covariance matrix estimate associated with the f-step ahead prediction of the state estimates. This estimate of the f-step ahead prediction variance is used to derive 95% (approximate) confidence bounds for the forecasts. This is under the approximating assumption that the prediction error can be characterised as a nonstationary Gaussian process (i.e. twice the square root of the variance at each time step is used to define the 95% confidence region).

16.4.1.3. *State adaption*

In the above KF Equations (16.11a) and (16.11b), the model parameters $\alpha_i, i = 1, 2$ and $\beta_j, j = 1, 2$, that define the state space matrices $\mathbf{F}$ and $\mathbf{G}$ in (16.10), are known initially from the model identification and estimation analysis based on the estimation data set. However, by embedding the model equations within the KF algorithm, we have introduced additional, unknown parameters, normally termed "hyperparameters" to differentiate them from the model parameters.[b] In this example, these hyperparameters are the elements of the NVR matrix $\mathbf{Q}_r$ and, in practical terms, it is normally sufficient to assume that this is purely diagonal in form. These two diagonal elements, in this example, are defined as $NVR_i = \sigma_{\zeta_i}^2/\sigma^2, i = 1, 2$. These specify the nature of the stochastic inputs to the state equations and so define the level of uncertainty in the evolution of each state (the quick

[b]Of course this differentiation is rather arbitrary since the model is inherently stochastic and so these parameters are simply additional parameters introduced to define the stochastic inputs to the model when it is formulated in this state space form.

and slow water level states respectively) relative to the variance of the measurement uncertainty σ^2. The inherent state adaption of the KF arises from the presence of the NVR parameters, since these allow the estimates of the state variables to be adjusted to allow for the presence and effect of the unmeasured stochastic disturbances that naturally affect any real system.

Clearly, the NVR hyperparameters have to be estimated in some manner on the basis of the data. One well known approach is to exploit Maximum Likelihood (ML) estimation based on prediction error decomposition (Schweppe, 1965; Young, 1999b). Another, used in the River Severn forecasting system described later, is to optimise the hyperparameters by minimising the variance of the multi-step-ahead forecasting errors. In effect, this optimises the memory of the recursive estimation and forecasting algorithm (Young and Pedregal, 1999) in relation to the rainfall-water level data. In this numerical optimisation, the multi-step-ahead forecasts are $\hat{y}_{k+f|k}$, where f is the forecasting horizon. The main advantage of this latter approach is, of course, that the integrated model-forecasting algorithm is optimised directly in relation to the main objective of the forecasting system design; namely the minimisation of the multi-step prediction errors.

16.4.1.4. *Alternative formulations of the Kalman filter*

The formulation of the KF used in the previous sections is not the only formulation that could be used in the present context. It has been assumed here that most of the uncertainty in the level forecasts is associated with the noise ξ_k on the output level measurement y_k in the state equations (16.10). This yields a particularly simple formulation of the KF but it is not necessarily the best one. An equally plausible assumption is that the level measurement is quite accurate and that the uncertainty derives mainly from the many stochastic influences affecting the state variables: i.e. it is the result of the "system input disturbance vector" $\boldsymbol{\zeta}_k$. In this connection, it should be recalled that the function of the KF is to attenuate the effects of the measurement noise ξ_k, whilst preserving the effects of these input disturbances, which are assumed to be real but unmeasurable inputs to the system.

This alternative "low measurement noise" formulation of the KF has the advantage that the state update will correct the state estimates so that the estimated output $\hat{y}_k = \mathbf{h}\hat{\boldsymbol{x}}_k$ will always be close to the measured level y_k, because the measurement noise ξ_k is constrained to be small. However, this is not such a simple formulation and its implementation introduces various complications. For instance, it would require constrained optimisation of the hyperparameters $NVR_i = \sigma^2_{\zeta_i}/\sigma^2, i = 1, 2$ under the prior assumption that the variance σ^2 of the noise ξ_k is known and is set at a relatively low value in relation to the magnitude of the level measurements: i.e. the specification and imposition of a high signal-noise ratio. Moreover, in order to introduce the required heteroscedastic features, all of the above hyperparameters will have to be made time variable or state-dependent in some manner.

Finally, as in the case of the measurement noise in the present formulation (see previous discussion in Section 16.4.1.1, above), it is quite likely that these input disturbances are serially correlated in time (i.e. they are coloured noise sequences). This is complicated because there are two major states, $x_{1,k}$ and $x_{2,k}$ for each rainfall-level component of the catchment model and so two input disturbances are required for each of these sub-models. Consequently, an extension to include coloured input disturbances would require a further extension of the state vector to allow for the additional stochastic states required in this situation. These latter considerations are the subject of current research at Lancaster in order to evaluate whether they introduce sufficient improvement in performance to justify the increased complexity (see also Vaughan and McIntyre, 2012).

16.4.1.5. *Parameter adaption*

Although the parameters and hyperparameters of the KF-based forecasting system can be optimised in the above manner, we cannot be sure that the system behaviour may not change sufficiently over time to require their adjustment. In addition, it is well known that the measurement noise ξ_k is quite highly heteroscedastic: i.e. its variance can change quite radically over time, with much higher variance occurring during storm events. For these reasons, it is wise to build some form of parameter adaption into the forecasting algorithm.

(i) Full Parameter Adaption. It is straightforward to update *all* of the parameters in the rainfall-water level (or flow) model since the parameter estimation algorithms in the CAPTAIN Toolbox can be implemented in a "real-time recursive" form that allows for sequential updating and the estimation of time-variable parameters: see Chapter 10 of Young (2011).

(ii) Gain Adaption. Full parameter adaption adds complexity to the final forecasting system and previous experience suggests that a simpler solution, involving a simpler scalar gain adaption is often sufficient. This is the

approach that was used successfully for some years in the Dumfries flood warning system (Lees *et al.*, 1993, 1994) and it involves the recursive estimation of the gain g_k in the following relationship:

$$y_k = g_k . \hat{y}_k + \epsilon_k, \tag{16.12}$$

where ϵ_k is a noise term representing the lack of fit and, in the case of the second order model (16.6),

$$\hat{y}_k = \hat{y}_{1,k} + \hat{y}_{2,k} = \frac{\hat{b}_0 + \hat{b}_1 z^{-1}}{1 + \hat{a}_1 z^{-1} + \hat{a}_2 z^{-2}} u_{k-\delta}, \tag{16.13}$$

where the "hats" denote the estimated values. In other words, the time variable scalar gain parameter g_k is introduced so that the model gain can be continually adjusted to reflect any changes in the steady state (equilibrium) response of the catchment to the effective rainfall inputs.

The associated recursive estimation algorithm for g_k is one of the simplest examples of the KF, in which the single state variable is the unknown gain g_k, which is assumed to evolve stochastically as a Random Walk (RW) process (Young, 2011)[c]:

$$p_{k|k-1} = p_{k-1} + q_g \tag{16.13a}$$

$$p_k = p_{k|k-1} - \frac{p_{k|k-1}^2 \hat{y}_k^2}{1 + p_{k|k-1} \hat{y}_k^2} \tag{16.13b}$$

$$\hat{g}_k = \hat{g}_{k-1} + p_k \hat{y}_k \{y_k - \hat{g}_{k-1} \hat{y}_k\}. \tag{16.13c}$$

Here, $\hat{g}_k$ is the estimate of g_k; while q_g is the NVR hyperparameter defining the stochastic input to the RW process, the magnitude of which needs to be specified. This NVR defines the "memory" of the recursive estimator, so that the higher its magnitude, the less the length of the memory: if the memory is too short, however, the estimate will tend to be too volatile; while if it is too long, it will not respond quickly enough to changes in the gain. The adapted forecast is obtained by simply multiplying the initially computed forecast by $\hat{g}_k$. Note that gain adaption of this kind is quite generic and can be applied to *any* model, not just the DBM model discussed here.

[c] It is also a scalar example of Dynamic Linear Regression (DLR) algorithm (Young, 1999b) available as the dlr routine in the CAPTAIN Toolbox.

(iii) Variance Adaption. The heteroscedasticity of the observational noise suggests that the noise variance should be updated in some manner. Sorooshian and Dracup (1980) present an approach to deal with correlated and heteroscedastic errors of flow measurements, based on a maximum likelihood approach; and an alternative procedure of either transforming the observation errors using an AR(1) model to obtain uncorrelated error, or introducing a Box–Cox transformation (Box and Cox, 1964) to deal with heteroscedasticity. Their methodology follows an *en bloc* estimation procedure with hydrological model and error transformation parameters treated as deterministic variables. Vrugt *et al.* (2005), on the other hand, apply a non-parametric, local difference-based estimator of the observation error variance, based on Hall *et al.* (1990) to model the error heteroscedasticity in their EnKF solution of the rainfall-flow model.

In order to account for the heteroscedasticity in the present study, a new approach is used, where the variance σ_k^2 of ξ_k is identified from the data as a state-dependent function of the simulated output, taking the following form:

$$\sigma_k^2 = \lambda_0 + \lambda_1 \hat{y}_k^2, \tag{16.14}$$

where λ_0 and λ_1 are new hyperparameters that are optimised together with the NVR hyperparameter that control the gain updating procedure discussed in (i). The estimate $\hat{\sigma}_k^2$ of σ_k^2 is fed back to the recursive KF engine (see previously, Section 16.4.1.2) as an estimate of the observational variance.

It is interesting to note, at this point, that an analysis of the heteroscedastic characteristics of the noise, in the above manner, has revealed the advantage of modelling in terms of the water level variable instead of the flow. In particular, the transformation of the water level y_k into the flow variable, Q_k through the rating curve, may be written in the form $Q_k = f(y_k)$. According to delta-method calculations (Box and Cox, 1964), for a smooth function $f(\cdot)$, the asymptotic variance $AV(Q_k)$ of the flow Q_k as $y \to y_k$ will have the form

$$AV(Q_k) = \dot{f}(y_k)^2 \mathrm{var}(y_k); \quad \dot{f}(y_k) = \left[\frac{df}{dy}\right]_{y=y_k}. \tag{16.15}$$

Here, the function denotes the rating curve relationship, which often can be approximated by a power function of water levels, with the power greater than 1. Hence, the variance of the flow is increased in comparison with the water level variance, in particular for larger flows. In this regard, it is interesting to note that Sorooshian and Dracup (1980) regarded error in the rating curve transformation of the water level observations as a main source of heteroscedasticity in the flow data.

16.4.1.6. *Single module summary*

The combination of a KF incorporation of the state space model with the gain and variance updating described in the previous sections, provides a tool for online DA. The NVR hyperparameters associated with the stochastic inputs to the state space equation (16.10) are estimated in the first place by maximum likelihood based on prediction error decomposition (see earlier, Section 16.4.1.3). Subsequently, in line with the current forecasting objectives, the hyperparameters minimising the variance of the f-step ahead prediction errors for the maximum water levels are estimated using an optimisation routine from the Matlab^{TM} computational software environment (based on the simplex direct search method (Nelder and Mead, 1965)). The optimisation is performed only during the estimation (calibration) stage and the optimal values obtained in this manner are then used during the application of the real-tine forecasting scheme.

In summary, the procedure for developing a single module of the forecasting system can be summarised as follows:

(1) Based on the available input-output data, estimate the complete nonlinear DBM model (16.6), using an optimisation procedure that exploits the rivbj routine in the CAPTAIN Toolbox to jointly optimise the nonlinear parameters $\{c_p, \gamma\}$ and the STF model parameters $\{a_1, a_2, b_0, b_1, \delta\}$, ensuring that the chosen model is physically meaningful, in accordance with DBM modelling requirements.
(2) Formulate the state space model and optimise the hyperparameters, without online updating, using maximum likelihood estimation based on prediction error decomposition.
(3) Optimise the updating and heteroscedastic variance parameters using a set of optimisation criteria based on the f-step ahead forecast error, where f is the required forecasting interval.

Three criteria are used during the identification of the model structure and estimation of its parameters. The first is a coefficient of determination associated with the water level predictions $R^2_{fp} = (1 - \sigma^2_{fp}/\sigma^2_y)$, with σ^2_{fp} and σ^2_y denoting the variances of f-step prediction error and observations, respectively. This is a multi-step-ahead forecasting efficiency measure, similar in motivation to the Nash–Sutcliffe efficiency measure used for model calibration (Nash and Sutcliffe, 1970). The others are the YIC (Young, 1989, 2001b) and AIC (Akaike, 1974) criteria, which are model order identification (identifiability) statistics: a low relative measure of these criteria ensures that the chosen model has a dynamic structure that reflects the information content of the data, so ensuring well identified parameters and no over-parameterisation.

16.4.2. *River water level routing*

Following previous practice with flow routing, the water level routing down the river can utilise linear dynamic models in a transfer function form. For instance, a first-order example of such a model takes the form:

$$y_{i,k} = \frac{b_0}{1 + a_1 z^{-1}} y_{i-1,k-\delta} + \xi_k, \tag{16.16}$$

where $y_{i-1,k}$ and $y_{i,k}$ are, respectively, the upstream and downstream water level measurements from the gauges an δ is the advective delay. If the noise ξ_k can be modelled well by and AR or ARMA process, then this can further enhance the model in forecasting terms. Model identification and estimation of this model is based on the same statistical procedures used in the rainfall-water level modelling, except that the added complexity of the SDP nonlinearity is now removed. Note that the Muskingum model, mentioned previously in Section 16.3, can be considered as a special case of this linear model (Young, 1986), so it would be identified from the data using the linear STF model (16.16) if it provided a suitable model of the level data.

Once again, if the STF model is of this first-order form, it represents a single linear store, whilst if it is higher order, it can be decomposed into a number of linear stores arranged in series or parallel, depending on the identification and estimation results. The incorporation of this model into the KF forecasting engine also involves the same procedure as in the rainfall-water level case, except that now the component of the KF devoted to this sub-model will be in its standard linear form, if necessary incorporating an SDP model for heteroscedastic noise. As we shall see in the Section 16.5, this approach to level routing is applied to the River Severn reach between Abermule and Welsh Bridge, where the model is first order with a pure advective time delay of $\delta = 23$ hours.

Further DBM modelling research has shown that, quite often, this linear approach to level routing can be improved by following the same SDP approach used in the case of rainfall-water level modelling. This leads again to a Hammerstein model form with an input nonlinearity that transforms the upstream water level before it enters the linear store. In the first-order case, this SDP model takes the form:

$$y_{i,k} = \frac{b_0}{1 + a_1 z^{-1}} f(y_{i-1,k-\delta}) + \xi_k, \tag{16.17}$$

where $f(y_{i-1,k-\delta})$ represents the nonlinearly transformed, advectively delayed upstream water level. Once again, if the noise ξ_k can be modelled well by and AR or ARMA process then this can be introduced to good effect. For forecasting, this model is introduced into the KF forecasting engine but now with the input nonlinearity present, as in the rainfall-water level case. In Section 16.5, this nonlinear approach to level routing is applied to the River Severn reaches between Buildwas and Bewdley.

16.4.3. *The complete catchment forecasting system*

Each rainfall-water level forecasting sub-system, together with the other sub-systems associated with the linear and nonlinear routing models, constitute modules in the complete forecasting system for the entire catchment. These are connected in a manner defined by the geographical location of the rainfall and flow sensors. The forecasts at each of the gauging stations are then based only on the upstream gauging station. The forecast lead times may be extended following the approach of cascading the predictions from upstream reaches as inputs to downstream reaches (Beven *et al.*, 2006; Romanowicz *et al.*, 2008) but the forecasting accuracy is affected as a result. Of course, the use of this kind of routing model relies on data acquisition along the river at as many points as is economically justified in order to obtain the required accuracy of the distributed forecasts along the river.

The complete model synthesised in this manner could be incorporated as a single large state space model within the modified KF forecasting engine. While this approach was evaluated and found to function satisfactorily, it was naturally simpler to disaggregate the system into a combination of suitably interconnected KF algorithms, each defined on the basis of the appropriate rainfall-water level or water level-water level model at the specified location. For the lead times not exceeding the natural sub-reach delay, these modules are effectively independent from each other. However, for larger lead times, when the forecasts of the rainfall-flow and routing modules are introduced instead of observed input, the modules become coupled. The influence of the uncompensated correlation between the modules on the forecasting performance does not seem to be large but it is a subject for future research.

In line with our decomposition of the KF, the optimisation of the hyperparameters for the whole forecasting system is based on the separate optimisation of the hyperparameters in each module. Overall optimisation was attempted but this failed because the optimisation hypersurfaces were too flat to allow for satisfactory convergence. Of course, this optimisation can be based on a number of different criteria. For example, it could seek to:

(1) maximise a statistical likelihood function (ML optimisation);
(2) minimise overall forecasting errors, or at specific forecasting horizons;
(3) minimise forecasting errors at flood peaks;
(4) satisfy multi-objective criteria (multi-objective optimisation), etc.

However, for simple illustrative purposes, in the present case study, optimisation was based on (2), namely, minimising the sum of the squares of the forecasting errors at specific forecasting horizons.

16.5. Case Study: The River Severn

The methodology outlined in the previous sections has been used to derive a real-time, adaptive forecasting system for the River Severn in the UK, as far downstream as the gauge at Bewdley. A map of the River Severn catchment is shown in Figure 16.1, where the locations considered are shown as black dots, with associated place names. The River Severn above Buildwas has one major tributary, the River Vyrnwy, which enters the main river upstream of the Montford gauging station. The data used in this study consist of hourly measurements of rainfall at the Dollyd, Cefn Coch, Vyrnwy Llanfyllin and Pen y Coed, in the Upper Severn and Vyrnwy catchments; and hourly water level measurements at Meifod, on the Vyrnwy, and Abermule, Welsh Bridge (Shrewsbury) and Buildwas, on the River Severn. The Severn catchment area at Abermule is 580 km^2, at Welsh Bridge 2325 km^2 and at Buildwas 3717 km^2. The Vyrnwy catchment area at Meifod is 675 km^2. The length of the River Severn between Abermule and Welsh Bridge is about 80 km; the distance between Welsh Bridge and Buildwas is about 37 km and the distance between Buildwas down to Bewdley is 40 km. There are 12 bridges on the 18 km long reach between Buildwas and Bridgenorth Bridge and a similar number on the 22 km long reach between Bridgenorth Bridge and Bewdley. There are also a number of rapids and weirs. It is worth noting that releases from the Llyn Clywedog

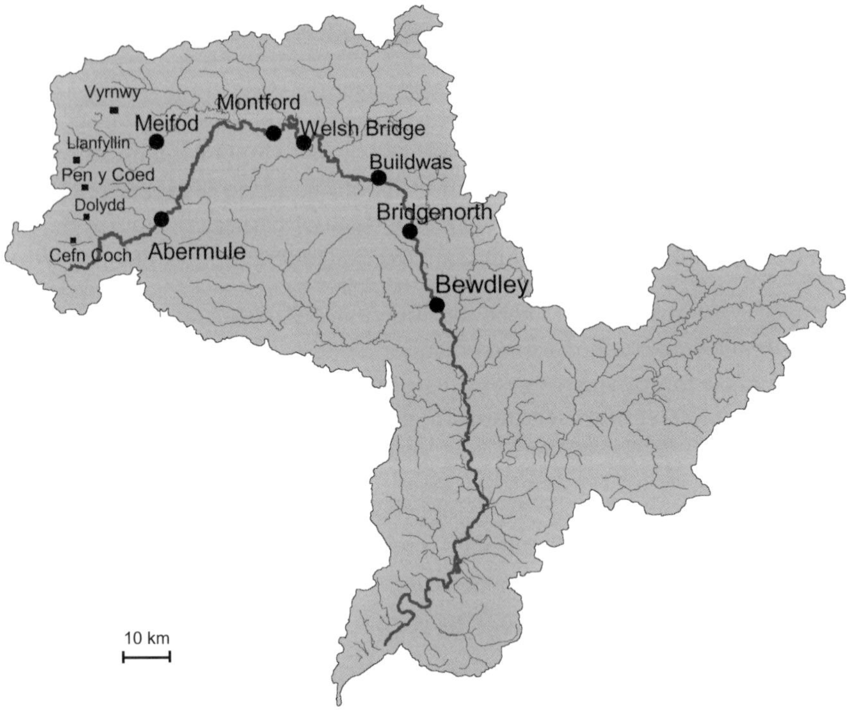

Figure 16.1. The River Severn catchment showing the water level/flow (black circles) and rainfall (small black squares) measurement sites used in the study.

reservoir, on Severn above Abermule, and Lake Vyrnwy, on the Vyrnwy tributary of Severn, are used to enhance low flows. Also, the water releases for water level maintenance may have an effect on low water level behaviour at Abermule and Meifod. As regards the data used for DBM modelling, model structure identification and estimation is based the October 1998 year flood event data; and the validation data starts at 9 am on the 24th of October 2000. In addition, the flood event in February 2002 is used for the validation of the sub-reaches between Buildwas and Bewdley.

The aim of the forecasting system design is the development of a relatively simple, robust, online forecasting system, with acceptable accuracy and the longest possible lead-time. In order to maximise the lead-times, rainfall measurements are used as an input to the rainfall-water level forecasts in the upper part of the Severn catchment. The sequential structure of the DBM forecasting model is shown as a block diagram in Figure 16.2, which also defines the nomenclature for the rainfall and water

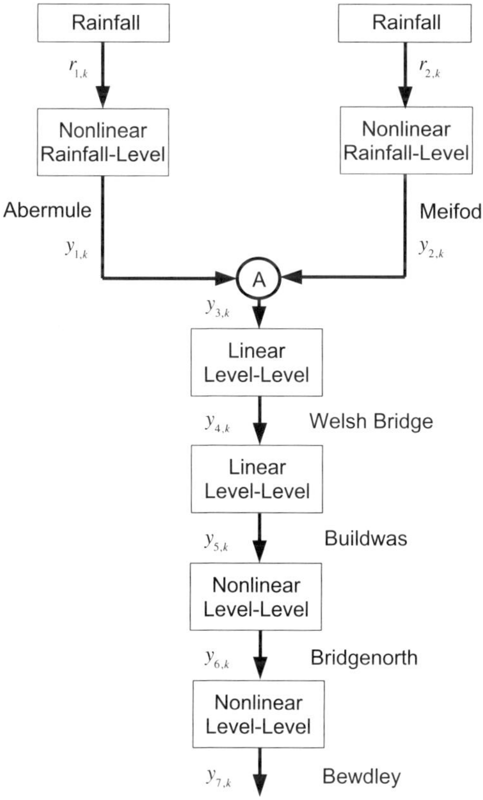

Figure 16.2. Block diagram of the River Severn Forecasting System DBM model.

level variables down the system. So far, the forecasting system has only been developed as far as Bewdley because, further downstream, there are numerous weirs controlling water levels locally along the river and the flow is no longer at all natural. We will only be able to extend the system to the lower part of the Severn when the additional information about the man-induced changes on the river are available.

In the following Sections 16.5.1.1 to 16.5.3.3, we present the rainfall-water level and water level routing models for the River Severn from Abermule down to Bewdley. The water level routing model for Welsh Bridge uses the averaged water levels from Abermule (Severn) and Meifod (Vyrnwy) as a single input variable: the use of these observations as two separate input variables was not possible due to their high cross-correlation and the associated poor identifiability of the model parameters

Table 16.1. Model structure identification results.

Model type	Location	Model structure	Forecast lead [h]	R_T^2 calibration	R_T^2 validation
Nonlinear	Abermule	[2 2 5]	5	0.954	0.945
Nonlinear	Meifod	[2 2 5]	5	0.970	0.953
Linear	Welsh Bridge	[1 1 23]	28	0.947	0.948
Linear	Buildwas	[2 2 7]	35	0.970	0.961

(multi-collinearity). All the DBM sub-models in this network are estimated using the statistical identification and estimation procedures discussed in Section 16.4. The results from this estimation and validation analysis for Abermule, Meifod, Welsh Bridge and Buildwas, as well as their diagnostic performance measures are summarised in Table 16.1.

16.5.1. *Rainfall-water level modelling*

The two nonlinear models described in this first section relate composite rainfall measures, obtained as a linear combination of rainfall gauge measurements with optimised weights, to the water levels measured at each location.

16.5.1.1. *Rainfall-water level model for Abermule*

The rainfall-water level model for Abermule uses rainfall measurements from three gauging stations in the Upper Severn catchment, Cefn Coch, Pen-y-Coed and Dolydd, as inputs. The weights for the rainfall measurements are derived through the least squares optimisation of the difference between the observed and modelled water levels at Abermule. The nonlinearity between the rainfall and water levels is identified initially using the non-parametric `sdp` estimation routine from the CAPTAIN toolbox. This non-parametric (graphical) relation is then parameterised using a step-wise power law parameterisation of the form given in (16.3) and the coefficients c_p and γ are optimised simultaneously with the estimation of the other STF model parameters. The model, after the decomposition into slow and fast components, has the form:

$$y_{1,k} = \frac{0.0038}{1 - 0.9955z^{-1}} u_{1,k-5} + \frac{0.134}{1 - 0.9304z^{-1}} u_{1,k-5} + \xi_k \qquad (16.18a)$$

$$u_{1,k} = \begin{Bmatrix} 1 & y_{1,k} < 3.36 \\ 0.68 & y_{1,k} \geq 3.36 \end{Bmatrix} y_{1,k}^{0.7} \cdot r_{1,k}. \qquad (16.18b)$$

The identified residence times for the fast and slow flow components of this model are 13 hours and 9 days, respectively. This model explains 93% of the variance of the observations, with the variance of the simulation modelling errors equal to $0.037\,\text{m}^2$.

The prediction error series shows both autocorrelation and heteroscedasticity, even though the latter is noticeably smaller than in the case of the model based on flow rather than water level measurements. In the case of the autocorrelation, an AR(3) model for the noise is identified based on the AIC criterion. The noise model could be incorporated into the forecasting model, as mentioned previously. However, in order to make the whole catchment model as simple as possible, and in compliance with the design objectives, it was decided not to do this in the upper reaches, down to Buildwas, but to retain the adaptive estimation of heteroscedastic variance of the predictions, with the option of introducing a noise model if the performance needed improvement and the added complexity could be justified at a later stage. As we shall see later in Section 6.2, however, AR noise models are incorporated in the nonlinear routing models from Buildwas to Bewdley. Note that, although the standard errors on the parameters of the STF model are generated during the estimation, the standard error estimates on the decomposed model parameters, shown here in (16.18a)–(16.18b), would need to be obtained by Monte Carlo simulation (e.g. Young, 1999a, 2001b, 2003) and this was not thought necessary in this illustrative example.

Based on these initial results, the gain and heteroscedastic variance updating procedures described earlier were applied to the rainfall-water level model (16.18a)–(16.18b). At the calibration stage, the resulting five-step-ahead forecasts of water levels explained 95.4% of the observational variance. The subsequent validation stage was carried out for the year 2000 floods and the results are shown in Figure 16.3, where the 5-hour-ahead forecast explains 94.5% of the variance of observations, only a little less than that obtained during calibration. The bank-full level at this site is at about 3.7 m.

The results obtained so far, of which those shown in Figure 16.3 are typical, suggest that the adaptive forecasting system predicts the high water levels well, which is the principal objective of the forecasting system design. The magnitude and timing of the lower peaks is not captured quite so well. This may be due to a number of factors, such as the changes in the catchment response times arising from a dependence on the catchment wetness. These possibilities have not been investigated so far, however,

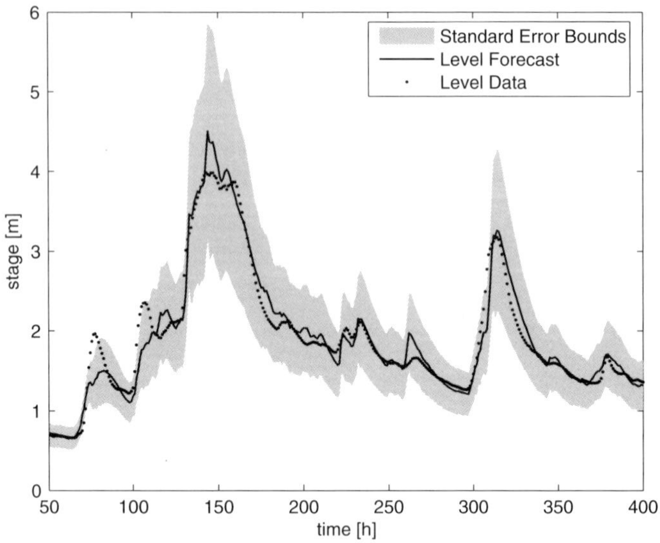

Figure 16.3. Abermule: 5-hour-ahead validation forecast, 24th October 2000. Note that here and in subsequent forecasting figures, the term "standard error" is used to mean twice the standard deviation, which is equivalent to the 95% uncertainty interval.

because the forecasting performance is considered satisfactory in relation to the current objectives.

16.5.1.2. *Rainfall-water level model for Meifod*

The same procedure used in the last section was applied to derive the DBM model for the Meifod gauging station on the River Vyrnwy. Amongst the rainfall measurement stations on the Vyrnwy catchment, three were chosen: Pen-y-Coed, Llanfyllin and Vyrnwy. The weights for the rainfall measurements were optimised, together with the parameters of the DBM model. The best rainfall-water level model, without any adaptive updating, has the same form as for Abermule, that is [2 2 5], and explains 92% of the variance of the observations. The full model equation, with the STF decomposition and the modified power-law transformation of the rainfall is given below:

$$y_{2,k} = \frac{0.0247}{1 - 0.9711z^{-1}} u_{2,k-5} + \frac{0.0424}{1 - 0.8491z^{-1}} u_{2,k-5} + \xi_k \quad (16.19a)$$

$$u_{2,k} = \begin{Bmatrix} 1 & y_{2,k} < 2.1 \\ 0.72 & y_{2,k} \geq 2.1 \end{Bmatrix} y_{2,k}^{0.1} \cdot r_{2,k}. \quad (16.19b)$$

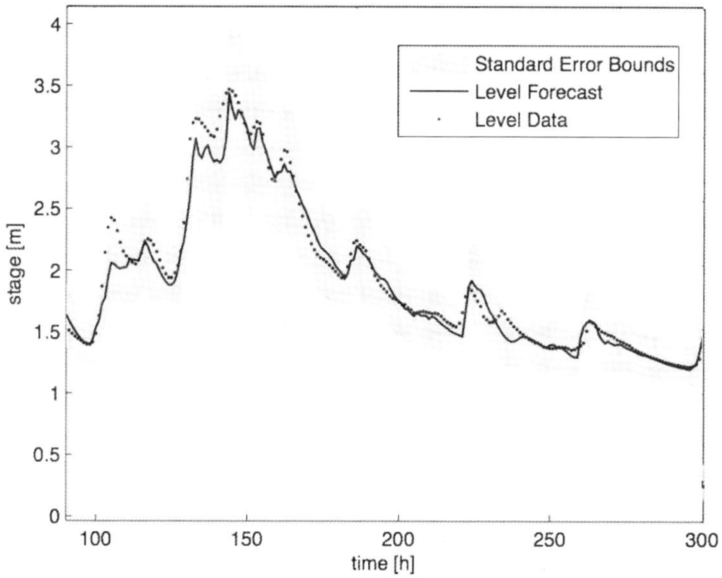

Figure 16.4. Meifod: 5-hour-ahead validation forecast based on upstream water level and rainfall inputs.

The identified residence times for the fast and slow flow components of this model are 6 hours and 34 hours, respectively.

The online updated 5-hour-ahead forecast for Meifod, over the calibration period in October 1998, explains 97% of the variance of observations. The validation was performed on the same time period in the year 2000 as the Abermule model and the results are shown in Figure 16.4. Here, the model explains 95.3% of the variance of observations. As in the case of Abermule, the results show some nonlinearity in timing. However, we do not have any information about the bank-full level for this site.

16.5.2. *Linear water level routing*

The part of the River Severn down from Meifod and Abermule to Bewdley is modelled using two linear routing models, each of the form outlined in Section 16.4.2. At each gauging station, the linear STF model is identified and estimated based on the water level measurements at each gauge and the nearest downstream gauging station. The first-order form of Equation (16.16) is identified for the two models using the rivbj routine in CAPTAIN for linear STF model estimation.

16.5.2.1. *Water level routing model for Welsh Bridge*

The best results for the water level data between Abermule, Meifod and Welsh Bridge were obtained when the STF model was calibrated on the water level data with the minimum water level at Welsh Bridge removed for identification and estimation purposes. It was then re-introduced for implementation of the model in the forecasting engine. The first-order model estimated in this manner is given below: it has a time delay of 23 hours and a residence time of about 28 hours.

$$y_{4,k} = \frac{0.0692}{1 - 0.9650z^{-1}} y_{3,k-23} + \xi_k, \tag{16.20}$$

where $y_{3,k-23}$ denotes denotes the average (circled A in Figure 16.2) of the water level measurements $y_{1,k}$ and $y_{2,k}$ at Meifod and Abermule, respectively, delayed by 23 hours; and $y_{4,k}$ denotes water level at Welsh Bridge. The 5 hour-ahead forecasts of water levels for Abermule and Meifod enable the Welsh Bridge water level forecasts to be extended to 28 hours, which should be more than adequate for flood warning purposes at Shrewsbury. The resulting, adaptively updated, water level forecast at Welsh Bridge explains 94.7% of the output variation.

Validation of the model at Welsh Bridge was performed using all of the upstream models: i.e. rainfall-water level models for Meifod and Abermule and water level routing model for Welsh Bridge, with summed water level observations/forecasts for Abermule and Meifod. The resulting online updated 28-hour-ahead forecasts are shown in Figure 16.5, validated on year 2000 floods. These forecasts explain 94.8% of the observed water level variance at Welsh Bridge, a little better than the calibration performance.

16.5.2.2. *Water level routing model for Welsh Bridge–Buildwas*

Identification and estimation of the water level routing model for Welsh Bridge–Buildwas resulted in a [2 2 7] model, which explains 98% of the variance of the observations. This model has the following decomposed STF form:

$$y_{5,k} = \frac{0.0034}{1 - 0.9866z^{-1}} y_{4,k-7} + \frac{0.6456}{1 - 0.4669z^{-1}} y_{4,k-7} + \xi_k, \tag{16.21}$$

where $y_{4,k-7}$ denotes denotes the water level measurement at Welsh Bridge, delayed by 7 hours and $y_{5,k}$ denotes water level at Buildwas. Here, the residence times are about 1 hour for the fast component and about 3 days for the slow component. The bank-full level at this site is at about 6 m.

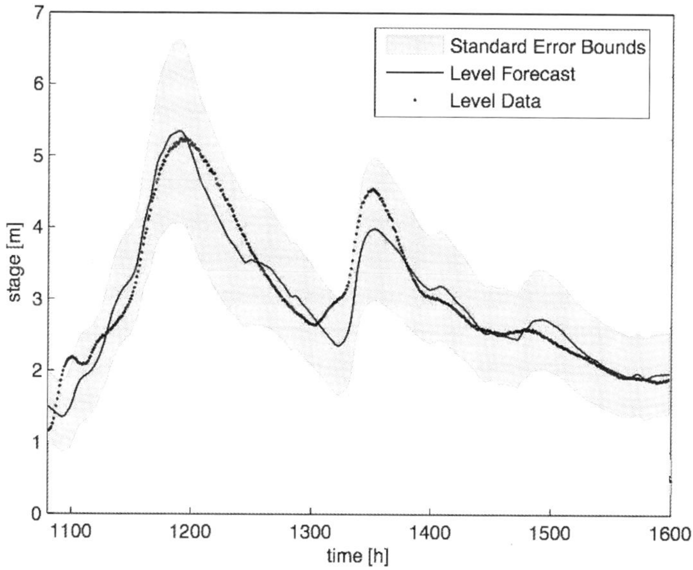

Figure 16.5. Welsh Bridge: 28-hour-ahead validation forecast based on upstream water level and rainfall inputs.

It is not clear what physical interpretation can be given to this identified parallel pathway model but it is probably a reflection of some complexity of the flow process in the river between Welsh Bridge and Buildwas and it would be interesting to investigate this further.

Finally, when the adaptive updating procedures were applied to the model and the 28 hour-ahead forecasts were used to extend the total forecast lead time of the final model to 35 hours, the forecast explains nearly 97% of the variance of observed water levels at Buildwas. The validation of the model was performed on the November floods from the year 2000 and the results are shown in Figure 16.6. The model explains 96% of the variance of the water level observations at Buildwas. The high peak values are predicted with reasonable accuracy, while the lower water level changes are over-predicted and there is a visible time difference between the simulated and observed water levels for the smaller peaks.

16.5.3. *Nonlinear water level routing*

The three, nonlinear routing models described in this section relate the upstream water levels to the downstream water levels measured at each location. The part of the River Severn down from Welsh Bridge to Bewdley

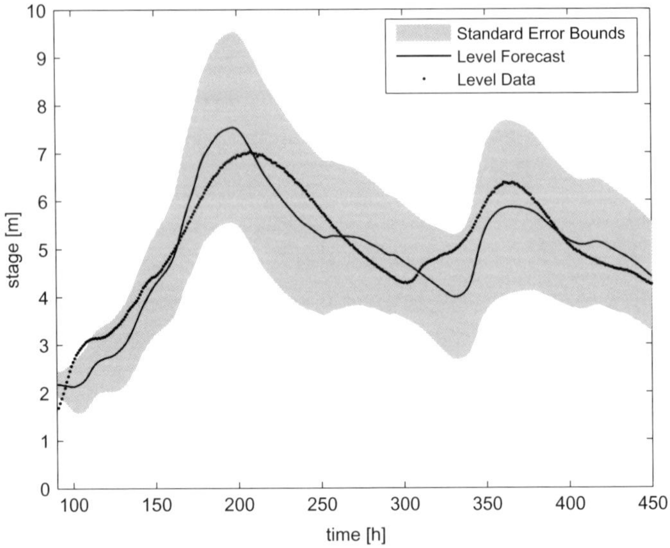

Figure 16.6. Buildwas: 35-hour validation forecast, 24th October 2000.

is modelled using three nonlinear routing models, each of the form outlined in Section 16.4.2. The Welsh Bridge–Buildwas reach was incorporated again in order to extend the forecast downstream using the same nonlinear modelling approach. At each gauging station, the nonlinear STF model is identified and estimated based on the water level measurements between this gauge and the nearest downstream gauging station.

The first-order form of Equation (16.17) is identified for all three of these models using the same statistical approach to that described for the rainfall-water level models discussed in previous sections. In each case, the identified non-parametric SDP nonlinearity is then parameterised using a Radial Basis Function (RBF) approximation (although any alternative, flexible parameterisation would be suitable[d]) and a ten element RBF is sufficient for all three models. In the final estimation phase of the modelling, the RBF parameters are re-optimised concurrently with the associated STF parameters, again using a similar optimisation procedure to that used for the rainfall-water level modelling. The results from the estimation analysis

[d]The RBF is often portrayed as a "neural" function but this can be misleading: as used here, the RBF is simply composed of Gaussian-normal shaped basis functions that, when combined linearly using optimised parameters ("weights"), can approximate well smooth nonlinear curves, such as the SDP-type nonlinearity.

Table 16.2. Model structure identification: Nonlinear routing.

Model type	Location	Model structure	Forecast lead [h]	R_T^2 calibration	R_T^2 validation
Nonlinear	Buildwas	[1 1 8]	8	0.997	0.996
Nonlinear	Bridgenorth	[1 1 2]	10	0.997	0.996
Nonlinear	Bewdley	[1 1 4]	14	0.998	0.989

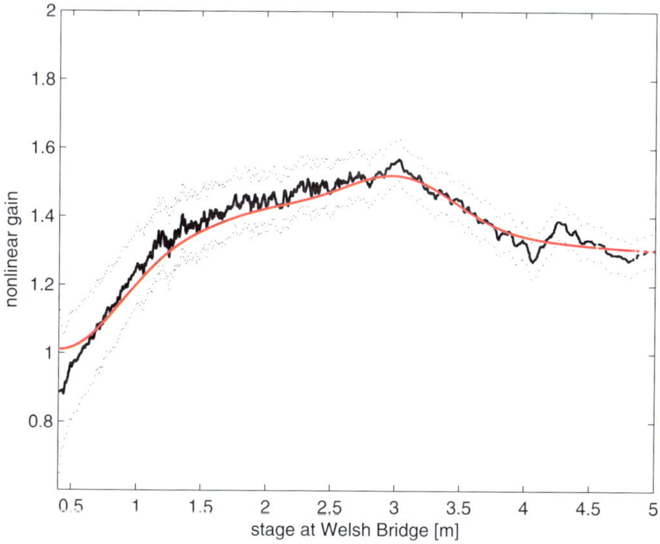

Figure 16.7. The estimated SDP nonlinearity for the Welsh Bridge–Buildwas routing model: the non-parametric estimate is shown as the black, full line, with its standard error bounds shown dotted; the red line is the optimal RBF estimate (see text).

for the DBM models for Buildwas, Bridgenorth and Bewdley, as well as their diagnostic performance measures are summarised in Table 16.2.

16.5.3.1. *Nonlinear water level routing for Welsh Bridge–Buildwas*

The linear water level routing model for Welsh Bridge–Buildwas has been presented in Section 16.5.2.2, but the same reach can be modelled using the nonlinearly transformed water levels at Welsh Bridge as an input to a linear first-order transfer function. The nonlinear relation is parameterised using ten radial basis functions as explained above and Figure 16.7 compares the RBF parameterised nonlinearity (red line) with the initial non-parametric estimate (black line). Note that this red line is not estimated as a RBF approximation to the non-parametric black line; it is the finally optimised RBF function and is plotted here to show that it compares well with the initial non-parametric estimate. Note also how the state-dependent gain defined by this nonlinear curve reduces substantially at water levels greater than 3 m, possibly suggesting that over-bank flow is occurring above this level.

The best identified model structure for Welsh Bridge–Buildwas reach, incorporating this nonlineaity, has the form:

$$y_{5,k} = \frac{0.4683}{1 - 0.5224z^{-1}} f_4(y_{4,k-8}) + \xi_k, \tag{16.22}$$

where $f_4(y_{4,k-8})$ denotes the nonlinearly transformed input water level at Welsh Bridge delayed by 8 hours and $y_{5,k}$ denotes water level at Buildwas. The noise ξ_k is identified by the AIC criterion as AR(13) process. The model residence time is equal to 1.5 hours.

The 8-hour-ahead forecast for Buildwas explains 99.42% of output water level variation for the validation period in November 2000. The results of validation are shown in Figure 16.8. The validation of the same model on the February 2002 flood data results in 99.62% explanation of the data.

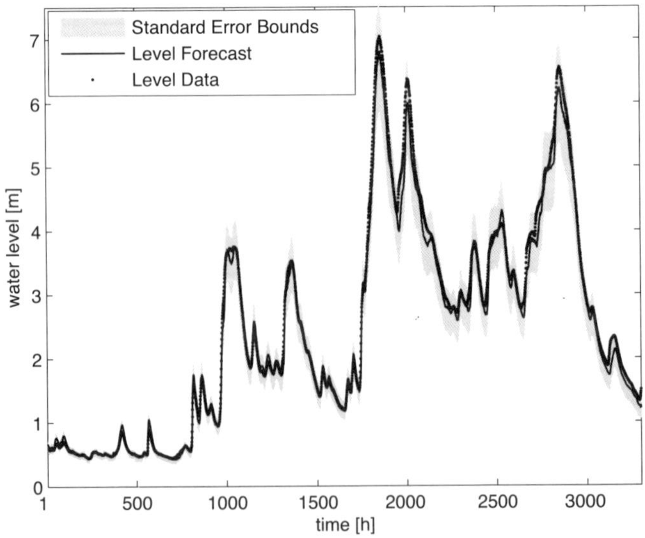

Figure 16.8. Buildwas: 8-hour-ahead validation forecast using the nonlinear SDP routing model.

16.5.3.2. *Nonlinear water level routing for Buildwas–Bridgenorth*

The best identified model structure for the Buildwas–Bridgenorth reach has the form:

$$y_{6,k} = \frac{0.6692}{1 - 0.3616z^{-1}} f_5(y_{5,k-2}) + \xi_k, \tag{16.23}$$

where $f_5(y_{5,k-2})$ denotes denotes the nonlinearly transformed input water level at Buildwas delayed by 2 hours and $y_{6,k}$ denotes water level at Bridgenorth. The noise ξ_k is identified by the AIC criterion as AR(11) process. The model residence time is equal to 1.0 hours.

The 2-hour-ahead forecast for Buildwas explains 99.99% of output water level variation for the validation period in 2002. However, as the forecast is only 2 hours ahead, it is useful to extend it using upstream forecasts. Also, the validation forecasts so far, as shown in Figures 16.3 to 16.8, have illustrated the overall performance of the forecasting system. In practice, however, the multiple-step-ahead forecasts, from a few hours ahead to the maximum forecast horizon, are required on demand at any time. Here, therefore, the results are presented in this operational mode. Figure 16.9 shows a series of six such operational forecasts over the major flood peak. The numbered circle with an interior cross mark is the forecasting origin in each case: all the forecasts are plotted as red lines over a period of between 3 and 14 hours ahead, together with their associated standard error bounds. The water level before the first forecast, at point 1, has declined from the previous flood peak and has levelled out. This is the first forecast where an increase in water level is predicted and occurs. At point 2, the forecast shows the water level continuing to rise sharply and so another, overlapping, forecast is shown at point 3, which predicts a still further rise but with the first sign of flattening out at the flood peak (these two forecasts tend to merge and are not easy to separate on the graph). Before point 4, the measured water levels have not been changing much around their peak values for about 12 hours, but the latest water level measurements at this point suggest a possible forthcoming decline and this is, indeed, confirmed by the forecast. Finally at points 5 and 6, the forecasts predict the continuing decline to the end of this particular data set.

Note that with the present implementation, the updated operational estimate of the level can be somewhat removed from the measurement because the measurement noise variance is not constrained to be small (compare the level forecast with the measured level at the start of the

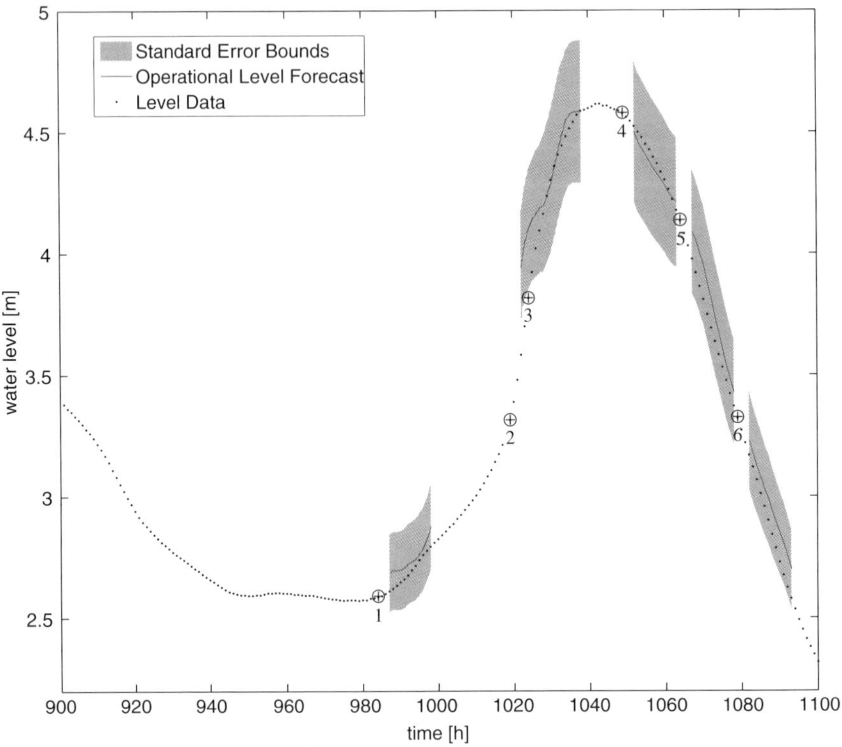

Figure 16.9. Bridgenorth: 14-hour-ahead operational forecasts using the nonlinear SDP routing model. The numbered circle with an interior cross marks the forecasting origin in each case (see text); the red line is the forecast; and the shaded bands are the estimated standard error bounds.

forecast period). If the alternative formulation outlined in Section 16.4.1.4 had been used, however, this updated estimate would be much closer to the measured level, with the closeness depending on the specified variance of the measurement noise σ^2.

16.5.3.3. *Nonlinear water level routing for Bridgenorth–Bewdley*

The best identified model structure for the Bridgenorth–Bewdley reach has the form:

$$y_{7,k} = \frac{0.4923}{1 - 0.4496z^{-1}} f_6(y_{6,k-4}) + \xi_k, \tag{16.24}$$

where $f_6(y_{6,k-4})$ denotes the nonlinearly transformed input water level at at Bridgenorth delayed by 4 hours and $y_{7,k}$ denotes water level at Bewdley. The noise ξ_k is identified by the AIC criterion as AR(11) process. The model residence time is just greater than one hour.

The 4-hour-ahead forecast for Bewdley explains 99.1% of ouput water level variations over the validation period in 2002 (the same 2002 flood event as at Bridgenorth, but considering a different peak wave). As in the previous reach, this forecast can be combined with the 10-hour-ahead forecast from Bridgenorth, thus obtaining a total 14 hours forecast lead time. Figure 16.10 shows this 14-hour-ahead forecast obtained for the same 2002 event, which explains 98.87% of the water level variations at Bewdley for this validation period.

It should be noted that the percentage of the variance explained provides a good measure of the overall performance and the figures cited in this and previous sections are based on the whole forecast period. In order to obtain a better idea of comparative performance, however, it is worth looking at shorter periods, for example over flood peaks. In the case of a 100 hour period covering the peak wave in 2002, for instance, the percentage explanation of the 14-hour-ahead forecast is 89.7%, while for the "naive" forecast (the current water level projected 14 hours ahead), this drops to 27.4%.

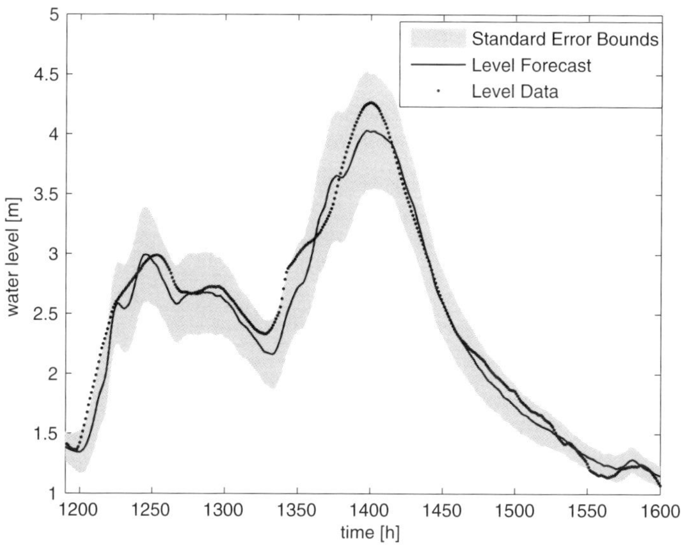

Figure 16.10. Bewdley: 14-hour-ahead validation forecast using the nonlinear SDP routing model.

16.6. Comments

The main aim of the Lancaster Real-Time Forecasting Study has been the derivation of a relatively simple and robust, online system for the forecasting of water levels in a river catchment, with the maximum possible lead-time and good real-time forecasting performance under high water level conditions. As a result, the analysis and forecasting system design is based on the water levels, rather than the more conventional flow variables, in order to obtain superior long-term forecasting performance. This removes the need to introduce the nonlinear water level-flow transformation, thus reducing the reliance on the prior calibration of this relationship and conveniently reducing the magnitude of the heteroscedasticity in the model residuals. In addition, improved forecasting performance is obtained by the introduction of real-time adaptive mechanisms that account for changes in both the system gain and the variance of the forecasting errors (heteroscedasticity).

An important contribution of the study is the development of an adaptive forecasting and DA system for a fairly large part of the River Severn catchment, based on a statistically estimated stochastic-dynamic, nonlinear DBM model composed of inter-connected rainfall-water level and water level routing modules. Although this adaptive forecasting system design is generic and could be applied to a wide variety of catchment systems, it has been developed specifically for the River Severn between Abermule and Bewdley, including the effects of the Vyrnwy tributary, using DBM models. As such, it is a quasi-distributed system: i.e. spatially distributed inputs but lumped parameter component models that consist of four main sub-models: two rainfall-water level models and four water level routing models for the river gauging stations downstream to Bewdley.

Without introducing external rainfall forecasts, the maximum length of the forecasting lead-time that can be handled successfully by this adaptive forecasting system depends on the advective delays that are inherent in the rainfall-water level dynamics; delays that depend on factors such as the changing speed (celerity) as the flood-wave moves along the river and the flood starts to inundate the floodplain. The gain adaptation can only off-set this nonlinearity in the effective time delay partially, but this still results in acceptable flood peak predictions. The result is a state/parameter updating system, where the parameter updating is limited to the minimum that is necessary to achieve acceptable long-term forecasting performance, combined with good potential for practical robustness and reliability. In this sense, the system represents a sophisticated development of the flood warning system for Dumfries (Lees *et al.*, 1993, 1994) that was operative during the 1990s. Sometimes, however, it may be necessary to adaptively update more parameters in the sub-models that comprise the catchment model. This is aided by the real-time recursive nature of the parameter estimation algorithms (Young, 2011) and a recent examples are given in Young (2010a,b).

The DA involves sequential updating at each measurement location. As a result, the forecast uncertainty of this decomposed model will be different from the uncertainty of the entire system, unless the correlation between the observations is negligible. However, we decided to use the decomposed system for the model identification and parameter estimation phase of the analysis due to much better identifiability of this decomposed form and the associated higher robustness of the resulting model. Further work is being undertaken on the propagation of uncertainty in the entire forecasting system (Beven *et al.*, 2006). However, in order to check the estimates of the uncertainty bands of the forecasts, we applied an empirical approach based on the observed behaviour of the past forecast (Gilchrist, 1978), which confirmed that they are reasonable.

In order to achieve the main aim of the study, namely acceptable long-horizon flood forecasting, the research reported here has concentrated on a design that achieves good forecast accuracy at high water levels for long lead times. Our results show that the changes of water level dynamics for these high peak values can be predicted adequately using the recursively updated adaptive gain and variance parameters. As a flood event continues, the same methodology can be used to produce accurate forecasts with much smaller prediction variance for shorter lead times. However, there remain some differences in the forecast responses for low flow values that we feel could be reduced further and research is continuing to investigate this possibility.

16.7. Conclusions

The main conclusions of the study described in this chapter are summarised below.

- The main objectives of a real-time flood forecasting system are to guide decision making about whether flood warnings should be issued to the stakeholders and the public and to help avoid the issuing of false warnings. To this end, the forecasting system proposed here is able to provide reasonably accurate forecasts, based on real-time adaptive mechanisms for state and parameter updating, together with the estimates of uncertainties. This will allow the decision-making process in the future to be based on an appropriate assessment of risk when issuing a warning.

- The Lancaster DBM model-based system is designed specifically to satisfy these objectives: its main aim is to process the online signals from the rainfall and water level sensors in a statistically efficient manner, in order to minimise forecasting errors.
- The DBM model-based state updating methods can be applied to any model that satisfies the KF requirements (e.g. models such as IHACRES (Jakeman *et al.*, 1990), PDM (Moore, 1985, 2007) and HYMOD (Moradkhani *et al.*, 2005b)). However, large, physically-based models are normally over-parameterised and so suffer from inherent ambiguity (equifinality). As a result, both parameter estimation and updating is problematic: a rather unsatisfactory solution requires the modeller to select a subset of identifiable parameters, with the remaining ones constrained, either deterministically or stochastically (e.g. using tightly defined Bayesian priors), to assumed known values.
- The DBM model/modified KF approach can match, or improve upon, the performance of the currently popular but computationally very intensive, methods of numerical Bayesian analysis (e.g. the ensemble KF), yet it is computationally much more efficient, with updates taking only a few seconds.
- Research and development is continuing to further improve the forecasting system performance. This includes the reduction of the autocorrelation in the multi-step-ahead forecasting errors by extending the stochastic parts of the DBM model in the upstream reaches down to Buildwas.
- A forecasting system based on the approach described in this chapter has now been incorporated as an option in the National Flood Forecasting System under the Environment Agency project *Risk-based Probabilistic Fluvial Flood Forecasting for Integrated Catchment Models.* Some details are available in the Phase 1 Report (Environment Agency, 2009).
- The forecasting system has been developed with, and exploits, computational tools from the CAPTAIN Toolbox (e.g. Taylor *et al.*, 2007) for Matlab^{TM}. For the latest information on CAPTAIN, see `http://captaintoolbox.co.uk/Captain_Toolbox.html`. A fully functional, version

of this Toolbox can be downloaded via this website or directly from `http://www.lancs.ac.uk/staff/taylorcj/tdc/download.php`.

Appendix 16.1. Recursive Estimation and the Kalman Filter

Recursive estimation is the embodiment of real-time updating. In recursive estimation, either the estimate of a model parameter vector, or the estimate of a state variable vector, are updated at the kth sampling instant based on the estimate obtained at the previous $(k-1)^{\text{th}}$ sampling instant. This update also depends upon the model that describes the relationship between the parameters or state variables and the latest data received at the kth sampling instant. So, for example, the state of a catchment model at a given k^{th} hour, as defined by the hourly flows at various locations in the catchment, is updated on the basis of the estimates of these flows at the previous $(k-1)^{\text{th}}$ hour, the catchment model, and the latest flow measurements, as received remotely from flow gauges at the various locations at the kth hour.

Recursive estimation was first developed by K. F. Gauss, sometime before 1826. A translation and an interpretation of his recursive linear least squares regression algorithm, in modern vector-matrix terms, is available in Appendix 2 of the first author's tutorial text on recursive estimation and time-series analysis (Young, 2011). Over one hundred and thirty years later, apparently without any knowledge of the Gauss algorithm, the systems theorist, Rudolf Kalman, developed a related algorithm for state variable estimation (Kalman, 1960). This Kalman Filter (KF) algorithm, as it has come to be known in many areas of science and social science, is a special form of recursive linear least squares estimation where the constant parameters considered by Gauss are replaced by state variables generated by a dynamic linear system, in the form of a set of linear, stochastic state equations (a vector Gauss–Markov process). Such equations could, for example, be those of a linear catchment model.

The simplest way to derive the linear least squares regression algorithm of Gauss is to consider the well known problem of estimating of the set of n unknown, constant parameters $x_j, j = 1, \ldots, n$, which appear in a "linear regression" relationship of the form:

$$x_k = h_{1,k}x_1 + h_{2,k}x_2 + \cdots + h_{n,k}x_n, \tag{A.1}$$

where the measurement y_k of x_k is contaminated by noise, i.e.

$$y_k = x_k + e_k = h_{1,k}x_1 + h_{2,k}x_2 + \cdots + h_{n,k}x_n + e_k, \tag{A.2}$$

while the $h_{j,k}; j = 1, 2, \ldots, n$, are exactly known, linearly independent variables which are also statistically independent of the measurement noise on y_k.

Most readers will recognise this problem and will know that the minimisation of the least squares criterion function for k samples, i.e.

$$J_2 = \sum_{i=1}^{k} \left\{ \sum_{j=1}^{n} [y_i - h_{j,i} x_j]^2 \right\} \tag{A.3}$$

requires that all the partial derivatives of J_2 with respect to each of the parameters $x_j, j = 1, 2, \ldots, n$, should be set simultaneously to zero. Such a procedure yields a set of n linear, simultaneous algebraic equations that are sometimes termed the "normal equations" and which can be solved to obtain the *en bloc* estimates $\hat{x}_j, j = 1, 2, \ldots, n$, of the parameters $x_j, j = 1, 2, \ldots, n$, based on the k data samples.

A simple and concise statement of the least squares results in the multi-parameter case can be obtained by using a vector-matrix formulation: thus by writing (A.2) in the alternative vector form,

$$y_k = \mathbf{h}_k^T \boldsymbol{x} + e_k, \tag{A.4}$$

where

$$\mathbf{h}_k^T = [h_{1,k}\ h_{2,k} \cdots h_{n,k}]; \quad \boldsymbol{x}^T = [x_1\ x_2 \cdots x_n]. \tag{A.5}$$

The superscript T denotes the vector/matrix transpose, and $\mathbf{h}_k^T \boldsymbol{x}$ is the vector inner product. With this notation, we are able to define J_2 as

$$J_2 = \sum_{i=1}^{k} [y_i - \mathbf{h}_i^T \boldsymbol{x}]^2 = \sum_{i=1}^{k} e_k^2. \tag{A.6}$$

Using the rules of vector differentiation, the partial derivatives of J_2 with respect to the vector of parameters $\boldsymbol{x}$ are given by,

$$\frac{\partial}{\partial \boldsymbol{x}} \sum_{i=1}^{k} [y_i - \mathbf{h}_i^T \boldsymbol{x}]^2 = -2 \sum_{i=1}^{k} \mathbf{h}_i [y_i - \mathbf{h}_i^T \boldsymbol{x}], \tag{A.7}$$

so that the normal equations are obtained from,

$$\frac{1}{2} \nabla_{\boldsymbol{x}}(J_2) = -\sum_{i=1}^{k} \mathbf{h}_i y_i + \left[\sum_{i=1}^{k} \mathbf{h}_i \mathbf{h}_i^T \right] \boldsymbol{x} = 0, \tag{A.8}$$

where $\frac{1}{2}\nabla_{\boldsymbol{x}}(J_2)$ denotes the gradient of J_2 with respect to all the elements of $\boldsymbol{x}$.

(*a*) *The Recursive Least Squares Algorithm*

Now, provided that the matrix $[\sum_{i=1}^{k} \mathbf{h}_i \mathbf{h}_i^T]$ is non-singular,[e] then the solution to the normal equations (A.8) takes the form

$$\hat{\boldsymbol{x}}_k = \left[\sum_{i=1}^{k} \mathbf{h}_i \mathbf{h}_i^T\right]^{-1} \sum_{i=1}^{k} \mathbf{h}_i y_i, \qquad \text{(A.9)}$$

where $\hat{\boldsymbol{x}}_k$ is the estimate of $\boldsymbol{x}$ based on the k data samples. This solution can be written as,

$$\hat{\boldsymbol{x}}_k = \mathbf{P}_k \mathbf{b}_k \qquad \text{(A.10)}$$

where,

$$\mathbf{P}_k = \left[\sum_{i=1}^{k} \mathbf{h}_i \mathbf{h}_i^T\right]^{-1}; \quad \mathbf{b}_k = \sum_{i=1}^{k} \mathbf{h}_i y_i \qquad \text{(A.11)}$$

and we see that $\mathbf{P}_k^{-1}$ and $\mathbf{b}_k$ can be obtained recursively from the equations:

$$\mathbf{P}_k^{-1} = \mathbf{P}_{k-1}^{-1} + \mathbf{h}_k \mathbf{h}_k^T; \quad \mathbf{b}_k = \mathbf{b}_{k-1} + \mathbf{h}_k y_k. \qquad \text{(A.12)}$$

A few lines of matrix manipulation (see page 30 *et seq* in Young, 2011) then produces the following Recursive Least Squares (RLS) algorithm for updating the estimate $\hat{\boldsymbol{x}}_k$ of $\boldsymbol{x}_k$ from the previous estimate $\hat{\boldsymbol{x}}_{k-1}$:

$$\hat{\boldsymbol{x}}_k = \hat{\boldsymbol{x}}_{k-1} + \mathbf{P}_{k-1}\mathbf{h}_k[1 + \mathbf{h}_k^T \mathbf{P}_{k-1}\mathbf{h}_k]^{-1}\{y_k - \mathbf{h}_k^T \hat{\boldsymbol{x}}_{k-1}\}, \qquad \text{(A.13)}$$

where $\mathbf{P}_k$ is obtained recursively from $\mathbf{P}_{k-1}$ by the equation:

$$\mathbf{P}_k = \mathbf{P}_{k-1} + \mathbf{P}_{k-1}\mathbf{h}_k[1 + \mathbf{h}_k^T \mathbf{P}_{k-1}\mathbf{h}_k]^{-1}\mathbf{h}_k^T \mathbf{P}_{k-1}. \qquad \text{(A.14)}$$

It is interesting to note that, following a little more matrix manipulation, an alternative form of the recursion (A.13) is the following:

$$\hat{\boldsymbol{x}}_k = \hat{\boldsymbol{x}}_{k-1} + \mathbf{P}_k\{\mathbf{h}_k y_k - \mathbf{h}_k \mathbf{h}_k^T \hat{\boldsymbol{x}}_{k-1}\}. \qquad \text{(A.15)}$$

---

[e]The need for linear independence between the x_j now becomes clear since linear dependence would result in singularity of the matrix $C_k = [\sum_{i=1}^{k} \mathbf{h}_i \mathbf{h}_i^T]$ with $det[C(k)] = 0$ and the inverse "blowing up".

This second form shows that the recursive update is in the form of a "gradient algorithm", since $\{\mathbf{h}_k y_k - \mathbf{h}_k \mathbf{h}_k^T \hat{\boldsymbol{x}}_{k-1}\}$ is proportional to the instantaneous gradient of the squared error function $[y_k - \mathbf{h}_k^T \hat{\boldsymbol{x}}]^2$. However, the form (A.13) is usually preferred in computational terms.

Equations (A.13) and (A.14) constitute the RLS algorithm. It is interesting to note the simplicity of the above derivation in vector-matrix terms which constrasts with the complexity of Gauss's original derivation (see Appendix A of Young, 2011) using ordinary algebra. Note also that the RLS algorithm provides a computational advantage over the stage-wise solution of the *en bloc* solution (A.9). In addition to the now convenient recursive form, which provides for a minimum of computer storage, note that the term $1 + \mathbf{h}_k^T \mathbf{P}_{k-1} \mathbf{h}_k$ is simply a scalar quantity. As a result, there is no requirement for direct matrix inversion even though the repeated solution of the equivalent classical solution (A.9) entails inverting an $n \times n$ matrix for each solution.

Since the RLS is recursive, it is necessary to specify starting values $\hat{\boldsymbol{x}}_0$ and $\mathbf{P}_0$ for the vector $\hat{\boldsymbol{x}}$ and the matrix $\mathbf{P}$, respectively. This presents no real problem, however, since it can be shown that the criterion function-parameter hypersurface is unimodal and so an arbitrary finite $\hat{\boldsymbol{x}}_0$ (say $\hat{\boldsymbol{x}}_0 = \mathbf{0}$), coupled with a $\mathbf{P}_0$ having large diagonal elements (say 10^6 in general), will yield convergence and performance commensurate with the stage-wise solution of the same problem. This setting is often referred to as a "diffuse prior", which exposes the fact that the algorithm has obvious Bayesian interpretation, in the sense that the *a priori* estimate $\hat{\boldsymbol{x}}_{k-1}$ at the $k-1^{\text{th}}$ sampling instant is updated to the *a posteriori* estimate $\hat{\boldsymbol{x}}_k$ at the kth sampling instant, based on the new data received at the kth sampling instant.

This Bayesian interpretation is enhanced still further if it is assumed that the noise e_k is a Gaussian white noise process with variance σ^2, for then the estimate $\hat{\boldsymbol{x}}_k$ will also have a Gaussian distribution and the statistical properties of the estimate, in the form of the covariance matrix, are conveniently generated by the algorithm if it is modified slightly to the form:

$$\hat{\boldsymbol{x}}_k = \hat{\boldsymbol{x}}_{k-1} + \mathbf{P}_{k-1}\mathbf{h}_k[\sigma^2 + \mathbf{h}_k^T \mathbf{P}_{k-1}\mathbf{h}_k]^{-1}\{y_k - \mathbf{h}_k^T \hat{\boldsymbol{x}}_{k-1}\} \tag{A.16a}$$

$$\mathbf{P}_k = \mathbf{P}_{k-1} + \mathbf{P}_{k-1}\mathbf{h}_k[\sigma^2 + \mathbf{h}_k^T \mathbf{P}_{k-1}\mathbf{h}_k]^{-1}\mathbf{h}_k^T \mathbf{P}_{k-1}, \tag{A.16b}$$

where now $\mathbf{P}_k$ is the covariance matrix, i.e.

$$\mathbf{P}_k = E\{\tilde{\boldsymbol{x}}_k \tilde{\boldsymbol{x}}_k^T\}; \quad \tilde{\boldsymbol{x}}_k = \boldsymbol{x}_k - \hat{\boldsymbol{x}}_k, \tag{A.16c}$$

so that the square root of the diagonal elements of $\mathbf{P}_k$ quantify the estimated standard errors on the elements of the estimate $\hat{\boldsymbol{x}}_k$.

(*b*) *The Kalman Filter*

In the RLS algorithm, there is an implict assumption that $\boldsymbol{x}_k$ is time invariant; an assumption that can be made explicit by stating that,

$$\boldsymbol{x}_k = \mathbf{F}\boldsymbol{x}_{k-1}; \quad \mathbf{F} = \mathbf{I}_n, \tag{A.17}$$

where $\mathbf{I}_n$ is the identity matrix (a purely diagonal $n \times n$ matrix with unity diagonal elements).

Using orthogonal projection theory, Kalman (1960) extended this assumption by considering the situation when $\boldsymbol{x}_k$ is not a vector of constant parameters, as in the RLS algorithm, but a vector of time-variable stochastic variables generated by the following set of stochastic state equations (a "Gauss–Markov process"):

$$\boldsymbol{x}_k = \mathbf{F}\boldsymbol{x}_{k-1} + \boldsymbol{\zeta}_k, \tag{A.18}$$

where $\mathbf{F}$ is a "state transition matrix" that defines the dynamic behaviour of $\boldsymbol{x}_k$ and is not, therefore, an identity matrix. This equation is obviously a generalisation of (A.17), allowing $\boldsymbol{x}_k$ to vary over time in a stochastic-dynamic manner, where the stochasticity arises from the input $\boldsymbol{\zeta}_k$, which is assumed to be a vector of zero mean, white noise inputs with covariance matrix $\mathbf{Q}$, which can be time-variable if there is information on this.

In practice, we might expect that, if $\boldsymbol{x}_k$ is a physically meaningful state vector, then it may be affected by one or more input variables. For instance, if the state variables are both observed and unobserved flows occurring at locations along a river, then these flows will be caused by the variations in the rainfall. We might also expect that not all the flows will be observed (e.g. some could be flows defined by the STF decomposition analysis discussed in the main text, which separates the observed flow into unobserved quick and slow components). Consequently, it is necessary to specify how the observed (measured) variables are related to the state variables by introducing a "measurement equation" that accounts also for any noise on these measured variables.

In the case of a single input variable $u_{k-\delta}$, delayed by δ samples (which is relevant to the rainfall-level models considered in this chapter, where $u_{k-\delta}$ is the delayed effective rainfall), these additional factors can be introduced quite straightforwardly by extending the model (A.18) to the form:

$$\boldsymbol{x}_k = \mathbf{F}\boldsymbol{x}_{k-1} + \mathbf{G}u_{k-\delta} + \boldsymbol{\zeta}_k \tag{A.19a}$$

$$y_k = \mathbf{h}^T\boldsymbol{x}_k + e_k. \tag{A.19b}$$

Here, the second equation (A.19b) is the measurement equation and it is very similar to the regression equation (A.4), except that it will be noted that the subsript k on $\mathbf{h}$ has been removed, showing that it is composed of constant elements. Although the elements of $\mathbf{h}$ could vary, if this is required, it is more usual, within this dynamic systems setting, that they will be constant since the relationship between the observed variable y_k and the state vector $\boldsymbol{x}_k$ will normally remain fixed. For example, in the case of state equations (16.10) in the main body of this chapter, the measured water level is sum of the quick and slow components, so that $\mathbf{h} = [1 \;\; 1]$.

Given the stochastic state equations (A.19a) and (A.19b), the recursive equations of the KF can be written in the following prediction-correction form:

***a priori* prediction:**

$$\hat{\boldsymbol{x}}_{k|k-1} = \mathbf{F}\hat{\boldsymbol{x}}_{k-1} + \mathbf{G}u_{k-\delta} \tag{A.20a}$$

$$\mathbf{P}_{k|k-1} = \mathbf{F}\mathbf{P}_{k-1}\mathbf{F}^T + \mathbf{Q} \tag{A.20b}$$

***a posteriori* correction:**

$$\hat{\boldsymbol{x}}_k = \hat{\boldsymbol{x}}_{k|k-1} + \mathbf{P}_{k|k-1}\mathbf{h}[\sigma^2 + \mathbf{h}^T\mathbf{P}_{k|k-1}\mathbf{h}]^{-1}\{y_k - \hat{y}_{k|k-1}\} \tag{A.20c}$$

$$\mathbf{P}_k = \mathbf{P}_{k|k-1} - \mathbf{P}_{k|k-1}\mathbf{h}[\sigma^2 + \mathbf{h}^T\mathbf{P}_{k|k-1}\mathbf{h}]^{-1}\mathbf{h}^T\mathbf{P}_{k|k-1} \tag{A.20d}$$

$$\hat{y}_k = \mathbf{h}^T\hat{\boldsymbol{x}}_k; \quad \hat{y}_{k|k-1} = \mathbf{h}^T\hat{\boldsymbol{x}}_{k|k-1}. \tag{A.20e}$$

The derivation of these equations is straightforward (Young, 2011) but, in any case, they make intuitive sense:

- Since the stochastic inputs $\boldsymbol{\zeta}_k$ and e_k both involve only zero mean, white noise variables, the state prediction equation (A.20a) provides a logical prediction of the state vector $\boldsymbol{x}_k$, given the equation (A.19a) for its propagation in time, because the expected values of these random inputs is zero and the dynamic system is assumed to be stable. Here, the subscript notation $k|k-1$ denotes the estimate at the kth sampling instant (in the main text one hour), based of the estimate at the previous $(k-1)^{\text{th}}$ sampling instant.

- The state update Equations (A.20c) and (A.20d) follow directly from their equivalent Equations (A.16a) and (A.16b), respectively, in the RLS algorithm.
- The covariance matrix prediction in Equation (A.20b) is not quite so obvious but, if we consider the estimation error $\tilde{\boldsymbol{x}}_k = \boldsymbol{x}_k - \hat{\boldsymbol{x}}_k$ then,

$$\tilde{\boldsymbol{x}}_k = \mathbf{F}\boldsymbol{x}_{k-1} - \mathbf{F}\hat{\boldsymbol{x}}_{k-1} + \boldsymbol{\zeta}_k = \mathbf{F}\tilde{\boldsymbol{x}}_{k-1} + \boldsymbol{\zeta}_k$$

because the effects of the input term $\mathbf{G}u_{k-\delta}$ cancel out. Consequently, the covariance matrix $\mathbf{P}_{k|k-1}$, which is the expected value of $\tilde{\boldsymbol{x}}_k\tilde{\boldsymbol{x}}_k^T$, is given by:

$$\mathbf{P}_{k|k-1} = E\{\tilde{\boldsymbol{x}}_k\tilde{\boldsymbol{x}}_k^T\} = E\{\mathbf{F}[\tilde{\boldsymbol{x}}_{k-1}\tilde{\boldsymbol{x}}_{k-1}^T]\mathbf{F}^T\} + E\{\boldsymbol{\zeta}_k\boldsymbol{\zeta}_k^T\} = \mathbf{F}\mathbf{P}_{k-1}\mathbf{F}^{\mathbf{T}} + \mathbf{Q}.$$

Here, use is made of the matrix identity $[\mathbf{F}\tilde{\boldsymbol{x}}_{k-1}]^T = \tilde{\boldsymbol{x}}_{k-1}^T\mathbf{F}^T$, associated with a matrix product; as well as the fact that the expected value of the cross-products between $\tilde{\boldsymbol{x}}_{k-1}$ and $\boldsymbol{\zeta}_k$ are zero because they are uncorrelated.
- The estimate $\hat{y}_k$ of the underlying "noise-free" output $\mathring{y}_k$ (see later comment), i.e. $\mathring{y}_k = \mathbf{h}^T\boldsymbol{x}_k$, is simply a combination of the state estimates defined by the known measurement equation vector $\mathbf{h}$ (e.g. in the case of the state space model (16.10) in the main text, $\mathbf{h}^{\mathbf{T}} = [1\ 1]$ and the estimate is the sum of the two state variables, in this case the "quick" and "slow" component water level estimates).
- Like its simple RLS relative, the KF can be considered in Bayesian terms. It must be emphasised, however, that the Gaussian assumptions utilised for this Bayesian interpretation of the algorithm are not essential to its success as an estimation algorithm. The algorithm is robust to the violation of these assumptions in the sense that the estimate $\hat{\boldsymbol{x}}_k$ is still the minimum covariance linear unbiased estimator of the state vector $\boldsymbol{x}_k$ (Norton, 1986).

In the context of the present chapter, the main reason for using the KF algorithm is forecasting water level or flow. The f-step-ahead forecasts are obtained by simply repeating the prediction steps in (A.20a) and (A.20b) f times, without correction (since no new data over this interval are available). The f-step ahead forecast variance is then given by:

$$\mathrm{var}(\hat{y}_{k+f|k}) = \hat{\sigma}_k^2 + \mathbf{h}^T\mathbf{P}_{k+f|k}\mathbf{h}, \tag{A.20f}$$

where $\mathbf{P}_{k+f|k}$ is the error covariance matrix estimate associated with the f-step ahead prediction of the state estimates. This estimate of the f-step ahead prediction variance is used to derive approximate 95% confidence bounds for the forecasts, under the approximating assumption that the prediction error can be characterised as a nonstationary Gaussian process (i.e. twice the square root of the variance at each time step is used to define the 95% confidence region).

It is important to note one important aspect of the stochastic formulation of the KF in relation to flood forecasting: namely the assumption that the measured water level (or flow) variable is subject to error and uncertainty. The objective of the KF is to provide an *estimate* of this variable so, at each recursion, the state update equation (A.20c) adjusts the estimate of the state vector $\hat{\boldsymbol{x}}_k$ to its most likely value, based on the past estimates and the latest noisy measurement. The associated estimate of the water level $\hat{y}_k$ in Equation (A.20e) is adjusted accordingly and it will not, in general, be equal to the measured water level. It is possible to over-ride this aspect of the KF and ensure that the estimate is updated so that the estimated water level is very close to the latest measured value by setting the noise variance σ^2 to a low value.

This means that the forecast generated by the KF should not be considered a forecast of the *measured* water level variable but an estimate of the underlying "noise-free" variable $\mathring{y}_k = \mathbf{h}^T \boldsymbol{x}_k$ (see A.4), together with a quantification of the uncertainty associated with this estimate. It is important to stress this point as there is a tendency to judge the quality of a forecast by how close it forecasts the measured water level. In fact, the forecast is the mean of an estimated random variable defined by this mean value and the associated confidence bounds, so that a good forecast is one in which, most of the time, the measured water level falls within the 95% confidence interval associated with the forecast.

Finally, it should be mentioned that other, more complex recursive algorithms have been developed for special problems: for instance, the recursive form of the Refined Instrumental Variable (RIV) algorithm discussed and used in the main text and Young (2011), which involves iteration as well as recursion.

Appendix 16.2. State-Dependent Parameter STF Models

Nonlinearity in rainfall-flow and water level models is very important because it defines the way in which the model responds under different catchment wetness conditions. Fortunately, it is straightforward to extend the linear STF model to allow for such nonlinear phenomena. In particular, the SDP STF model class can characterise a wide variety of nonlinear, stochastic, dynamic phenomena, including chaotic systems.

In the discrete-time case, the general SDP form of the STF model between rainfall r_k and level y_k is one where all of the parameters can be functions of other variables. In the case where each SDP is a function of only one variable, the model can be written as follows (*cf* Equation (16.1)):

$$y_k = \frac{B(\mathbf{w}_k, z^{-1})}{A(\mathbf{v}_k, z^{-1})} r_{k-\delta} + \xi_k \tag{A.21}$$

where,

$$\begin{aligned} A(\mathbf{v}_k, z^{-1}) &= 1 + a_1(v_{1,k})z^{-1} + a_2(v_{2,k})z^{-2} + \cdots + a_n(v_{n,k})z^{-n}, \\ B(\mathbf{w}_k, z^{-1}) &= b_0(w_{0,k}) + b_1(w_{1,k})z^{-1} + b_2(w_{2,k})z^{-2} + \cdots + b_m(w_{m,k})z^{-m}. \end{aligned}$$

In other words, it is assumed that any parameter in this SDTF model (A.21) may vary over time because it is a function of other ("state") variables $v_{i,k}, i = 1, 2, \ldots n$ and $w_{i,k}, i = 0, 1, 2, \ldots m$ that are the elements of the vectors $\mathbf{v}_k$ and $\mathbf{w}_k$. Typical variables used in this regard are the input r_k or output y_k and their past values, which are the state variables associated with the Non-Minimal State Space (NMSS) representation of the STF model (see Taylor *et al.* (2000) and the prior references therein).

The general SDP model subsumes a simple nonlinear model that is particularly important in a hydrological context. This is the "Hammerstein" model, where a single SDP nonlinearity affects the numerator parameters in the SDTF model (A.21) while the denominator coefficients are time-invariant: i.e.

$$\begin{aligned} A(z^{-1}) &= 1 + a_1 z^{-1} + a_2 z^{-2} + \cdots + a_n z^{-n}, \\ B(w_{0,k}, z^{-1}) &= b_0(w_{0,k}) + b_1(w_{0,k})z^{-1} + b_2(w_{0,k})z^{-2} + \cdots + b_m(w_{0,k})z^{-m}. \end{aligned}$$

Furthermore, the numerator SDPs $b_i(w_{0,k}), i = 0, 1, 2, \ldots m$ are defined so that the SDP nonlinearity $w_{0,k}$ can be factored out of the model, producing a single, nonlinearity operating on the input variable $u_{k-\delta}$. In the hydrological context considered here, this nonlinearity is found to be a function of the flow or level variable, in this case either the level y_k, for the rainfall-level model (16.5); or the upstream level $y_{i-1,k-\delta}$, for the routing model (16.17). In the rainfall-flow case, this converts the measured rainfall into an "effective rainfall" (see Section 16.4.1). This model can be written in the form,

$$y_k = \frac{B(z^{-1})}{A(z^{-1})} u_{k-\delta} + \xi_k, \tag{A.22}$$

where the effective rainfall $u_{k-\delta} = f(r_{k-\delta})$ is the SDP-transformed rainfall.

The SDP nonlinearity in the Hammerstein model is normally inferred from the data in two stages. First, the sdp algorithm in the CAPTAIN Toolbox is used to obtain a non-parametric (graphical) estimate that identifies the basic characteristics of the nonlinearity. Then, a parametric form that is able to model the non-parametric curve is selected and the parameters of this parameterised nonlinearity and the associated linear STF are estimated concurrently by optimisation based on either nonlinear least squares or maximum likelihood. A typical example of such an effective rainfall nonlinearity is shown in Figure 16.11, where the initial non-parametric estimate is shown as a full line, while the final optimised, parametric estimate

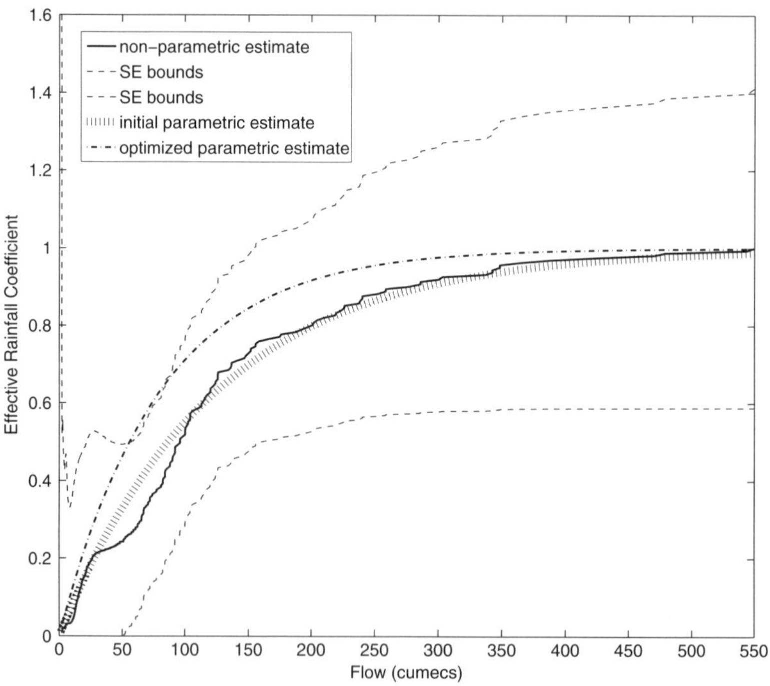

Figure 16.11. Typical SDP effective rainfall nonlinearity estimation results.

is shown as a dash-dot line. Also shown are the Standard Error (SE) bounds on the non-parametric estimate (dashed lines) and the initial parametric fit to the non-parametric estimate. In this case, the parametric function is of the form $u_k = (1 - e^{\gamma y_k}) r_k$.

References

Akaike, H. (1974). A new look at statistical model identification, *IEEE T. Automat. Contr.*, **19**, 716–723.

Bertino, L., Evensen, G. and Vackernagel, H. (2003). Sequential data assimilation techniques in oceanography, *Int. Stat. Rev.*, **71**, 223–242.

Beven, K.J., Leedal, D.T., Smith, P.J. *et al.* (2011). "Identification and Representation of State Dependent Non-Linearities in Flood Forecasting using DBM Methodology" in: Wang, L. and Garnier, H. (eds), *System Identification, Environmetric Modelling and Control*, Springer-Verlag, Berlin.

Beven, K.J., Romanowicz, R.J., Pappenberger, F. *et al.* (2006). "The Uncertainty Cascade in Flood Forecasting" in: Balbanis, P., Lambroso, D. and Samuels, P.P. (eds), *Innovation, Advances and Implementation of Flood Forecasting Technology: Proceedings of the ACTIF meeting*, Tromsø, Norway.

Box, G.E.P. and Cox, D.R. (1964). An analysis of transformations, *J. Roy. Stat. Soc. B* (Methodological), **26**, 211–252.

Box, G.E.P. and Jenkins, G.M. (1970). *Time Series Analysis Forecasting and Control*, Holden-Day, San Francisco.

Cluckie, I.D., Xuan, Y. and Wang, Y. (2006). Uncertainty analysis of hydrological ensemble forecasts in a distributed model utilising short-range rainfall prediction, *Hydrol. Earth Syst. Sci. Discuss.*, **3**, 3211–3237.

Environment Agency UK (2009). *Risk-Based Probabilistic Fluvial Flood Forecasting for Integrated Catchment Models: Phase 1 Report*, Technical Report Science Report — SR SC080030/SR1, Environment Agency, Bristol, UK.

Evensen, G. (1994). Sequential data assimilation with a nonlinear quasi-geostrophic model using monte carlo methods to forecast error statistics, *J. Geophys. Res.*, **99**, 10143–10162.

Gilchrist, W. (1978). *Statistical Forecasting*, Wiley, Chichester.

Goswami, M., O'Connor, K.M., Bhattarai, K.P. *et al.* (2005). Assessing the performance of eight real-time updating models and procedures for the Brosna river, *Hydrol. Earth Syst. Sci.*, **9**, 394–411.

Hall, P., Kay, J.W. and Titterington, D.M. (1990). Asymptotically optimal difference-based estimation of variance in nonparametric regression, *Biometrika*, **77**, 521–528.

Holder, L. (1985). *Multiple Regression in Hydrology*, Technical report, Institute of Hydrology, Wallingford, U.K.

Jakeman, A.J., Littlewood, I.G. and Whitehead, P.G. (1990). Computation of the instantaneous unit hydrograph and identifiable component flows with application to two small upland catchments, *J. Hydrol.*, **117**, 275–300.

Kalman, R.E. (1960). A new approach to linear filtering and prediction problems, *J. Basic Eng.-T. ASME*, **82–D**, 35–45.

Khan, M.H. (1993). Muskingum flood routing model for multiple tributaries, *Water Resour. Res.*, **29**, 1057–1062.

Krzysztofowicz, R. (1999). Bayesian theory of probabilistic forecasting via deterministic hydrologic model, *Water Resour. Res.*, **35**, 2739–2750.

Krzysztofowicz, R. (2002). Bayesian system for probabilistic river stage forecasting, *J. Hydrol.*, **268**, 16–40.

Lees, M.J., Young, P.C. and Ferguson, S. (1993). "Adaptive Flood Warning" in: Young, P.C. (ed.), *Concise Encyclopedia of Environmental Systems*, Pergamon, Oxford.

Lees, M.J., Young, P.C., Ferguson, S. *et al.* (1994). "An adaptive flood warning scheme for the River Nith at Dumfries" in: White, W.R. and Watts, J. (eds), *International Conference on River Flood Hydraulics*, Wiley, Chichester, pp. 65–77.

Madsen, H., Rosbjerg, D.J., Damgård, J. *et al.* (2003). "Data Assimilation in the MIKE 11 Flood Forecasting System using Kalman Filtering Water Resources Systems" in: *Proceedings of Symposium HS02b*, IUGG, Sapporo, Japan, IAHS Publ. no. 281, pp. 75–81.

Madsen, H. and Skotner, C. (2005). Adaptive state updating in real-time river flow forecasting — a combined filtering and error forecasting procedure, *J. Hydrol.*, **308**, 302–312.

Moore, R.J. (1985). The probability-distributed principle and runoff production at point and basin scales. *Hydrol. Sci. J.*, **30**, 273–297.

Moore, R.J. (2007). The PDM rainfall-runoff model, *Hydrol. Earth. Syst. Sci.* **11**, 483–499.

Moradkhani, H., Hsu, K.L., Gupta, H.V. *et al.* (2005a). Uncertainty assessment of hydrologic model states and parameters: sequential data assimilation using the particle filter, *Water Resour. Res.*, **41**, W05012, doi:10.1029/2004WR003604.

Moradkhani, H., Sorooshian, S., Gupta, H.V. *et al.* (2005b). Dual state-parameter estimation of hydrological models using ensemble Kalman filter, *Adv. Water Resour.*, **28**, 135–147.

Nash, J.E. and Sutcliffe, J.V. (1970). River flow forecasting through conceptual models: discussion of principles, *J. Hydrol.*, **10**, 282–290.

Nelder, J.A. and Mead, R. (1965). A simplex method for function minimization, *Comput. J.*, **1**, 308–313.

Norton, J.P. (1986). *An Introduction to Identification*, Academic Press, New York.

Pappenberger, F., Beven, K.J., Hunter, N.M. *et al.* (2005). Cascading model uncertainty from medium range weather forecasts (10 days) through a rainfall-runoff model to flood inundation predictions within the European flood forecasting system (EFFS), *Hydrol. Earth Syst. Sci.*, **9**, 381–393.

Park, J., Obeysekera, J. and Zee, R.V. (2005). Prediction boundaries and forecasting of nonlinear hydrologic stage data, *J. Hydrol.*, **312**, 79–94.

Porporato, A. and Ridolfi, L. (2001). Multivariate nonlinear prediction of river flows, *J. Hydrol.*, **248**, 109–122.

Refsgaard, J.C. (1997). Validation and intercomparison of different updating procedures for real-time forecasting, *Nord. Hydrol.*. **28**, 65–84.

Romanowicz, R.J. and Beven, K.J. (1998). Dynamic real-time prediction of flood inundation probabilities, *Hydrol. Sci. J.*, **43**, 181–196.

Romanowicz, R.J., Beven, K.J. and Young, P.C. (2004a). "Data Assimilation in the Identification of Flood Inundation Models: Derivation of On-Line Multi-Step Ahead Predictions of Flows" in: Webb, B., Arnell, N., Onf, C., *et al.* (eds), *BHS International Conference: Hydrology, Science and Practice for the 21st Century*, pp. 348–353.

Romanowicz, R.J., Young, P.C. and Beven, K.J. (2004b). "Assessing Uncertainty in Assessing Flood Risk" in: *Proceedings of First International Conference on Flood Risk Assessment*, Bath, U.K., pp. 127–138.

Romanowicz, R.J., Young, P.C. and Beven, K.J. (2006). Data assimilation and adaptive forecasting of water levels in the River Severn catchment, *Water Resour. Res.*, **42**(W06407), doi:10.1029/2005WR004373.

Romanwicz, R.J., Young, P.C., Beven, K.J. *et al.* (2008). A data based mechanistic approach to nonlinear flood routing and adaptive flood level forecasting, *Adv. Water Resour.*, **31**, 1048–1056.

Schweppe, F. (1965). Evaluation of likelihood functions for Gaussian signals, *IEEE T. Inform. Theory*, **11**, 61–70.

Sørensen, R.J. and Madsen, H. (2004). Efficient sequential techniques for the assimilation of tide gauge data in three dimensional modelling of the North Sea and Baltic Sea system, *J. Geophys. Res.*, **109**, 10.1029/2003JC002144.

Sorooshian, S. and Dracup, J.A. (1980). Stochastic parameter estimation procedures for hydrologic rainfall-runoff models: correlated and heteroscedastic error cases, *Water Resour. Res.*, **16**, 430–442.

Taylor, C.J., Chotai, A. and Young, P.C. (2000). State space control system design based on non-minimal state variable feedback: further generalization and unification results, *Int. J. Control.*, **73**, 1329–1345.

Taylor, C.J., Pedregal, D.J., Young, P.C. *et al.* (2007). Environmental time series analysis and forecasting with the CAPTAIN toolbox, *Environ. Modell. Softw.*, **22**, 797–814.

Thirumalaiah, K. and Deo, M.C. (2000). Hydrological forecasting using neural networks, *J. Hydro. Eng.*, **5**, 180–189.

Vaughan, M. and McIntyre, N. (2012). "An assessment of DBM Flood Forecasting Models" in: *Proceedings of the Institution of Civil Engineers — Water Management*, **165**, 105–120.

Vrugt, J.A., Diks, C.G.H., Gupta, H. *et al.* (2005). Improved treatment of uncertainty in hydrologic modelling: combining the strengths of global optimization and data assimilation, *Water Resour. Res.*, **41**, W01017, doi:10.1029/2004WR003059.

Wheater, H.S., Jakeman, A.J. and Beven, K.J. (1993). "Progress and Directions in Rainfall-Runoff Modelling", in: Jakeman, A.J., Beck, M.B. and McAleer, M.J. (eds), *Modelling Change in Environmental Systems*, Wiley, Chichester.

Young, P.C. (1986) "Time-Series Methods and Recursive Estimation in Hdrological Systems Analysis" in: Kraijenhoff, D.A. and Moll, J.R. (eds), River flow Modelling and Forecasting, Dordrecht, Netherlands.

Young, P.C. (1989). "Recursive Estimation, Forecasting and Adaptive Control" in: Leondes, C.T. (ed.), *Control and Dynamic Systems*, Academic Press, San Diego, pp. 119–166.

Young, P.C. (1993). "Time Variable and State Dependent Modelling of Nonstationary and Nonlinear Time Series" in: Subba Rao, T. (ed.), *Developments in Time Series Analysis*, Chapman and Hall, London, pp. 374–413.

Young, P.C. (1999a). Data-based mechanistic modelling, generalised sensitivity and dominant mode analysis, *Comput. Phys. Commun.*, **117**, 113–129.

Young, P.C. (1999b). Nonstationary time series analysis and forecasting, *Prog. Environ. Sci.*, **1**, 3–48.

Young, P.C. (2000). "Stochastic, Dynamic Modelling and Signal Processing: Time Variable and State Dependent Parameter Estimation" in: Fitzgerald, W.J., Walden, A., Smith, R. *et al.* (eds), *Nonlinear and Nonstationary Signal Processing*, Cambridge University Press, Cambridge, pp. 74–114.

Young, P.C. (2001a). *Advances in Real-Time Flood Forecasting*, Technical Report TR/176, Centre for Research on Environmental Systems and Statistics, Lancaster University.

Young, P.C. (2001b). "Data-Based Mechanistic Modelling and Validation of Rainfall Flow Processes" in: Anderson, M.G. and Bates P.D. (eds), *Model Validation: Perspectives in Hydrological Science*, John Wiley, Chichester, pp. 117–161.

Young, P.C. (2001c). "The Identification and Estimation of Nonlinear Stochastic Systems" in: Mees, A.I. (ed.), *Nonlinear Dynamics and Statistics*, Birkhauser, Boston, pp. 127–166.

Young, P.C. (2002). Advances in real-time flood forecasting, *Philos. T. R. Soc. A.*, **360**, 1433–1450.

Young, P.C. (2003). Top-down and data-based mechanistic modelling of rainfall flow dynamics at the catchment scale, *Hydrol. Process.*, **17**, 2195–2217.

Young, P.C. (2004). "Identification and Estimation of Continuous-Time Hydrological Models from Discrete-Time Data" in: Webb, B., Arnell, N., Onf, C. *et al.* (eds), *Hydrology: Science and Practice for the 21st Century*, Vol. 1, British Hydrological Society, London, pp. 406–413.

Young, P.C. (2005). "Rainfall-runoff Modeling: Transfer Function Models" in: Anderson, M.G. (ed.), *Encyclopedia of Hydrological Sciences*, Volume 3, Part II, Wiley, Hoboken, N.J., pp. 1985–2000.

Young, P.C. (2006). "Data-Based Mechanistic Modelling and River Flow Forecasting" in: *Proceedings 14th IFAC Symposium on System Identification*, Newcastle, Australia, pp. 756–761.

Young, P.C. (2008). The refined instrumental variable method: unified estimation of discrete and continuous-time transfer function models, *Journal Européen des Systèmes Automatisés*, **42**, 149–180.

Young, P.C. (2009). A unified approach to environmental systems modeling, *Stock. Env. Res. Risk A.*, **23**, 1037–1057.

Young, P.C. (2010a). "Real-Time Updating in Flood Forecasting and Warning" in: Pender, G.J. and Faulkner, H. (eds), *Flood Risk Science and Management*, Wiley-Blackwell, Oxford, pp. 163–195.

Young, P.C. (2010b). Gauss, Kalman and advances in recursive parameter estimation, *J. Forecast. (Special issue celebrating 50 years of the Kalman Filter)*, **30**, 104–146.

Young, P.C. (2011). *Recursive Estimation and Time-Series Analysis: An Introduction for Student and Practitioner*, Springer-Verlag, Berlin.

Young, P.C. (2013). Hypothetico-inductive data-based mechanistic modeling of hydrological systems, *Water Resour. Res.*, **49**, 2, 915–935.

Young, P.C. and Beven, K.J. (1994). Data-based mechanistic modelling and the rainfall-flow nonlinearity, *Environmetrics (Special Issue: Environmental Time Series Analysis)*, **5**, 335–363.

Young, P.C. and Garnier, H. (2006). Identification and estimation of continuous-time, data-based mechanistic models for environmental systems, *Environ. Modell. Softw.*, **21**, 1055–1072.

Young, P.C., McKenna, P. and Bruun, J. (2001). Identification of nonlinear stochastic systems by state dependent parameter estimation, *Int. J. Control.*, **74**, 1837–1857.

Young, P.C. and Pedregal, D.J. (1999). Recursive and en-bloc approaches to signal extraction, *J. App. Stat.*, **26**, 103–128.

Young, P.C., Romanowicz, R.J. and Beven, K.J. (2006). Updating algorithms in flood forecasting. Technical Report Report UR5, Flood Risk Management Research Consortium, U.K., December, 2006. Available at www.floodrisk.org.uk.

Young, P.C. and Tomlin, C. (2000). "Data-Based Mechanistic Modelling and Adaptive Flow Forecasting" in: Lees, M.J. and Walsh, P. (eds), *Flood Forecasting: What Does Current Research Offer the Practitioner?*, Centre for Ecology and Hydrology on behalf of British Hydrological Society, Wallingford, U.K. pp. 26–40.

CHAPTER 17

Uncertainty Estimation in Fluvial Flood Forecasting Applications

Kevin Sene
Atkins, UK

Albrecht H. Weerts
Deltares, Delft, Netherlands

Keith Beven
Lancaster Environment Centre,
Lancaster University, UK

Robert J. Moore
Centre for Ecology & Hydrology,
Wallingford, UK

Chris Whitlow
Edenvale Young, Bristol, UK

Stefan Laeger and Richard Cross
Environment Agency, Bristol, UK

17.1. Introduction

Flood warning is a well established way to mitigate the damage from flooding. Given sufficient lead-time, valuables and other property can be moved, flood resilience measures invoked (e.g. flood boards, sandbags) and people evacuated from areas at risk. Flood control structures can also be operated to reduce the risk from flooding, and emergency works performed to clear watercourses of debris, and to reinforce flood defences and embankments.

Flood warning is usually most effective when a long lead-time can be provided, and a flood forecast provides a way of extending the time available as well as increasing the accuracy of the warning, and reducing the false alarm rate. For river level and flow forecasting, the approaches used can range from simple correlation techniques through to complex integrated catchment models combining rainfall-runoff, flow routing and coastal forecasting components. However, there are many uncertainties associated with the generation of forecasts, and these are often characterised as arising from initialisation, modelling and forcing errors (e.g. Beven, 2009; Laeger *et al.*, 2010).

It has long been recognised that an estimate of the uncertainty in a flood forecast should be obtained to assist with operational decision making (e.g. Buizza, 2008; Krzysztofowicz, 2001; Schaake *et al.*, 2007) although with much debate about the extent to which this information should be disseminated. Some potential operational benefits include being right more often when issuing warnings, providing a more realistic representation of capabilities in communicating with other organisations and the public, and being more likely to bracket what actually happens in terms of flood warning threshold crossings and timings.

The resulting uncertainty estimates may be used qualitatively, such as through simple graphical displays, or in a more quantitative way, as inputs to a user's flood incident management procedures or decision-support system, perhaps using cost-loss and risk-based techniques to assist in decision making (e.g. Hall and Solomatine, 2008; Krzysztofowicz, 1999; Roulin, 2007; Todini, 2004). Options for communicating uncertainty include making the estimates available only to technical experts, such as flood forecasting duty officers, through to making them widely available to the public in radio and television broadcasts, and via the internet.

For real-time use, the operational requirement is usually to derive estimates of the forecast uncertainty from all sources, for the locations where these are required, which are often called "forecasting points". The two main options are (i) to assess and take account of forecast errors through comparison with real-time and/or historical observations and (ii) to make prior assumptions about the magnitude of individual sources of uncertainty (in rating curves, rainfall forecasts, etc.), and to propagate these estimates through the model cascade. Examples of techniques in the first category include probabilistic data assimilation and forecast calibration methods, whilst the second approach is often called forward uncertainty propagation (e.g. Beven, 2009; Pappenberger *et al.*, 2007).

Probabilistic data assimilation techniques make use of real-time observations to improve the forecast and, in addition to assessing the overall uncertainty in a forecast, also aim to reduce the uncertainty. One of the earliest operational applications was to estimate uncertainty for a forecasting system for the River Nith in Scotland, using a Data-Based Mechanistic (DBM) approach (Lees *et al.*, 1994). Other approaches include Ensemble Kalman Filtering (EKF) (Evensen, 1994) and Particle Filtering (PF) (Moradkhani *et al.*, 2005).

By contrast, the aim of (probabilistic) forecast calibration techniques is to make an off-line assessment of the uncertainty over a hindcasting period, by comparing previous forecasts with the subsequently observed river levels and flows. The resulting statistical model is then assumed to apply during real-time operation. Post-processing techniques of this type include Bayesian uncertainty processors (Krzysztofowicz and Herr, 2001; Krzysztofowicz and Kelly, 2000; Reggiani *et al.*, 2009; Reggiani and Weerts, 2008; Todini, 2008), quantile regression (Koenker, 2005; Weerts *et al.*, 2011), percentile matching (Wood and Schaake, 2008), hydrological model output statistics processors (Montanari and Brath, 2004; Seo, *et al.*, 2006), and Autoregressive Moving Average (ARMA) error prediction (Moore, 2007; Moore *et al.*, 2010). However, this is a developing area (Wilks and Hamill, 2007), and operational applications have only recently started to be implemented.

By contrast, simple forward uncertainty propagation techniques such as "what-if" approaches have been used by operational forecasters for decades; for example, to explore the sensitivity of forecasts to assumptions about future rainfall, and to other factors, such as flood defence failures and gate operations. These methods are also widely used off-line to explore model performance, and to assess the needs for improvements to models and the underlying data.

More sophisticated approaches have also been developed, such as the Ensemble Streamflow Prediction (ESP) technique which was introduced into operational use in the 1980s for forecasting river flows at medium to long lead-times (Day, 1985; Wood and Schaake, 2008). In its original form, this approach was based on sampling of individual years from the historical meteorological record, and propagating these values through the forecasting model. However, in more recent developments, the estimates are conditioned on categorical rainfall forecasts of the next month's rain (Moore *et al.*, 1989) or short-range rainfall forecasts and other information. Also, with the introduction of ensemble meteorological forecasting techniques in the 1990s, much of the focus in recent years in Europe, the US and elsewhere has been on the operational use of ensemble rainfall forecasts in improving flood forecasts at longer lead-times, and for flash flood events (e.g. Cloke and Pappenberger, 2009; Environment Agency, 2010a; Golding, 2009; Thielen *et al.*, 2009; Webster and Hoyos, 2004).

The current situation is therefore that there are many uncertainty estimation techniques potentially available or under development, which capture and/or reduce the overall uncertainty to various extents, and which perform best at different lead-times. Varying levels of investment may also be required in terms of staff time and software development, and possibly also in computing infrastructure to overcome model run-time constraints.

Here, initial work is described on developing a decision framework and guidelines to help potential users through these various choices, taking account of typical hydrological and operational constraints, such as flood warning lead-time requirements and model run-times. Some key design principles were that a risk-based approach should be used, matching the effort expended and accuracy of the technique to the level of flood risk, and that the approach should be generic and capable of being extended in the future as new techniques become available.

The overall aim is to help practitioners to select and implement appropriate real-time techniques which can add value to the forecasting process and are robust enough to use operationally (Laeger *et al.*, 2010). Also, the focus was on techniques for the generation of uncertainty estimates, rather than the communication of uncertainty. Whilst this is an important topic, it remains an active area for research (e.g. Demeritt *et al.*, 2007; Demuth *et al.*, 2007; Nobert *et al.*, 2010).

The resulting framework builds on similar guidelines developed for related areas, such as for deterministic flood forecasting (Environment Agency, 2002), and flood risk modelling generally (Beven *et al.*, 2010). Several examples are presented for application of the framework within the Environment Agency in England and Wales, for whom the original version was developed (see Laeger *et al.*, 2010), together with a number of uncertainty estimation techniques for integrated catchment forecasting models. The examples are included to illustrate key concepts, rather than necessarily being the recommended approaches to use in any given situation. A brief look to the future of probabilistic forecasting is also provided, together with some key priorities for research.

17.2. A Proposed Framework for Selecting Operational Uncertainty Estimation Techniques

17.2.1. *Operational requirements*

When estimating uncertainty in forecasts in real-time, two key questions are: (i) how will the information be used?; and (ii) what minimum lead-time is required for decision making?

Probabilistic information can be used in several ways, ranging from just a qualitative assessment, through to the requirement for calibrated probability distributions for input to a decision-support system (see Section 17.2.4). The level of effort required to develop techniques, and interpret outputs, varies between approaches, and should ideally be linked to the level of risk. For example, a more sophisticated approach is probably required for a tidal barrier, protecting many thousands of people from flood risk, than for isolated properties in a rural area.

Here, risk is defined as the combination of probability and consequence, where consequence is typically defined in terms of the number of properties at risk from flooding (although other choices are sometimes made, such as the likely economic damage). In the framework, the level of risk influences both the choice of technique and the extent of the analysis work performed in collating the information required to apply the framework. For example, for a low risk location, if run-times are likely to be an issue, then a simpler, faster technique might be used whereas, for a high risk location, additional investment in computer processing power might be found to be worthwhile, following extensive testing of prototype configurations to assess the impacts on system performance.

Table 17.1 provides an indication of the types of probabilistic outputs which might be required (Environment Agency, 2010b). Each approach requires a different level of detail in terms of the probabilistic information provided. Although it is difficult to generalise, the requirement for the probabilistic content of forecasts increases moving down the table, with higher risk applications requiring a more detailed approach.

Flood warning authorities are also increasingly adopting performance targets and levels of service agreements for the warning, forecasting and dissemination components of the overall flood warning service (e.g. Andryszweski *et al.*, 2005). For probabilistic forecasts, in addition to objectives for deterministic performance measures such as maximum false alarm rates, and the accuracy of peak levels, further measures might also

Table 17.1. Summary of typical operational requirements for probabilistic flood forecasts (adapted from Environment Agency, 2010b).

Category	Requirement	Description
Qualitative	Visualisation	"Eyeball" assessments of the spread of ensemble members with forecast lead-time, and relative to threshold values, and of other factors such as the clustering of ensembles.
	Persistence-based approaches	Techniques which compare the number of threshold exceedances between successive model runs.
Intermediate	Threshold-frequency approaches	Calibration of thresholds based on the historical model performance, obtained over a calibration period (e.g. based on flow return periods) (e.g. Reed *et al.*, 2007).
Quantitative	Physical-threshold approaches	As for threshold-frequency approaches but using actual threshold values in decision making (e.g. flood defence levels).
	Cost-loss approaches	Assessment of appropriate actions based on consideration of the economic value or utility of forecasts, and the optimum probability thresholds for taking action.
	Bayesian uncertainty processors (whole-system versions)	Similar to cost-loss approaches, considering the predictive uncertainty taking account of all information available up to the time of the forecast, and including economic and subjective views of flood warning decision criteria.

be introduced for the probabilistic component. Again, a higher standard of performance may be required for high risk locations. For example, some key measures which quantify different aspects of the forecast include the following items (e.g. Weerts *et al.*, Chapter 15 of this volume):

- *Reliability* — which is a measure of the agreement between the forecast probability and the observed frequency over many flood events (e.g. the forecast bias).
- *Resolution* — which indicates the ability of the forecast to discriminate between true events and true non-events among different events.
- *Sharpness* — which is a measure of the tendency to forecast with a concentration of large probabilities around some value, as opposed to small probabilities spread over a wide range of values.

Incremental improvements in existing measures might also be assessed; for example, the reduction in false alarm rates.

As with deterministic forecasts, the issue then arises of how to assess performance at the design stage, before a system has been implemented. However, often an initial estimate may be obtained from the performance of the same technique on a similar catchment or — for higher risk locations — by exploratory modelling using a pre-operational model configuration.

Another consideration is the lead-time requirement for flood warning. Many studies have shown that the effectiveness of flood warnings increases as the lead-time is increased, although eventually a point of diminishing returns is reached, and the risk of false alarms also increases (e.g. World Meteorological Organisation, 1994). Typically, the severity of warnings is escalated as the forecast time of the event approaches, ranging from an initial alert in the early stages, through to a full flood warning when flooding seems imminent. Also, longer lead-times may be required for higher risk locations; for example, where large numbers of properties need to be evacuated if flooding seems likely, decisions need to be made about deploying demountable defences, or critical infrastructure is at risk.

Lead-time requirements vary between organisations. For example, in the Environment Agency, a minimum lead-time of 2 hours is specified for flood warnings, where technically feasible, with longer lead-times for high risk locations (e.g. to trigger a Major Incident Plan) whilst Outlook Statements are currently provided for 2–5 days ahead. For small fast-response catchments, rainfall forecast inputs are typically required to achieve this flood warning lead-time whilst, for larger catchments, an adequate lead-time may be possible based on rainfall alone estimated from rain gauge or weather radar observations, or by routing flows from river gauging stations further upstream. However, for Outlook Statements, for the typical scale of catchments in the UK, forecasts usually rely solely on rainfall forecasts as the main forcing input at longer lead-times.

When translating these requirements into the minimum lead-time required (ideally) for flood forecasts, the additional time required for decision making and issuing the warning also needs to be considered (e.g. Carsell *et al.*, 2004; Environment Agency, 2002; Sene, 2008). Approaches to issuing warnings can include loud hailers, sirens, door knocking, telephone calls and — more recently — automated multi-media dissemination systems. The time taken to issue a warning, from taking the initial decision to issue a warning, to contacting the final person, varies between approaches, and often depends on the number of properties at risk.

17.3. Hydrological Constraints

Having defined the main operational requirements, in practice it may not always be possible to achieve these due to a combination of hydrological and operational constraints. The main hydrological constraints are:

- The catchment response time.
- The lead-time provided by meteorological forecasts (where used).

The catchment response time is the time taken for a rainfall input to translate into flows at the required forecasting point(s) in an integrated catchment model. Obviously this is not a fixed value, and can depend on factors such as antecedent conditions, the location in the catchment, rainfall depths and distributions, and artificial influences. However, the notion of a typical response time under flooding conditions is useful when deciding on the most appropriate uncertainty estimation techniques to use.

For meteorological forecasts, in addition to rainfall inputs, some other operational uses in flood forecasting can be for air temperature forecasts (for snowmelt forecasting), and air pressure and wind field forecasts (for coastal surge forecasting). For short- to medium-range forecasting, the current "state of the art" is to use high resolution non-hydrostatic Numerical Weather Prediction (NWP) models with a grid resolution as small as 1–2 km embedded in regional and global-scale models, combined with nowcasting models for the next few hours ahead (e.g. Golding, 2009; Roberts *et al.*, 2009).

Typically, at these length scales, forecast skill is exhibited for lead-times up to several days ahead, although obviously decreases with increasing lead-time and is significantly less for some types of event, such as convectively driven storms. Lower resolution forecasts may show skill weeks to months ahead in some parts of the world, although the model length scales used (typically 25–100 km at present) are less suitable for use with hydrological forecasting models except in very large catchments. Meteorological services therefore often only issue forecasts up to a fixed lead-time, which is presently typically 0–6 hours ahead for the nowcasting component, 2–5 days for the current generation of high-resolution local area forecasting models, but 10–15 days ahead or longer for ensemble forecasts derived at a global scale (e.g. Cloke and Pappenberger, 2009; Weerts *et al.*, Chapter 15 of this volume; Werner *et al.*, Chapter 19 of this volume).

When using meteorological forecasts in a flood forecasting model, the maximum forecast lead-time that can be provided consists of the sum of the catchment response time and the maximum lead-time for which meteorological forecasts are available (or are considered to add value).

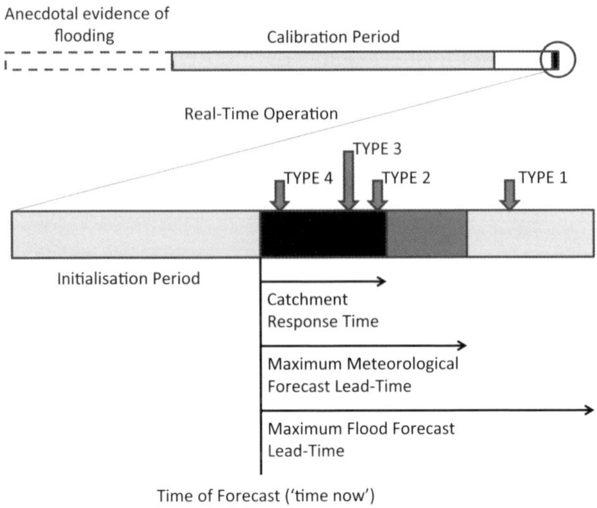

Figure 17.1. Illustration of some key timescales in real-time hydrological forecasting.

The catchment response time is also an important factor when considering the value of data assimilation since, at longer lead-times, the information content of the observed river levels or flows diminishes, reducing the value of this approach.

Figure 17.1 summarises these various timescales and lead-times, together with a number of other timescales which are important to the forecasting process.

The uppermost bar shows the time available for model operation in real-time, relative to an example period of historical data for calibration of the techniques. Normally some time elapses between calibrating the technique and using it operationally, as shown by the white bar. There may also be anecdotal evidence from before the start of observations, which can sometimes be used during model development as part of the model validation process, such as evidence of peak water levels or flood extents in major historical flood events.

The inset shows a breakdown of some typical times for real-time operation, with the maximum flood forecast lead-time possible consisting of the sum of the hydrological response time and the lead-time provided by the meteorological forecast. Some probabilistic techniques, such as Bayesian model averaging (see Section 17.6.3), also require an initialisation or training period of recent observed data during real-time operation.

In some organisations, a "climatological" forecast may also be generated for longer lead-times — for example assuming zero rainfall or mean values for the time of year — although this is not shown in the figure.

The figure also shows four catchment scenarios for the lead-time requirement for flood warnings, indicated as Types 1, 2, 3 and 4. For a Type 1 catchment, the required flood warning lead-time can only be obtained by using a flood forecast which relies solely on meteorological forecasts. This might be the situation for a small fast response catchment, for example, or for a large catchment in which a very long lead-time is ideally required for issuing warnings (e.g. for evacuating large numbers of properties). By contrast, for the Type 2 situation, the forecast relies mainly on rainfall observations by rain gauge, weather radar or satellite. Types 3 and 4 place even less reliance on rainfall data, and correspond to larger catchments in which the required lead-time can be achieved simply by routing flows from river gauging station observations further upstream (for Type 4 catchments).

These catchment types are a development of those proposed by Lettenmaier and Wood (1993) and — for the Environment Agency situation of relatively short flood warning lead-time requirements — it proved convenient to adopt the catchment type classifications shown in Table 17.2 (Environment Agency, 2010b). Here, a distinction is made between the hydrological (catchment runoff) and hydraulic (river channel flow routing) response times.

The advantage of introducing this classification scheme is that, in the absence of other information (for example, performance statistics for an existing model), the catchment type can be linked to the main types of errors which cause uncertainties in a forecasting model, which are (Environment Agency, 2010b):

- *Initialisation Errors* — due to errors in the observations used to estimate precipitation; potential evaporation and temperature; discharge or other boundary conditions used in calculating the model states in historical mode of operation to initialise a forecast into the future (and which will normally be updated as new observations are received).
- *Modelling Errors* — arising from an uncertain model structure and its parameters, model resolution, uncertain river control structure operating/management rules etc.
- *Forcing Errors* — errors which occur in the forecast-mode of operation when a model component is forced with an input derived from another

Table 17.2. Links between lead-time requirements and catchment response (adapted from Lettenmaier and Wood, 1993).

Type	Catchment	Criterion	Description and key forcing inputs for flood warnings
1	Very fast responding basins	$T_{warning} \gg T_{total}$	The desired lead-time is such that the warning or outlook must be issued on the basis of water that has not yet fallen as rain. In this case a rainfall forecast is the only means to provide a timely warning when using a flood forecasting model.
2	Small-to-medium-size basins	$T_{warning} < T_{total}$ and $T_{catchment} \gg T_{river}$.	The warning or outlook will be issued on the basis of water that is already in the catchment and is mainly determined by the hydrological travel time.
3	Medium-size basins	$T_{warning} < T_{total}$ and $T_{catchment} \sim T_{river}$	The warning or outlook will be issued on the basis of water that is already in the catchment and river and the response time is determined by the hydrological response time and the hydraulic response time.
4	Large river basin	$T_{warning} < T_{river}$ or $T_{catchment} \ll T_{river}$	The warning or outlook will be issued on the basis of water that is already in the main channel; or the hydrological response time is insignificant compared to the hydraulic response time.
5	Coastal/tidal zone	$T_{warning} \gg T_{surge}$	The desired lead-time is such that the warning or outlook may be issued on the basis of wind conditions that have not yet occurred. In this case wind and pressure forecasts are necessary for a timely warning.

$T_{warning}$ = desired warning time, T_{total} = total response time, T_{river} = travel time through main river, $T_{catchment}$ = hydrological response time, T_{surge} = coastal forecast lead-time

model with its own initialisation and modelling errors; for example, a NWP model or the hydrological or flow routing outputs at a flow forecasting point further upstream.

Note that, whilst this classification is often used, in practice the initialisation and modelling errors are often inter-related; in particular, the initial conditions for a model run are often derived from a previous model run rather than, say, from direct observations of soil moisture.

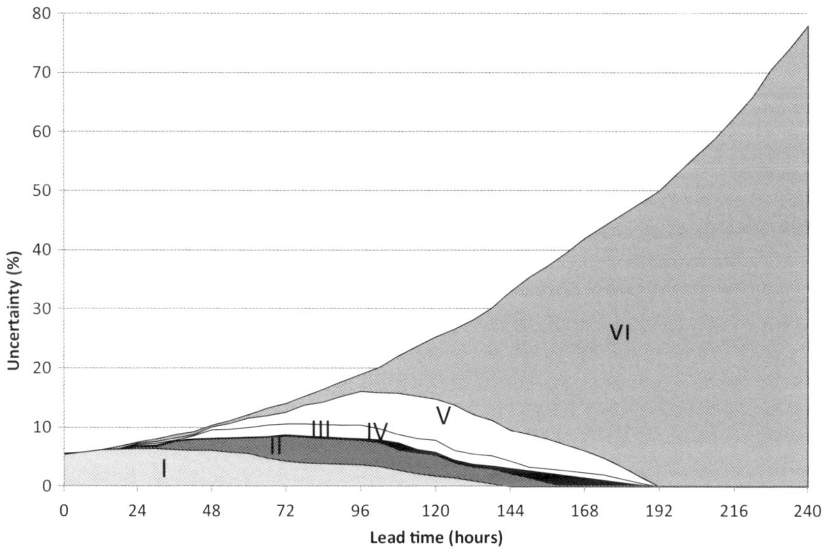

Figure 17.2. Example of the contributions to total uncertainty as a function of forecast lead-time for a specific downstream location in a large river. Key to sources of uncertainty: I = rating curve(s) for upstream station(s), II = flow routing or hydrodynamic component, III = runoff from major tributaries and headwaters, IV = initial soil moisture conditions, V = areal rainfall estimates from observations (rain gauges, weather radar), VI = rainfall forecasts (adapted from Werner *et al.*, Chapter 19 of this volume).

Figure 17.2 illustrates how the main sources of error in a flood forecast can vary with the forecast lead-time. The example is for a forecasting point at the downstream end of a large river catchment, and shows how there is a transition from initialisation and modelling errors to forcing errors as the lead-time increases. Of course, a similar figure for a small, fast response catchment might show almost complete reliance on rainfall forecasts even at short lead-times.

For this example, Figure 17.3 illustrates how these different sources of uncertainty vary in both magnitude and type with a selection of increasing lead-times. Although these results are specific to the catchment and forecasting model considered, plots of this type can assist with focussing effort on which methods to use, and the sources of uncertainty to consider.

Table 17.3 summarises how initialisation, modelling and forcing errors typically arise for each of the five catchment types defined in Table 17.2. This type of assessment can provide a focus for off-line forward uncertainty propagation analyses to assist with model development and — as discussed

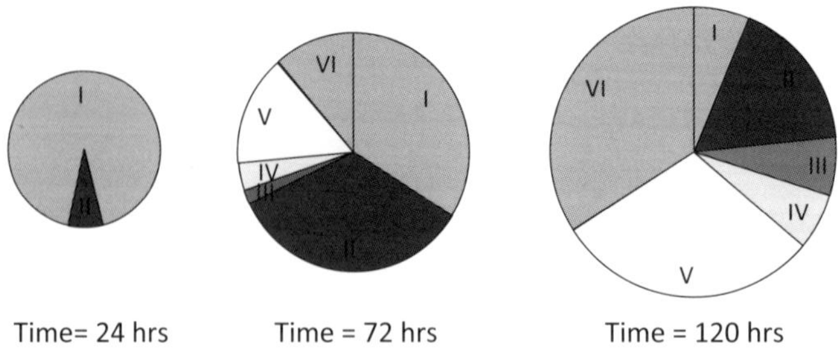

Figure 17.3. Illustration of the main sources and magnitude of uncertainty for three lead-times, for the example shown in Figure 17.2 (key as for Figure 17.2).

later — can also help with selecting an appropriate uncertainty estimation technique for real-time use.

17.4. Operational Constraints

A number of operational constraints may also limit the choice of uncertainty estimation technique. For probabilistic forecasting applications, these typically relate to the time and budget available, data volumes and record lengths, and model run-time issues.

As in other areas of flood risk management, the first of these items is often linked to the level of risk. For example, for higher risk locations, a higher investment might be justified, as shown by a formal cost–benefit analysis, based on the estimated cost of implementation and the likely damages which could be avoided through improved forecasts. However, it is worth noting that, although flood warning benefit estimation techniques are well-established for deterministic forecasts (e.g. Carsell *et al.*, 2004; Environment Agency, 2002), techniques for estimating the additional incremental benefits from probabilistic forecasts are still under development and an active area for research.

The issue of data volumes and record lengths is one which, to some extent, can be addressed through additional computing power, and other operational measures, such as collecting more data. Data volumes are most likely to be an issue when using grid-based forcing inputs, such as ensemble rainfall forecasts, and may limit the amount of data which can be stored online. This can influence the length of the initialisation period which can be provided for some techniques, and can also cause bottlenecks in data transfer with potential run-time issues.

The record length requirements for calibrating probabilistic techniques are in many ways similar to those required for deterministic forecasting

Table 17.3. Dominating uncertainties for each forecasting situation (Environment Agency, 2010b).

Type	Catchment	Source
1	Very fast responding basins	**Initialisation Errors**: — Past Areal Mean Rainfall — Potential Evaporation **Modelling Errors**: — Rainfall-Runoff Model Parameters — Rainfall-Runoff Model Structure **Forcing Errors:** — Rainfall Forecasts
2	Small-to-medium-size basins	**Initialisation Errors**: — Past Areal Mean Rainfall — Potential Evaporation **Modelling Errors**: — Rainfall-Runoff Model Parameters — Rainfall-Runoff Model Structure **Forcing Errors:** — Rainfall Forecasts (for longer lead-times/outlook statements)
3	Medium-size basins	A mixture/combination of 2 and 4
4	Large river basins	**Initialisation Errors**: — High Flow Ratings — Ungauged Lateral Inflows — Tidal Boundary **Modelling Errors**: — Hydraulic/Routing Model Parameters — River Channel/Floodplain Survey — River Control Structures **Forcing Errors:** — High Flow Ratings (forecast inflows) — Forecast Tidal Boundary — Forecast Lateral and other Inflows (for longer lead-times/outlook statements) — Rainfall Forecast (for even longer lead-times/outlook statements)
5	Coastal/tidal zone	**Initialisation Errors**: — Water levels in the coastal zone **Modelling Errors**: — Bathymetry — Model domain/resolution **Forcing Errors:** — High Flow Ratings (forecast inflows/levels) — Boundary Conditions — Wind and Pressure Forecast

techniques. Ideally, the observational record should include one or more extreme events to provide reassurance that the model can cope with low probability events although, due to limited record lengths, often this is not possible (hence the key role of data assimilation in helping to cope with the unexpected, as discussed later).

Calibration records should also be homogeneous, in the sense of correcting for, or excluding, any periods in which there were changes in instrumentation, catchment characteristics, model components, or the forcing inputs used to operate the model. Whilst this is generally straightforward to achieve for the hydrological component, if deterministic or ensemble meteorological forecasts are used, then hindcasting for this component can be a considerable undertaking, which is best performed by the meteorological service providing the information (e.g. Uppala *et al.*, 2005). Alternatively, for the ensemble forecasting component, an older form of the current operational NWP model might be used instead, alongside the latest operational model for the deterministic component. For example, modern high-end personal computers are now fast enough to operate some models from 5–10 years ago in real-time, allowing hindcasting exercises to be performed quickly and easily, at the expense of some reduction in performance compared to using the latest operational models.

However, of the various operational constraints, model run-times are perhaps the main concern, since for some techniques multiple model runs may be required for each forecast that is issued, placing an additional load on the forecasting system. This may also limit the number of ensemble members which can be considered (leading to sampling issues), or may even mean that a probabilistic forecast cannot be obtained in an operationally useful time. The time available for model runs typically consists of the interval between forecast runs, less any time for pre- and post-processing of inputs and outputs. For some forecasting services, model runs might only be performed hourly, or less, whereas, in the case of the Environment Agency for example, forecast runs are performed every 15 minutes in the time leading up to, and immediately following, a flood event. However, it is also important to note that there are a number of probabilistic forecasting techniques for which the additional run-time requirement is minimal, compared to that of the underlying, deterministic flood forecasting model: several examples are provided later.

The most likely situations in which forecast times may become prohibitive are with forward uncertainty propagation techniques for models that include a hydrodynamic component, and for some ensemble data assimilation and hydrological uncertainty processing techniques. In this situation, some approaches to reducing run-times can include (Laeger *et al.*, 2010):

- Computational improvements, e.g. parallel processing, faster processors.
- Model configuration changes, e.g. nested models, model simplification or rationalisation.
- Statistical approaches, e.g. sampling or grouping of ensembles.
- Model emulators, e.g. simpler models to emulate the behaviour of more complex models.

Typically, computational improvements would normally only be considered for high-risk locations, or performed as part of an organisational upgrade to computer systems.

Model configuration changes are a cheaper option and, in Environment Agency practice, for example, deterministic hydrodynamic models are usually required to complete a model run of a specified duration, on a specified computer processor, within a time of less than six minutes. To achieve this it is usual to simplify the model using a coarser discretisation in areas away from the forecasting point(s) of interest, and to make other changes which improve the model stability and convergence. Run-time reductions of 1–2 orders of magnitude are sometimes possible without sacrificing model performance at the key forecasting points (e.g. Chen *et al.*, 2005). However, if multiple model runs are required, the overall time to derive a forecast may still be prohibitive (depending on the overall forecast run interval).

The third option, of statistical sampling approaches, is one which has been considered extensively in meteorology, and more recently in ensemble hydrological forecasting (e.g. Cloke and Pappenberger, 2009). However, results tend to be model and situation specific, requiring extensive testing of the impact of reduced sample sizes on the overall model outputs.

The final option, of emulators, requires development of a simpler, faster running model to emulate the behaviour of the more complex model at specific cross-sections of interest in flood warning. The objective is to generate a compact, parametrically efficient model to recreate the behaviour of the parent distributed model. The resulting emulation can be very accurate and fast for a wide range of flow conditions (e.g. Beven *et al.*, 2008) while still allowing a mapping of level or flow forecasts back to physical cross-section data to provide real-time inundation estimates.

Of course, emulation of this type depends on the accuracy of the original hydrodynamic model. Where the hydrodynamic model is used to simulate the effects of different control strategies, or complex backwater effects, different emulators might be needed for different conditions. The state-space form of the emulator is readily incorporated into existing Kalman filter-based data assimilation algorithms such as those described in Romanowicz *et al.* (2006, 2008), if level data can be made available at the emulation site.

17.5. Decision Tree

Having defined the main requirements and constraints, the overall framework is summarised in a series of key questions and worksheets, supported by a range of flowcharts and tables. This section presents several examples of these items for application to flood forecasting in real-time. The full framework is discussed in Laeger *et al.* (2010) and Environment Agency (2010b), which also considers off-line applications of uncertainty estimation techniques.

The framework is based around the following overall classification of techniques:

- *Forward Uncertainty Propagation* — methods which seek to estimate the uncertainty from individual sources requiring prior assumptions based on experience and/or analysis of historical data.
- *Probabilistic Data Assimilation* — methods which use real-time observations to improve a forecast, including providing an estimate of uncertainty.
- *Probabilistic Forecast Calibration* — methods which aim to calibrate the probabilistic content of forecasts by evaluation of the recent or historical performance over a hindcasting period (sometimes called conditioning, or statistical post-processing).

For example, Table 17.4 shows suggestions for the most appropriate general category of technique to use for some of the operational requirements summarised in Table 17.1.

Given ongoing research on this topic, this table in particular is intended just as an initial guide on choice of techniques, and it is anticipated that it will be improved and extended in future issues of the framework. Also, it is worth noting that the categories of probabilistic data assimilation and forecast calibration — although convenient to use here — sometimes

Table 17.4. Initial indication of uncertainty estimation requirements for several approaches to decision making using probabilistic forecasts (adapted from Environment Agency, 2010b).

Operational Requirement	Forward Uncertainty Propagation	Probabilistic Data Assimilation	Probabilistic Forecast Calibration
Visualisation	Sensitivity tests may be sufficient	Optional	Not required
Persistence-based approaches	Sensitivity tests may be sufficient	Optional	Not required
Threshold-frequency approaches	Sensitivity tests may be sufficient	Recommended	Recommended
Physical-threshold approaches	Sensitivity tests may be sufficient	Essential	Essential
Cost-loss approaches	Sampling from probability density function or physically-based reasoning	Essential	Essential

mean different things to different people (with alternative names including updating and conditioning, sometimes with qualifiers such as "...with respect to historical data").

Another point to note is that, for some forecasting situations, a combination of data assimilation and forecast calibration techniques may be useful. This takes advantage of the reductions in uncertainty provided by data assimilation techniques for lead-times within the catchment response time, whilst still providing an overall estimate of uncertainty at longer lead-times using forecast calibration techniques (see Figure 17.1).

To provide an overall summary of the technique selection process, Figure 17.4 presents a decision tree summarising the main choices between approaches, whilst taking account of the main hydrological and operational constraints. The grey shaded area indicates forward uncertainty propagation techniques, for which perhaps the most commonly considered sources of uncertainty in integrated catchment models are those illustrated, i.e. rainfall inputs (e.g. rain gauges, weather radar), rainfall-runoff model parameters, rating curves and ensemble rainfall forecasts. For illustration, two possible model run-time reduction approaches are shown: these might also need to be considered for some probabilistic data assimilation and forecast calibration techniques (although this is not illustrated here).

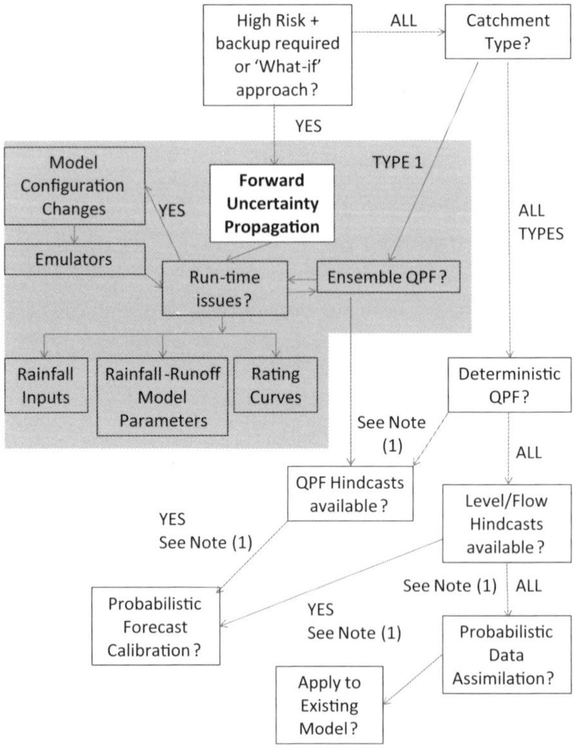

Figure 17.4. Example of a decision tree providing an initial guide to the selection of techniques. Note (1) — depends on the operational requirement, including the level of risk (QPF = Quantitative Precipitation Forecast).

Table 17.5 also provides an overall summary of techniques which might be used for a range of potential applications. Of course, it is not possible to summarise all aspects of a complex decision-making process in a single diagram or table, but this is at least a starting point for deciding on which techniques should be considered in each situation.

The framework is intended to be adapted and improved as further information becomes available. Also, the discussion so far has been mainly in terms of the generic classifications of forward uncertainty propagation, data assimilation and forecast calibration techniques. The following section describes examples of specific techniques which might be included in the framework for each of these categories, together with the rationale for some of the decisions which are shown in Figure 17.4. The focus is on real-time uncertainty estimation techniques, rather than off-line techniques.

Table 17.5. Suggestions for techniques which might be used for different applications (adapted from Environment Agency, 2010b).

Potential Application	Description
Qualitative assessments of uncertainty	Inclusion of plumes, spaghetti plots and other graphical ways of displaying uncertainty in the outputs to flood forecasting and warning duty officers. These estimates could be derived using data assimilation and/or forecast calibration techniques, depending on the operational requirement and lead-time, and/or forward uncertainty propagation of ensemble rainfall forecasts.
Resilience to telemetry failure	Provision of backup uncertainty estimation techniques in real-time in case of telemetry failure at one or more forecasting points where data assimilation is used, typically for high-risk locations. By definition, the absence of real-time data means that forward uncertainty propagation techniques are used. Typically this would be based on the latest update of total error before failure; however, that uncertainty may expand rapidly in reach-to-reach model cascades if there are multiple failures but will be constrained again as soon as there is a site with working telemetry.
Reduction of uncertainty	Use of real-time data to both constrain uncertainty, and to provide estimates of uncertainty, at a range of lead-times, for use in operational decision making, and possibly decision-support systems/cost-loss approaches, etc. This requires the use of data assimilation techniques, possibly combined with forecast calibration techniques for higher risk applications, and at longer lead-times.
High risk locations	For obtaining calibrated probabilistic forecasts suitable for input to decision-support systems, cost-loss approaches, and other types of operational decision making, both data assimilation and forecast calibration techniques may be required, depending on the level of risk, and the operational requirement.
Flash flooding applications	For providing flood warnings at lead-times beyond the hydrological response time of the catchment, estimates of uncertainty can be derived using data assimilation and/or forecast calibration techniques for lead-times up to the response time, then ensemble rainfall forecasts or deterministic rainfall forecasts with a forecast calibration technique.
Outlook statements	As for flash flooding, for long lead-times, beyond the hydrological response time of the catchment, this requires the use of forecast calibration techniques, and/or ensemble rainfall forecasts, and could involve data assimilation and/or forecast calibration techniques at shorter lead-times, depending on the level of risk, model run-times and the operational requirement.

(*Continued*)

Table 17.5. (*Continued*)

Potential Application	Description
Probabilistic inundation mapping	Mapping flood depths and extent in real-time with estimates of uncertainty, based on estimates for levels both at forecasting points, and intermediate locations (e.g. at nodes in hydrodynamic models). The use of emulators is an option where full hydrodynamic models are computationally prohibitive for real-time forecasting using multiple/ensemble model runs (as an alternative to reduced complexity models), particularly where forecasts are only required at cross-sections of particular interest for flood warning purposes.
Rainfall-level models	Calibration and implementation of new types of models, such as rainfall-level models (with estimates of uncertainty) without introducing uncertainty through the rating curve in estimating discharge.
Multi-model ensembles	Use of ensembles consisting of the outputs from several types of model to investigate model structural issues together with techniques such as Bayesian model averaging.

17.6. Examples of Uncertainty Estimation Techniques

17.6.1. *Forward uncertainty propagation*

In forward uncertainty propagation techniques, prior assumptions are made about the nature of the individual sources of initialisation, modelling and forcing errors in a cascade of flood forecasting models.

Some examples might include uncertainties in the catchment averaging scheme for rainfall data (when rain gauges are used), rating curves at high flows and individual parameter values in a model, such as the roughness coefficients in a hydrodynamic model. Model structural errors might also be considered. Depending on the sources of uncertainty considered, sampling techniques can range from qualitative estimates, such as assuming a likely range of values, through to Monte Carlo, Bayesian, fuzzy set, multi-model and other approaches (e.g. Beven, 2009; Beven, Chapter 3 of this volume).

Although an important tool for off-line use, such as for investigating model performance issues, for real-time use, some significant limitations of this approach can include:

- The run-times can quickly become prohibitive if large sample sizes are required and/or the model includes a hydrodynamic component.

- The resulting uncertainty estimates and/or probability distributions are necessarily based on the prior assumptions, and usually will not account for all sources of uncertainty, and the interactions between them.

The issue of parameter interdependence can be particularly important, since for many types of model the individual parameters, or sets of parameters, cannot be considered in isolation; for example, for the parameters which define the surface stores in a rainfall-runoff model. Where there is some understanding of the degree of independence, this can be estimated by multivariate sampling schemes such as Latin hypercube or Copula sampling (Beven, 2009).

So, although forward uncertainty propagation can be a useful technique for off-line use, for real-time use there are more powerful data assimilation and forecast calibration techniques which provide estimates for the overall uncertainty, and avoid these potential run-time and sampling issues (see Sections 17.6.2 and 17.6.3). However, the approach is well-suited for real-time use in some situations, including the following two applications:

- In situations where there is an obvious dominating source of uncertainty, such as when using rainfall forecasts at long lead-times, or for flash flood situations.
- As a back-up to data assimilation and other techniques which rely on a feed of real-time data, in the event of telemetry or instrumentation failure.

In the first example, the main value is for the use of ensemble rainfall forecasts at lead-times beyond the catchment response time. Here, the effectiveness of probabilistic data assimilation approaches reduces, and typically the overall uncertainty is dominated by the uncertainty in rainfall forecasts. Of course, ensemble meteorological forecasts themselves are normally generated as part of a sampling process for initial conditions with the inherent uncertainties and limitations that this causes, such as dependence on prior assumptions, partial sampling of overall uncertainty etc. Due to computational limitations, there are also necessarily limitations on the number of ensemble members which can be provided. From a hydrological forecaster's perspective, the sampling approach is normally decided by the meteorological service and — as for the maximum lead-time available — is not a factor which can be changed when using a specific meteorological forecasting product. However, it is worth noting that this too is an active area for research, with the latest operational

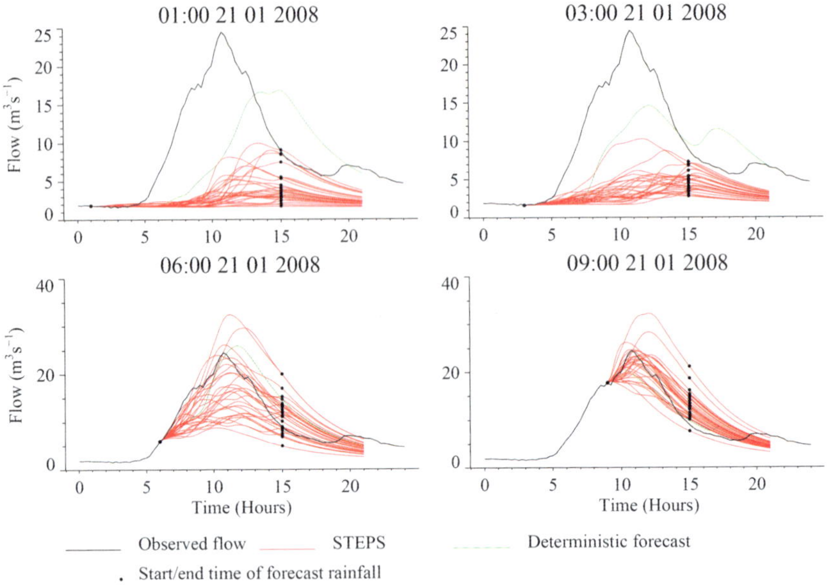

Figure 17.5. Example of propagation of STEPS ensemble rainfall forecasts through a PDM rainfall-runoff model for the Upper Calder catchment to the town of Todmorden. Note that, for this example, the STEPS forecasts ran to 15:00 on 21 January 2008, as indicated by the black dots, with a further 6 hours of zero rainfall appended to each (Environment Agency, 2010b).

numerical weather prediction models allowing for uncertainty both in initial conditions and model parameters and grid resolution. Also, some nowcasting approaches for short-range forecasts are now being formulated within an overall stochastic framework to provide ensemble nowcasts, as illustrated in Figure 17.5.

The figure shows an example of the propagation of a 30-member ensemble rainfall forecast through a rainfall-runoff model using the STEPS nowcasting approach (Bowler *et al.*, 2006). Multi-model NWP ensembles are also increasingly used for medium- to long-range meteorological forecasting.

The second application — as a backup to more sophisticated techniques — might be used in situations where it is essential to provide an estimate of uncertainty, even in the event of telemetry or instrumentation failures. In this case, data assimilation techniques which rely on those real-time feeds of data may fail. One example might be when uncertainty estimates are used as inputs to a decision-support system for a high-risk location, such as a tidal barrier. However, even in this case, as described in the following section, it is worth noting that some probabilistic data assimilation techniques can continue to provide uncertainty bounds which reflect the lack of real-time data, and so may also be useful in this situation.

17.6.2. *Probabilistic data assimilation*

Data assimilation techniques are widely used to improve the accuracy of forecasting model outputs, and deterministic approaches include input updating, state updating, parameter updating and output updating (e.g. Moore, 1999; Refsgaard, 1997; Serban and Askew, 1991).

For some approaches, uncertainty estimates can be derived as part of the assimilation process. The term probabilistic data assimilation is used here for approaches of this type, which typically involve some form of filtering scheme. These include Kalman filtering approaches (Kalman, 1960), including extended and ensemble versions, and particle filters.

Figure 17.6 shows some possible locations in an integrated catchment model at which different interventions can be made to reduce and quantify forecast uncertainty.

As for deterministic data assimilation approaches, the great value of probabilistic approaches is the ability to adapt to new, unexpected

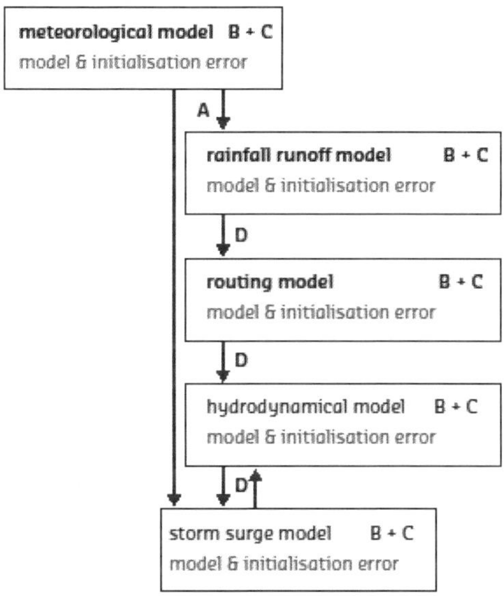

Figure 17.6. Illustration of the sources of errors, and possible types of improvement, in an end-to-end integrated catchment flood forecasting model (A = forecast calibration, B = state updating, C = parameter updating, D = output updating and/or forecast calibration) (Environment Agency, 2010b).

situations as they arise, as reflected by the real-time data. In particular, from an operational perspective, the next flood event is the most important, and will usually be different from all previously observed events, due to factors such as:

- Differences in the rainfall pattern and depths around the catchment.
- Rainfall/radar observation uncertainty.
- Rainfall forecast uncertainty.
- Antecedent condition and runoff generation uncertainty.
- River channel effects such as roughness/velocity uncertainty and rating curve uncertainty.
- Event-specific factors, such as flood defence failures and telemetry failures.

The use of real-time data (data assimilation) is therefore important in updating model outputs to allow — to some extent — for unforeseen factors as they occur, whilst noting that the information content of observed levels and flows decreases with increasing lead-time, particularly for lead-times beyond the catchment response time.

For probabilistic approaches, real-time data can be used in three main ways:

- Use a probabilistic data assimilation technique which generates estimates of uncertainty as part of the assimilation process, as an addition to an existing deterministic forecasting model.
- Use a forecasting model capable of both modelling river levels or flows and updating values in real-time, with estimates of uncertainty.
- Apply a deterministic data assimilation approach and then use a forecast calibration technique to estimate the remaining uncertainty in forecasts (see Section 17.6.3).

Ensemble Kalman filtering provides one example of a technique which can be used with an existing model (Burgers *et al.*, 1998; Butts *et al.*, 2005; Evensen, 1994; Weerts and El Sarafy, 2006; Weerts *et al.*, Chapter 15 of this volume). The method has the advantage of making few assumptions about the nature of the relationship between inputs and outputs, but for operational use there are two important factors to consider:

- As for forward uncertainty propagation techniques, the probabilistic interpretation of the outputs depends strongly on the validity of the prior assumptions, and whether all sources of uncertainty and their interactions are included.

- There may be run-time limitations on the number of ensemble members which can be considered, again affecting the probabilistic interpretation of outputs due to the reduced sample size.

Particle filtering approaches (Moradkhani *et al.*, 2005) require fewer prior assumptions, but may still have run-time constraints, and may be more sensitive to assumptions about the sources of model and observation errors (Weerts and El Sarafy, 2006).

One simpler approach which avoids both of these potential problems is to use a method which considers total uncertainty, at the forecasting points of interest, and only requires a single run of the underlying deterministic model.

For example, in the adaptive gain approach (Lees *et al.*, 1994), the basis of the method is to apply a scalar multiplicative gain to correct the model forecast at specific locations where observed data are available. The gain is presumed to evolve stochastically, for example as a random walk. The change in gain over time can then be evaluated using a Kalman filter, with the gain being updated as pairs of observations and model predictions become available. Forecasts are generated by multiplying the deterministic forecast of the hydrological or hydraulic model by the forecast value of the gain. Since the forecasting of the gain is done in a stochastic rather than deterministic fashion, uncertainty bounds on the prediction can be provided.

In addition, in the event of failure of an instrument or telemetry feed, an estimate for the uncertainty is still provided although, as illustrated in Figure 17.7, the bounds widen out for the time that the data feed is lost. The example is for a trial application of the approach to estimating the uncertainty in river level forecasts at the Linstock gauging station in the River Eden catchment in England for a flood event in January 2005 when there were several "drop-outs" in the data available for this gauge.

This approach is one component in the DBM modelling methodology (Beven *et al.* 2011; Leedal *et al.*, 2008; Lees *et al.*, 1994; Romanowicz *et al.*, 2006, 2008; Smith *et al.*, 2012; Young, 2002; Young *et al.*, Chapter 16 of this volume). The full DBM methodology provides an example of the use of a stochastic model, capable of both providing a forecast, and an estimate for the probability distribution of levels or flows.

The basic elements of a DBM model are a non-linear transformation of the input time series which feeds into a linear transfer function that

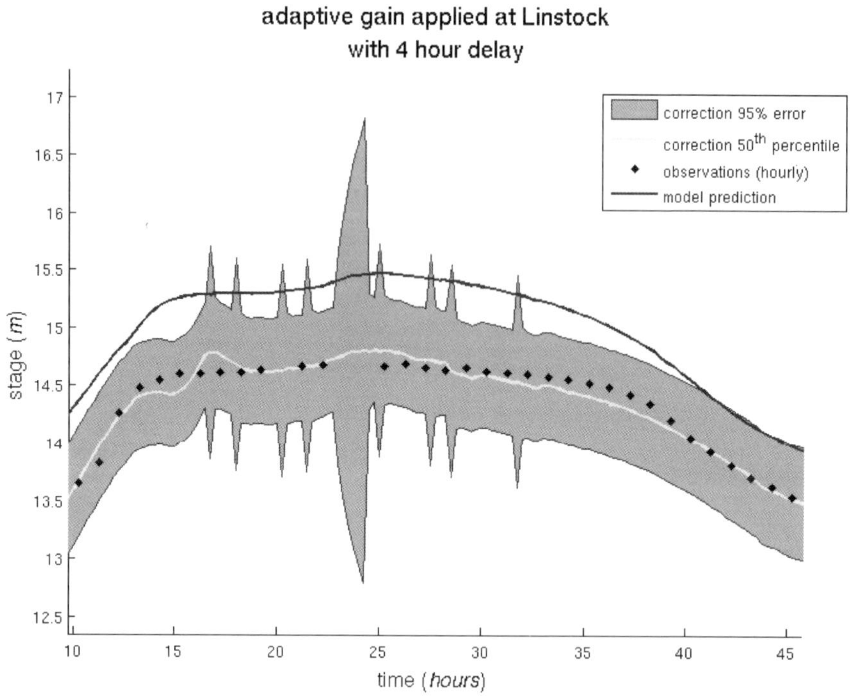

Figure 17.7. Adaptive Gain applied to outputs from ISIS model for Linstock, River Eden. Solid line: hydrodynamic model forecasts. Dots: hourly observed water levels. Grey: 95% prediction quantiles for 4 hour-ahead forecasts.

controls the distribution of the predicted outputs in time. In a cascade of model components where observations are available in real-time at a number of sites, the propagation of uncertainty through the cascade can be constrained by the use of data assimilation using a modified Kalman filter technique.

The resulting model might be used as the primary forecasting approach, or as part of a multi-model ensemble, combining several modelling approaches, as described in the next section. One particular advantage of the DBM approach is that relationships can be derived directly between rainfall and levels, avoiding the usual need for a valid high flow rating equation when developing rainfall-runoff models. Figure 17.8 shows an example of the 95% confidence limits derived using the DBM approach for another location in the River Eden catchment, this time for a high flow event in January 2009.

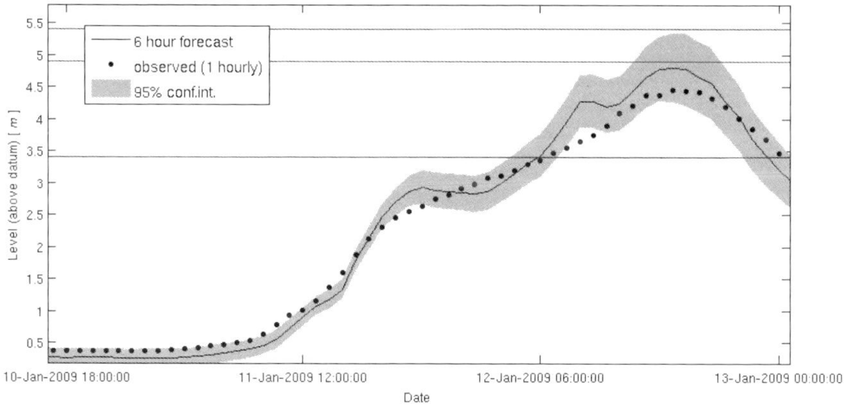

Figure 17.8. Example of outputs from a DBM model cascade for the River Eden, January 2009. Continuous plot of six hour ahead forecasts at Sheepmount using hourly time-steps and updating.

17.6.3. *Probabilistic forecast calibration*

Probabilistic forecast calibration techniques aim to assess the uncertainty in model forecasts, over a hindcasting period. Post-processing techniques of this type are often an addition to an existing model, and so are quick to operate since they do not require multiple model runs. They are often developed as a purely statistical "catch-all" (ensemble) post-processor, in order to reduce and account for the integrated hydrologic uncertainty (Seo *et al.*, 2006).

Quantile regression (Koenker, 2005) provides one example of this approach. Here, the uncertainty is assessed by applying a statistical model which has been calibrated off-line, typically using several years or more of observations and hindcasts. The basis of the method is to derive regression relationships for the transformed forecast errors, for a user-selected set of quantile values (e.g. 5%, 25%, 50%, 75% and 95%) and lead-times (e.g. 1, 2, 3, 6, 12, 24 and 48-hours ahead). Various formulations can be considered, and Figure 17.9 shows an example of the output for one case study, in which the best results were obtained when the transformed forecast errors were conditioned on the value of the predicted water level or flow (Weerts *et al.*, 2011).

In this example, the forecast extends to 48-hours ahead, and at the longer lead-times relies on rainfall forecasts derived from a deterministic NWP model. To derive the quantile regression model, an archive of four years of rainfall forecasts was used, in addition to archived values

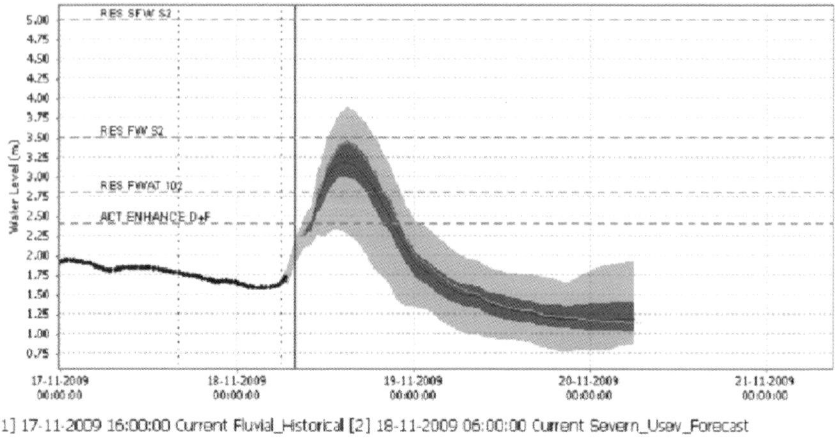

Figure 17.9. Example of quantile regression outputs for the Welshbridge forecasting location in the Upper Severn catchment during November 2009.

for weather radar, rain gauge, river level and river flow data over that period.

As indicated earlier, an alternative approach to deriving uncertainty estimates at longer lead-times is to use ensemble rainfall forecasts. In principle the use of real-time ensembles should be better able to cope with the unexpected, such as previously unobserved rainfall distributions and intensities, compared to the use of uncertainty estimates calibrated on a historical period. Ideally, rather than using the raw outputs (as provided by the meteorological service), calibrated estimates would be used. However, this remains an active research area, and the raw outputs continue to be used in most hydrological forecasting applications to date (e.g. Weerts *et al.*, Chapter 15 of this volume; Wilks and Hamill, 2007). Note also that, at long lead-times, most well-designed ensembles will tend towards a climatological estimate of future conditions.

One other consideration is that, due to computer processing limitations, the area resolution of ensemble meteorological forecasts is usually an order of magnitude less than for deterministic forecasts, although this gap continues to narrow. For example, a typical situation is that, for a grid resolution of 1 km for the deterministic product, the resolution for the ensemble product might be 4–5 km or more. Also, an archive of hindcast ensemble forecasts is required for calibration, which may not necessarily be available, or may require a time-consuming hindcasting or re-analysis task. As an alternative, an archive can be built up over time by saving forecasts as they are used operationally. However, the resulting records may not be homogeneous due to changes to model parameterisations, data assimilation procedures, model resolution, etc., which meteorological services often make over time to improve forecast accuracy. This in turn is an argument for using a system of version control and controlled upgrades to new products as these become available. A third approach, which was discussed earlier, is to use an older, less computationally intensive operational NWP model for the ensemble component, in parallel with the latest operational model for the deterministic component.

Another approach is to combine a probabilistic forecast calibration approach to assess the hydrological model error with an ensemble rainfall forecasting approach to assess the rainfall forecast error (Moore *et al.*, 2010), the latter dominating the forecast uncertainty at longer lead-times. Figure 17.10 shows an example of this approach combining a parametric ARMA approach with the STEPS ensemble rainfall forecast estimates shown in Figure 17.5.

In this example, the ARMA model is applied to the logarithm of errors from a PDM rainfall-runoff model (Moore, 2007) simulation, and theoretical quantiles employed to obtain uncertainty bands, under limiting assumptions

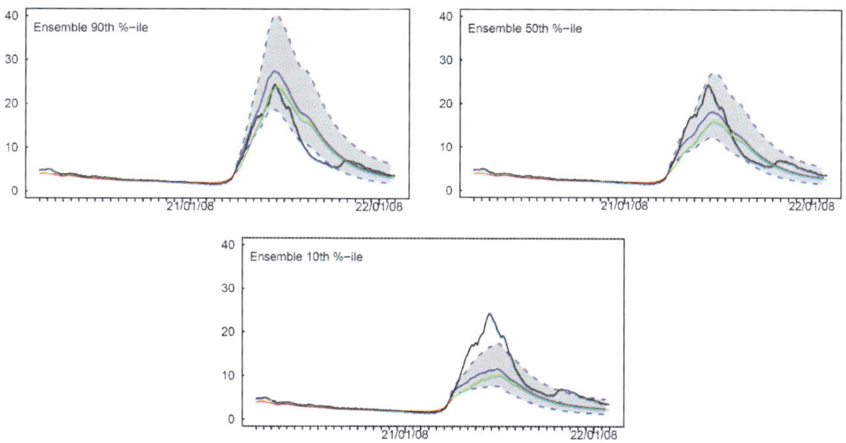

Figure 17.10. Uncertainty of flow forecasts using a STEPS ensemble rainfall forecast with time-origin 06:00 21 January 2008. Model uncertainty is indicated by the 95% probability band in grey. Rainfall forecast uncertainty is indicated across the graphs by the high, medium and low percentile flows (90th, 50th and 10th percentiles). Observed river flow: black line. Simulated river flow: red line. Percentile flows obtained using ensemble rainfall forecasts: green line-simulated. The blue line is the ARMA updated forecast.

of normality and constant variance of the residual errors. In this case, the "real-time" raw ensemble rainfall forecast inputs may be expected to better reflect the uncertainty of a specific "unusual" storm, rather than storm conditions "typical" over the period of calibration. The ARMA uncertainty bands provide an estimate of total model uncertainty, but are conditional on the use of rainfall observations rather than forecasts. Used together as shown in Figure 17.10, they provide a way of jointly assessing uncertainty due to model error and rainfall forecast error. The use of calibrated ensemble rainfall forecasts might lead to further improvements, once techniques for this approach become available.

Finally, the use of multi-model ensembles provides another route to assessing model uncertainty, and in particular model structural errors. In this approach, several hydrological models are operated in parallel, and techniques such as Bayesian model averaging (Beckers *et al.*, 2008; Raftery *et al.*, 2005; Todini, 2008; Vrugt and Robinson, 2007) can be used to provide an overall estimate of uncertainty, based on the performance of the individual models in the recent past. Ideally, models should be of different types or "brands" to provide a truly competing set of outputs. Of course, if the models include a hydrodynamic component, there may be run-time issues with this approach. Also, there is no guarantee that any of the models in the ensemble provides the "correct" forecast in a given flood event, although the resulting forecast would be expected to be more robust than the usual situation of relying on the outputs from a single forecasting model.

17.7. Current Research Themes in Probabilistic Flood Forecasting

In the past decade, there has been a huge increase in interest in probabilistic flood forecasting techniques. Many organisations with flood warning responsibilities are starting to implement operational or trial applications. Here, initial ideas have been outlined for a risk-based framework for categorising and selecting techniques, depending on the operational requirement and the hydrological and operational constraints.

Whilst much of this interest has been prompted by recent improvements in the resolution of ensemble rainfall forecast products, this has also extended to consideration of other sources of uncertainty, such as in tidal surge forecasts, and model-related errors. There has also been a resurgence of interest in Kalman filtering and related techniques which were first applied to flood forecasting applications in the 1970s (Chiu 1978; Wood and Szollosi-Nagy, 1980).

The overall expectation is that the use of probabilistic information should lead to better risk-based decision making in issuing flood warnings and other operational decisions during flood events. From a practitioner's point of view, the principle benefits are (Laeger *et al.*, 2010):

- Providing a more structured and transparent approach for assessing uncertainties and their effects on flood forecasts regardless of the experience of individual forecasters. This should provide a clear audit trail of the information available and how it has (or has not) influenced operational decisions.
- Increasing lead-times through using rainfall forecast ensembles, albeit often with additional uncertainty at longer lead-times (however, work still needs to be done to assess the readiness of duty officers and professional partners to utilise these forecasts).
- Allowing for calculated precautionary actions to be taken in high-risk locations in response to low probability forecasts.
- Providing additional information to support marginal decisions.
- Targeting model, telemetry and data improvements to the areas where they matter most. Off-line assessments of model performance after floods events can play a major role in this.

The uncertainty framework described here attempts to provide an overall structure to guide the selection of appropriate techniques to meet some of these different requirements, whilst recognising that some of these methods still remain active areas for research. In particular, the development of forecast calibration techniques for ensemble meteorological forecasts has been noted as an area where further research is required.

Of course, beyond the purely technical issues, there are a number of questions which remain active areas for research, including the issues of:

- The operational applications of uncertainty information, particularly for low probability/high impact events.
- Improved understanding of the assumptions and limitations in the probabilistic estimates provided.
- Communication of uncertainty information to professional partners and to the public.
- Assessing the incremental economic benefits from the availability of uncertainty information.
- Training and raising awareness for technical experts and professional partners.

- Techniques for reducing model run-times, such as emulators and statistical sampling.
- Performance measures and validation techniques for probabilistic flood forecasts.

Major advances have been made in these areas in recent years, with an ever-widening set of case studies and applications available to inform practitioners on the benefits of a probabilistic approach.

Acknowledgements

The uncertainty framework described here was originally developed for the Environment Agency in England and Wales as part of the R&D project "Risk-based Probabilistic Fluvial Flood Forecasting for Integrated Catchment Models". Several of the figures are based on case studies which made use of data and models provided by the Environment Agency during that project. Thanks are due to the following people who produced some of the findings and results shown here: Joost Beckers, Alex Minett, Hessel Winsemius and Jan Verkade from Deltares; Paul Smith and Dave Leedal of Lancaster University for Figures 17.7 and 17.8; Steve Cole, Alice Robson and Phil Howard from CEH Wallingford; Andrew Craig from Edenvale Young; and, Marc Huband and Neil Breton from Atkins. The uncertainty framework was also partly developed under the Flood Control 2015 programme (http://www.floodcontrol2015.com/).

References

Andryszewski, A., Evans, K., Haggett, C. *et al.* (2005). "Levels of Service Approach to Flood Forecasting and Warning" *Proc. Int. Conf. Innovation, Advances and Implementation of Flood Forecasting Technology*, Tromso, Norway, October 17–19 2005.

Beckers, J.V.L., Sprokkereef, E. and Roscoe, K. (2008). "Use of Bayesian Model Averaging to Determine Uncertainties in River Discharges and Water Level" *4th International Symposium on Flood Defence: Managing Flood Risk, Reliability and Vulnerability*, Toronto, Ontario, Canada, May 6–8 2008.

Beven, K.J., Young, P.C., Leedal, D.T. *et al.* (2008). "Computationally Efficient Flood Water Level Prediction (With Uncertainty)" in: Samuels, P., Huntingdon, S., Allsop, W. *et al.* (eds), *Flood Risk Management: Research and Practice*, CRC Press, Boca Raton.

Beven, K.J. (2009). *Environmental Modelling: an Uncertain Future?*, Routledge, London.

Beven, K.J., Leedal, D. and McCarthy, S. (2010). Guidelines for assessing uncertainty in flood risk mapping, data and modelling. Flood Risk Management Research Consortium, Work Package 1.7.

Beven, K.J., Leedal, D.T., Smith, P.J. *et al.* (2011) "Identification and Representation of State Dependent Non-Linearities in Flood Forecasting Using the DBM Methodology" in: Wang, L. and Garnier H. (eds), *System Identification, Environmetric Modelling and Control*, Springer-Verlag, Berlin, pp. 341–366.

Bowler, N.E., Pierce, C.E. and Seed, A.W. (2006). STEPS: a probabilistic precipitation forecasting scheme which merges an extrapolation nowcast with downscaled NWP, *Q. J. Roy. Meteor. Soc.*, **132**, 2127–2155.

Buizza, R. (2008). The value of probabilistic prediction, *Atmos. Sci.Lett.*, **9**, 36–42.

Burgers, G., van Leeuwen, P.J. and Evensen, G. (1998). Analysis scheme in the ensemble Kalman filter, *Mon. Weath. Rev.*, **126**, 1719–1724.

Butts, M.B., Falk, A.K., Hartnack, J. *et al.* (2005). "Ensemble-Based Methods for Data Assimilation and Uncertainty Estimation in the FLOODRELIEF Project" *Proc. Int. Conf., Innovation, advances and implementation of flood forecasting technology*, Tromso, Norway, October 17–19 2005.

Carsell K.M., Pingel, N.D. and Ford, D.T. (2004). Quantifying the benefit of a flood warning system, *Nat. Haz. Rev.*, **5**, 131–140.

Chen, Y., Sene, K. and Hearn, K. (2005). "Converting 'Section 105' or SFRM Hydrodynamic River Models for Real Time Flood Forecasting Applications" *40th Defra Flood and Coastal Management Conference, University of York*, July 5–7 2005, Paper 6b-3.

Chiu, C. (ed.) (1978). *Applications of Kalman Filters to Hydrology, Hydraulics and Water Resources*, University of Pittsburgh Press, Pennsylvania.

Cloke, H.L. and Pappenberger, F. (2009). Ensemble flood forecasting: a review, *J. Hydrol.*, **375**, 613–626.

Day, G.N. (1985). Extended streamflow forecasting using NWSRFS, *J. Water Res. Pl.-ASCE*, **111**, 157–170.

Demeritt, D., Cloke, H., Pappenberger, F. *et al.* (2007). Ensemble predictions and perceptions of risk, uncertainty and error in flood forecasting, *Environ. Haz.*, **7**, 115–127.

Demuth, J., Morss, R.E., Lazo, J.K. *et al.* (2007). "Assessing how the US Public Understands and uses Weather Forecast Uncertainty Information" American Meteorological Organisation, *16th AMS Conf. on Applied Climatology*, San Antonio, Texas, Paper 3.3.

Environment Agency (2002). *Real Time Modelling Guidelines*, R&D project WSC013/5.

Environment Agency (2010a). *Hydrological Modelling using Convective Scale Rainfall Modelling*, Science Report SC060087. Authors: Schellekens, J., Minett, A.R.J., Reggiani, P., Weerts, A.H. (Deltares); Moore, R.J., Cole, S.J., Robson, A.J., Bell, V.A. (CEH Wallingford), Environment Agency, Bristol, UK.

Environment Agency (2010b). *Risk-Based Probabilistic Fluvial Flood Forecasting for Integrated Catchment Models*, Science Report SC080030, Environment Agency, Bristol, UK.

Evensen, G. (1994). Sequential data assimilation with a non-linear quasi-geostrophic model using Monte Carlo methods to forecast error statistics, *J. Geophys. Res.*, **97**, 905–924.

Golding, B. (2009). Long lead time flood warnings: reality or fantasy?, *Meteorol. Appl.*, **16**, 3–12.

Hall, J. and Solamatine, D. (2008). A framework for uncertainty analysis in flood risk management decisions, *Int. J. River Basin Manage.*, **6**, 85–98.

Kalman, R. (1960). New approach to linear filtering and prediction problems, *J. Basic Eng.-T. ASME*, **82D**, 35–45.

Koenker, R. (2005). *Quantile Regression*, Cambridge University Press, Cambridge.

Krzysztofowicz, R. (1999). Bayesian theory of probabilistic forecasting via deterministic hydrologic model, *Water Resour. Res.*, **35**, 2739–2750.

Krzysztofowicz, R. (2001). The case for probabilistic forecasting in hydrology, *J. Hydrol.*, **249**, 2–9.

Krzysztofowicz, R. and Herr, H.D. (2001). Hydrologic uncertainty processor for probabilistic river stage forecasting: precipitation-dependent model, *J. Hydrol.*, **249**, 46–68.

Krzysztofowicz, R. and Kelly, K.S. (2000). Hydrologic uncertainty processor for probabilistic river stage forecasting, *Water Resour. Res.*, **36**, 3265–3277.

Laeger, S., Cross, R., Sene, K. *et al.* (2010). "Risk-Based Probabilistic Fluvial Flood Forecasts for Integrated Catchment Models" British Hydrological Society, *BHS Third International Symposium*, Role of Hydrology in Managing Consequences of a Changing Global Environment, Newcastle University, 19–23 July 2010.

Leedal, D.T., Beven, K.J., Young, P.C. *et al.* (2008) "Data Assimilation and Adaptive Real-Time Forecasting of Water Levels in the River Eden Catchment, UK" in: Samuels, P., Huntingdon, S., Allsop, W. *et al.* (eds), *Flood Risk Management: Research and Practice*, CRC Press, Balkema.

Lees, M. J., Young, P.C., Ferguson, S. *et al.* (1994). "An Adaptive Flood Warning Scheme for the River Nith at Dumfries" in: White, W.R. and Watts, J. (eds), *2nd International Conference on River Flood Hydraulics*, John Wiley and Sons, Hoboken, NJ.

Lettenmaier, D.P. and Wood, E.F. (1993). "Hydrologic Forecasting", in: Maidment, R.D., *Handbook of Hydrology*, McGraw-Hill, New York, Section 26.1–26.30.

Montanari, A. and Brath, A. (2004). A stochastic approach for assessing the uncertainty of rainfall-runoff simulations, *Water Resour. Res.*, **40**, W01106, doi:10.1029/2003WR002540.

Moore, R.J. (1999). "Real-time Flood Forecasting Systems: Perspectives and Prospects" in: Casale, R. and Margottini, C. (eds), *Floods and landslides: Integrated Risk Assessment*, Springer, London, pp. 147–189.

Moore, R.J. (2007). The PDM rainfall-runoff model, *Hydrol. Earth Syst. Sci.*, **11**, 483–499.

Moore, R.J., Jones, D.A. and Black, K.B. (1989). Risk assessment and drought management in the Thames Basin, *Hydrol. Sci. J.*, **34**, 705–717.

Moore, R.J., Robson, A.J., Cole, S.J. *et al.* (2010). Sources of uncertainty and probability bands for flood forecasts: an upland catchment case study, *Geophys. Res. Abstr.*, **12**, EGU2010-151609.

Moradkhani, H., Hsu, K.-L., Gupta, H. *et al.* (2005). Uncertainty assessment of hydrologic model states and parameters: sequential data assimilation using the particle filter, *Water Resour. Res.*, **41**, W05012, doi:10.1029/2004WR00 3604.

Nobert, S., Demeritt, D. and Cloke, H.L. (2010). Informing operational flood management with ensemble predictions: lessons from Sweden, *J. Flood Risk Manage.*, **3**, 72–79.

Pappenberger, F., Beven, K.J., Frodsham, K. *et al.* (2007). Grasping the unavoidable subjectivity in calibration of flood inundation models: a vulnerability weighted approach, *J. Hydrol.*, **333**, 275–287.

Raftery, A.E., Gneiting, T., Balabdaoui, F. *et al.* (2005). Using Bayesian model averaging to calibrate forecast ensembles, *Mon. Weather Rev.*, **133**, 1155–1174.

Reed, S., Schaake, J. and Zhang, Z. (2007). A distributed hydrologic model and threshold frequency-based method for flash flood forecasting at ungauged locations, *J. Hydrol.*, **337**, 402–420.

Refsgaard, J.C. (1997). Validation and intercomparison of different updating procedures for real-time forecasting, *Nord. Hydrol.*, **28**, 65–84.

Reggiani, P. and Weerts, A.H. (2008). Probabilistic quantitative precipitation forecast for flood prediction: an application, *J. Hydrometeorol.*, **9**, 76–95.

Reggiani, P., Renner, M., Weerts, A.H. *et al.* (2009). Uncertainty assessment via Bayesian revision of ensemble streamflow predictions in the operational River Rhine forecasting system, *Water Resour. Res.*, **45**, W02428, doi:10.1029/2007WR006758.

Roberts, N.M., Cole, S.J., Forbes, R.M. *et al.* (2009). Use of high-resolution NWP rainfall and river flow forecasts for advanced warning of the Carlisle flood, *Meteorol. App.*, **16**, 23–34.

Romanowicz, R.J., Young, P.C. and Beven, K.J. (2006). Data assimilation and adaptive forecasting of water levels in the River Severn catchment, United Kingdom, *Water Resour. Res.*, **42**, W06407, doi:10.1029/2005WR004373.

Romanowicz, R.J., Young, P.C., Beven, K.J. *et al.* (2008). A data-based mechanistic approach to nonlinear flood routing and adaptive flood level forecasting, *Adv. Water Resour.*, **31**, 1048–1056.

Roulin, R. (2007). Skill and relative economic value of medium-range hydrological ensemble predictions, *Hydrol. Earth Syst. Sci.*, **11**, 725–737.

Schaake, J.C., Hamill, T.M., Buizza, R. *et al.* (2007). HEPEX the Hydrological Ensemble Prediction Experiment, *B. Am. Meteorol. Soc.*, **88**: 1541–1547.

Sene, K.J. (2008). *Flood Warning, Forecasting and Emergency Response*, Springer, Dordrecht, p. 303.

Seo, D.-J., Herr, H.D. and Schaake, J.C. (2006). A statistical post-processor for accounting of hydrologic uncertainty in short-range ensemble streamflow prediction, *Hydrol. Earth Syst. Sci. Discuss.* **3**, 1987–2035.

Serban, P. and Askew, A.J. (1991). Hydrological forecasting and updating procedures, *IAHS-AISH P.* Publication No. 201, 357–370.

Smith, P.J., Beven, K.J. and Horsburgh, K. (2012) Data-based mechanistic modelling of tidally affected river reaches for flood warning purposes: an example on the River Dee, UK, *Quart. J. Roy. Meteor. Soc.*, doi:10.1002/qj.1926.

Thielen, J., Bartholmes, J., Ramos, M.-J. *et al.* (2009). The European Flood Alert System — Part I: concept and development, *Hydrol. Earth Syst. Sci.*, **13**, 125–140.

Todini, E. (2004). Role and treatment of uncertainty in real-time flood forecasting, *Hydrol. Process.*, **18**, 2743–2746.

Todini, E. (2008). A model conditional processor to assess predictive uncertainty in flood forecasting, *Int. J. River Basin Manage.*, **6**, 123–137.

Uppala, S.M., Kållberg, P.W., Simmons, A.J. *et al.* (2005). The ERA-40 re-analysis, *Q. J. Roy. Meteor. Soc.*, **131**, 2961–3012.

Vrugt, J.A. and Robinson, B.A. (2007). Treatment of uncertainty using ensemble methods: comparison of sequential data assimilation and Bayesian model averaging, *Water Resour. Res.*, **43**, W01411, doi:10.1029/2005WR004838.

Webster, P.J. and Hoyos, C. (2004). Prediction of monsoon rainfall and river discharge on 15–30 day time scales, *B. Am. Meteorol. Soc.*, **85**, 1745–1765.

Weerts, A. and El Serafy, G.Y.H. (2006). Particle filtering and ensemble Kalman filtering for state updating with hydrological conceptual rainfall-runoff models, *Water Resour. Res.*, **42**, W09403, doi:10.1029/2005WR004093.

Weerts, A.H., Winsemius, H.C. and Verkade, J.S. (2011). Estimation of predictive hydrological uncertainty using quantile regression: examples from the National Flood Forecasting System (England and Wales), *Hydrol. Earth Syst. Sci.*, **15**, 255–265, doi:10.5194/hess-15-255-2011.

Wilks, D.S. and Hamill, T. (2007). Comparison of ensemble-MOS methods using GFS reforecast, *Mon. Weather Rev.*, **135**, 2379–2390.

Wood, A.W. and Schaake, J.C. (2008). Correcting errors in streamflow forecast ensemble mean and spread, *J. Hydrometeorol.*, **9**, 132–148.

Wood, E.G. and Szollosi-Nagy, A. (eds) (1980). *Real-Time Forecasting/Control of Water Resource Systems*, Pergamon Press, Oxford.

World Meteorological Organisation (1994). *Guide to Hydrological Practices 5th Edition*, WMO No. 168, Geneva.

Young, P.C. (2002). Advances in real-time flood forecasting, *Philos. T. R. Soc. A.*, **360**, 1433–1450.

CHAPTER 18

Case Study: Decision Making for Flood Forecasting in the US National Weather Service

Robert Hartman and John Schaake
NOAA National Weather Service, USA

18.1. Introduction

River and flood forecasts for the US are produced by 13 National Weather Service (NWS) River Forecast Centers (RFCs). The RFCs have four basic functions:

1. Continuous hydrometeorological data assimilation, river basin modelling, and hydrologic forecast preparation.
2. Technical support and interaction with supported and supporting NWS offices.
3. Technical support and interaction with outside water management agencies and users.
4. Applied research, development, and technological implementation to facilitate and support the above functions.

RFCs provide hydrologic guidance for timescales that vary from *hours* (flash flood guidance and support to local flood warning systems), to *days* (traditional flood forecasts), to *weeks* (snowmelt forecasts), to *months* (reservoir inflow and seasonal water supply). The following is a brief description of how guidance for each of these timescales is produced.

18.2. Flash Flooding and Rapid Hydrologic Response

The NWS defines a flash flood as a flood event which occurs within six hours of the causative event. The rapid and often localised development of these events requires real-time assessment and immediate public warning. As such, the National Weather Service responsibility for public flash flood watches and warnings rests with 122 continuously staffed Weather Forecast Offices (WFOs). RFCs support the WFO flash flood warning system by providing flash flood guidance (amount of rain required to cause flash flooding) based on current watershed conditions. Flash flood guidance is provided in gridded form for durations of 1, 3 and 6 hours.

Many cities and counties with serious flash flood problems have installed and actively maintain automated local flood warning systems. The California–Nevada RFC pioneered the concept, design, development, and application of ALERT (Automated Local Evaluation in Real Time). The systems consist of automated event-reporting rain and river gauges and a computer system that analyses current and developing hydrologic conditions. Data from the river and rain gauges are generated in response to changing conditions (e.g. 0.04 inches of rain or 0.1 foot change in river level) and transmitted to the computer base station using line-of-sight radio operated on a hydrologic warning frequency. All data are automatically passed to the nearest WFO as well as the supporting RFC. Collaboration among the operators of these systems is facilitated by organisations such as the Alert Users Group (http://www.alertsystems.org) and the National Hydrologic Warning Council (http://hydrologicwarning.org).

18.3. Flood Forecasting

Flood forecasts are provided for more than 4000 locations in the US. Operational flood forecasting is an exercise in system integration as shown in Figure 18.1. Data collection, meteorological forecasting, hydrologic modelling, co-ordination, decision support, product generation, and dissemination components must all come together in a robust environment that works efficiently and reliably.

Operational flood forecasting is highly dependent upon reliable and timely data. The RFCs rely on data collected by many other agencies and groups. It is important to note that the flood forecasting function at all RFCs relies heavily upon automated data as opposed to manual observations. Automated data are transmitted from field sites using a variety of techniques that include line-of-sight radio, microwave, satellite,

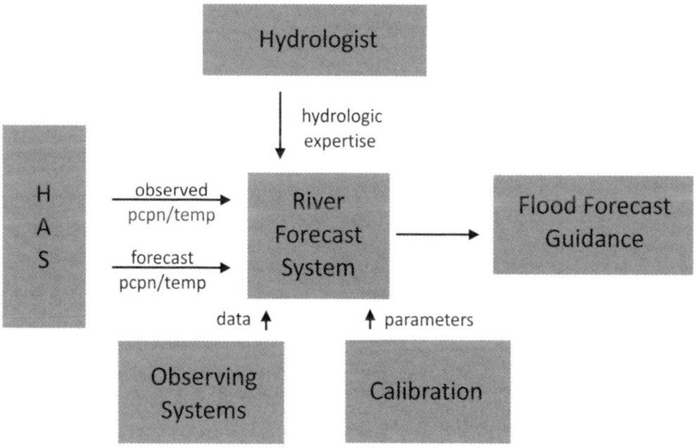

Figure 18.1. Operational flood forecasting process.

telephone, and meteor burst. Agencies that collect and provide real-time hydrometeorological data include the US Bureau of Reclamation (USBR), US Army Corps of Engineers (USACE), US Geological Survey (USGS), US Forest Service (USFS), US Natural Resources Conservation Service (NRCS), state agencies, cities and counties with local flood warning systems, numerous irrigation districts, and many utility companies involved in hydropower generation. In combination, the network consists of thousands of measurements from rain gauges, air temperature sensors, river gauges, and gauges that measure reservoir elevation. Many other variables such as wind speed, wind direction, relative humidity, and barometric pressure are also available. Data are collected in a local database at each RFC and screened for quality before being passed along to the river forecasting model. A significant proportion of RFC resources are dedicated to data collection, storage, processing and quality control.

The RFCs use the CHPS (Community Hydrologic Prediction System, www.nws.noaa.gov/oh/hrl/chps/index.html) to simulate and project river flows and stages. CHPS is based on the Deltares FEWS (Flood Early Warning System, www.deltares.nl) infrastructure with extensions to meet RFC modelling, computation, interface, and data requirements. Many of the forecast watersheds in the US require the use of a snow model. The snow model accumulates solid precipitation, manages the heat content of the snowpack, and delivers meltwater to the soil model when the snowpack begins to melt. The temporal delay caused by this natural process is important to both flood management and water resources interests. If all precipitation in the US fell as rain, flooding would be significantly worse and more frequent and agricultural irrigation would be significantly less reliable.

CHPS is operated at each RFC in an interactive fashion through a graphical user interface. The interface allows the forecaster to run, evaluate, and modulate the operation at each forecast point. Model performance is evaluated over the last several days to several weeks and the streamflow is projected for the next 2–10 days (varies by RFC) using current model conditions as well as forecasts of temperature, precipitation, and reservoir regulation.

Forecasts of streamflow necessarily include forecasts of precipitation and air temperature. These weather forecasts are critical to the NWS's ability to provide adequate public warning time in the event of a flood. Quantitative Precipitation Forecasts (QPF), temperature forecasts and snow levels for the next two to ten days in six hour blocks are integrated into RFC hydrologic forecasts by the RFC HAS (Hydrometeorological Analysis and Support) unit (Figure 18.2).

The HAS unit is a staff of professional meteorologists specifically tasked with ensuring that forecast weather is included in the hydrologic forecast process. The HAS unit integrates guidance from the NCEP's (National Center for Environmental Prediction) Hydrometeorological Prediction Center (HPC) and the WFOs in the RFC area. Some HAS units also operate mesoscale atmospheric models and orographic procedures that utilise both synoptic and mesoscale model information. On the RFC operations floor, the interaction between the HAS unit meteorologists and forecast hydrologists contributes to the quality of NWS hydrologic forecasts and guidance.

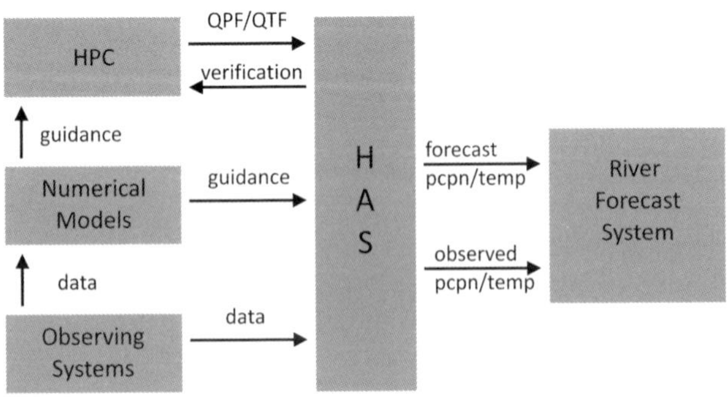

Figure 18.2. Operational hydrometeorological analysis and support function.

Each RFC breaks up its forecast workload into several geographic pieces. This allows forecasters to simultaneously concentrate on specific problem areas. The art of flood forecasting involves a delicate and skilled balance of experience, analysis of current conditions, and analysis of model guidance. Traditionally, RFC flood forecast hydrologic models extend several days into the future. But, flood warnings released to the general public may only extend one-to-two days due to excessive uncertainty. The traditional flood forecasting approach has been deterministic and has not attempted to quantify the uncertainty. Hydrologic forecasters balance lead time and accuracy with each forecast issued. Frequent updates during flood events are used to keep up with of changing conditions.

Hydrologic forecasting requires a great deal of co-ordination. Many federal, state, local and private entities are involved in storing, releasing and diverting streamflow. The integration of water management activities into the forecast process is critical to the accuracy and utility of resulting forecasts.

While the RFCs generate the flood forecast guidance, it is the WFOs that issue public flood warnings. The 122 WFOs provide decision support for local customers and issue public advisories, watches and warnings that include known impacts associated with forecast water levels.

18.4. Reservoir Inflow Forecasting

In addition to flood forecasts, RFCs provides reservoir inflow forecasts to the appropriate water management agencies. These forecasts allow reservoir operators to anticipate regulation changes required by very wet or very dry weather. Reservoir forecasts are produced in the flood forecast modelling process and are updated once a day and more often if required as conditions change. Reservoir operators inform the RFCs of regulation plans and changes so that planned reservoir releases can be incorporated in downstream river forecasts.

One of the major challenges to extending river forecast lead times to beyond a few days is to account for the effects of future reservoir operations and their uncertainty.

18.5. Spring Snowmelt Forecasting

Snowmelt forecasts are provided each spring to assist water and emergency management entities in dealing with the timing and magnitude of the spring snowmelt peak. These may be made for lead times ranging from a week to several months. To generate these forecasts, RFCs use the flood forecast model described above in an ensemble mode. This approach, called Ensemble Streamflow Prediction (ESP), provides a distribution of future streamflow outcomes that can be sampled in any way desired. The process starts with the current model conditions (or states) and then develops a set of forecast scenarios based on the historical observations of temperature and precipitation. The approach is extremely flexible in that any time window and any attribute of streamflow can be analysed. The most typical use is for weekly, monthly, and seasonal volumes and peaks. In addition, the system can be used to describe a whole host of information such as the distribution of low flow stages during the upcoming summer. This technique has been very successful in quantifying spring flood risks in the Upper Midwest of the US weeks in advance.

18.6. Seasonal Water Supply Forecasting

Water supply forecasts predate flood forecasting efforts in most parts of the western US. These forecasts are extremely important to water managers interested in providing water for agricultural, municipal, environmental, and commercial purposes. Winter precipitation stored in the mountain snowpack makes seasonal water supply forecasting possible. In many basins of the western US as much as 70% of the annual streamflow arises from the melting of the seasonal snowpack each spring. Two modelling approaches are used to prepare seasonal runoff estimates.

Statistical models have been used to forecast spring runoff for many years and they remain the tool of choice for many involved in this process. Statistical models are simple to use and make direct use of monthly snow course, snow pillow, and precipitation data. As they are optimised to predict the seasonal volume, they are among the most accurate and dependable. The statistical model resembles a multiple linear regression where the independent variables include monthly snow course, snow pillow, precipitation and, occasionally, streamflow observations. The models are calibrated in advance using an average of 30 years of observed data. Principal components analysis is used to deal with correlation among the input variables. Cross-validation errors are computed to find equations that forecast as well as fit the observed data.

The same process (ESP) used to generate the weekly snowmelt forecasts can be extended to estimate seasonal volumes. ESP forecasts for each watershed are developed and compared with the results from the statistical models. Often the approaches complement each other lending insight to the process. Small adjustments are sometimes made to ensure that the ESP (both weekly snowmelt and seasonal) and statistical models are consistent. Over the past decade, ESP procedures have gained popularity as they can provide updated forecasts on a daily basis and are less susceptible to the central tendency issues exhibited by statistical regression models when conditions are far from average.

Seasonal volume forecasts are commonly coordinated with other agencies. The exchange of information in this process improves the quality of forecasts and minimises conflicting public forecast information.

18.7. Forecast Uncertainty

A major limitation of the existing operational hydrologic forecast system in the US is that much more information about forecast uncertainty is needed. Accordingly, a new approach is being developed that will use ensemble techniques to quantify uncertainty. This will enable longer-lead times and more informative flood forecasts. Weather and climate forecast information will be used to generate ensemble precipitation and temperature forecasts for each hydrologic forecast sub-area. New techniques for ensemble data assimilation will be developed to account for uncertainty in initial conditions. In addition to this, statistical post-processing techniques will be developed to ensure that the final hydrologic forecasts are unbiased and statistically reliable. A significant effort will be required to educate potential users and ensure that they have decision support tools that can make effective use of the uncertainty information.

Other hydrologic forecast system improvements under development include applications of distributed modelling, improved methods to estimate hydrologic model parameters and account for their uncertainty, and methods to provide spatially distributed hydrologic forecast information.

CHAPTER 19

Quantifying and Reducing Uncertainties in Operational Forecasting: Examples from the Delft FEWS Forecasting System

Micha Werner
Deltares and UNSESCO-IHE, Delft, Netherlands

Paolo Reggiani
Deltares and RWTH Aachen University, Germany

Albrecht H. Weerts
Deltares, Delft, Netherlands

19.1. Introduction

Flood forecasting, warning and response are generally held as an effective non-structural approach to reduction of flood risk, primarily by addressing the potential of reduction of flood-related losses with the availability of a timely and effective warning (UN ISDR, 2004; WMO, 2006), as well as the potential of reduction of the flood hazard through flood alleviation measures (e.g. temporary defences, inundation areas). The recognition of this strength has led to relative widespread implementation of flood forecasting and warning services, particularly where the legal and institutional settings of such systems are well established to ensure a flood warning is responded to effectively.

Real-time flood forecasting makes use of a cascade of inter-linked hydrological and in some cases hydrodynamic models, embedded in a data-management environment. These model chains are run in two principal operational modes; (i) a historical mode, and (ii) a forecast mode. In the first mode the models are forced by hydrological and meteorological observations over a limited time period prior to the onset of the forecast. In the second mode, the models are run over the required forecast lead time, whereby the internal model states at the end of the historic run are taken as initial conditions for the forecast run. Often the models will be forced using quantitative forecasts of precipitation and temperature. The forecasting lead-time depends primarily on the lead-time requirement posed by flood warning, and may extend from as little as 2–3 hours to as much as up to 240 hours ahead. In the latter case the shorter lead-times may be used in the actual operational warning, while forecasts at the longer lead-times are used as guidance.

In each of the steps in the model and data processing chain, uncertainties can be attributed to the model inputs, the model structure, internal model states and model parameterisation, with the total predictive uncertainty accumulating in the output product (Pappenberger *et al.*, 2007). Depending on the lead-time at which forecasts are issued in comparison to the hydrological response time, the dominant uncertainties will lie in the inputs derived from observations, the hydrological runoff and routing models, or if applicable, the hydraulic models. As the process of forecasting is geared primarily at providing timely and accurate information for the flood-warner when deciding on the issuing of a flood warning, the uncertainties within the process will need to be considered. These uncertainties are generally recognised and in most operational forecasting systems these are indeed addressed.

In this chapter, several examples of how uncertainties are addressed and constrained within operational forecasting systems are given, with the examples selected to cover a wide range of approaches with which this can be done. These examples are taken from operational forecasting systems that use the Delft FEWS (Werner *et al.*, 2004, 2013) operational forecasting system to manage data and models in the real-time environment. This system has been applied in several operational forecasting systems, including forecasting systems for the Rhine basin in the Netherlands (Sprokkereef, 2001) and Switzerland (Bürgi, 2002), across England and Wales and Scotland (Cranston *et al.*, 2007; Lardet *et al.*, 2007; Werner, *et al.*, 2009; Whitfield, 2005). In the first section we give an overview of the Delft FEWS data-management environment. Key to this data management system is that it is centred on the data instead of the models used for forecasting, allowing it to be flexible to the data and models used. Secondly, we describe how the uncertainties/errors in the model chain can be reduced through real-time data assimilation (error correction and/or state updating) in an operational setting. Thirdly, we focus on how input uncertainty can be quantified either by applying what-if scenarios and/or through using probabilistic weather forecasts. Finally, we end this chapter with a look forward at how a more complete description of the predictive uncertainty can be derived and reduced using Bayesian revision of ensemble predictions. Examples of the methods described taken from the operational systems also illustrate some of the methods used in disseminating the probabilistic forecast information.

19.2. The Delft FEWS Operational Forecasting System

Operational flood forecasting systems have been under development at forecasting centres for several decades. Particularly where these have been used operationally for significant time, the systems have often developed gradually, and in many cases have been centred on the particular use of a given model and the data it requires (Werner and Whitfield, 2007). While this model-centric approach leads to a system that can adequately provide forecasts using the model and data it was designed for, it offers distinct disadvantages in the face of changing model and data requirements.

The Delft FEWS flood forecasting system (Werner *et al.*, 2004, 2013) takes a different approach. Rather than setting the models central stage, the system gives data and data processing a central position. To fulfil the data requirement in providing a forecast at the lead time needed in flood warning, the system allows a sequence of data handling and processing steps to be defined. These steps could include the import of data from a variety of external sources such as observed hydrological and meteorological data or numerical weather prediction models, transformation of data across spatial and temporal scales, and where required, the running of hydrological and hydraulic process models. As in any other step in the data processing sequence, models are considered to each have a data input requirement, a specific role, and a data output product. Models linked to the system can be connected through the use of an open data exchange interface (Werner and Heynert, 2006), and some 30+ process models are currently capable of being used. This open approach to integration of models and data has several advantages. The relative simplicity with which a wide variety of modelling systems can be configured to run operationally within the framework enables easier migration of new techniques from the academic to the operational arena. Additionally, the use of the system by operational forecasting staff can be decoupled from the specifics of

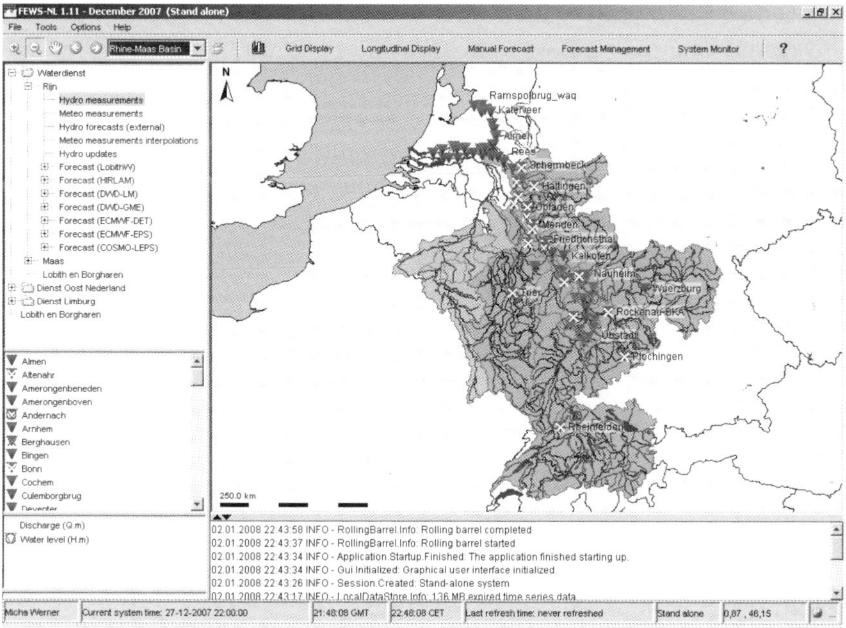

Figure 19.1. Main user display of the Delft FEWS system, as configured for the Rhine and Meuse basins. The stations selected are the hydrological gauging stations, where observed water levels are available. Many of these sites are also the primary warning locations.

the models themselves. This may be of significant benefit as it will result in lower efforts in re-training large numbers of staff involved in operational forecasting following a change of the models and data (Werner and Whitfield, 2007). Since its introduction, the Delft FEWS framework has been applied in several operational flood forecasting systems, as well as a number of pilot, pre-operational, and research systems (see e.g. Figure 19.1, depicting the front end of the Delft FEWS system as configured for the Rhine basin).

A key element to how time-series data are handled within the framework is that the ensemble dimension is included, thus allowing ensembles of time series data to be easily stored and processed. When running through a cascade of data processing and modelling steps, or workflow, there are essentially two methods available in running through an ensemble:

(i) Each ensemble member is run individually, meaning the coordination of the ensemble run is within the forecast processor in Delft FEWS which is used to run the required sequence of steps. As a result, a model used in this way need not be aware that it is run within an ensemble, and from the perspective of the model it is used in exactly the same way as in a deterministic run.

(ii) The ensemble is passed to the data processing step or the process model as a whole. In this case, the responsibility of dealing with the ensemble is vested in the data processing step itself. Typically this method is used in establishing statistical properties of the ensemble, where the complete ensemble must be considered.

The two methods in running ensembles allow flexible definition of how ensembles are used in the forecast process. In many cases, the same cascade of process models and data handling steps will be configured once to run in deterministic mode, and then again to run using input data that has an ensemble dimension, in which case the cascade will be run as an ensemble.

19.3. Reducing Uncertainty through Data Assimilation

Within the cascade of processing steps and process models that constitute a forecast, uncertainties accumulate from the various sources to form the total uncertainties in the forecast (Pappenberger *et al.*, 2007). The availability of real-time data up to the start time of the forecast can, however, be utilised to reduce the uncertainties. Where observed precipitation and temperature data will typically be used in establishing (hydrological) model boundary conditions, observations of water levels and discharge can be effectively used in quantifying error in the modelled water levels and discharges from the process models. Through data assimilation the quantification of these errors can be used to improve the reliability of the forecasts. Refsgaard (1997) describes four basic approaches in data assimilation: (i) updating of the inputs to the model; (ii) updating of the internal model states; (iii) updating of the model parameters; and (iv) correction of errors in the model outputs. Of these the second and fourth are currently the most applied in operational forecasting systems using the Delft FEWS framework.

19.3.1. *Error correction*

Error correction (also known as output correction) is one of the oldest and most versatile methods of data assimilation (Broersen *et al.*, 2005;

Madsen *et al.*, 2000), and it is applied in almost all operational forecasting systems. The basic principle of error correction is that systematic errors (i.e. structural differences between model and discharge measurements) during the historical mode (up until the start of forecast or forecast T0) are used to derive a statistical model of that systematic error. This statistical model can then be used during the forecast mode to predict the error when there are no measurements available.

A good example where the use of this error correction method leads to a better historical simulation (and as a consequence also a better forecast) is the operational forecasting system for the Rivers Rhine and Meuse (Sprokkereef, 2001). Interpolated temperature and rainfall fields, derived from synoptic meteorological measurement stations, are used as inputs to hydrological model of the Rivers Rhine and Meuse. Subsequently, discharges calculated by the hydrological model are input into the hydrodynamic model of the Rhine and Meuse. In Figure 19.2 the measured discharge of the Meuse at the gauge of Borgharen is shown by the thick grey line. The dashed line shows the results of the hydrological-hydraulic model cascade without any form of data assimilation. If error correction is applied at

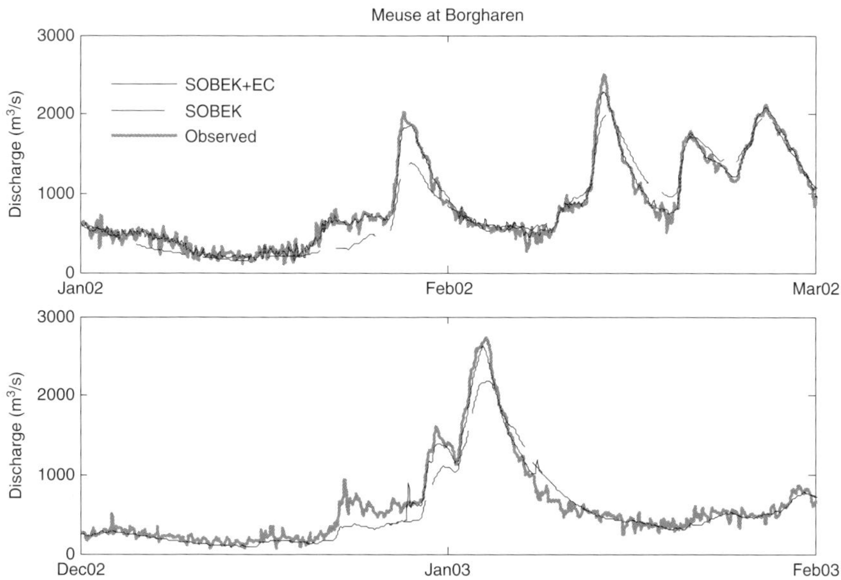

Figure 19.2. Simulated flows at the gauge at Borgharen on the Meuse River for two flood events. Both the flow at Borgharen simulated using corrected inputs (SOBEK+RE) and using simulated inputs (SOBEK) are shown.

the main tributaries (Ourthe, Vesdre, Ambleve, Sambre and Lesse) as well as at the upper model boundary (Chooz) the simulations with the hydrodynamic model of the Meuse improve dramatically, resulting in the solid line which is much closer to the measurements than the dashed line. It is clear that error correction will result in much better forecasts during operational forecasting because the initial state of the hydraulic model is closer to the "true" state of the modelled system at the start of the forecast.

The effect of error correction on the forecast is shown for two locations along the Rhine in Figure 19.3. Here the Root Mean Square Error (RMSE) of the HBV forecasts at Maxau (upper boundary hydraulic model) and Cochem (tributary) are shown as a function of lead time for a hindcast period (December 2002–January 2003). Figure 19.3a shows that error correction results in a large improvement (lower RMSE) for the first 24 hours. The RMSE values, however, increase rapidly with increasing lead time. After 32 hours only a small improvement of the forecasts is visible at the gauge at Cochem (Mosel). For lead times longer than in the order of 72 hours, the forecasts at Maxau without error correction even perform slightly better than with error correction. Figure 19.3b shows an example of how the error correction works during a forecast at Maxau, the upper boundary of the hydrodynamic model.

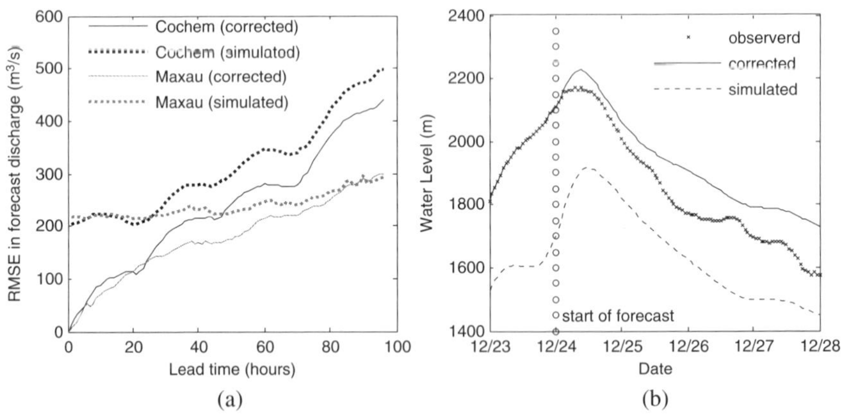

Figure 19.3. (a) Lead-time accuracy of the discharge forecast expressed as RMSE at the gauging stations of Cochem on the Mosel River, and Maxau on the Rhine River. Both the accuracy with and without error correction are shown. (b) Shows an example of the corrected and simulated flows at the gauge of Maxau in the Rhine for the forecast of 24th of December 2002.

19.3.2. *State updating*

Another approach to data assimilation with the objective of improving the forecast is to update the state of the hydrological model or the hydraulic models through sequential data assimilation. With sequential data assimilation the prior Probability Density Function (pdf) of the model state is estimated (forecasted). This prior estimate of the pdf of the model state is subsequently updated by using the available measurements resulting in a posterior pdf of the model state. This can then be used during the next forecasting step.

Operational sequential data assimilation through ensemble Kalman filtering in meteorological, hydrological and hydraulic forecasting (El Serafy *et al.*, 2007; Evensen, 1994; Evensen, 2003; Madsen and Canizares, 1999; Weerts and El Serafy, 2006) or residual re-sampling filtering (van Leeuwen, 2003; Weerts and El Serafy, 2006) is becoming more and more feasible through enhanced computing power and the availability of generic data assimilation packages (El Serafy *et al.*, 2007) which can be used within the Delft FEWS framework. The main advantage of state updating (filtering) over output correction is that it is possible to explicitly take both model and data uncertainties into account. However, the specification of the model and data uncertainties is also the main difficulty in implementing such a filter because these uncertainties are generally poorly known.

Besides the issue of estimating the model and data uncertainties, a compromise between the computational costs to run an ensemble filter (the required computational costs correspond to N times a normal model run, where N is the ensemble size) and the accuracy of the outcome of that filter (the larger N the more accurate the estimate of the prior and posterior model pdf becomes) must be reached. El Serafy *et al.* (2004) showed that with an ensemble size of in the order of 30 members acceptable results could be obtained for the Rhine.

In the operational forecasting system for the Rhine, an ensemble Kalman filter using 32 ensemble members is implemented for the hydraulic SOBEK-RE model of the Rhine using the generic data assimilation module DATools, one of the components available in the Delft FEWS system. Correlated Gaussian noise on the calculated water level using a linear correlation function with a correlation length of about 40 km is generated using the data uncertainty engine (Brown and Heuvelink, 2007). This noise accounts for system and input errors. Observations at 14 gauging (each at an intermediate distance of about 20–40 km) along the main river are being used to update the system state (water level and discharge).

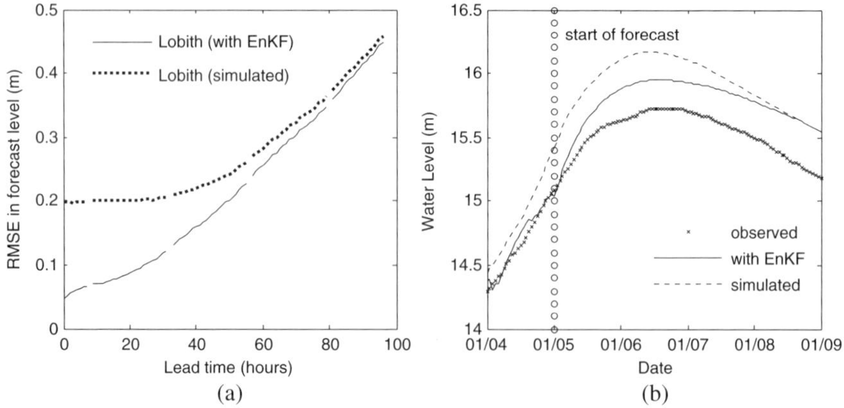

Figure 19.4. (a) RMSE of the water level forecast at the gauge of Lobith on the Rhine with EnKF and without assimilation as a function of lead time determined over a two month hindcast (December 2002/January 2003). (b) Observed water level together with the mean of the EnKF water level forecast and the water level forecast without assimilation at Lobith for an event in January 2003.

The observation error is thought to be uncorrelated Gaussian noise with a standard deviation of 0.05 m. During the forecast no noise is applied and the mean of the 32 ensembles will converge to the reference run (no assimilation).

Figure 19.4a shows an example of the EnKF filtering on the forecast at the measurement station Lobith at the Dutch–German border. From this figure, it becomes clear that the EnKF forecast of the water level starts much closer to the measurements and slowly converges to the forecast without assimilation because the influence of system and input noise is not filtered out during the forecast. The maximum error at the start of the forecast without assimilation can be as large as 0.45 m against 0.13 m for the ensemble mean forecast with assimilation.

Figure 19.5a shows an example of the effect of EnKF for the gauge at IJsselkop, which is located downstream of Lobith. The observations here are used for verification and not used in assimilation, confirming that EnKF can improve the estimate of the states along the river. Obviously the results will depend on the selected correlation scales of the input and system noise, the selected observation error and the amount and the location of gauges with observations used in updating.

Figure 19.4b and Figure 19.5b show the RMSE at Lobith and IJsselkop respectively for a hindcast with and without EnKF over a two month period (December 2002–January 2003). It shows that the RMSE of the

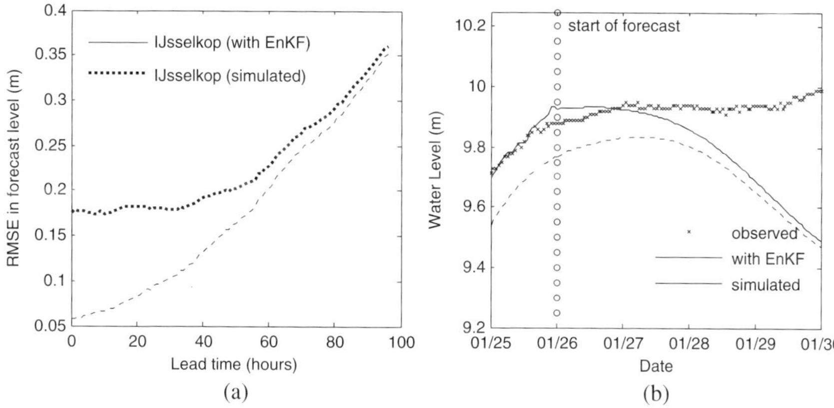

Figure 19.5. (a) RMSE of the water level forecast at the gauge of IJsselkop on the Rhine with EnKF and without assimilation as a function of lead time determined over a two month hindcast (December 2002/January 2003). (b) Observed water level together with the mean of the EnKF water level forecast and the water level forecast without assimilation at IJsselkop for an event in January 2003.

forecasts with EnKF at T0 (start of forecast) is almost equal to the assumed measurement error. It takes more than two days before the RMSE converges to the RMSE of the forecast without assimilation. For locations upstream of Lobith this is less (in the order of one day) and for locations further downstream (for example IJsselkop) this convergence equally takes more than two days.

19.4. Quantifying Input/System Uncertainty

Application of data assimilation has been shown to significantly reduce uncertainties at the shorter lead-times. However, as lead times increase, the innovation from using real-time data in data assimilation will clearly decrease. Figure 19.6 shows a schematic view of the relative contribution of different sources of uncertainty at varying lead-times. At short lead-times, where the innovation from data assimilation is the most apparent, the dominant source of uncertainty is that from the gauged data used in data assimilation. As lead-times increase, uncertainties in routing observed data from upstream gauges to the forecast point of interest will start to influence the total uncertainty. For lead-times beyond the time of travel between the upstream gauging stations and the point of interest, errors in modelling the runoff response from the catchments, as well as incorrect representation of the initial state of the runoff models, will begin to

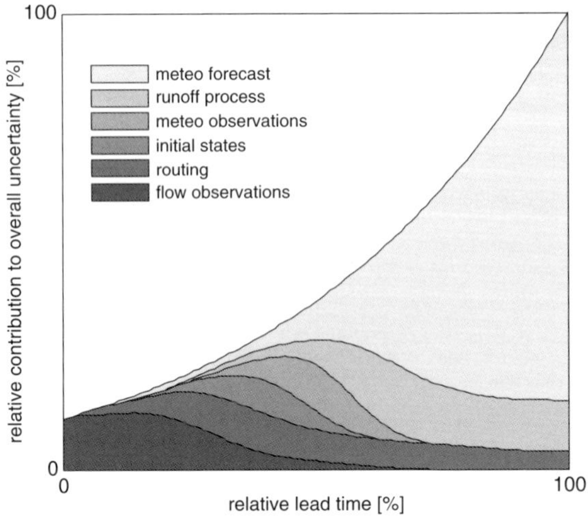

Figure 19.6. Schematic representation of relative contributions of differing sources of uncertainty to the total uncertainty in the forecast at different lead times.

dominate the uncertainty in the forecast. These model uncertainties are combined with the uncertainty of the temperature and more importantly precipitation observations used in determining the boundary conditions to the runoff response models. The uncertainty in the observations themselves will be exacerbated by methods used to derive, for example, catchment average precipitation and temperature.

At lead-times greater than the response time from observed rainfall, uncertainties in the prediction of rainfall and temperature will dominate the uncertainties in the response, increasing in importance as the lead time of the forecast increases. It is clear that how important each of these sources of uncertainty is will depend on the relation between the desired warning lead-time and the response times of each of the contributions (Lettenmaier and Wood, 1993; Werner *et al.*, 2005), but the desire for extending lead times either for use in operational warning or as guidance on the trend will mean that the uncertainties in inputs to the forecast cascade derived from numerical weather prediction will increase in significance.

For the forecaster to make use of forecast values at, in particular, the longer lead-times, it is important that the influence of uncertainty in the inputs is quantified. A logical approach to this quantification is to explore the sensitivity of the forecast values due variable inputs. Within the operational forecasting systems utilising the Delft FEWS framework, several approaches are used, ranging from ad-hoc application of differing input scenarios, to probabilistic ensemble forecasting.

19.4.1. *Quantifying input uncertainties through what-if scenarios*

A common approach to exploring the effect of uncertain inputs into the forecast cascade is through the use of what-if scenarios. In these, the forecaster applies ad-hoc transformations to the inputs, such as multiplication and time shifts, and re-runs the forecast to investigate the impact. Within the Delft FEWS framework any number of forecasts can be made with differing what-if scenarios applied to the inputs, and the resulting forecasts compared to explore the influence of variation in the inputs. In several cases, the what-if scenarios to be applied have been standardised, creating what could be referred to as an empirical ensemble. Table 19.1 provides an overview of some of the empirical methods used in operational forecasting systems utilising Delft FEWS. It is clear that these examples take a very pragmatic line in indicating uncertainty to both the forecaster and, sometimes, to the end user. In some of the cases the request for implementation of such an indication of uncertainty, however crude from a statistical point of view, was made explicitly by the users of the forecasts to provide some indication of uncertainty (Bürgi, 2002; Halquist, 2006).

All of the approaches described in Table 19.1 are constructed within the process of making a forecast. Where a standardised set of scenarios is

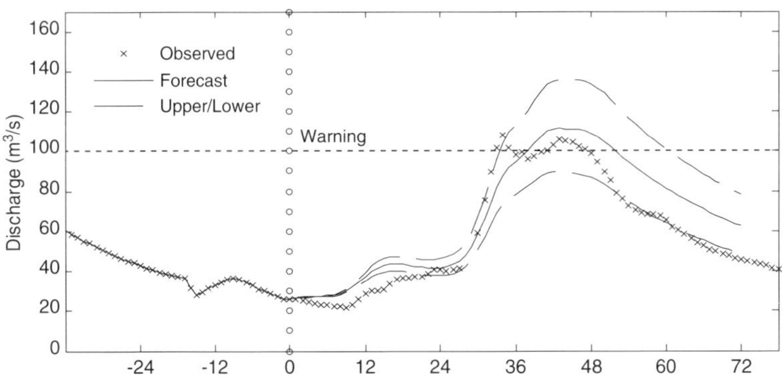

Figure 19.7. Forecast for the Emme at Emmenmat in Switzerland, showing the default forecast, as well as upper and lower scenarios using user defined multipliers on flow, precipitation and temperature.

Table 19.1. Examples of empirical methods for considering input uncertainties applied in operational forecasting systems.

Approach	Forecasting system	Description
What-if scenarios	Several	Support of what-if scenarios is taken in almost all operational forecasting systems to investigate the influence of uncertain (meteorological inputs). Scenarios are implemented during the forecast process and typically apply multipliers to the forecast rainfall, or allow the forecaster to input a user defined forecast rainfall profile based on own judgement.
Standardised multipliers on meteorological and hydrological inputs	FEWS-Rhine Federal Office for the Environment, Switzerland	In the current[a] operational forecasting system used for the Rhine basin in Switzerland (Bürgi, 2002), a standardised what-if scenario is applied. The normal forecast derives temperature and precipitation inputs from the Meteo-Suisse 7 km NWP model for the shorter lead times, falling back to ECMWF deterministic forecast for the longer lead times. A mini ensemble is created by setting user-defined multipliers on forecast precipitation and temperature. An additional multiplier is set for post-processing the hydrograph to allow for an indication of increasing error with lead time when there is neither snow nor precipitation. Figure 19.7 gives an example of the results of this three member ensemble.
Combinations of meteorological input products	National Flood Forecasting System (NFFS) Environment Agency, UK	In NFFS two deterministic rainfall forecast products are available, a radar nowcast product and a NWP forecast product. Under normal forecasting conditions these time series are merged, with the radar gaining priority until it runs out, then falling back to the NWP forecast and finally to a zero rainfall profile. To explore the influence of each of these inputs, a set of standard what-if scenarios have been defined, using different combinations of input products. The combinations explored include the default merged rain profile, a radar-only forecast, an NWP-only forecast, as well as a forecast using zero future rainfall (Figure 19.8).

(*Continued*)

[a]The use of this standardised scenario has since been superseded by the operational use of a hydrological ensemble forecast using a meteorological ensemble input.

Table 19.1. (*Continued*)

Approach	Forecasting system	Description
Re-sampling best guess and 5% and 95% confidence interval precipitation forecast	HPC QPF Sampling NCRFC, National Weather Service, USA.	The Hydrometeorological Prediction Centre (HPC) in the US produces both a best-guess quantitative precipitation forecasts, as well as 5% and 95% confidence interval forecasts. These are sampled using the first 24, first 48 and first 72 (60 in the case of CI forecasts) hours of precipitation data to obtain a nine member empirical ensemble precipitation input. This is used in the operational forecast as an input to a nine member hydrological ensemble forecast (Halquist, 2006).

combined in an empirical ensemble, this is generally sampled in one of the first steps of the data processing cascade, with the following steps being configured to simply loop over the available samples.

19.4.2. *Quantifying uncertainty through ensemble weather prediction*

In a special issue dedicated to advances in quantitative precipitation forecasting through numerical weather prediction, Collier and Krysztofowicz (2000) pointed out the importance of the availability of a forecast of the short-term precipitation to avoid falling back to an assumption of zero future rain in the middle of a storm. The utility of such a precipitation forecast has since been widely demonstrated (e.g. Golding, 2000; Habets, 2004; Ibbit *et al.*, 2000).

Whilst the use of numerical weather prediction for deriving boundary conditions in the forecast has been a significant development in extending the lead time of hydrological forecasting, it is recognised that there are also considerable uncertainties in these weather predictions. To address these, ensemble prediction systems have been established, with the first of these systems becoming operational in the early 1990s. Examples are the ECMWF-EPS system operated by the European Centre for Medium Range Weather Prediction (Molteni *et al.*, 1996) and the GEFS system operated by the US National Centre for Environmental Protection (Tracton and Kalnay, 1993). These are global ensemble prediction systems, predicting

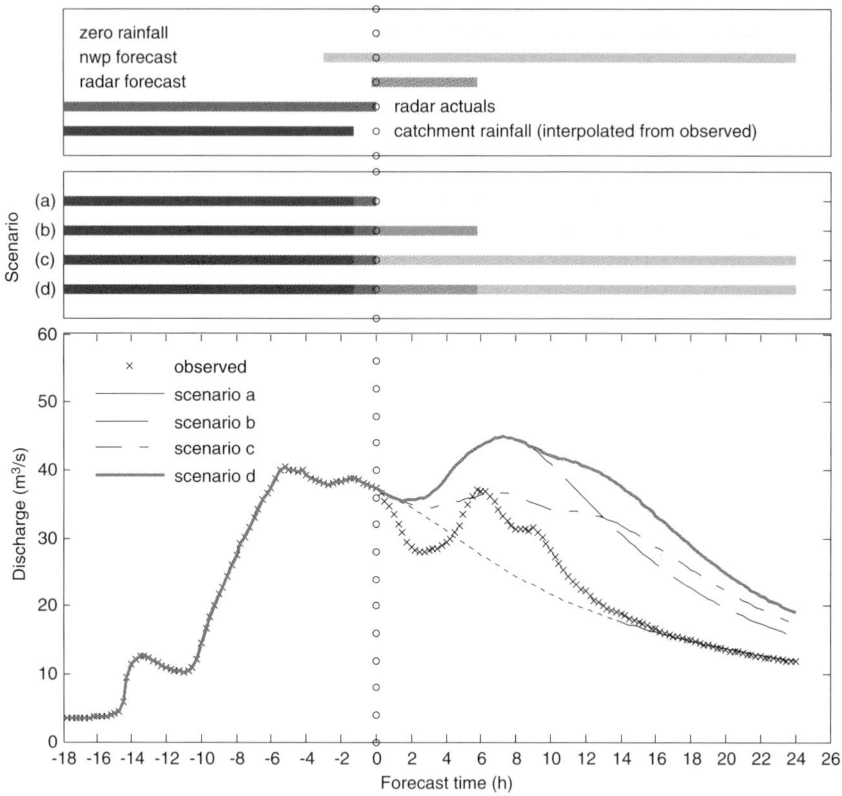

Figure 19.8. Standard set of scenarios used by the Environment Agency in England and Wales. The different forecast inputs are shown in the top figure. These include observed catchment rainfall, radar actuals, radar forecasts (lead time 6 hours), NWP forecast (lead time 36 hours) and a zero rainfall profile. The middle figure shows four scenarios created using differing combinations of these inputs. The lower input shows the response to these scenarios at the gauge of Gargrave on the Aire, for a forecast on the 7th of January 2005 at 23:00.

the evolution of the weather with an emphasis on medium-term predictions (5–15 days lead-time). Ensemble forecasts are generated by perturbing the initial conditions, assumed *a priori* to be equally likely, and computing the evolution of the meteorology due to these perturbed initial conditions. The availability of predicted ensemble parameters such as precipitation and temperature, makes the use of these ensembles in creating ensemble hydrological forecasts a logical next step (Bartholmes and Todini, 2005), with the European Community Sponsored EFAS system (De Roo *et al.*,

2003; Gouweleeuw *et al.*, 2005) being an example, but these ensembles have also found application in several operational fluvial forecasting systems.

Predictions from these global ensembles may be applied in forcing the hydrological models in two ways, either directly where the boundary conditions (usually temperature and precipitation) are obtained directly from the meteorological ensemble, or indirectly where these are obtained following a downscaling approach (Buizza, 2004). Although skill is shown in using the direct approach for medium-sized basins (Roulin, 2007), these are generally better applicable to larger-scale basins. For smaller basins, the indirect approach may be applied, where the large-scale predictions are downscaled using for example regression techniques (Regimbeau *et al.*, 2007), possibly complemented with re-sampling techniques to ensure spatial correlations are correctly maintained (Clark *et al.*, 2004). Alternatively, a limited area ensemble prediction system such as COSMO-LEPS (Marsigli, 2005) may be used, which is nested on the members of the ECMWF global ensemble. The direct approach in deriving hydrological forcing can then be followed again (Diomede *et al.*, 2006).

In contrast to the creation of an ensemble through re-sampling within the forecast process, the complete set of ensemble inputs are now simply imported in Delft FEWS using a suitably configured import process. During this import the ensemble size and available lead-time are identified from the import source, allowing it to dynamically adjust to variations in lead-time and ensemble size.

Subsequent to the import of the ensemble, the cascade of data processing and process models can simply be run for each ensemble member to derive a hydrological forecast ensemble. In most cases the process models are unaware that they are running as a part of an ensemble. Currently the ensembles used in Delft FEWS apply the direct approach to model forcing, where catchment precipitation and temperature inputs are obtained through averaging the NWP grid cells or through an appropriate spatial interpolation approach, which may include an elevation correction. Besides this simple scaling due to altitude, no downscaling methods are currently applied. If available, this would, however, simply require integration of an additional step in the cascade run for each ensemble member.

Table 19.2 provides some examples of ensemble prediction systems used in driving the hydrological ensemble forecast from selected operational forecasting systems utilising the Delft FEWS framework. This shows that in several cases, multiple ensembles are even considered. The first two of these EPS are relatively straightforward, with all ensemble members

Table 19.2. Overview of meteorological ensembles applied in operational forecasting systems using the Delft FEWS framework.

Ensemble system	Forecasting system	Description
ECMWF-EPS	FEWS-NL Rhine and Meuse Catchments, Institute for Inland Water Management and Waste Water Treatment, the Netherlands	Global Ensemble Prediction System (EPS). The current T255L40 EPS has a horizontal resolution of about 80 km, and has 51 members, of which the first member is the control run (Molteni *et al.*, 1996). The lead time of the re-sampled EPS used here is 240 hours at a resolution of 12 hours.
COSMO-LEPS	FEWS-NL *See above* FEWS-DE Rhine Catchments, Federal Institute of Hydrology, Germany FEWS-Po, ARPA-SIM, Bologna, Italy FEWS-Rhine (experimental), Federal Office for the Environment, Switzerland	Limited-area Ensemble Prediction System. This 16 member EPS is obtained by running a non-hydrostatic limited area model, nested on the members of the ECMWF-EPS ensemble. The ECMWF-EPS ensembles used in providing the 16 member forecast are obtained through a cluster analysis of the full 51 EPS members for three ensuing forecasts. The resulting 10 km resolution ensemble is much better suited to resolving severe weather at small scales (Marsigli *et al.*, 2005).
SRNWP-PEPS	FEWS-Rhine (experimental) *See above*	This ensemble of short range NWP products is actually a multi-model, or poor man's ensemble (Quiby and Denhard, 2003). The ensemble is constructed using the deterministic high resolution NWP models from participating Meteorological agencies across Europe, with up to 21 ensemble members being available at any one time (depending on how many contributing deterministic forecasts are available). The lead time of each member varies, as well as the resolution and the spatial domain.

being uniform in length, and with uniform spatial and temporal resolution. The SRNWP-PEPS ensemble is an exception to this. This is in effect a multi-model ensemble, with differing spatial and temporal resolutions, as well as differing domains for each ensemble member. This variability is

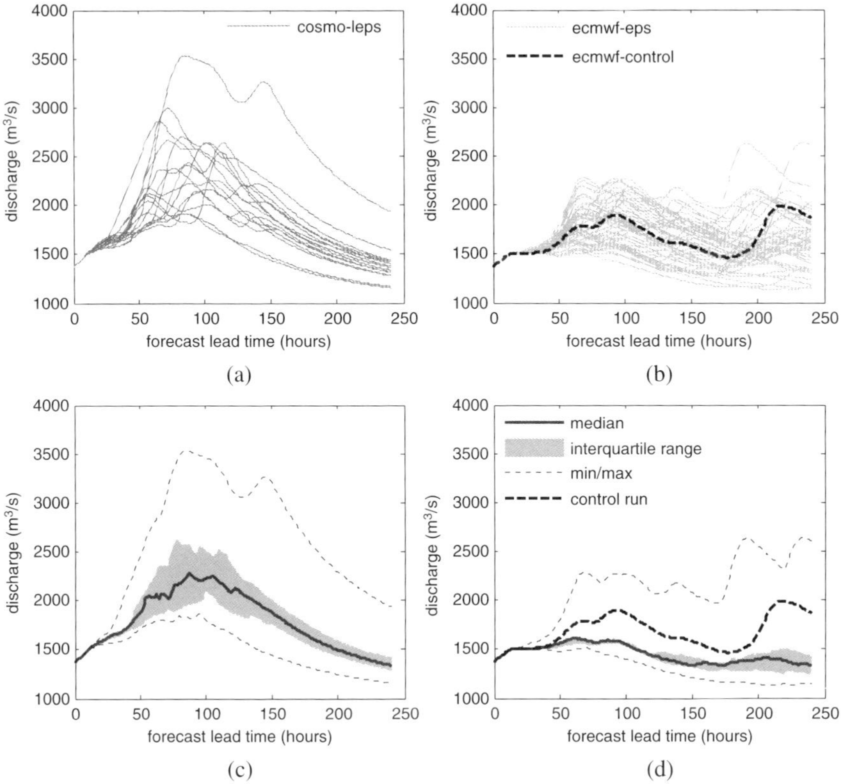

Figure 19.9. Example of two NWP ensemble forecasts for the River Rhine at Maxau, COSMO-LEPS and ECMWF-EPS. Both forecasts have the base time of 1st March 2007 06:00 UTC; (a) and (b) show the raw ensemble outputs, while (c) and (d) shows the parameterised ensemble outputs.

quite challenging to run, with the ran properties for each member being adapted dynamically by Delft FEWS depending on the properties of each of the contributing deterministic NWP models. The variability in lead-time also creates difficulties in interpreting the results using standard statistical parameterisations as the number of ensemble members to consider will differ with lead-time.

Figure 19.9 provides an example of outputs from both the ECMWF-EPS ensemble and the COSMO-LEPS ensemble for the Rhine at Maxau for the same forecast base time. In the upper two plots the raw ensemble outputs are shown, 16 members for the COSMO-LEPS ensemble and 50+ members control run for the ECMWF-EPS ensemble. The lower plots show the parameterised ensemble results, showing the minimum and maximum of all members, the median and the inter-quartile range. The ensemble results can be seen to be quite different for these two NWP ensembles. These differences are attributed to scale effects, as this gauge is towards the upper end of the catchment. For gauges further down in the catchment these differences are less pronounced. There is for both also an under-representation of spread at the shorter lead-times. This is an obvious consequence of considering only the meteorological uncertainties, as described through the meteorological ensemble forecast. The meteorological uncertainties become significant only at lead times greater than the response time of the runoff and routing processes in the catchment (see Figure 19.6).

19.4.3. *Quantifying uncertainty through multi-model ensembles*

Considering only the meteorological uncertainties as discussed above shows that at the shorter lead times there is an under-representation of uncertainty. At these lead-times, uncertainties in the process models will have a significant contribution to the overall uncertainty (see Figure 19.6) and when considering uncertainties in the metrological predictions, only these are clearly not addressed. It is widely accepted that in process models such as runoff response and routing models there may be multiple-model structures and model parameterisations that simulate the behaviour of the true system to an equally acceptable degree (Beven and Freer, 2001; Duan *et al.*, 2007). An approach to addressing this is to apply an ensemble of models, either as a multi-model ensemble where optimal calibrations of models of differing structure are used in the forecast cascade (Duan *et al.*, 2007; Georgekakos *et al.*, 2004; Regonda *et al.*, 2006), or through running multiple parameter sets of the same model structure (Weerts *et al.*, 2003). Predictions made with the differing models can be post-processed through, for example, Bayesian model averaging to improve the forecasting skill (Duan *et al.*, 2007).

Application of such multiple-model ensembles requires significant computational resources, as well as extensive knowledge and support requirements for multiple process models, making the approach organisationally difficult to implement in an operational environment. A few cases of operational forecasting systems implementing the Delft FEWS framework do, however, apply a more pragmatic approach to running multiple model ensembles. In the National Flood Forecasting System in England and Wales (Whitfield, 2005) and in FEWS Scotland (Cranston *et al.*, 2007), there are several forecast locations where multiple models are used to predict flows and levels at the same point. Typically the multiple models are chosen to improve forecasting resilience. In several reaches of the River Severn, for example, a hydrodynamic model is applied as the primary forecast model. However, should this fail, a simple hydrological routing model is also run, and as a last option a correlation model relating observed upstream flows and levels to the downstream point may also be applied. Under most conditions all models will be available, thus providing some indication of the variability in the forecast due to differing model structures, and providing a sanity check on the forecasts made with the primary model.

19.4.4. *Quantifying uncertainty in seasonal predictions*

Most of the applications of both the empirical ensembles and the meteorological and hydrological ensembles described above consider forecasts for the short and medium ranges. For water resources planning, volumetric forecasts with even longer lead-times may be desirable. Such seasonal forecasts can clearly be derived through extending the lead-time of the meteorological ensemble forecasts, using for example downscaled GCM predictions (Luo *et al.*, 2007) and processing these through the model cascade. Another approach available within the Delft FEWS framework follows the empirical ensemble streamflow prediction procedure proposed by Day (1985). In this procedure an empirical ensemble of precipitation and temperature inputs is sampled from the validated long-term time series of catchment average temperature and precipitation. A sample is drawn from each available year in the historical series using the current day as the starting point. The ensemble thus created is a representation of the climatology of catchment temperature and precipitation, and can be run through the forecast model cascade, resulting in climatology-based seasonal forecast, conditional on the states of the system at the time of forecast.

Figure 19.10 shows an example of a seasonal forecast for the gauge at Dilworth on the Buffalo River in the northern US. Figure 19.10a shows the mean catchment precipitation series from 1948–2004. This is re-sampled to create an ensemble of 54 members. Figure 19.10b shows the cumulative precipitation of each ensemble member, as well as the ensemble mean. Figure 19.10c shows the resulting hydrological forecast, conditioned on the initial states from January 2005. The ensemble mean is shown as well as the 10% and 90% confidence limits. Note that spring snowmelt dominates flooding in this basin.

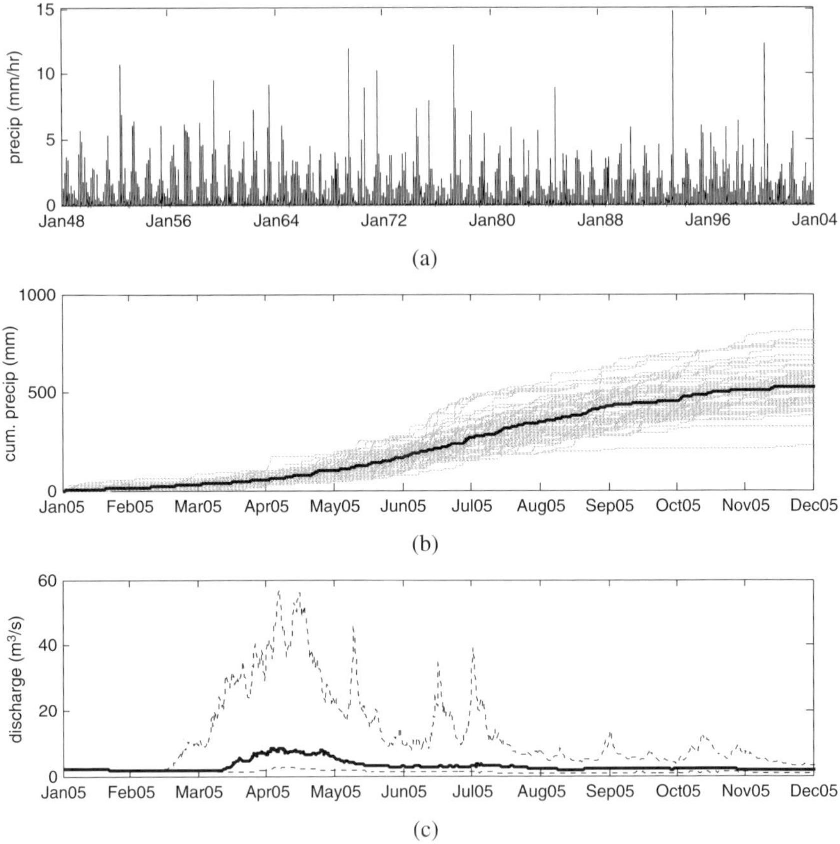

Figure 19.10. Seasonal forecast for Dilworth on the Buffalo River, MN, USA. (a) Mean catchment precipitation series from 1948 through 2004. (b) Cumulative precipitation of each ensemble member, as well as the ensemble mean. (c) Ensemble mean and 10–90% confidence interval of the forecast discharge.

19.5. Quantifying and Reducing Predictive Uncertainty with the Bayesian Forecasting System (BFS) Approach

The previous section discussed several approaches of addressing uncertainties in either the inputs or in the model structure. Particularly when only uncertainties in the meteorological inputs are considered, the representation of the uncertainty at the short lead-times is seen to be too low. As there will be model error at these short lead-times, this under-representation of uncertainty will lead to lower skill at these shorter lead-times. It is at these shorter lead-times, however, that the prime interest of flood warning lies, with the longer lead-times being used more as guidance. A more comprehensive approach to gain a description of the overall uncertainty of a flow forecast available within the Delft FEWS framework is through the application of Bayesian revision. Krzysztofowicz (1999) proposed the Bayesian Forecasting System (BFS) theory for stream flow predictions. This theory constitutes a general framework for Bayesian inference on the uncertainty of a flow forecast, while using deterministic hydrological and/or hydraulic models.

The basic concept of the BFS is to derive the uncertainty of a forecast by "revising" *prior* knowledge on the behaviour of the system over a historical period of operation. If performed correctly, the result of the revision process, referred to as *posterior* density function, represents a reliable assessment of the uncertainty of the forecast, which is conditional on a whole range of information available at the begin of a forecast. It is referred to as *predictive* uncertainty (Krzysztofowicz, 2001; Todini, 2007).

A simple Bayesian uncertainty processor for a flood forecasting system, based on the application of Bayes theorem, can be formulated in terms of the random variables s_n,h_n,h_0 (Krzysztofowicz and Kelly, 2001) as follows:

$$\phi_n(h_n|s_n, h_0) = \frac{f_n(s_n|h_n, h_0)g_n(h_n|h_0)}{k_n(s_n|h_0)}, \tag{19.1}$$

where s_n is the flow forecasted at lead time n (expressed in the number of hours or days), h_n is the flow which has been retroactively observed for day n, and h_0 is the flow observed at the onset of the forecast. We note that we have limited the conditioning to the random variables h_0 and h_n. However, the formulation can be arbitrarily expanded to include additional conditioning variables, if necessary.

The conditional probability density $f_n(g|g)$ is a likelihood of actual discharges, given a model forecast and conditioning observation(s). The conditional density $g_n(g|g)$ is a prior probability density function on the flow predicted for day n, conditional on the observation h_0. The denominator $k_n(s_n|h_0)$ is the expected density on the forecasted flow s_n given by the total probability law:

$$k_n(s_n|h_0) = \int_{-\infty}^{\infty} f_n(s_n|h_n, h_0)g_n(h_n|h_0)dh_n. \tag{19.2}$$

The prior knowledge on the system, which is stochastically described by the conditional probability density function $g_n(h_n|h_0)$, is based on assumed probability distributions on water levels. A proper specification of the prior density is essential in obtaining an informative posterior. Failing to do so may compromise the performance of the processor. The determination of an adequate prior function constitutes a challenge, especially in the absence of a sufficiently long historic time series of observations, as is the case in poorly monitored basins.

In the most simplistic case, the prior density is assumed as constant function, $g_n(h_n) = c$, whereby all possible flow rates are regarded as equally probable (uninformative prior). Such a prior will, however, not deliver a posterior density with adequate informative content about the expected uncertainty of the flow forecast. However, with sufficient observations available, more elaborate prior densities can be proposed. The densities can be subsequently conditioned on additional information, such as multiple stream flow observations at locations further upstream and different points in time.

The revision of the prior density is performed by means of a combination with a likelihood function, as stated in Equation (19.1). The likelihood function is a probability density on future flow rates for a given lead-time, conditional on forecasts produced by the flow forecasting model. The likelihood can also be conditioned on additional information, such as multiple flow observations at locations further upstream, as shown in Reggiani and Weerts (2007, 2008). In fact it is a stochastic specification of the forecasting model error, and enters the processor via Equation (19.1). The likelihood function is ultimately responsible for importing model bias or systematic peak time lag errors into the Bayesian revision process.

Another challenge in applying Equation (19.1) consists in the estimation of parameterised expressions for the density functions $f_n(g|g)$ and $g_n(g|g)$. Performing the underlying statistical analysis online during a real-time flow forecast is generally undesirable due the computational efforts required by processing large amounts of data. Extended calculation times are incompatible with the time constraints imposed by forecasting operations.

A solution to this problem has been proposed by Krzysztofowicz and Kelly (2001). They first carry out a statistical analysis of long series of observations against forecasts off-line, and subsequently parameterise empirical probability distributions via suitable parametric models. These models, for example Weibull or gamma distributions, imply fitting two or three parameters to the data points. In more complex situations, modelled

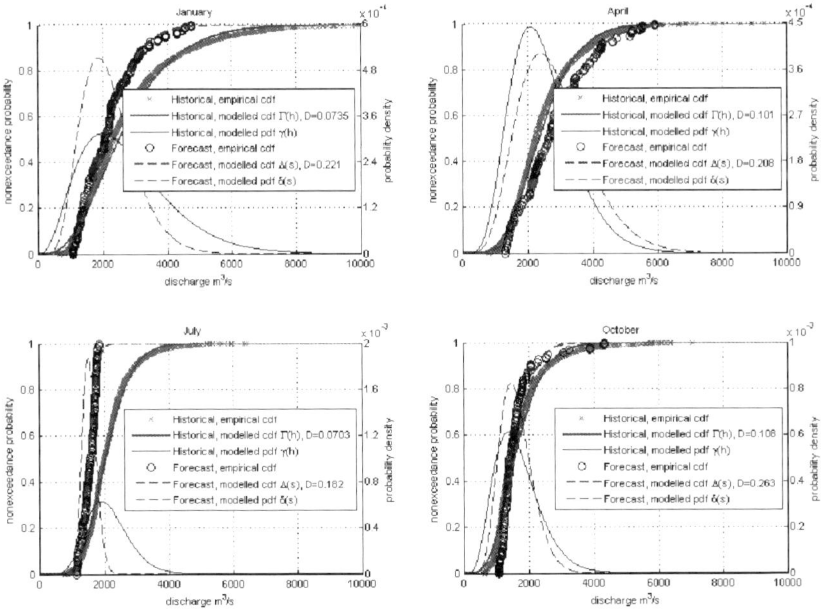

Figure 19.11. Empirical and modelled probability distributions and densities for selected months, established at the gauge of Lobith on the River Rhine.

distributions obtained by piecewise fits can also be considered. These can then be used online through rapid evaluation of algebraic expressions, with a significant reduction of the calculation time involved.

Figure 19.11 shows an example of empirical distributions of observed and modelled water levels at the gauging station Lobith on the River Rhine. The four sub-plots refer to probability distributions for selected months, representing the seasons winter, spring, summer and winter, respectively. The empirical distributions on the observed water levels are performed over a 100+ year series of observations from 1st January 1901 to 1st June 2007. The distribution of the modelled flows refers to a period from 1st June 2004 to 1st June 2007, for which continuous forecasts were available.

Figure 19.12 presents a sequence of plots showing the evolution of the predictive uncertainty for an ensemble stream-flow forecast produced with the ECMWF ensemble weather prediction system (Molteni and Buizza, 1996) applied to the River Rhine. The resulting ensemble forecast includes 50 member forecasts plus 1 control run. The lead-time ranges from 1–6 days. The boxes show the observed and forecasted discharges at the gauging station at Lobith on the 9th February 2007. The vertical lines in the lower

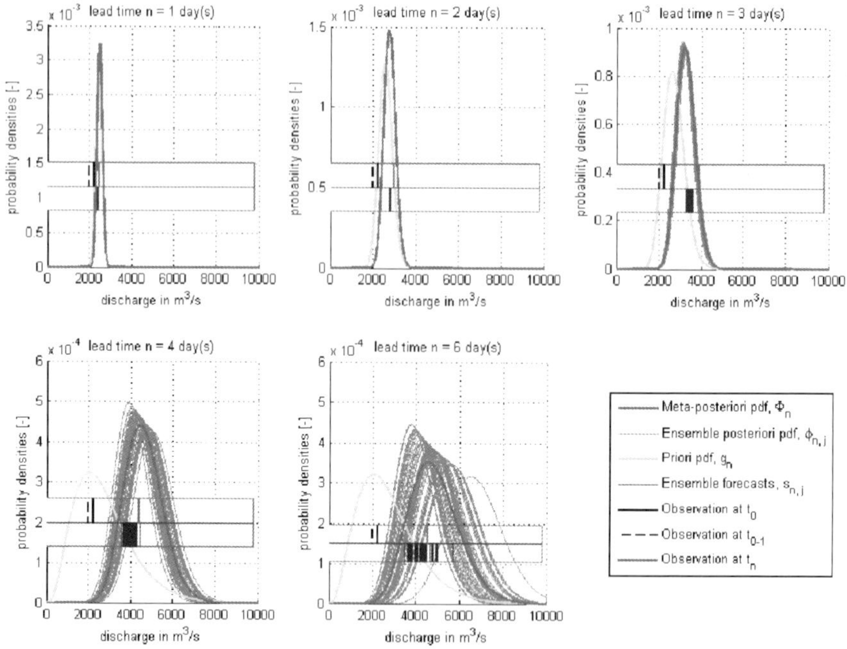

Figure 19.12. Prior and revised posterior distributions of an ensemble stream flow prediction for a flood event on the 9th February 2007 at the gauge of Lobith on the River Rhine.

box are the forecast discharge ensembles; the dashed vertical line in the upper box is the discharge that has been observed on that particular forecasting day. The continuous black line is the discharge observed at Lobith at the onset of the forecast. It can be observed that for lead-times of 1 and 2 days the forecast discharge and the conditional probability distributions collapse onto one single curve. This can be explained through the mutual resemblance of the weather forecast ensemble members during the first 24 hours. From day 3 onwards the forecasted discharges and respective density functions begin to diverge.

The effect of the Bayesian revision of the prior density is evident. The processor, which has been trained on the basis of additional information from a forecasting model and upstream observations, learns from historical experience and delivers a revised posterior distribution. With increasing lead time the posterior densities are shifted, whereby the peak of the distribution (corresponding to maximum probability of occurrence) moves closer towards the (retrospectively) observed discharge.

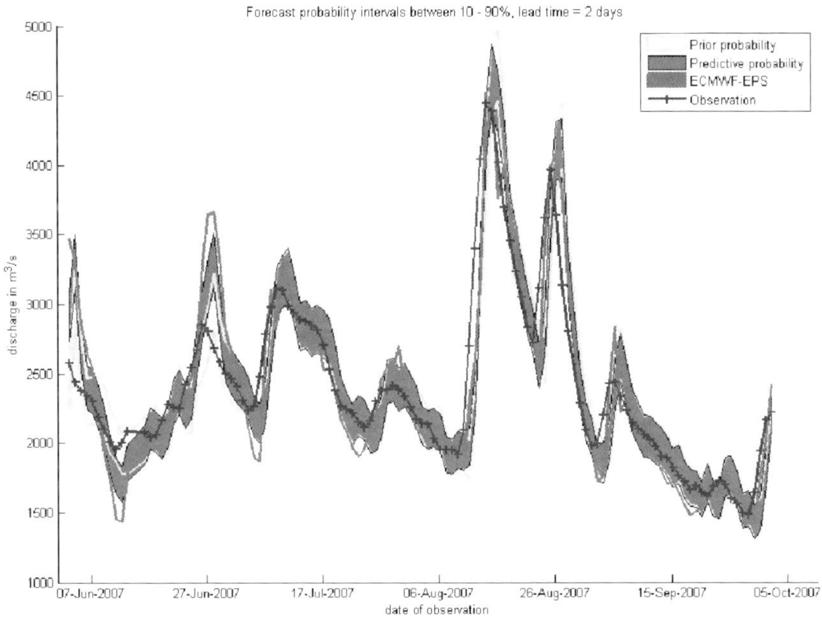

Figure 19.13. Verification of the Bayesian processor for the period 1st June 2007–1st October 2007 on the River Rhine at Lobith.

Figure 19.13 shows a verification of the uncertainty processor for the period between June to September 2007 and a forecasting lead time of 2 days. The continuous line indicates the observed discharge, the lighter shadowed area shows the 10%–90% probability interval for the prior, and the darker area represents the respective probability interval for the posterior density. The prior probability density function has been conditioned on upstream observations.

The envelope determined by the lightest grey shaded area indicates the uncertainty band for the unprocessed ensemble forecast, derived by simple probability ranking. It represents the 10%–90% probability interval for the total uncertainty of the model chain, expressed by the total probability $P(s_n > s^*)$. While $P(s_n > s^*)$ represents the unconditioned probability of the model chain, the Bayesian processor delivers a posterior probability density on the expected stream flow, h_n, conditional on all information (observations and model forecast) available at the start of the forecast.

The reliable estimation of the predictive uncertainty provided through the framework described offers significant possibilities for objective uncertainty assessment. At the time of writing this framework was yet to be applied in an operational forecasting system, but verification of the results for the Rhine basin show a improvement of forecast skill at both short and long lead times (Reggiani *et al.*, 2008).

19.6. Discussion and Conclusions

The range of methods in quantifying and reducing uncertainties in hydrological forecasts as applied in the operational forecasting systems described above clearly illustrate the importance of explicit consideration of the uncertain forecast when used as an element in the flood warning decision process.

The empirical ensemble methods described have often been adopted out of pragmatic considerations of the forecasters involved in the operational forecasting processes, to gain insight into forecast uncertainty prior to more complex approaches, such as that provided through Bayesian revision becoming operational (Halquist, 2006). Although these empirical methods lack the statistical formalism of the more advances ensemble approaches, their value to the forecaster should not be underestimated. A good example is given in the running of different combinations of input data, including a forecast with zero forecast rainfall (Figure 19.8). Whilst this zero rainfall forecast may not be very useful in indicating a potential threshold crossing at the onset of a flood event, it does clearly indicate to the forecaster the degree to which the predicted hydrograph is affected by the forecast precipitation. The confidence the forecaster has invested in the precipitation forecast will in turn determine how confident that forecaster will be on issuing a warning that is predominantly in the domain of the hydrograph, influenced either by the observed (albeit also uncertain) precipitation or the uncertain forecast precipitation.

It is clear that the use of ensembles of meteorological inputs, combined with advanced methods of post-processing, have significant benefits in providing a more complete and statistically sound quantification of uncertainty in the forecast. The utility of ensemble forecasts in increasing the skill of hydrological predictions has been shown by several authors (Regimbeau *et al.*, 2007; Roulin, 2007), and Roulin (2007) showed that using the ensemble forecast to underpin flood warning has more relative economic value than using the deterministic forecast alone, or even the ensemble mean. Such translation of the probabilistic forecast into an economical value that balances the expected damage due to a flood event against the expected loss due to a false warning (and possible ensuing evacuation) can thus provide stakeholders with an instrument for rational decision making under conditions of uncertainty (Raiffa and Schlaifer, 1961).

The examples described from operational systems utilising the Delft FEWS forecasting framework also show that many of the technological constraints once apparent in running ensemble forecasts operationally have now largely been overcome. This allows methods that provide more complete analyses of the full predictive uncertainty, such as that provided in the Bayesian revision approach to be applied in the operational forecasting domain as opposed to being used only in academic study. Despite this potential in the operational use of the ensemble methods, and increasing adoption in operational forecasting systems, significant challenges remain in the dissemination of these ensemble results. Operationally, a warning issued on the basis of a (probabilistic) forecast is never a fully automated process, but is done following interpretation of the forecast results by the duty forecaster. Surveys of both lay persons (Handmer and Proudley, 2007) and hydrologists (Demeritt *et al.*, 2007) and their response to probabilistic forecast products suggest that difficulty in interpretation of such probabilistic forecasts is not to be underestimated. The examples of ensemble forecasts shown in the various figures in this chapter demonstrate some of the methods available with the Delft FEWS framework for dissemination of ensemble forecasts. The visual information obtained from the different methods is quite clear in Figure 19.12, when the raw "spaghetti" output from an ensemble forecast is compared against the same forecast represented as a statistical representation. Other methods of summarising probabilistic forecast information to support interpretation by the forecaster are being explored, including examples such as those presented in Ramos *et al.* (2007) and Werner *et al.* (2005).

For all methods of quantifying uncertainty in the forecast, it is important that once the forecast has been made these are suitably verified (Schaake *et al.*, 2007). Methods for verification of forecasts are well-established (Wilks, 1995), and such verification provides clear insight into the value and the skill of the ensemble predictions at different lead times, giving invaluable information to the forecaster in interpreting the forecast products. This again reflects the importance of the forecaster in the flood warning process. The primary objective of reducing and quantifying uncertainties in this process is to increase the confidence with which warnings are issued, and as a result, increase the reliability of these flood warnings.

References

Bartholmes, J. and Todini, E. (2005). Coupling meteorological and hydrological models for flood forecasting, *Hydrol. Earth Syst. Sci.*, **9**, 333–346.

Broersen, P.M.T. and Weerts, A.H. (2005). "Automatic Error Correction of Rainfall-Runoff Models in Flood Forecasting Systems" in: *Proceedings IEEE/IMTC Conference*, Ottawa, Canada, pp. 1531–1536.

Brown, J.D. and Heuvelink, G.B.M. (2007). The data uncertainty engine (DUE): a software tool for assessing and simulating uncertain environmental variables, *Comput. Geosci.*, **33**, 172–190.

Buizza, R. (2004). "Use of Ensemble Forecasts for Hydrological Applications" in: De Roo, A. and Thielen, J. (eds), *2nd European Flood Alert System Workshop*, Ispra, European Commission, DG Joint Research Centre.

Bürgi, T. (2002). "Operational Flood Forecasting in Mountainous Areas — An Interdisciplinary Challenge" in: Spreafico, M. and Weingartner, R., *International Conference in Flood Estimation*, CHR Report II-17. Bern, CH, pp. 397–406.

Clark, M., Gangopadhyay, S., Hay, L. *et al.* (2004). The Schaake shuffle: a method for reconstructing space-time variability in forecasted precipitation and temperature fields, *J. Hydrometeorol.*, **5**, 243–262.

Collier, C. and Krzysztofowicz, R. (2000). Quantitative precipitation forecasting, *J. Hydrol.*, **239**, 1–2.

Cranston, M., Werner, M., Janssen, A. *et al.* (2007). "Flood Early Warning System (FEWS) Scotland: An Example of Real Time System and Forecasting Model Development and Delivery Best Practice" in: *Defra Conference on Flood and Coastal Management*, Paper 02–3, York, UK.

Day, G.N. (1985). Extended streamflow forecasting using NWSRFS, *J. Water Res. Pl.-ASCE*, **111**, 157–170.

Demeritt, D., Cloke, H., Pappenberger, F. *et al.* (2007). Ensemble predictions and perceptions of risk, uncertainty and error in flood forecasting, *Environ. Haz.*, **7**, 115–127.

De Roo, A., Gouweleeuw, B., Thielen, J. *et al.* (2003). Development of a European flood forecasting system, *Intl. J. River Basin Manage.*, **1**, 49–59.

Diomede, T., Marsigli, C., Nerozzi, F. *et al.* (2006). Quantifying the discharge forecast uncertainty by different approaches to probabilistic quantitative precipitation forecast, *Adv. Geosci.*, **7**, 189–191.

Duan, Q., Ajami, N.K., Gao, X. *et al.* (2007). Multi-model ensemble hydrologic prediction using Bayesian model averaging, *Adv. Water Res.*, **30**, 1371–1386.

El Serafy, G.Y., Gerritsen, H., Hummel, S. *et al.* (2007). Application of data assimilation in portable operational forecasting systems — the DATools assimilation environment, *Ocean Dynam.*, **57**, doi 10.1007/s10236-007-0124-3.

El Serafy, G.Y. and Mynett, A.E. (2004). "Comparison of EKF and EnKF in SOBEK-River: Case Study Maxau-IJssel" in: Liong, S.-Y., Phoon, K.-K. and Babovic, V. (eds), *Proceedings of the 6th international conference on Hydroinformatics*, World Scientific, Singapore, pp. 513–520.

Evensen, G. (1994). Sequential data assimilation with a non-linear quasi-geostrophic model using Monte Carlo methods to forecast error statistics, *J. Geophys. Res.*, **97**, 905–924.
Evensen, G. (2003). The ensemble Kalman filter: theoretical formulation and practical implementation, *Ocean Dynam.*, **53**, 343–367.
Georgakakos, K.P., Seo, D.-J., Gupta, H. *et al.* (2004). Towards the characterization of streamflow simulation uncertainty through multimodel ensembles, *J. Hydrol.*, **298**, 222–241.
Golding, B.W. (1990). Quantitative precipitation forecasting in the UK, *J. Hydrol.*, **239**, 286–305.
Habets, F., LeMoigne, P. and Noilhan, J. (2004). On the utility of operational precipitation forecasts to served as input for streamflow forecasting, *J. Hydrol.*, **293**, 270–288.
Halquist, J.B. (2006). "Use of HPC QPF Confidence Interval Forecasts to Produce an Ensemble of River Forecasts" in: *20th Conference on Hydrology, 86th Annual Meeting of the American Meteorological Society*, Atlanta, US.
Handmer, J. and Proudley, B. (2007). Communicating uncertainty via probabilities: the case of weather forecasts, *Environ. Haz.*, **7**, 79–87.
Ibbitt, R.P., Henderson, R.D., Copeland, J. *et al.* (2000). Simulating mountain runoff with meso-scale weather model rainfall estimates: a New Zealand experience, *J. Hydrol.*, **239**, 19–32.
Krzysztofowicz, R. (1999). Bayesian theory of probabilistic forecasting via deterministic hydrologic model, *Water Resour. Res.*, **35**, 2739–2750.
Krzysztofowicz, R. (2001). The case for probabilistic forecasting in hydrology, *J. Hydrol.*, **249**, 2–9.
Krzysztofowicz, R. and Kelly, K.S. (2000). Hydrologic uncertainty processor for probabilistic river stage forecasting, *Water Resour. Res.*, **36**, 3265–3277.
Lardet, P., Cranston, M. and Werner, M. (2007). "Flood Early Warning System (FEWS) for South West Scotland: A State-of-the-Art Real Time Forecasting System" in: *Charting The Course: New Perspectives in Floodplain Management*, ASFPM 31st Annual Conference, Norfolk, USA.
Lettenmaier, D.P. and Wood, E.F. (1993). "Hydrologic Forecasting" in: Maidment, R.D. (ed.), *Handbook of Hydrology*, McGraw-Hill, New York.
Luo, L., Wood, E.F. and Pan, M. (2007). Bayesian merging of multiple climate model forecasts for seasonal hydrological predictions, *J. Geophys. Res.*, **112**, D10102, doi:10.1029/2006JD007655.
Madsen, H., Butts, M., Khu, S. *et al.* (2000). "Data Assimilation in Rainfall Runoff Forecasting", in: *Proceedings of the 4th Hydroinformatics Conference, Iowa, USA*, IAHR, Cedar Rapids, IA, US.
Madsen, H. and Canizares, R. (1999). Comparison of extended and ensemble Kalman filters for data assimilation in coastal area modelling, *Int. J. Numer. Meth. Fl.*, **31**, 961–981.
Marsigli, C., Boccanera, F., Montani, A. *et al.* (2005). The COSMO-LEPS mesoscale ensemble system: validation of the methodology and verification, *Nonlinear Proc. Geoph.*, **12**, 527–536.

Molteni, F., Buizza, R., Palmer, T.N. *et al.* (1996). The ECMWF ensemble prediction system: methodology and validation, *Q. J. Roy. Meteor. Soc.*, **122**, 73–119.

Pappenberger, P., Matgen, P., Beven, K.J. *et al.* (2007). Influence of uncertain boundary conditions and model structure on flood inundation predictions, *Adv. Water Resour.*, **29**, 1430–1449, doi:10.1016/j.advwatres.2005.11.012.

Quiby, J. and Denhard, M. (2003). "SRNWP-DWD Poor-Man Ensemble Prediction System: The PEPS Project", *Eumetnet News*, 8, 1–6. Available at http://www.eumetnet.eu.org/ (Accessed on 08/01/2008).

Raiffa, H. and Schlaifer, R. (1961). *Applied Statistical Decision Theory*, MIT Press, Cambridge, MA.

Ramos, M.H., Bartholmes, J. and Thielen-del Pozo, J. (2007). Development of decision support products based on ensemble forecasts in the European flood alert system, *Atmos. Sci. Lett.*, **8**, 113–119.

Refsgaard, J.C. (1997). Validation and intercomparison of different updating procedures for real-time forecasting, *Nord. Hydrol.*, **28**, 65–84.

Reggiani, P., Renner, M., Weerts, A.H. *et al.* (2008). Uncertainty assessment via Bayesian revision of ensemble stream flow predictions in the operational River Rhine forecasting system, *Water Resour. Res.*, **45**, W02428–W02442.

Reggiani, P. and Weerts, A.H. (2007). Probabilistic quantitative precipitation forecast for flood prediction: an application, *J. Hydrometeorol.*, DOI:10.1175/2007JHM858.1.

Reggiani, P.M. and Weerts, A.H. (2008). A Bayesian approach to decision making under uncertainty: an application to real-time forecasting in the River Rhine, *J. Hydrol.*, **356**, 56–69.

Regimbeau, F., Habets, F., Martin, E. *et al.* (2007). Ensemble streamflow forecasts over France, *ECMWF Newsletter*, **111**, 21–27.

Roulin, E. (2007). Skill and relative economic value of medium-range hydrological ensemble predictions, *Hydrol. Earth Syst. Sci.*, **11**, 725–737.

Schaake, J.C., Hamill, M., Buizza, R. *et al.* (2007). HEPEX: the hydrological ensemble prediction experiment, *B. Am. Meteorol. Soc.*, **88**, 1541–1547.

Sprokkereef, E., Buiteveld, H., Eberle, M. *et al.* (2001). Extension of the Flood Forecasting Model FloRIJN, *NCR Publication* 01/12/2001. Available at www-ncr-web.org (Accessed on 08/01/2008).

Todini, E. (2007). Hydrological modeling: past, present and future, *Hydrol. Earth Syst. Sci.*, **11**, 468–482.

Tracton, M.S. and Kalnay, E. (1993). Operational ensemble prediction at the National Meteorological Center, *Weather Forecast.*, **8**, 379–398.

UN ISDR (2004). *International Strategy for Disaster Reduction: Guidelines for Reducing Flood Losses*, UN- ISDR, Bonn, Germany.

van Leeuwen, P.J. (2003). A variance minimizing filter for large-scale applications, *Mon. Weather Rev.*, **131**, 2071–2084.

Weerts, A.H., Diermanse, F., Reggiani, P. *et al.* (2003). Assessing and quantifying the combined effect of model parameter and boundary uncertainties in model based flood forecasting, *Geophys. Res. Ab.*, **5**, 14564.

Weerts, A. and El Serafy, G.Y. (2006). Particle filtering and ensemble Kalman filtering for state updating with hydrological conceptual rainfall-runoff models, *Water Resour. Res.*, **42**, W09403, doi:10.1029/2005WR004093.

Werner, M., Cranston, M., Harrison, T. *et al.* (2009). Recent developments in operational flood forecasting in England, Wales and Scotland, *Meteorol. Appl.*, **16**, 13–32.

Werner, M. and Heynert, K. (2006). "Open Model Integration — A Review of Practical Examples in Operational Flood Forecasting" in: Gourbesville, P., Cunge J., Guinot, V. *et al.* (eds), *7th International Conference on Hydroinformatics: Volume 1*, Research Publishing, Nice, pp. 155–162.

Werner, M., Reggiani, P., De Roo, A. *et al.* (2005). Flood forecasting and warning at the river basin and at the European scale, *Nat. Haz.*, **36**, 25–42.

Werner, M., Schellekens, J., Gijsbers, P. *et al.* (2013). The Delft-FEWS flow forecasting system, *Environ. Modell. Softw.*, **40**, 65–77.

Werner, M., van Dijk, M. and Schellekens, J. (2004). "Delft-FEWS: An Open Shell Flood Forecasting System" in: Liong, S.-Y., Phoon, K.-K. and Babovic, V (eds), *6th International Conference on Hydroinformatics*, World Scientific, Singapore, pp. 1205–1212.

Werner, M. and Whitfield, D. (2007). On model integration in operational flood forecasting, *Hydrol. Process.*, **21**, 1519–1521.

Whitfield, D. (2005). "The National Flood Forecasting System (NFFS) of the UK Environment Agency" in: Balabanis, P., Lumbroso, D. and Samuels, P. (eds), *International Conference on Innovation, Advances and Implementation of Flood Forecasting Technology*, Tromsø, Norway. Available at http://www.actif-ec.net (Accessed on 07/08/2008).

Wilks, D.S. (1995). "Statistical Methods in the Atmospheric Sciences: An Introduction", *International Geophysics Series*, **59**, Academic Press, San Diego.

WMO (2006). *Flood Forecasting Initiative*, report on the technical conference on improved meteorological and hydrological forecasting, Geneva, Switzerland, 20–23 November 2006.

CHAPTER 20

Real-Time Coastal Flood Forecasting

Kevin Horsburgh
National Oceanography Centre, Liverpool, UK

Jonathan Flowerdew
Met Office, Exeter, UK

20.1. Elements of Coastal Flood Forecasting

The coastal flood warning system for the UK was established as a direct result of the worst natural disaster to affect the country in recent times — the 1953 North Sea storm surge. During the night of 31 January, coastal flooding caused the loss of 307 lives in East Anglia and a further 1,795 fatalities in the Netherlands (McRobie *et al.*, 2005). Storm surges are the sea level response to wind stress and atmospheric pressure gradient (Pugh, 1987). Surges and high tides are critical components of total sea level during coastal flood events. Together they may exceed coastal defence thresholds directly, or raise the still water level such that storm wave action can cause significant overtopping or breaching. The operational warning system that was created is called the UK Coastal Monitoring and Forecasting Service (UKCMF) and is summarised diagrammatically in Figure 20.1. UKCMF provides a primary alert service for coastal flood risk to the Environment Agency, which is in turn responsible for local interpretation and warning the public.

This section considers the various sources of uncertainty that may contribute to forecast inaccuracies for coastal flooding. It conveys the limitations of the numerical methods employed to generate sea level predictions, and the means of error quantification. For coastal flooding there are a number of key areas of uncertainty: defence failure; overtopping

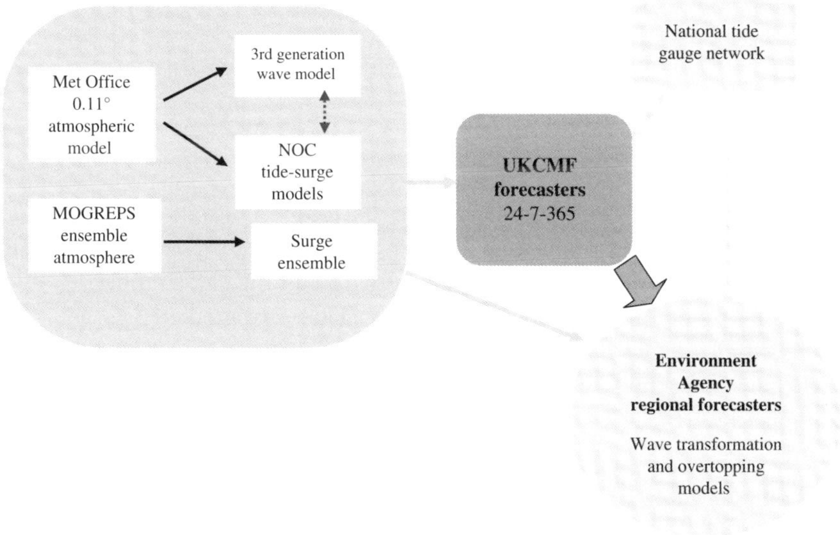

Figure 20.1. Components of the UK Coastal Monitoring and Forecasting service.

rates; accuracy of tidal predictions; accuracy of surge generation; and uncertainty in the meteorological forcing that causes storm surges and waves. In an estuarine environment water levels will also be affected by river flow, and any attempts at forecasting must take fluvial and pluvial processes into account. This level of complexity has been explored in probability forecasting exercises as part of the Flood Risk Management Research Consortium (Xuan, 2007) but here the focus is limited to the marine processes of surge and tide.

Tidal heights for coastal forecasts are derived from tide tables generated using harmonic analysis, since even at the finest resolution (e.g. Jones and Davies, 1996) numerical models do not give comparable accuracy. Storm surges are forecast with a hydrodynamic ocean model which is run once with tide and meteorological forcing, and then again with tides only; the surge is obtained by subtracting the two. Harmonically-predicted tide and computer-modelled surge are then added to provide a time series of water level upon which wave effects are added. The most significant source of uncertainty for storm surge magnitude is the causal meteorology, specifically the wind strength at the sea surface and the horizontal gradients of atmospheric pressure at sea level. This uncertainty can be managed effectively using a suite of surge models forced by a meteorological ensemble.

A case study is presented here that demonstrates the utility of ensemble surge forecasting.

20.2. Uncertain Inputs

20.2.1. *Tidal predictions*

Tidal predictions are obtained from a harmonic analysis of tide gauge observations (usually a dataset of one year or longer). Practically, this involves a least-squares fitting procedure to obtain the amplitudes and phases of a number of distinct frequencies. Tides are predicted to great — but not 100% — accuracy. The observed sea level also contains the effects of atmospheric forcing (the surge) which is noise as far as harmonic analysis is concerned. In this chapter we only use the word "surge" to imply a genuine meteorological contribution to sea level. Generally, the difference between observed sea levels and tidal predictions is properly referred to as the residual.

Some examples of the stability (over time) of tidal constituents are given in Pugh (1987). Tides at most ports are predicted using either 60 or 100 constituents, and the combined instantaneous error can amount to several centimetres of amplitude, as well as phase errors (in the time of high water for instance) of tens of minutes. These errors are proportional to the tidal amplitude so are typically largest in regions of high tidal range such as the Bristol Channel. Horsburgh and Wilson (2007) showed mathematically how small phase advances with respect to the predicted tide create a residual with tidal period that peaks halfway up the rising tide, as illustrated in Figure 20.2. This phenomenon may reflect the limit of accuracy of the harmonic method, or it may result from an increase in water depth due to surge. The latter is one aspect of tide-surge interaction. Tides and surges are (to first order) shallow water waves, with phase speeds of $(gh)^{1/2}$ where h is the water depth and g is the acceleration due to Earth's gravity. A positive surge will therefore increase the water depth and thus the phase speed of both tide and surge as they travel. It is impossible, generally, to distinguish a genuine surge that alters the phase of the tide from a harmonic error. In fact, many properties of a non-tidal residual time series stem from the choice of definition (i.e. subtraction). In an arithmetic sense this is perfectly correct but it can be misleading if one wishes to quantify sea level change due to real meteorological drivers.

One method of estimating the error inherent in harmonic analysis is to run a numerical model with no meteorological forcing. To demonstrate this,

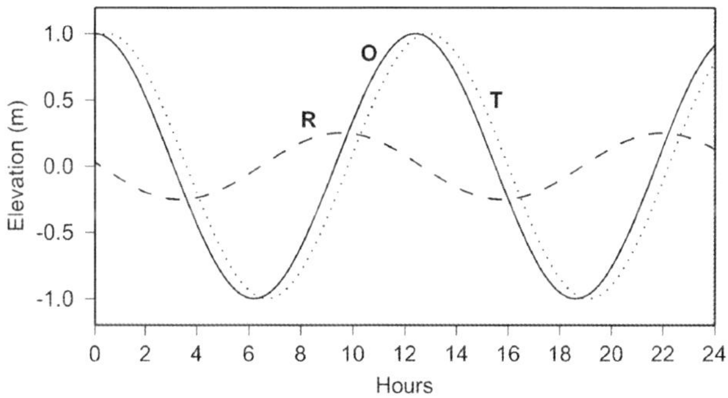

Figure 20.2. Illustration of how a simple phase alteration, with no change to the amplitude, can give rise to a residual with tidal period. The solid line (O) represents observations, the dotted line tidal predictions (T), and the dashed line the corresponding residual (R = O-T).

the NOC CS3X surge model was run for the year 2004, forced only by its lateral tidal boundary conditions (as described in Section 20.2.2). On the basis of that run, a 50-constituent harmonic analysis was performed for the 44 port locations that make up the UK strategic tide gauge network. Using these constituents, the tide at every port was then predicted for 2005. The time series at each site were finally compared with the directly simulated values when the model was restarted for 2005. The Root Mean Square (RMS) error, averaged over all gauge sites, was 7 cm with a maximum value of 29 cm at Newport, in the Bristol Channel. This experiment is unrealistically noise free; if one were to add some white noise to the analysed time series (to represent the effect of weather on the basic process of harmonic analysis) it seems reasonable to conclude that a typical harmonic prediction error for the UK coastline is of the order 10 cm.

During very light winds it is possible to infer harmonic errors from tide gauge data. In the absence of significant wind stress the surge component of any sea level observation can be approximated by a pressure correction, the so-called inverted barometer effect (see Pugh, 1987). This equates a supposed slope of the sea surface to the horizontal pressure gradient, and yields a rule of thumb where sea level changes by 1 cm for each millibar of deviation from the mean sea level pressure of 1012 mb. Figure 20.3 shows such an occurrence where the pattern of tides at Avonmouth was well predicted, yet mean high water errors of 30 cm were recorded over the four day period. Winds over the south-west of the UK were less than 10 m/s during this time, and an inverse barometer correction has been applied to the observed data.

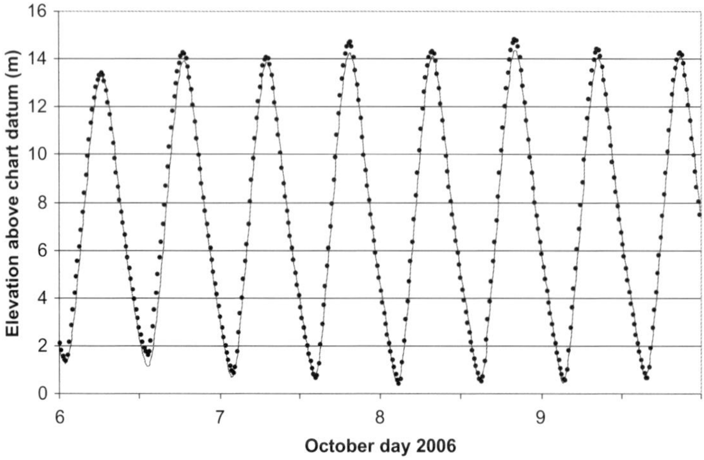

Figure 20.3. Tidal predictions (solid line) and observed inverse-barometer corrected sea level (dots) at Avonmouth tide gauge for the period 6–10 October 2006.

Table 20.1. High water residual error at Avonmouth for 6–10 October 2006.

	RMS error at all high waters for October 2006	Error at 19.30 on 7 October 2006
Standard tide table supplied to STFS	30 cm	44 cm
Analysis based on data for October 2006	20 cm	25 cm
Analysis based on previous 30 days measurements	—	13 cm

In an attempt to improve the predictions a new analysis was performed based on data for the month of October 2006 only. The tide was re-predicted (necessarily using fewer constituents) and both RMS and maximum errors were reduced as seen in Table 20.1. A further improvement was obtained by analysing 30 days of data terminating at 00 UTC on 7 October 2006, and then generating predictions for just 24 hours. Short-term predictions based on recent real-time data offer prospects of an automated analysis-prediction approach to optimise the operational forecasting of the tidal component of sea level.

Obviously this technique is not appropriate for making predictions any further into the future. Furthermore, when stronger winds give rise to any appreciable surge then the short period analyses are likely to be less accurate. The use of artificial intelligence techniques (e.g. fuzzy systems or neural networks) to improve tidal forecasting is an active area of research, having shown some utility in real-time forecasting of sea levels in the North Sea (Randon *et al.*, 2007).

20.2.2. *Uncertainty in the surge models*

Storm surges are the sea level response to wind stress and atmospheric pressure gradient, and they are a critical component of total sea level during coastal flood events. Numerical models for storm surge prediction solve the governing equations of fluid flow on regular grids. A good review of storm surge modelling is given by Gonnert *et al.* (2001), and operational coastal flood warning systems for European seas are described in detail by Flather (2000). To achieve finer resolution around coastlines, it is possible to nest finer grids (e.g. Greenberg, 1979) or use finite element techniques (e.g. Jones and Davies, 2006). Since the wind and atmospheric pressure gradient are spatial fields that give rise to a free surface gradient, then two-dimensional depth-averaged models of both surges and wind-driven currents have been very successful as predictive tools in shelf seas. A recent review of operational capabilities (Ryabinin *et al.*, 1996) found that 75% of models used internationally were two-dimensional and depth-averaged, and that their forecast skill was comparable to 3D models that resolved the vertical structure of current. All 2D models typically solve the following equations in discretised form, where Cartesian formulations are used here for simplicity (the equations are more usually expressed in spherical coordinates for regular latitude-longitude grids):

$$\frac{\partial U}{\partial t} + U\frac{\partial U}{\partial x} + V\frac{\partial U}{\partial y} - fV = -g\frac{\partial \eta}{\partial x} - \frac{1}{\rho}\frac{\partial P_a}{\partial x} + \frac{1}{\rho H}(\tau_{sx} - \tau_{bx}), \quad (20.1)$$

$$\frac{\partial V}{\partial t} + U\frac{\partial V}{\partial x} + V\frac{\partial V}{\partial y} + fU = -g\frac{\partial \eta}{\partial y} - \frac{1}{\rho}\frac{\partial P_a}{\partial y} + \frac{1}{\rho H}(\tau_{sy} - \tau_{by}), \quad (20.2)$$

$$\frac{1}{H}\frac{\partial \eta}{\partial t} + \frac{\partial U}{\partial x} + \frac{\partial V}{\partial y} = 0, \quad (20.3)$$

where U and V are vertically averaged velocity fields in the x and y directions, η is the free surface elevation, H is the total depth, f is the Coriolis parameter, g is Earth's gravity, ρ is the uniform density of water, P_a is the atmospheric pressure, τ_{sx} and τ_{bx} are respectively surface and bed stresses in the x direction, with similar terms for the y direction in Equation (20.2).

The current UK operational surge model, CS3X, covers the entire north-west European continental shelf at 12 km horizontal resolution. Its surface boundary conditions are the sea level pressure and 10 m wind fields from the Met Office North Atlantic Extended (NAE) atmospheric model, at a similar spatial resolution (0.11° longitude by 0.11° latitude). Tidal input at the model open boundaries consists of the largest 26 constituents. Finer resolution models of the Bristol Channel and south coast are nested within this outer domain. The model suite runs four times each day and simulations consist of a six hour hindcast portion (where the model is forced with meteorological reanalysis) followed by a 48 hour forecast. The modelled surge is derived by subtracting a tidal model run from one forced by both tide and atmosphere. Validation of the models is performed monthly by comparison with observed sea level data from the UK national tide gauge network. Typical monthly mean RMS errors in the model accuracy are of the order 10 cm, but maximum instantaneous errors at the critical time of high water can be up to 60 cm (Wortley *et al.*, 2007) although such an error is very rare.

The wind stress on the sea surface is usually parameterised with a quadratic formulation:

$$\tau_{sx} = C_D \rho_a W_x |W|, \tag{20.4}$$

where ρ_a is the density of air, $|W|$ is wind speed 10 m above the sea surface, and W_x is its component in the x coordinate direction. The drag coefficient, C_D, is not generally constant but is prescribed to have some dependence on wind speed, and occasionally surface roughness (i.e. the wave field) in coupled wave-surge models (see Wolf, 2008, for a review). Any error in wind strength or direction supplied by the atmospheric model will result in corresponding surge errors. The performance of NOC model codes (effectively the numerics and choice of parameters) is known to be comparable with surge prediction models from other European nations (Flather *et al.*, 2003). Nevertheless, all deterministic models are constrained by the limitations in their forcing, and the non-linear response to that forcing. Forecasters and managers can control for this by understanding the magnitude of such errors, and the circumstances in which they tend to occur, as well as by employing complementary probabilistic forecasting techniques.

Sensitivity studies based on two events illustrate the dependency of surge models on fairly small changes in the wind field. Event 1 was a surge affecting all North Sea ports on 31 October 2006–1 November 2006, where the residual peaked at around 2.2 m at Sheerness. Event 2, on 2–3 December 2006, mainly affected the west coast and produced an observed residual of almost 1 m at Liverpool. On both occasions the models performed well, predicting the surge to within 20 cm at all times, implying that the modelled meteorology was accurate for those events. Figure 20.4a shows the impact of a 10% increase in wind speed over the simulation period. The maximum difference obtained between the standard run and the modified wind run was 40–50 cm over the southern North Sea with larger values in the Thames Estuary, even for this modest perturbation to the wind speed. Larger differences are obtained when the wind error is larger, or when the wind direction also contains systematic inaccuracy as in Figure 20.4b. The time series for Sheerness (Figure 20.4c) illustrates the significance of these differences at a local level.

Figure 20.5 shows the equivalent diagrams for the surge of 2–3 December 2006. This represents a different type of weather pattern where the most significant surge was along the west coast and in the Irish Sea. In this case, when the winds forcing the model were artificially increased by 10%, there was a difference of approximately 30 cm in parts of Liverpool Bay with values up to 50 cm in estuaries and inlets. The selected time series plot is for Liverpool, where the residual is shown. Although the perturbations used in these trials were arbitrary (and the perturbation chosen was time invariant), the results nevertheless demonstrate the consequences of small inaccuracies in the forcing data. Closed marine basins like the North Sea or the Irish Sea effectively act as integrators for any sea level error. These sensitivity studies make a strong case for ensemble-based forecasting of surge, where realistic uncertainties in the meteorology are explicitly handled. In contrast, changes in key parameter choices within the surge model (e.g. the drag coefficients for both bed friction and air-sea coupling) had a less significant effect on the surge response in these trials.

20.3. Ensemble Surge Forecasting

The non-linearity of both meteorological and ocean models means that any deterministic forecast is strongly affected by its initial conditions, as well as choices for those parameters used to describe unresolved physical processes (e.g. bed friction in hydrodynamic models). Another argument for the ensemble forecasting of surges is that a single, deterministic forecast (which contains implicit error) makes it difficult for a forecaster to accurately determine the risk of a particular critical threshold or warning level being

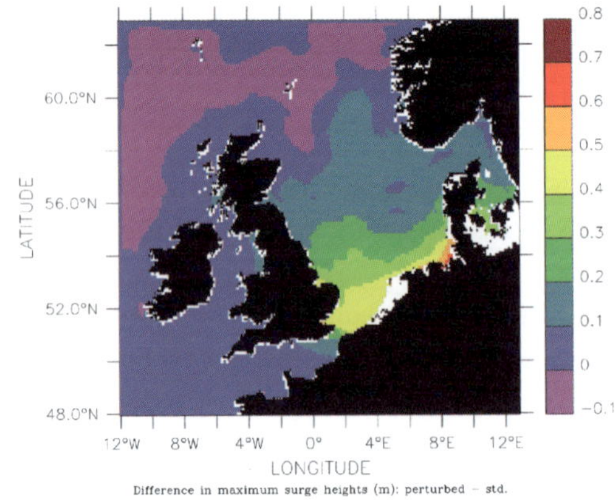

(a)

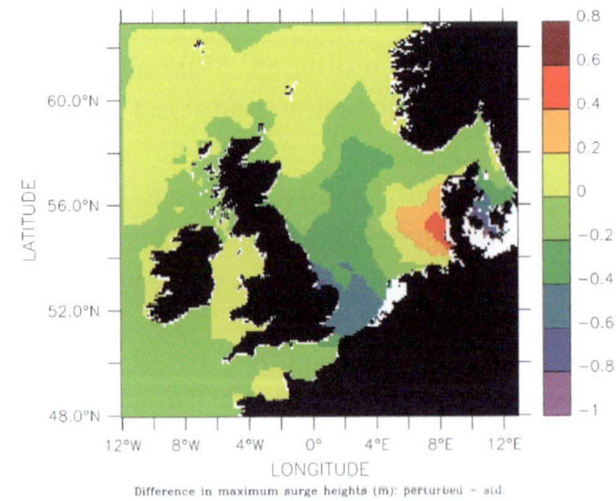

(b)

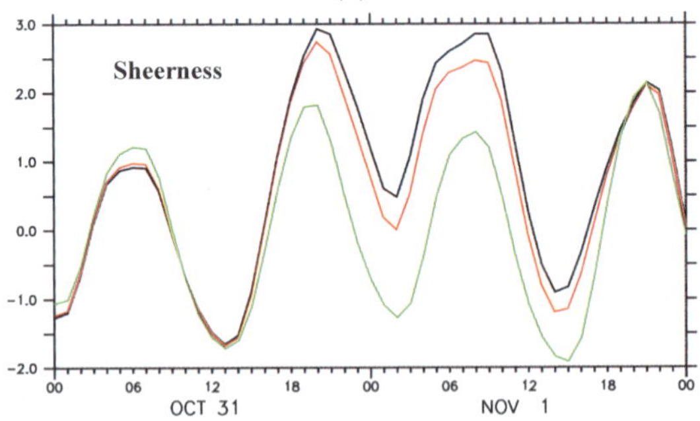

(c)

Figure 20.4. Sensitivity of Event 1 (31 October–1 November 2006) to differences in wind forcing. (a) Maximum elevation difference with 10% increase in wind. (b) Maximum elevation difference with 30° change in wind direction. (c) Sheerness time series for tide (green), total level (red), total level with 10% increased winds (black).

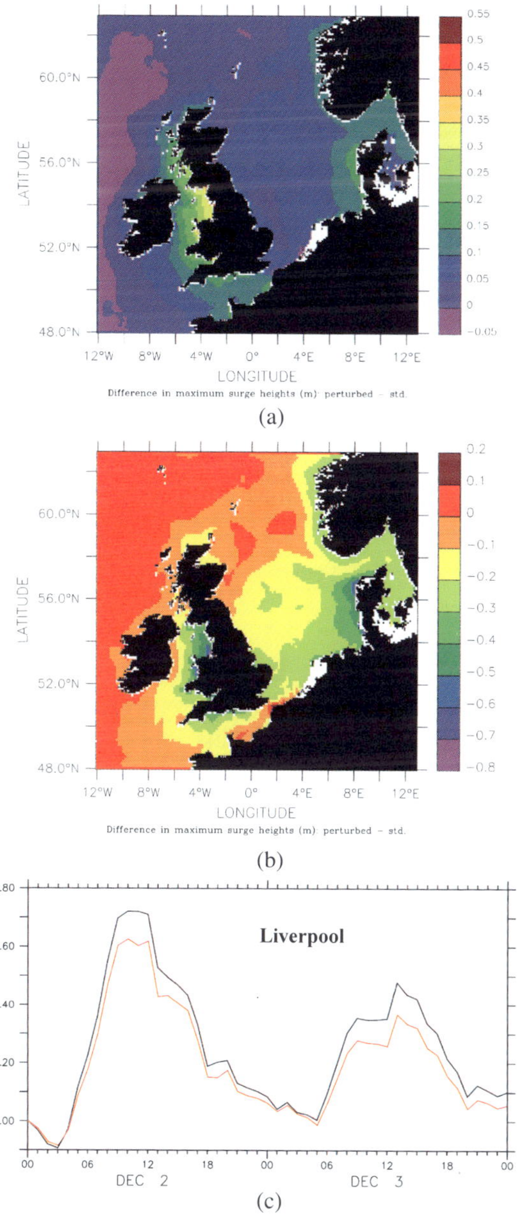

Figure 20.5. Sensitivity of Event 2 (2–3 December 2006) to differences in wind forcing. (a) Maximum elevation difference with 10% increase in wind; (b) Maximum elevation difference with 30° change in wind direction; (c) Liverpool residual time series for unperturbed level (red); level with 10% increased winds (black).

exceeded. Ensemble forecasting quantifies the uncertainty by making many numerical simulations using different choices of initial states and key parameters. It is a popular method for handling the uncertainty inherent in short-term weather prediction and climate models (e.g. Buizza *et al.*, 2005). The computational expense of numerical weather and hydrodynamic models rules out a Monte Carlo approach, where the uncertainty in model input is randomly sampled over thousands of trials. Instead, ensemble members are chosen very carefully to efficiently sample the uncertainty. A recently developed ensemble-based surge forecasting system has been validated over the period 2006–2008 (Flowerdew *et al.*, 2010). The meteorological input to the NOC CS3X surge model is taken from the Met Office Global and Regional Ensemble Prediction System (MOGREPS, Bowler *et al.* 2008). MOGREPS is based on two ensembles: the regional North Atlantic and European (NAE) model provides the meteorological data. During 2006–2008 it ran with double the grid spacing (0.22°) of the deterministic model. A lower resolution global model provides boundary conditions for the regional system. Each forecast cycle starts with a single meteorological analysis based on observations, to which perturbations are applied to generate 24 ensemble members (23 perturbed controls, plus one unperturbed control).

The meteorological perturbations are generated using the Ensemble Transform Kalman Filter (ETKF). This scheme uses estimates of the observation error to scale and mix differences between the individual members and the ensemble mean, taken from the T + 12 state of the previous forecast cycle (Wang and Bishop, 2003). The ETKF is based on data assimilation theory and aims to quantify the uncertainty in the meteorological analysis: the perturbations thus generated are valid from the initial time, which is essential for a short-range ensemble system. The MOGREPS system currently runs twice per day, with global forecasts at 0000 and 1200 GMT, and regional runs at 0600 and 1800. The regional simulations forecast out to T + 54 hours and the output is available to forecasters approximately six hours after the run begins. The 24 MOGREPS regional ensemble members then drive 24 separate simulations of the surge model, which is initialised from the deterministic model suite.

During the development of the ensemble system it became apparent that the initial condition of the surge model was a far less important constraint on the surge development than the forcing meteorology. Figure 20.6 shows the results of a sensitivity experiment based around a significant surge event in the Irish Sea. Initial conditions for the surge model were

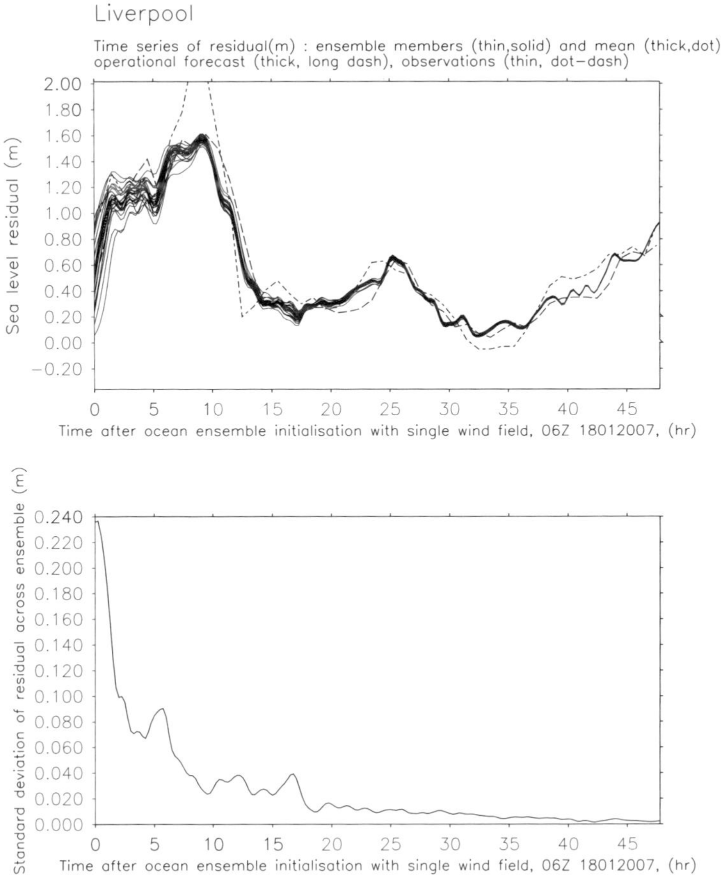

Figure 20.6. Upper panel: sea level residual curves from the surge event at Liverpool on 18 January 2007. The significance of the different curves is explained above the diagram. Lower panel: standard deviation from the ensemble of 24 residuals at Liverpool.

taken from each ensemble member at 0600 on 18 January 2007; the complete ensemble was then run forward using the atmospheric forcing from just one ensemble member. The figure shows the spread of residual values from the 24 runs at Liverpool, where the surge was greatest. The lower plot demonstrates that the memory of the initial state is lost rapidly, with the standard deviation halving in less than three hours. These results imply that for surge models the surface boundary conditions (i.e. meteorological forcing) are dominant over the initial condition, whose influence decays quickly into the next forecast.

The ensemble surge model output can be processed to provide a variety of useful graphical products that are available to forecasters. The upper panel in Figure 20.7 shows the mean and spread of the ensemble residual at midday, on 20 October 2008. We define spread as the standard deviation of all ensemble members, including the control run. The white contours show a surge of approximately 1 m in Liverpool Bay, and the red shading indicates an uncertainty of typically 6 cm in the forecast.

The narrow spread in this instance implies only small uncertainty in the forecast. The lower panel in Figure 20.7 is a probability plot showing the fraction of surge members that exceed 0.6 m at the forecast time. This is a useful graphical means of translating the ensemble output into some estimate of risk, or threshold exceedance, at critical locations.

20.4. Surge Ensemble Verification

Verification of the ensemble surge forecasting system took place over the winters of 2006/2007 and 2007/2008. The observational data were taken from 36 tide gauges of the UK national network, available from the British Oceanographic Data Centre (BODC) both as real-time (raw) data and quality-controlled data after a three month processing delay. To maximise the significance of the verification statistics, and integrate any local effects into meaningful diagnostics, the verification measures combined the results from all 36 locations. Quality control was applied to the raw observations, including a filter that selected from multiple observations whichever surge most closely matched the preceding data. The results use model forecasts at the corresponding times to all available observational data, at 15 minute intervals; it is probably better practice to average both the model and observed data onto hourly values to remove sea level variability (specifically seiching) which the hydrodynamic model cannot resolve.

It was found that one of the most useful indicators of forecast accuracy was the ensemble spread (which is a readily available diagnostic for forecasters, as we have seen). This is demonstrated in Figure 20.8, which shows a good relationship between the spread of the ensemble and the actual forecast errors. In the case of low spread, which occurs most frequently, there is an error of approximately 10 cm that is not spread-dependent (i.e. the intercept on the y-axis in Figure 20.8). This value is consistent with the limits of accuracy on tidal prediction, from which the observed residuals are derived. Figure 20.8 shows that the individual perturbed members have the greatest RMS error, since they are deliberately adjusted

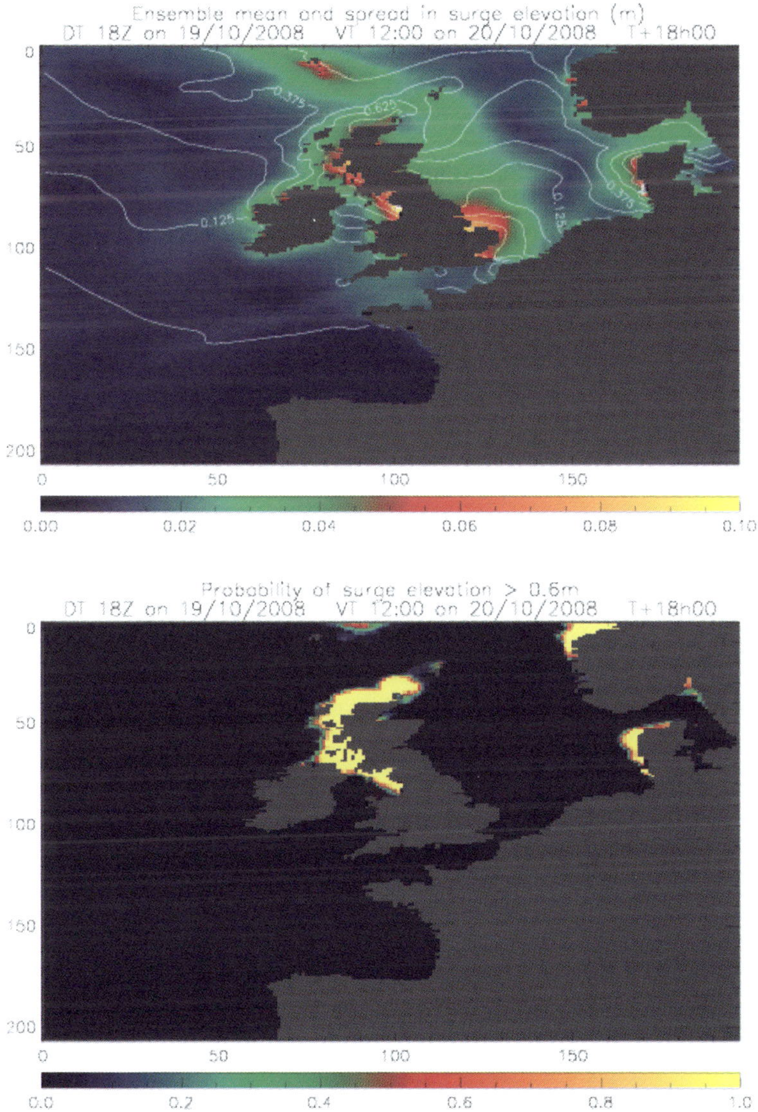

Figure 20.7. Upper panel: ensemble mean and spread of sea level residual for 1200 on 20 October 2008 from forecast on 1800 19 October 2008. Lower panel: probability of residual exceeding 0.6 m at 1200 on 20 October 2008 from the same ensemble forecast.

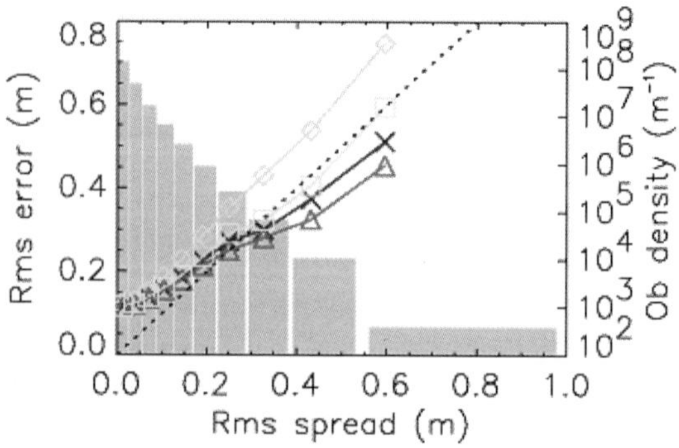

Figure 20.8. RMS error of the forecasts for all observations, binned as a function of ensemble spread. The grey histograms indicate the relative population of each spread value, given by the right-hand scale. Error-spread performance is shown for the ensemble mean (triangles), individual perturbations (diamonds), the control member (crosses) and the corresponding deterministic model run (squares).

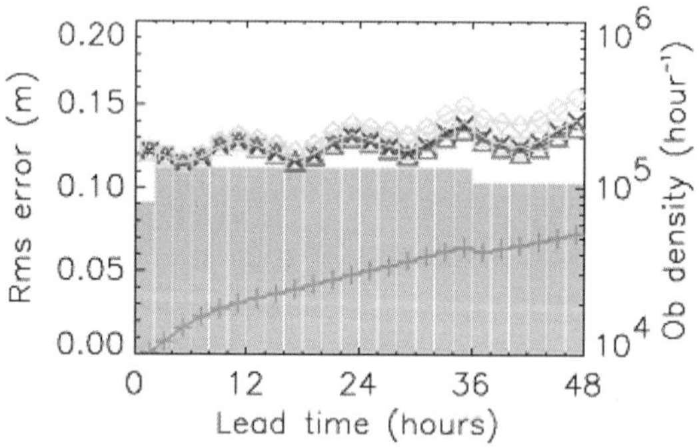

Figure 20.9. Spread (lower) and error quantities (symbols as in Figure 20.8) in the residual as a function of forecast lead time.

away from the best estimate (the analysis) of the state of the atmosphere. The ensemble mean consistently has the lowest error, particularly at high spreads, which implies that the sampling of uncertainty provided by the system allows the mean to produce, on average, a better forecast than any perturbed member (or the control member). The ensemble mean also has lower RMS error than the corresponding deterministic forecast over this relatively short verification period.

Figure 20.9 shows the spread and the errors of the various forecast quantities, as a function of lead-time. The ensemble variance is approximately linear with lead-time, as would be expected of a random walk process. The forecast errors are dominated by the initial 12 cm error,

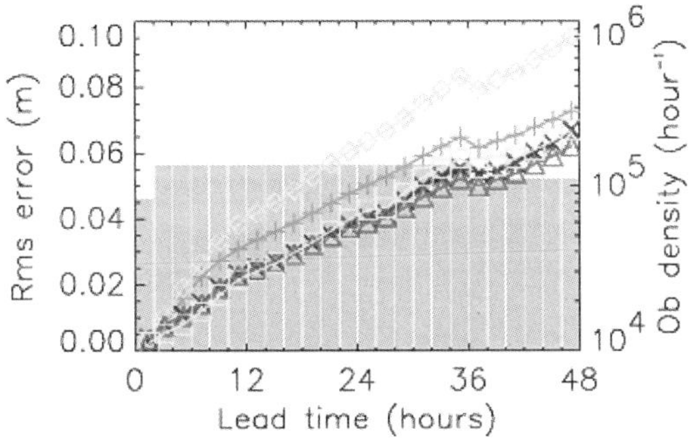

Figure 20.10. Spread (vertical dashes) and error quantities (symbols as in Figure 20.8) in the residual compared with a modelled hindcast residual, as a function of forecast lead-time.

which reduces to 7–8 cm when the five ports with highest tidal range are excluded. There is some growth of error with lead-time, and at the full extent of the forecast period (T + 48), the ensemble mean (triangles) again performs best. The semidiurnal oscillation in the error quantities can be attributed to errors in tidal predictions, and is virtually eliminated when residual RMS errors are recalculated with respect to the hindcast residual (as in Figure 20.10) rather than the observed residual. The hindcast residual is derived from a surge model forced with reanalysed meteorology; a tide-only run is subtracted to yield a modelled residual that has no dependency on observations, and thus contains no errors due to limitations of the harmonic method. The ensemble spread is a much better match to this meteorologically-focused aspect of surge forecast error.

A common application in forecasting is to make a yes/no decision (e.g. close a barrier, evacuate a district, etc.) and a large number of empirical measures have been devised to assign skill to the forecasts that inform such decisions. One useful verification measure is the Brier Skill Score (BSS). This is a variant of the Brier Score (BS) which is the mean squared error of probability forecasts:

$$\mathrm{BS} = \frac{1}{N}\sum_{k=1}^{N}(P_k - O_k)^2. \tag{20.5}$$

In the case of storm surges, P could be the ensemble probability of either the residual or the total water level exceeding a particular threshold. The observations (O) of the event take the value 1 if the threshold was exceeded and 0 if it was not. However, for rare events like flooding it is possible to get a reasonable BS in the absence of any real skill. The BSS measures the improvement of the probabilistic forecast relative to a reference forecast based on climatology (in our case, how frequently the same threshold is exceeded over the verification period).

$$\mathrm{BSS} = 1 - (\mathrm{BS}/\mathrm{BS}_{\mathrm{ref}}). \tag{20.6}$$

A perfect BSS is 1. If the ensemble BSS is no better than a guess based on the climatological frequency of occurrence then the best available BSS is 0. Figure 20.11 presents the BSS of our ensemble prediction of residuals greater then 1 m, as a function of lead-time, aggregated over all ports and the full validation period. These results have been calculated against hindcasts rather than observed residuals because of the greatly reduced statistical noise. Corresponding results based on the tide gauge data are discussed in Flowerdew *et al.* (2010). In these calculations, an event was defined as the hindcast residual exceeding 1 m at some point within a 12 hour window. The ensemble forecast probability was the fraction of members with residuals that exceeded 1 m within the same window (although not necessarily at precisely the same time). The choice of a

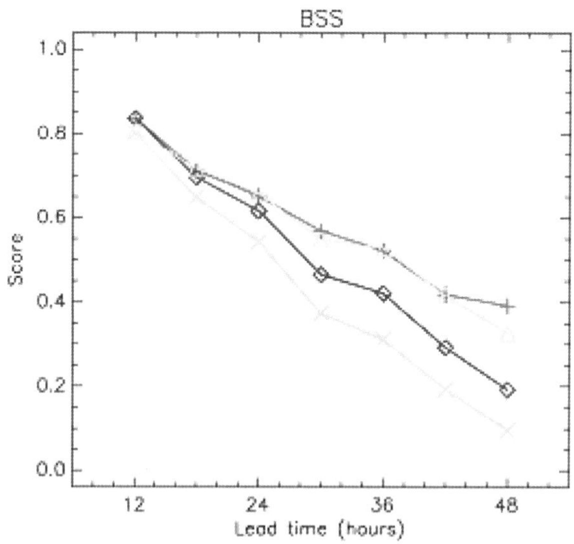

Figure 20.11. Brier skill score (relative to hindcast) for ensemble prediction of residuals greater than 1 m. For comparison with the full ensemble (plus symbols) we show the control member (crosses, "ndress'), and two dressings of the control member: "odress" (diamonds) and "mdress" (triangles).

12 hour window is appropriate since tide is a critical component when total water levels are considered.

The comparison traces in Figure 20.11 show the results from three different ways of converting ("dressing") a single deterministic forecast into a probabilistic one. Here they are applied to the ensemble control member. The simplest scheme ("ndress") assumes the deterministic forecast to have no error, giving p = 0 if the forecast is below the threshold and p = 1 if it exceeds it. The "odress" scheme assumes a constant Gaussian error with a standard deviation equal to the overall RMS forecast error (4 cm when verifying against hindcasts). In this case the forecast probability is the fraction of this assumed distribution that exceeds the threshold. The last method ("mdress") uses a Gaussian error distribution whose standard deviation is the overall RMS error plus and a fraction of the individual forecast magnitude. These comparisons quantify the benefit obtained from the ensemble's dynamical representation of forecast error over and above simpler climatological expressions of forecast error. As Figure 20.11 shows, for this threshold the ensemble and magnitude-based dressing have jointly the best skill, ahead of the constant dressing, and far better then the undressed control. The improvement is larger at longer lead-times (which hints at the usefulness of longer forecasts).

The BSS for total water levels (including harmonically predicted tides) exceeding port-specific alert levels is illustrated in Figure 20.12. The ensemble now clearly performs better than either method of climatological dressing, showing the value of its dynamic simulation over the simple statistical approaches. Extrapolation implies that the ensemble has increased skill over climatologically dressed forecasts out to about six days.

20.5. Case Study: 9 November 2007

On the 9 November 2007, the east coast of the UK experienced the worst storm surge for approximately 50 years. It was initially feared that the event would be as bad as the North Sea storm surge of January 1953, but fortunately the forecasts for both winds and surges decreased over the 24 hours preceding the event. There was minor flooding in East Anglia, and some precautionary evacuations, but surge levels from the wash to the Thames were typically 20 cm lower than the long-range predictions. Although the maximum residual of 2.4 m (above Ordnance Datum Newlyn (ODN)) was the third highest seen in the Sheerness tide gauge record, this is not the most useful metric. The effects shown in Figure 20.2,

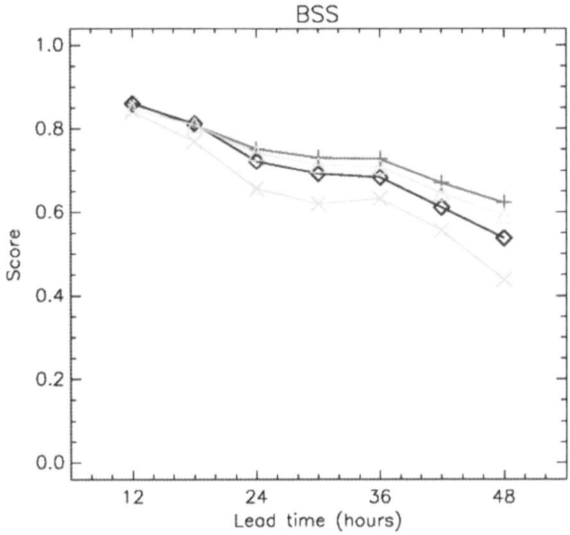

Figure 20.12. BSS (relative to hindcast) for ensemble prediction of total water level exceeding the port-specific alert level. Lines and symbols as for Figure 20.11.

combined with the fact that wind stress is most effective at generating surge in shallow water, resulted in the maximum residual being obtained typically 3–5 hours prior to high water at this location (Horsburgh and Wilson, 2007). A more practical measure is the skew surge, which is the difference between the elevation of the predicted astronomical high tide and the nearest experienced high water (e.g. de Vries *et al.*, 1995). In terms of skew surge (or indeed the total water level) the 2007 surge elevations in the Thames were approximately 1 m less than in 1953.

A full account of the event and performance of the deterministic model is given by Horsburgh *et al.* (2008). Their analysis shows that the storm surge along the east coast of the UK was modelled accurately throughout the entire period. The deterministic forecast available immediately prior to the event (at midnight on 9 November 2007) was within 9 cm of the observed skew surge level at Lowestoft and within 21 cm at Sheerness (see Figure 20.14 for locations). The accuracies of the surge hindcast using re-analysed meteorological forcing were 1 cm and 8 cm respectively. The MOGREPS-driven surge ensemble system was on trial at the time of the storm, and Figure 20.13 shows the ensemble of sea level residuals at Sheerness from the run at 1800 GMT on 8 November 2007. In this figure, the red curves show the warning thresholds for individual ports; if the

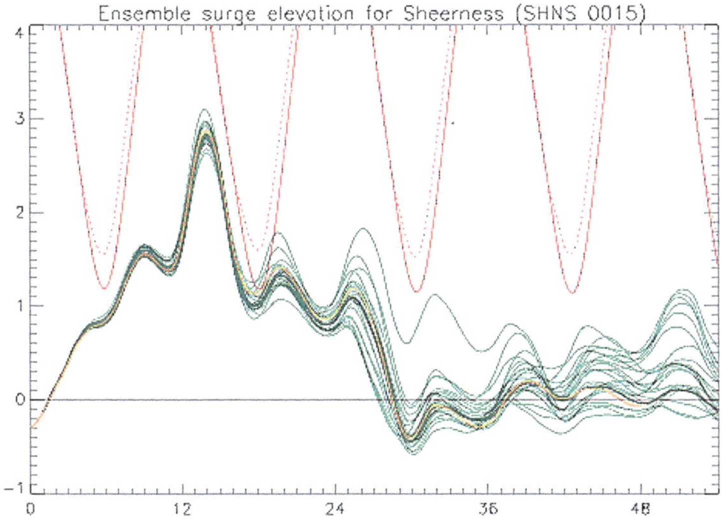

Figure 20.13. Ensemble surge elevation (m) for Sheerness from the 1800 GMT run of MOGREPS on 8 November 2007. The green lines are the residuals from the 24 individual ensemble members and the yellow line is the forecast from the deterministic model. The red curves denote the warning thresholds acted on by UKCMF forecasters.

threshold is predicted to be exceeded then the green ensemble curve will intersect the red curve. Sea level forecasts at Sheerness are of importance because when combined with river flow measurements they determine the closure of the Thames Barrier. High water at Sheerness on 9 November 2007 can be identified by the minimum in the solid red curve in Figure 20.13, approximately 18 hours into this forecast. As expected, the residual peaks in the complete ensemble (green curves) are found approximately four hours earlier. By high water, only three of the 24 ensemble members (and the deterministic model, shown by the yellow line) predicted that the threshold would be exceeded. In the event it was not, however, the Thames Barrier was closed as a precaution.

Table 20.2 shows values of the skew surge from the earlier ensemble ran (at 0600 GMT) for both Sheerness and Lowestoft, and compares them with observations and successive runs of the deterministic model. The ensemble mean forecast for Sheerness was superior to either of the deterministic forecasts (although this was not the case for Lowestoft). The best accuracy was obtained from the re-analysis model runs. These are not available to forecasters, of course, but it is their superior accuracy that makes them so useful in statistical verification.

Table 20.2. Values (m) of the observed skew surge for Sheerness and Lowestoft at the first high water on 9 November 2007, from forecasts of the deterministic surge model, and from the ensemble forecast run at 0600 GMT on 8 November 2007.

	Lowestoft	Sheerness
Tide gauge observations	1.66	0.83
1200 deterministic forecast on 8 November 2007.	1.85	1.11
0000 deterministic forecast on 9 November 2007.	1.57	1.04
Reanalysis (hindcast) model run	1.65	0.91
MOGREPS ensemble mean	1.90	0.98
MOGREPS ensemble minimum	1.37	0.32
MOGREPS ensemble maximum	2.27	1.28

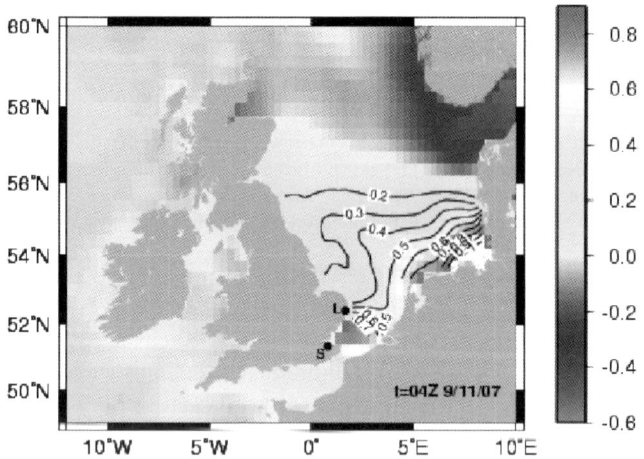

Figure 20.14. Difference (m) between the sea level residual from the ensemble member giving the largest skew surge along the UK east coast and that of the deterministic model, at 0400 GMT on 9 November 2007. The locations of Lowestoft (L) and Sheerness (S) are shown.

The higher sea level residuals exhibited by a minority of ensemble members would have had implications for the flood hazard in eastern England. Regional forecasters are interested in the sea levels that might be achieved in the most extreme case. To demonstrate the spatial properties of the ensemble that gave the largest skew surge along the UK east coast, Figure 20.14 depicts the difference between the residuals of that member and the deterministic model. In the case of the most extreme forecast, sea levels would have been higher by up to 80 cm along the affected coastline for a period of more than three hours spanning high water. Although that

Figure 20.15. Potential inundation (without defences or in the event of a breach) for one of the most affected areas on 9 November 2007. The brown shading indicates the flood extent with sea levels of 2.86 m above ODN (as occurred); red areas are the additional inundation with sea levels of 3.24 m above ODN as predicted by the most extreme ensemble member. The black arrow (bottom right) shows the location of a power station.

particular scenario had only a 1 in 24 chance of occurrence, it is valid to explore the potential consequences with detailed inundation mapping.

A map of the flooding that would have occurred in the absence of defences is given in Figure 20.15. It is not currently feasible to include coastal defences in this type of mapping. Therefore, Figure 20.15 could be thought of as showing the inundation extent in the event of a serious breach.

The brown shading indicates the (undefended) flooding that would have occurred on 9 November 2007 with the actual sea levels; the red shading shows additional flooding had the most extreme forecast been realised. In this instance, the area affected by the higher water levels is not significant overall, although a power station is specifically affected when the higher level is imposed. The same spread of values at other vulnerable locations may have had more significant implications. Figure 20.15 implies a step-like topography that exposes all low lying areas to risk at moderate extreme levels. There is presumably a higher threshold which, if exceeded, would place larger areas at risk. This scenario demonstrates a particular use of ensemble predictions. Wherever a series of locally critical thresholds can be identified, inundation mapping using simple GIS-techniques or more complex dynamic models (e.g. Bates *et al.*, 2005) can set multiple warning levels to which the probabilistic output of a surge ensemble system can be applied. Ensemble forecasts of sea level combined in this way with locally determined, risk-based thresholds can form part of a cost/loss decision making tool that provides a robust scientific and economic basis for mitigation decisions.

During this event, which had a high media profile, the ensemble predictions gave confidence to those responsible for emergency response because they provided a measure of uncertainty at critical times in the forecast period. Event-based results such as these, and the outcomes of the broader verification process described earlier, endorse the utility of probability-based forecasts for coastal flood risk management. As mean sea levels rise by 20–90 cm towards the end of this century (IPCC, 2007), coastal flood managers will become more reliant on operational warning systems in order to protect lives and infrastructure. Forecasting systems must become more sophisticated in response to this challenge. Although it is clear that the current forecasting process performs well, future improvements to all aspects of the modelling system (e.g. wave-surge coupling) will lead to better risk management and long-term decisions on coastal defence. Ensemble surge modelling is a key component of this improved system.

Acknowledgements

We are grateful to Mr Ian Bowler of the Environment Agency for the mapping that produced Figure 20.15. The ensemble surge development work was funded by the Defra/EA R&D programme under contract SC050069. The surge models within UKCMF are developed and maintained by the National Oceanography Centre, funded by the Environment Agency, and run at the Met Office.

References

Bates, P.D., Dawson, R.J., Hall, J.W. *et al.* (2005). Simplified two-dimensional numerical modelling of coastal flooding and example applications, *Coast. Eng.*, **52**, 793–810.

Bowler, N.E., Arribas, A., Mylne, K.R. *et al.* (2008). The MOGREPS short-range ensemble prediction system, *Q. J. Roy. Meteor. Soc.*, **134**, 703–722.

Buizza, R., Houtekamer, P.L., Toth, Z. *et al.* (2005). A comparison of the ECMWF, MSC and NCEP global ensemble prediction systems, *Mon. Weather Rev.*, **133**, 1076–1097.

de Vries, H., Breton, M., Demulder, T. *et al.* (1995). A comparison of 2D storm-surge models applied to three shallow European seas, *Environ. Softw.*, **10**, 23–42.

Flather, R.A. (2000). Existing operational oceanography, *Coast. Eng.*, **41**, 13–40.

Flather, R.A., Williams, J., Blackman, D.J. *et al.* (2003). *Investigation of Forecast Errors at Sheerness during 2002*, POL Internal Document, No. 151, p. 46.

Flowerdew, J., Horsburgh, K., Wilson, C. *et al.* (2010). Development and evaluation of an ensemble forecasting system for coastal storm surges, *Q. J. Roy. Meteor. Soc.*, **136**, 1444–1456.

Gonnert, G., Dube, S.K., Murty, T.S. *et al.* (2001). *Global Storm Surges: Theory, Observations and Applications*, German Coastal Engineering Research Council, p. 623.

Greenberg, D.A. (1979). A numerical model investigation of tidal phenomena in the Bay of Fundy and Gulf of Maine, *Mar. Geod.*, **2**, 161–187.

Horsburgh, K.J., Williams, J.A., Flowerdew, J. *et al.* (2008). Aspects of operational model predictive skill during an extreme storm surge event, *J. Flood Risk Manage.*, **1**, 213–221.

Horsburgh, K.J. and Wilson, C. (2007). Tide-surge interaction and its role in the distribution of surge residuals in the North Sea, *J. Geophys. Res. Oceans*, **112**, C08003, doi:10.1029/2006JC004033.

IPCC (2007). *Climate Change 2007*: *Synthesis Report. Contribution of Working Groups I, II and III to the Fourth Assessment Report of the Intergovernmental Panel on Climate Change* (core writing team, Pachauri, R.K and Reisinger, A. (eds)), IPCC, Geneva, Switzerland, p. 104.

Jones, J.E. and Davies, A.M. (1996). A high-resolution three dimensional model of the M2, M4, M6, S2, N2, K1 and O1 tides in the eastern Irish Sea, *Estuar. Coast. Shelf S.*, **42**, 311–346.

Jones, J.E. and Davies, A.M. (2006). Application of a finite element model (TELEMAC) to computing the wind induced response of the Irish Sea, *Cont. Shelf Res.*, **26**, 1519–1541.

McRobie, A., Spencer, T. and Gerritsen, H. (2005). The big flood: North Sea storm surge, *Philos. T. R. Soc. Lond. A*, **363**, 1263–1270.

Pugh, D.T. (1987). *Tides, Surges and Mean Sea-Level: A Handbook for Engineers and Scientists*, Wiley-Blackwell, Chichester, p. 477.

Randon N., Lawry J., Horsburgh K.J. *et al.* (2008). Fuzzy Bayesian modelling of sea-level along the east coast of Britain, *IEEE Transac. Fuzzy Syst.*, **16**, 725–738.

Ryabinin, V.E., Zilberstein, O.I. and Seifert, W. (1996). *Storm Surges*, World Meteorological Organisation, Marine Meteorology and Related Oceanographic Activities Report No. 33, WMO/TD-No.779, p. 121.

Wang, X. and Bishop, C.H. (2003). A comparison of breeding and ensemble transform Kalman filter ensemble forecast systems, *J. Atmos. Sci.*, **60**, 1140–1158.

Wortley, S., Crompton, E., Orrell, R. *et al.* (2007). *Storm Tide Forecasting Service: Operational Report to the Environment Agency for the period 1st June 2006 to 31st May 2007*, Met Office, Exeter, United Kingdom, p. 67.

Xuan, Y. (2007). *Uncertainty Propagation in Complex Coupled Flood Risk Models using Numerical Weather Prediction and Weather Radars*, PhD thesis, University of Bristol.

SECTION V

UNCERTAINTIES IN LONG-TERM CHANGE IN FLOOD RISK

CHAPTER 21

Detecting Long-Term Change in Flood Risk

Cíntia B. Uvo
TVRL, University of Lund, Sweden

Robin T. Clarke
Instituto de Pesquisas Hidráulicas, Brazil

21.1. Risk and Probability

We need first to consider the terms "flood risk" and "probability of flooding". They are often taken as synonymous, but in the context of decision theory, they mean different things, as the following example illustrates. Suppose that a river may overtop its banks, but that there is some uncertainty as to which of the following three events may occur: there is no flooding (denoted by E_1); the river overtops its banks by up to 50 cm (E_2); or, the flood depth is greater than 50 cm (E_3). One, and only one, of the three events may occur, and their probabilities of occurrence will be denoted by $p(E_1)$, $p(E_2)$ and $p(E_3)$, with $p(E_1) + p(E_2) + p(E_3) = 1$; for the moment, we do not consider how these probabilities are calculated. Now suppose that the engineer responsible for a city's flood defences must decide which of three courses of action he/she will take: from his/her experience in dealing with similar flood threats he/she may decide not to issue a flood warning (a decision denoted by d_1: "Green"), because similar meteorological and hydrological conditions in the past have not caused flooding. Alternatively, he/she may decide to take some provisional steps to deal with possible flooding (d_2: "Orange"), or to go the step further and issue a full flood warning with mobilisation of civil defences and evacuation of people (d_3: "Red"). One, and only one of the decisions d_1, d_2, d_3 can be

Table 21.1. Decision table for a hypothetical example on flood warning.

Event:	E_1:	E_2:	E_3:
Probability:	$p(E_1)$	$p(E_2)$	$p(E_3)$
Decision:			
d_1: Green	$C(1,1)$	$C(1,2)$	$C(1,3)$
d_2: Orange	$C(2,1)$	$C(2,2)$	$C(2,3)$
d_3: Red	$C(3,1)$	$C(3,2)$	$C(3,3)$

taken. Then to each combination of decision d_i with event E_j $(i, j = 1,\ 2,\ 3)$ there will be a set of consequences denoted by $C(i, j)$; for example with $i = 1$ and $j = 3$, corresponding to the issue of no flood warning and the occurrence of a severe flood, the consequence $C(1,3)$ is likely to include the component "engineer loses job". Suppose, for the sake of simplicity, that the consequences $C(i, j)$ can be measured in terms of financial loss (almost always, the consequences must be measured in terms of "utility", measured on a scale from zero to one and which has a one-to-one relationship with the ordered set of consequences $C(i, j)$). Then the events, probabilities, decisions and consequences can be put in the form shown in Table 21.1.

A final column may be added to this table, giving the "expected utility" (or, when utility can be measured in terms of cost, the less general "expected cost"). In terms of expected cost, the values in this additional column are the sums of the $C(i, j)$ in each row, weighted with their corresponding probabilities: giving, for the first row, $p(E_1)\,C(1,1) + p(E_2)\,C(1,2) + p(E_3)\,C(1,3)$ with similar expressions for the second and third rows. Decision theorists refer to these "expected costs" as the "risks" associated with the different decisions; note that, with this definition, a "flood risk" involves not only the probabilities of flooding $p(Ej)$, but also their consequences $C(i, j)$. This example, although greatly oversimplified, illustrates another important point: whilst the hydrological modeller may be able to estimate the probabilities of flood events occurring, these probabilities by themselves are of no use for decision making, without the essential contribution supplied by the decision-maker: namely, (i) the list of decisions from which he/she must choose the "best" (in this case, the three decisions "Green", "Orange" and "Red"), and the consequences $C(i, j)$ associated with the combinations of decisions with uncertain events. This is why collaboration between hydrological modeller and decision-maker is so important: the risk associated with each of the possible decisions cannot be determined by either one acting independently of the other.

The title of this chapter is "Detecting Long-Term Change in Flood Risk". In light of the preceding discussion, we therefore interpret this in terms of changes in the probabilities of flooding.

21.2. The Meaning of "Long-Term"

A further comment is needed concerning the nature of "long-term" changes. A long-term change in flood probabilities may be interpreted as meaning either (i) that the probability of flooding shows a trend, whether increasing or decreasing; or (ii) that because of year-to-year persistence in geophysical phenomena, there are very slow changes in the probability of flooding over long periods, without any overall trend being evident. Distinguishing between (i) and (ii) is obviously very difficult, perhaps even impossible, since what may appear as a trend over a rather short period of record may, over a much longer period of record, come to be seen as part of a slow fluctuation. This is clear, for example, in the very long sequence of Nile water levels (Figure 21.1) used as an example in the book by Beran (1994), and in other long geophysical records.

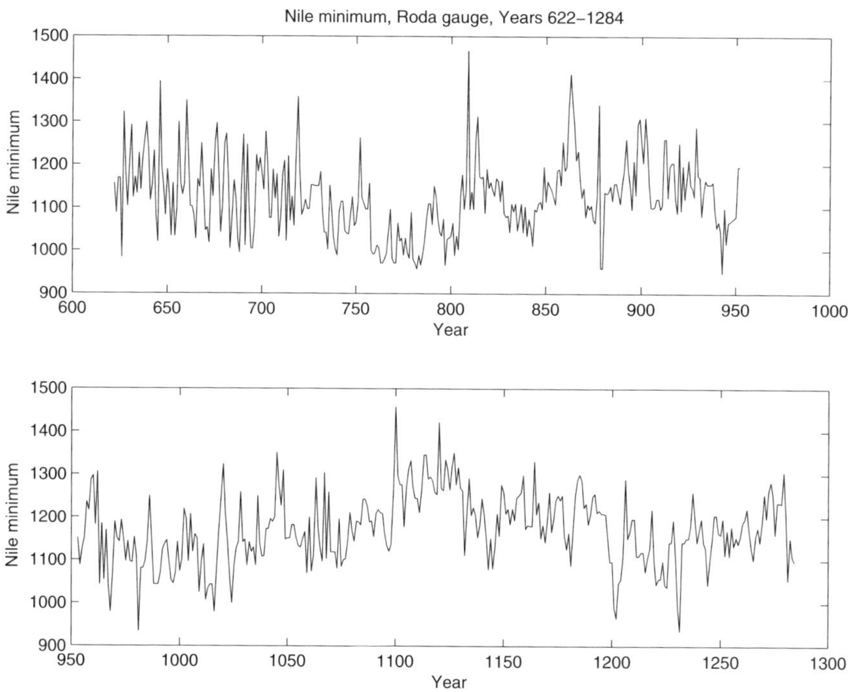

Figure 21.1. Annual minimum water levels 622–1284: Nile River at the Roda gauge, Cairo (Tousson, 1925, p.366–385: in Beran, 1994).

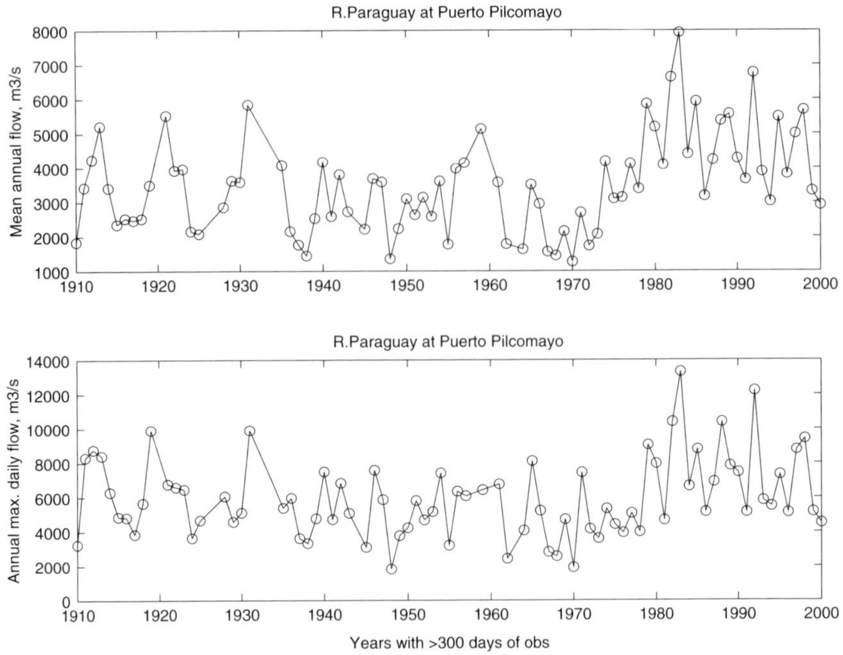

Figure 21.2. Mean annual (top) and maximum mean daily (bottom) flows for the Paraguay River at Puerto Pilcomayo for the period 1910–2000.

In practice, however, the questions about flood frequency that hydrologists are called upon to answer are based on flood records that usually began some 50 years ago, and which commonly do not exceed 100 years' duration. Changes in flood probabilities cannot be established over periods longer than the period of flood record, and for the purpose of this contribution we therefore take "long-term" to mean "extending over a period of 50 to 100 years' duration". Figure 21.2 (bottom) shows the annual maximum mean daily flows of the Paraguay River at Puerto Pilcomayo for the period 1910–2000; the mean annual flows (top figure) appear to show an increasing trend since about 1970, and this trend is possibly reproduced, although less strongly, in the annual maxima of the bottom figure. As another example, Figure 21.3 shows the number of closures of flood barrier on the River Thames, since its construction in 1983. There appears to be a positive trend up to about 2001; thereafter, the data may suggest a negative trend, or it may simply be that this is an artefact of the year-to-year variability in number of closures.

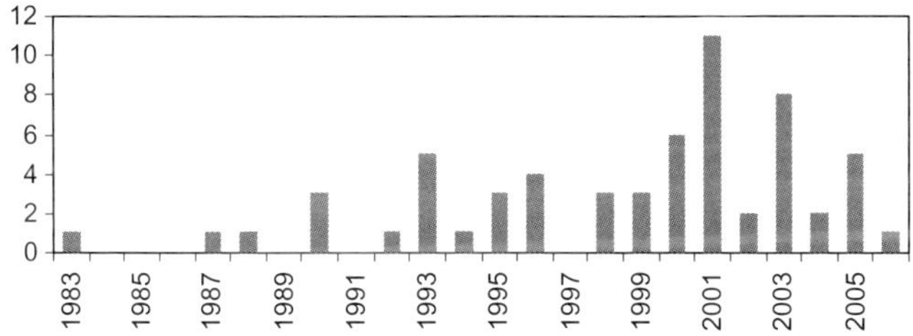

Figure 21.3. Number of closures of flood barrier per year on the River Thames. (Source: UK Environment Agency.)

21.3. The Need for Care in Selecting a Hypothesis for Testing Trends

When analysing changes in flood phenomena, whether in terms of the numbers of floods occurring per decade (N say) or the magnitudes of floods (Q), it is common to begin with plots of the variables N or Q against time (i.e. year number). If a trend looks apparent in the plot, then a formal test of significance, whether parametric or non-parametric, is made, and a conclusion is drawn as to whether the observed trend is "statistically significant", or whether it could reasonably be ascribed to chance. Although common, this procedure is incorrect because the hypothesis being tested is suggested by the same data that are used to test it. The statistical argument is therefore circular. For a statistical test of a null hypothesis (such as the null hypothesis that no trend exists in a flood record) to be valid, the hypothesis should be independent of the data on which it is to be tested.

A further caution is necessary where trends in several flood records collected over a region, such as in a large drainage basin, are to be tested for trend. It is common for the record at each site to be tested individually and separately from the others. However, regional flood records are likely to be correlated: extreme floods at one gauging site are likely to occur in the same years as extreme floods at other gauging sites not too distant from it. Statistical tests for trends in which the record from each site is treated individually are likely to overestimate the significance of a trend. The effects of cross-correlation between flood flows is illustrated by the Douglas *et al.* (2000) in their analyses of trends in extreme flows in the USA; they reported no evidence of significant trends in flood flows, but noted that:

> "A dramatically different interpretation would have been achieved if regional cross-correlation had been ignored. In that case, statistically

significant trends would have been found... in two thirds of the flood flow analyses."

Instead of using statistical tests of whether flood characteristics have changed over time during a period of record using time t as an independent variable (particularly after looking at the data), it is better to relate the flood characteristic to the possible variables which might be driving the change. Changes in land use, particularly deforestation and urbanisation, or long-term fluctuation in climate, may give rise to changes in flood characteristics. It is then appropriate to look for causative variables, such as growth in impervious area within a drainage basin, or crop yields as measures of the growth of land under cultivation, to which flood characteristics may be related. Where changes in flood characteristics may be a consequence of fluctuating climate, possible explanatory variables may be suggested by anomalies in sea surface temperatures or climatic indices such as the North Atlantic Oscillation Index (Hurrell *et al.*, 2003), the Pacific Decadal Oscillation Index (Mantua *et al.*, 1997), and others: see, for example, Allasia (2007).

Thus instead of relating flood characteristics to time t, it is preferable to relate them to variables which are possible drivers of change, denoted generically by x. Relation to such variables may also be a surer way of extrapolating flood characteristics into the future. If deforestation, as measured by the variable x, is expected to continue into future years at a rate that can be estimated, however crudely, or if the variability of a chosen climate related index can be estimated, this may provide a better guide to future flood characteristics than extrapolation of a trend on a time-variable t. However, there may also be advantage in looking for relationships between flood characteristics and both variables, x and t; if adding the time-variable t to a relation between flood characteristics and a causative variable x adds no further information, we can assume that x is a good predictor of future changes in flood characteristics. If, however, information is added by the inclusion of t, then x does not by itself explain the observed changes, and additional causative variables must be explored.

The above paragraph refers to the extrapolation of trends in flood characteristics into the future, but there may also be scope for extending them into the past, either by correlation with longer flood records at other sites, or by using explanatory variables correlated with the flood characteristic of interest. Other proxies used for the extension of annual maximum discharges time series into the past are tree rings (*Schongart et al.* 2005) and laminated sediments (Sander *et al.* 2002; Werritty *et al.*, 2006, among others).

Time-trends in variables describing flood magnitudes, and the relationships between them and explanatory variables, can be explored using either parametric or non-parametric methods; we discuss some of the details below. The point of departure for parametric methods is a probability distribution for the flood variable (such as annual maximum instantaneous discharge in $m^3\,s^{-1}$) with a small number of parameters, one or more of which are related to the explanatory variable(s). The parameters in this model are then estimated from whatever flood record is available. The starting point when non-parametric methods are used is commonly the null hypothesis that no time-trend (or relationship with an explanatory variable) exists. This hypothesis is then tested using the flood record and if it is rejected, a specific form is assumed for the relationship and the flood record is used to estimate its parameters.

21.4. Parametric Methods for Flood Frequency Analysis that Allow for Long-Term Trend Detection

Setting aside for the moment the possibility of trends, it is well known (e.g. Coles, 2001) that the analysis of extremes such as flood magnitudes by parametric methods can be approached in two ways, generally termed (i) the "block" approach, and (ii) the Peaks-Over-Threshold (POT) approach. Another alternative is to analyse trends by examining the number of events by unit time (iii).

21.4.1. *"Block" methods*

In the block approach, units of time (usually years) are taken, and the maximum event (such as peak discharge) observed in each time-unit (block) is abstracted and used to form a derived sequence of annual maxima. Thus if a record of daily flow in a river is of length N years (assuming the years complete), the derived sequence will have N values of annual maximum discharge. Except perhaps in very large drainage basins, it is often reasonable to assume that the N annual maxima are statistically independent, being a sample from the Generalised Extreme Value (GEV) distribution (Coles, 2001) with Cumulative Distribution Function (cdf) given by:

$$G(q) = exp\{-[1 + \xi(q - \mu)/\sigma]^{-1/\xi}\}, \tag{21.1}$$

where μ, σ and ξ are parameters which determine the distribution position, dispersion and shape. Coles (2001) gives a full account of how to estimate the parameters μ, σ, ξ from the available flood record; see also Tawn and Coles (Section 2.5, this volume). Where a trend in flood magnitude Q is thought to be present, the parameter μ in (21.1) may be related to time t by including another parameter β to give a linear relationship of the form:

$$\mu(t) = \alpha + \beta t, \tag{21.2}$$

or, preferably, a linear relationship with a causative variable x instead of t: see Coles (2001) who gives an example in which annual maximum sea level at Fremantle, Australia, is related to the value of the Southern Oscillation Index (SOI) SOI(t) in year t, $\mu(t) = \alpha + \beta\,\mathrm{SOI}(t)$. Alternative models would relate the dispersion and shape parameters σ and ξ to these variables (Coles, 2001, Chapter 6). In the case of a model using (21.1), with the parameters σ, ξ not regarded as time-variant, the four parameters $\alpha, \beta, \sigma, \xi$ can be estimated by maximum likelihood, and likelihood theory provides a large sample test, if needed, of the null hypothesis of no trend, $\beta = 0$.

21.4.2. *POT methods*

The second POT approach uses a threshold discharge, denoted by u (Coles, 2001, Chapter 4) and all discharges that exceed this threshold are retained for analysis. Thus within any given year, the number of extreme events may be zero (if no event exceeds the threshold), one, or any greater number. Provided that the threshold is set sufficiently large, the excess over the threshold $y = Q - u$, conditional on $Q > u$ is the generalised Pareto distribution with cumulative distribution function,

$$H(y) = 1 - (1 + \xi y/\tilde{\boldsymbol{\sigma}})^{-1/\xi}, \tag{21.3}$$

where $\tilde{\boldsymbol{\sigma}} = \sigma + \xi(u - \mu)$ and the parameters μ, σ, ξ define the distribution position, dispersion and shape, as in the case of the GEV distribution. Similar devices to those described for GEV distribution can be used to introduce time-trends into the parameters, or to explore relationships with causative variables such as the extent of deforestation, or climatic indices.

21.4.3. *Methods for analysing trends in the number of flood events per unit of time*

The number of times per year, per decade, or any other time unit, that a river overflows its banks or exceeds some other threshold, may also be of interest, and in this case the variable to be analysed — the number of flood events per unit of time — will constitute a sequence of integers, possible small and including zeros. To analyse such a record for time-trend, or to explore a relationship with explanatory variables, simple regression methods — with the usual assumptions that observations of the response variable are continuous, normally distributed and with constant variance — are unlikely to be satisfactory and an approach which describes adequately the non-negative-integer character of the observations is to be preferred. Generalised Linear Models (GLMs) are appropriate for the purpose, and a plausible point of departure is that the number of events in the j-th time interval, say N_j, is a Poisson variable with time-variant mean. A GLM relating this flood variable to an explanatory variable, say time x, using a linear model would be defined as follows:

$$\begin{aligned} \mathrm{E}[N] &= \mu(x) \\ g[\mu(x)] &= \alpha + \beta x, \end{aligned} \tag{21.4}$$

where $g[\cdot]$ is a known link function (McCullagh and Nelder, 1989). One appropriate form for the link function when data are Poisson distributed is the logarithm, so that the second of the two equations in (21.4) becomes

$$log_e[\mu(x)] = \alpha + \beta x. \tag{21.5}$$

Table 21.2 shows the number of closures of the River Thames Flood Barrier, to prevent tidal surges, over the period 1983–2006. Assuming that the number of closures follows a Poisson distribution with mean linear related to year number, fitting the model defined by (21.4) and (21.5) by maximum likelihood shows that the parameters α and β are estimated by $\hat{\alpha} = \textit{-185.5} \pm \textit{41.6}$ and $\hat{\beta} = \textit{0.0934} \pm \textit{0.0208}$, and Figure 21.4 shows the relation between the observed and fitted frequencies of closures.

21.5. Non-Parametric Methods for Long-Term Trend Detection in Flood Records

A good account of non-parametric statistical methods is given in the book by Hipel and McLeod (1994), and the Mann–Kendall test in particular has been widely used for trend analysis in environmental time series. An advantage often advanced in support of this test is that, in contrast to the parametric methods described above, there is no requirement to specify the probability distribution of the data to which it is applied: the basic

Table 21.2. Number of closures of the Thames Flood Barrier against tidal surges, 1983–2006 (Source: UK Environment Agency).

Year	1983	1984	1985	1986	1987
N	1	0	0	0	1
Year	1988	1989	1990	1991	1992
N	1	0	3	0	1
Year	1993	1994	1995	1996	1997
N	5	1	3	4	0
Year	1998	1999	2000	2001	2002
N	3	3	6	11	2
Year	2003	2004	2005	2006	
N	8	2	5	1	

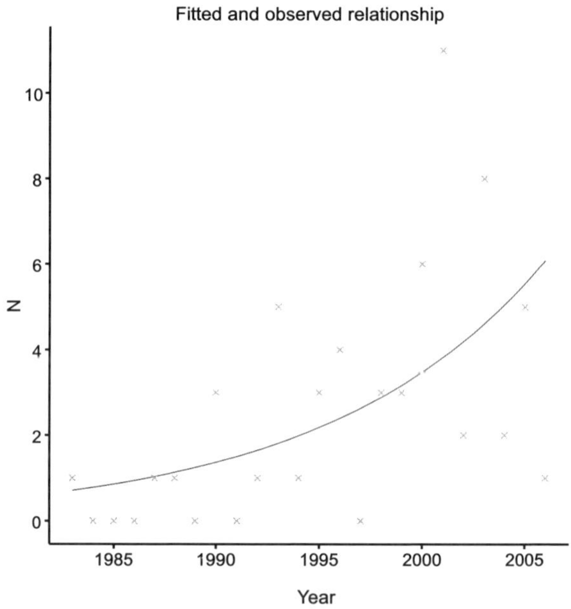

Figure 21.4. Relationship between observed numbers (crosses) of closures of the River Thames Flood Barrier and estimated numbers (line) by GLMs.

requirement is only that values in the data sequence be statistically independent, a not-unreasonable assumption in the case of annual maximum flood discharges (except possibly in very large basins with year-to-year storage).

However, as pointed out by Cox and Hinkley (1974), there are a number of general considerations which limit the practical importance of non-parametric methods. Where the Mann–Kendall test is used to identify a trend in time, the null hypothesis of time-trend absent is tested against the alternative hypothesis of monotonic time-trend, so that no explicit mathematical form of the nature of any trend is specified. If the null hypothesis is rejected and it is required to say something about the form of the trend (whether, for example, it is linear), parameters must be introduced to describe it. If the parameters must be estimated from the flood record, they will be subject to uncertainty, and the Mann–Kendall test does not supply measures of this uncertainty. A similar comment applies where, say, annual floods are to be related to an explanatory variable such as the percentage of natural forest remaining in a drainage basin or the intensity of Sea Surface Temperature (SST) anomalies. The Spearman rank correlation test (Hipel and McLeod, 1994) can be used to test whether the ranked series (of annual floods and percentage forest or SST) are correlated, but the test says nothing about the form of the relationship between the two variables.

As a test for whether time-trends exist in annual flood series, the Mann–Kendall test has other limitations. Where annual flood series are available from several gauging stations with correlated records, it is natural to ask whether any time-trend is similar at the several sites, so that a regional estimate of the time-trend could be calculated. Although multivariate versions of the Mann–Kendall test exist (Hirsch *et al.*, 1982; Hirsch and Slack, 1984; Lettenmaier, 1988) it is not obvious how they could be adapted to provide estimates of a regional trend in flood magnitude, and of its uncertainty. The same comment holds for the case where several correlated flood records are to be related to explanatory variables X.

We conclude this section with a quotation from Cox and Hinkley (1974) in their discussion of non-parametric methods: "... the main objective in analysing data, especially fairly complicated data, is to achieve a description and understanding of the system under investigation in concise and simple terms. This nearly always involves a model having a small number of unknown parameters as possible, each parameter describing some important aspect of the system."

21.6. Normal Scores Regression

Like the Mann–Kendall test for trend, normal scores linear regression is a statistical test of the hypothesis that no trend exists, without providing estimates of parameters that describe the form of any trend, where one is shown to exist, nor of their standard errors. Like the Mann–Kendall test, it requires no assumption about the nature of the probability distribution of values in the annual flood sequence. The procedure is to put the values in the flood series in ascending order of magnitude, and then to replace each value (say the i-th) in this derived sequence by the expected value of the i-th value in a sample from a normal distribution. A linear regression is calculated and the significance of its slope is tested in the usual way.

The normal scores test, and the Spearman rank correlation test, were among the tests used to test for trend in over 1000 flood records analysed and presented in the UK Flood Estimation Handbook (Robson and Reed, 1999).

21.7. Conclusion

The analysis of limited flood records for the detection and explanation of trend, whether in flood magnitudes or in numbers of flood occurrences per unit of time, requires particular attention to the causes of the trend and to the spatial correlation that may exist in regional flood records.

References

Allasia, D. (2007). "Assessment of Hydro-Climatic Forecast for the High River Paraguay" PhD thesis, Instituto de Pesquisas Hidráulicas da Universidade Federal do Rio Grande do Sul [in Portuguese].

Beran, J. (1994). *Statistics for Long-Memory Processes,* Monograph on Statistics and Applied Probability, Vol. 61, Chapman & Hall, London.

Coles, S. (2001). *An Introduction to Statistical Modeling of Extreme Values*, Springer Series in Statistics, Springer, London.

Cox, D.R. and Hinkley, D.V. (1974). *Theoretical Statistics*, Chapman & Hall, London.

Douglas, E.M., Vogel, R.M, Kroll, C.N. (2000). Trend in floods and low flows in the United States: impact of spatial correlation, *J. Hydrology*, **240**, 90–105.

Hipel, K.W. and McLeod, A.I. (1994). *Time Series Modelling of Water Resources and Environmental Systems.* Available from www.stats.uwo.ca/faculty/aim/1994Book/ (Accessed 10/09/2012).

Hirsch, R.M. and Slack, J.R. (1984). A non-parametric trend test for seasonal data with serial dependence, *Water Resour. Res.*, **20**, 727–732.

Hirsch, R.M., Slack, J.R., Smith, R.A. (1982). Techniques of trend analysis for monthly water quality data, *Water Resour. Res.*, **18**, 107–121.

Hurrell, J.W., Kushnir,Y., Ottersen, G. *et al.* (2003). "An Overview of the North Atlantic Oscillation" in: Hurrell, J.W., Kushnir, Y., Ottersen, G. *et al.* (eds), *The North Atlantic Oscillation: Climate Significance and Environmental Impact*, AGU Geophysical Monograph Series, p. 134.

Lettenmaier, D.P. (1988). Multivariate non-parametric tests for water quality, *Water Resour. Bull.*, **24**, 505–512.

Mantua, N.J., Hare, S.R., Zhang, Y. *et al.* (1997). A Pacific interdecadal climate oscillation with impacts on salmon production, *Bull. Amer. Meteorol. Soc.*, **78**, 1069–1079.

McCullagh, P. and Nelder, J.A. (1989). "Generalized Linear Models" *Monographs on Statistics and Applied Probability*, **37**, Chapman & Hall, New York.

Robson A, Reed, D.W. (1999). *Flood Estimation Handbook*, (5 Volumes), UK Natural Environment Research Council.

Sander, M., Bengtsson, L., Holmkvist, B. *et al.* (2002). The relationship between annual varve thickness and maximum annual discharge (1909–1971), *J. Hydrol.*, **263**, 23–35.

Schongart, J, Piedade, M.T.F., Wittmann, F. *et al.* (2005). Wood growth patterns of *Macrolobium acaciifolium* (Benth.) Benth. (Fabaceae) in Amazonian black-water and white-water floodplain forests, *Oecologia*, **145**, 454–461.

Tousson, O. (1925). Mémoire sur l'Histoire du Nil, Mémoires de l'Institut d'Egypte.

Werritty, A, Paine, J.L., Macdonald, N. *et al.* (2006). Use of multi-proxy flood records to improve estimates of flood risk: Lower River Tay, Scotland, *CATENA*, **66**, 107–119.

CHAPTER 22

Detecting Changes in Winter Precipitation Extremes and Fluvial Flood Risk

Robert L. Wilby
Department of Geography,
Loughborough University, UK

Hayley J. Fowler
School of Civil Engineering & Geosciences,
Newcastle University, UK

Bill Donovan
Environment Agency, Bristol, UK

22.1. Introduction

There is a widely held perception that flood risk has increased across Europe during the last decade (EEA, 2005). Following extensive flash flooding in England, the Pitt Review (2008) concluded that "[t]he summer 2007 floods cannot be attributed directly to climate change, but they do provide a clear indication of the scale and nature of the severe weather events we may experience as a result." The review further asserted that, "timely decisions will allow organisations the flexibility to choose the most cost-effective measures, rather than being forced to act urgently and reactively. Early action will also avoid lock-in to long-lived assets such as buildings and infrastructure which are not resilient to the changing climate." This echoes the position held by the Stern Review (2006) that early action on adaptation will bring clear economic benefits by reducing the potential costs of climate change to people, property, ecosystems and infrastructure. The UK Government's Climate Change Act places a statutory requirement upon competent authorities to undertake risk assessments as part of their duties covering adaptation to climate change. The Environment Agency (2007) also called on key utilities and public services to take responsibility for "climate proofing" critical infrastructure, facilities and services.

Following the flooding of the summer 2007, the UK Government's Environment Secretary at the time, Hilary Benn, announced that budgets for flood risk management would reach at least £650 million in 2008, increasing to £700 million in 2009 and eventually attaining £800 million by 2010. Inevitably, higher spending on flood defence infrastructure will prompt questions about *when* and *where* to prioritise investment. This chapter addresses three complementary issues. First, we define the terminology and then summarise the latest scientific evidence on detection of changes in climate and their attribution to human influence. Although we are primarily concerned with climate change detection we explain the current limits to attribution. Second, we provide an assessment of trends in rainfall and flood metrics across the UK and then review the factors that determine the potential detectability of changing climate extremes at regional scales. Third, we describe a methodology and provide a case study of detection times for projected changes in seasonal precipitation extremes in north-west (NWE) and south-east (SEE) England. Results for other regions of the UK are reported in Fowler and Wilby (2010).

22.2. Detection and Attribution Studies

Detection is the process of demonstrating that climate has changed in some defined statistical sense, without providing the reason(s) for the change (Hegerl *et al.*, 2007). A change is detected in observations when the likelihood of an observation (such as an extreme temperature) lies outside the bounds of what might be expected to occur by chance. Changes may not be found if the underlying trend is weak compared with the "noise" of natural climate variability (e.g. McCabe and Wolock, 1991; Wolock and Hornberger, 1991). Conversely, there is always a small chance of spurious detection (perhaps due to an outlier event at one end of the observational record). The change might not necessarily occur linearly, indeed, there is evidence of abrupt, step changes in global climate such as the widespread changes in precipitation that occurred in the 1960s (Baines and Folland, 2007; Narisma *et al.*, 2007). We should also keep in mind that climate records from neighbouring sites are often strongly correlated, a feature that is exploited by studies that pool data to maximise the strength of the signal (e.g. Fowler and Kilsby, 2003a; Pujol *et al.*, 2007). However, trends are highly susceptible to false tendency and/or causation (see Legates *et al.*, 2005; Sparks and Tryjanowski, 2005). This can arise because data are not homogeneous, being affected by a host of non-climatic influences such as encroachment of urban areas, changes in observer, instrumentation, monitoring network density, station location or exposure (see Davey and Pielke, 2005; Kalnay and Cai, 2003).

Attribution is the process of establishing the most likely cause(s) of detected changes at a defined level of statistical confidence (Hegerl *et al.*, 2007). The problem for climate change studies is that there is no physical control, unlike epidemiological studies of causation. So attribution ultimately comes from comparing earth system observations against an earth systems model. Thus, attribution of observed climate change to human influence is accepted when there is consistency between a modelled response to combined human plus natural forcing and the observed pattern of climate behaviour. Climate models are used to isolate the unique "fingerprints" of different external forcings such as greenhouse gas concentrations, variations in solar radiation, or sulphate aerosols originating from volcanic eruptions. In one of the classic studies of this kind, Stott *et al.* (2001) showed that rising global mean temperatures in the second half of the twentieth century could only be explained by combining the human influence on atmospheric composition with natural forcings by solar output and volcanism.

Both detection and attribution require accurate characterisation of internal climate variability, ideally over several decades or longer. Internal climate variability is normally defined by long control simulations of coupled ocean–atmosphere global climate models without external forcing because instrumental records are too short. Alternatively, climatologies of the last millennium may be reconstructed from proxy evidence such as tree rings or lake sediments. Whenever an observation lies outside the range of internal variability, climate change may be detected; whenever a pattern of climate anomalies is only explained by inclusion of human influences in model simulations, attribution is established. However, uncertainties in the representation of physical processes in climate models, in reconstructing climate from proxy data, and in the effects of the forcing agents mean that attribution results are most robust when based on several models (Zhang *et al.*, 2006).

Evidence of human influences on the climate system has been accumulating steadily since the first detection studies published in 2007 (Table 22.1). Early studies focused on large-scale, long-term changes in variables such as seasonal or annual global mean temperatures (e.g. Stott and Tett, 1998; Tett *et al.*, 1999). However, since the IPCC Third Assessment Report (IPCC, 2001) there has been a proliferation of studies covering

Table 22.1. Examples of climate change detection and attribution studies.

Realm	Spatial/Temporal domain	Sources
Temperature	Mean temperature trends for North America	Karoly *et al.* (2003)
	Mean temperature trends in Australia	Karoly and Braganza (2005)
	Annual mean central England temperatures	Karoly and Stott (2006)
	Four indices of temperature extremes over US	Meehl *et al.* (2007)
	Decadal mean temperatures over six continents	Stott (2003)
	European summer temperature anomaly of 2003	Stott *et al.* (2004)
	Global mean temperature changes	Stott *et al.* (2001)
Precipitation	Volcanic influence on global mean	Gillett *et al.* (2004b)
	Precipitation over land areas	Lambert *et al.* (2004)
	Twentieth century annual precipitation trends	Zhang *et al.* (2007)
	Decreasing totals over SW Australia	Timbal *et al.* (2005)
	Trends in winter rainfall over SW Australia	Cai and Cowan (2006)
Troposphere	Sea level pressure	Gillett *et al.* (2003)
	Tropopause height changes	Santer *et al.* (2003)
	Atmospheric moisture content	Santer *et al.* (2007)
	Near surface humidity	Willett *et al.* (2007)
Oceans	Heat content of oceans	Barnett *et al.* (2005)
	Arctic sea ice extent	Gregory *et al.* (2002)
	Wave conditions in the North Sea	Pfizenmayer and von Storch (2001)
Hydrosphere	Trend in global drought	Burke *et al.* (2006)
	Global hydrological cycle since 1550	Tett *et al.* (2007)
	Trend in arctic river discharges	Wu *et al.* (2005)
Biosphere	Length of the growing season	Christidis *et al.* (2007)
	Warming over different vegetation types	Dang *et al.* (2007)
	Canadian forest fires	Gillett *et al.* (2004)

a much wider range of variables, and some are now even attributing climate changes at sub-continental scales (see Barnett *et al.*, 2005). For example, Karoly and Stott (2006) assert that the observed warming in annual mean central England temperatures since 1950 is very unlikely due to natural factors but is consistent with the expected response to anthropogenic forcing.

With the possible exception of rainfall reductions over south-west Australia (Timbal *et al.*, 2005), attribution of rainfall trends to human influence is not yet possible below the scale of the global land area, broken into latitudinal zones (Zhang *et al.*, 2007). However, changes in moderately extreme precipitation events are, in theory, more robustly detectable than changes in mean precipitation (Hegerl *et al.*, 2004; 2006) because as precipitation increases (under higher temperatures and greater water holding capacity of the atmosphere) a greater proportion falls in heavy and very heavy rainfall events (Katz, 1999; Pall *et al.*, 2007; Trenberth *et al.*, 2003). Disproportionate increases in heavy rainfall have been widely reported for the observed climate record (Groisman *et al.*, 2005) but rates of change and/or regional patterns of observed and simulated rainfall extremes show little similarity in early studies (e.g. Kiktev *et al.*, 2003). This is partly due to the inability of climate models to adequately resolve extreme precipitation at sub-grid box scales, the scale mismatch between point observations and gridded climate model output, and the difficulty of defining statistically robust "extreme" indices (Hegerl *et al.*, 2006).

Attributing individual extreme weather events to human activities presents additional challenges. It is often said that *this kind of event is consistent with expected climate changes or might become more commonplace in the future*. Formal attribution can only be accomplished in a probabilistic sense. In much the same way that smoking increases the risk of lung cancer by X%, the risk of an extreme meteorological event is said to be Y% more likely as a consequence of past greenhouse gas emissions than without. For example, the exceptional European heat wave in the summer of 2003 resulted in an area-average temperature anomaly of 2.3°C compared with the 1961–1990 average. By simulating probability distributions of summer temperatures with and without anthropogenic emissions it was shown that the risk of a 2003 heat wave has increased by a factor of at least two due to human influence (Stott *et al.*, 2004). Attributing changes in flood risk to carbon dioxide emissions is highly problematic because of the very large uncertainty in modelled precipitation changes at the river catchment scale (Prudhomme *et al.*, 2003). For example, using results from the EU PRUDENCE project, Fowler and Ekström (2009) showed the extent to which the present generation of Regional Climate Models (RCMs) is unable to reproduce properties of observed extreme summer rainfalls.

22.3. Changes in UK Flood Risk Indices

Long-term changes in UK fluvial flood risk have been reviewed elsewhere (see Wilby *et al.*, 2008). Several studies report increasing winter precipitation, larger multi-day rainfall totals, and higher contributions from intense daily events since the 1960s (Table 22.2). Others suggest strong

Table 22.2. Observed changes in UK precipitation and runoff (from Wilby *et al.*, 2008).

Region	Period (metrics)	Key findings	Sources
9 homogeneous rainfall regions	1766–2000 (monthly rain)	Significantly wetter winters in west Scotland; 1990s unusually wet	Alexander and Jones (2001)
102 stations across Scotland	1914–2004, 1961–2004 (daily, seasonal rain and snow cover)	Increased winter rainfall since 1960s across all regions but up to 70% in north; increase in spring rainfall in west since 1910s; large reductions in autumn/spring snow cover; increase in heavy winter rain in north and west	Barnett *et al.* (2006)
28 rainfall stations and 15 gauging stations	1881–onwards	Increase in annual rainfall maxima in since 1960s; no long-term trend in rainfall maxima, flood peaks, overall flood volume or duration	CEH & UKMO (2001)
56 gauging stations across western UK	1962–2001 (seasonal, annual flow)	Significant increases in mountainous west during autumn and winter	Dixon *et al.* (2006)
204 stations across UK	1961–2000 (daily rain)	Significant increase in annual maxima, 5- and 10-day rainfall events over Scotland and northern England, especially in the 1990s	Fowler and Kilsby (2003a;b)
97 undisturbed catchments across UK	1963–2002, 1973–2002 (annual flow)	Increasing runoff trend for Scotland, and maritime western areas of England and Wales	Hannaford and Marsh (2006; 2007)
56 stations in the English Lake District	1971–2000 (daily, monthly rain)	Increased heavy precipitation at high elevation sites; weakening of the Cumbrian rain shadow	Malby *et al.* (2007)
689 stations across UK	1961–2006 (daily rain)	Heavy precipitation events are making a greater contribution to winter totals and less in summer	Osborn *et al.* (2000); Maraun *et al.* (2008)

(*Continued*)

Table 22.2. (*Continued*)

Region	Period (metrics)	Key findings	Sources
890 gauging stations	1870–onwards (seasonal rain, peak flows)	Increase in winter rainfall since 1960s; flood rich and flood poor periods but no overall trend	Robson (2002), Robson *et al.* (1998)
47 gauging stations across UK	1970–2002 (seasonal flow)	Significant increases during autumn and winter at <25% sites; no clear regional pattern	Wade *et al.* (2005)
13 meteoro-logical and 38 gauging stations across Scotland	1970–1996 (seasonal rain, flow)	Significant increase in runoff for one third of sites	Werritty (2002)
15 rivers in England and Wales	1865–2002 (reconstructed seasonal flow)	Significant long-term increase in winter flow in only three rivers; increasing annual flows since 1970s	Wilby (2006)
Stations with snowfall >13 cm in Britain	1861–1996 (catalogue of heavy snowfalls)	Decadal variability in the frequency of heavy snowfall peaking in the 1860s, 1870s and the 1970s; no overall trend	Wild *et al.* (2000)

regional gradients with larger winter increases in the north and west of the UK, and at higher elevation sites. These patterns translate into increased winter mean runoff, especially in the upland areas of western England and Wales. The most widely accepted explanation is that the changes were forced by a strongly positive phase of the North Atlantic Oscillation (NAO) since the 1960s. This displaced storm tracks northwards and strengthened westerly moisture advection from the Atlantic over north-west Europe (Haylock and Goodess, 2004). Whether or not recent variability in the NAO is itself a manifestation of anthropogenic forcing remains an open question (Hegerl *et al.*, 2007). However, when longer UK rainfall-runoff records are analysed, many of the trends found in shorter series cease to be significant (CEH and UKMO, 2001; Robson, 2002; Robson *et al.*, 1998; Wilby *et al.*, 2008). As mentioned above, trend detection for extreme events is far from straightforward because the outcome can depend on the chosen metric, period of record, power of the statistical test, and confounding factors:

- *Flood risk metrics.* A variety of indices have been used to detect changes in flood risk. For example, the sample studies in Table 22.2 employ series of monthly, seasonal and annual rainfall/river flow, annual maxima of daily rainfall intensities/river flows or N-day rainfall totals/maximum flows, proportional contribution of heavy events to total rainfall, and annual counts of peaks over threshold flows. Analyses may be performed using point or area average data, individual records, data from networks of stations, via pooling of (rainfall) maxima by region or elevation, and using gridded extreme indices (e.g. Alexander *et al.*, 2006). In some cases, more exotic indicators may be applied such as frequencies of flood generating atmospheric circulation patterns (e.g. Bárdossy and Filiz, 2005), changes in the timing of extreme events (e.g. Fowler and Kilsby, 2003b); or standardised precipitation indices (e.g. Seiler *et al.*, 2002). In any event, the chosen metric should be meaningful in terms of flood generation mechanisms and interpretable from a policy perspective (e.g. sub-daily data are needed for flash flooding). Fortuitously, model experiments show that changes in indices of extreme precipitation are stronger than corresponding changes in mean precipitation (Hegerl *et al.*, 2004). Widely employed indicators of extreme events (e.g. number of days with precipitation greater than 10 mm, or the fraction of total rainfall due to events exceeding the 95th percentile amount) are acknowledged to be "not as extreme" as they could be (Frich *et al.*, 2002; Tebaldi *et al.*, 2006). Here, the trade-off is between having extremes that are severe enough to have repercussions for society, yet are not so rare that there are insufficient events to enable detection of change.
- *Period of record.* When longer rainfall and river flow records are analysed, many trends found in shorter series cease to be significant. This can be due to the influence of outliers (at the start or the end of the record), or simply the large multi-decadal variability of the UK rainfall regime. It is widely recognised that extreme statistics derived from conventional 30 year climatological periods are subject to large uncertainty (Kendon *et al.*, 2008; Lane, 2008). Following a detailed review of the 2000/01 UK flooding, it was recommended that records of at least 50 years duration should be used for investigating possible climate change signals in rainfall and river flows (CEH and UKMO, 2001; Kundzewicz and Robson, 2004). Unfortunately, continuous river flow records of this length, with accompanying meta-data, are rare. Out of more than 1000 flow records

held by the Global Runoff Data Centre, less than 200 worldwide have continuous daily flow series longer than 40 years that extend into the late 1990s (Kundzewicz *et al.*, 2005).

- *Statistical tests.* As noted, climate variability affects the detectability of a trend within a finite series of observations. The possibility of erroneous trend detection (Type I errors) are controlled by choosing an appropriate statistical test, and level of confidence. Widely used methods for extremes include (logistic) linear regression, "change point" tests, and the non-parametric Spearman rank correlation and Mann–Kendall tests (Lanzante, 1996). Alternatively, an actual trend might be overlooked because it is overshadowed by short-term climate fluctuations (Type II error). In this case, the detectability depends on the power of the test as a function of record length, magnitude of the trend and rarity of event(s). Even when presented with the same data, different (flood) indices may reveal different numbers of significant trends (e.g. Svensson *et al.*, 2005). Detectability of trends in very rare events is particularly difficult for small sample sizes but can be improved through regional pooling of data (Frei and Schär, 2001). For most practical situations, this points to the necessity of analysing "less extreme" extremes.
- *Confounding factors.* Unerring faith in detected trends without a supporting conceptual framework can lead to invalid statistical inference (Sparks and Tryjanowski, 2005). Creeping or sudden changes in meteorological records can arise for a range of reasons such as changing instrumentation or instrument location, observing or recording practices, site characteristics, or sampling regime (Zwiers and von Storch, 2004). Some claim that the entire observational record has been contaminated to some extent by human influences on climate and hydrology (Tett *et al.*, 2007). Discharge records may respond to a host of non-climatic influences including land cover and management, urbanisation, river regulation, water abstraction and effluent returns, or flood flows may bypass gauging structures (Archer, 2007; Legates *et al.*, 2005). Occasionally false trends and biases arise from the statistical method used to divide data, such as percentile-based indices for temperature and precipitation extremes (Michaels *et al.*, 2004; Zhang *et al.*, 2005).

In summary, the choice of index, spatial and temporal scale of aggregation, statistical test (including significance testing), and the account taken of confounding factors, all require careful justification in a detection study. The next section briefly describes earlier work on climate change detection at river catchment scales. Then we outline a method for detecting changes in heavy precipitation using RCM output. This involves calculating *when* and *where* changes in downscaled rainfall metrics are not likely to have been entirely the result of (model-estimated) natural internal variability.

22.4. Detection Time for Changes in Heavy Precipitation

Detection times for climate change *trends* in hydroclimatic data depend on the strength of the assumed trend, the sample variance of the time series in question, and the probabilities of making errors of Type I or II. Preliminary estimates using data for river catchments in the US and UK suggest that statistically robust trends in seasonal runoff are unlikely to be found until the second half of the 21st century (Wilby, 2006; Ziegler *et al.*, 2005).

The same statistical relationship can be inverted to estimate the strength of trend required for detection by specified time horizons. For example, Figure 22.1 shows that, on average, a 60% change in winter rainfall is needed for detection by 2025 given long-term (L) inter-annual variability over the UK catchments used by Wilby (2006), whereas the change needed for detection by 2055 is 44%. The smallest changes needed for detection in winter rainfall over the same periods are in the River Tyne catchment (54% and 39% respectively). Even smaller changes are needed for detection in annual rainfall totals by 2025 and 2055 (25% and 18% respectively).

Analysis of UK *winter* and *annual* rainfall totals suggests that changes of ~25% will be needed for early detection (by year 2025) in the best

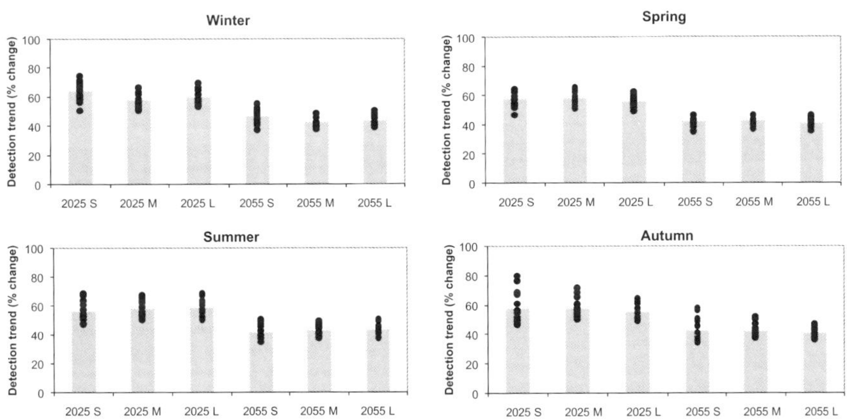

Figure 22.1. Detectable trends (% change) in seasonal mean rainfall by 2025 and 2055 for sample variances derived from short (S, 1961–1990), medium (M, 1921–1990) and long (L, 1865–1990) records. Each point represents a different river catchment; the bars show the mean detection trend by time horizon and variance estimate.

case catchment (River Tyne). Although increases (or decreases) in seasonal totals could affect overall flood risk, it is suspected that more useful flood-risk information can be extracted from daily precipitation indices. Furthermore, as noted before, projected trends in indices of heavy precipitation may be detectable earlier than trends in seasonal means (Hegerl *et al.*, 2004). However, more complex flood events involving rainfall and snowmelt, or fluvial flooding and tidal surge (as in Svensson *et al.*, 2002) are clearly not captured by "rain only" indices.

22.5. Case Study: North-West and South-East England

Here, we provide a case study for detecting changes to extreme rainfall metrics based on a combination of Regional Frequency Analysis (RFA) and pattern scaling. The estimation of the frequencies of extreme events is difficult as extreme events are, by definition, rare and observational records are often short. In RFA, data from the "region" are assumed to share the same frequency distribution and only differ in their magnitude (mean or median values). Therefore, RFA trades "space for time" and pools standardised data from several different sites within a "region" to fit a single frequency distribution (Hosking and Wallis, 1997).

22.5.1. *Data*

We employed RCM output from the PRUDENCE (Prediction of Regional scenarios and Uncertainties for Defining European Climate change risks and Effects) ensemble, which contains daily data for a range of climatic variables for control (1961–1990) and future (2071–2100) time slices (Christensen *et al.*, 2007). We used 13 RCM integrations from the PRUDENCE ensemble. All experiments yield daily precipitation totals for control (1961–1990) and future (2071–2100) time periods (Christensen *et al.*, 2007) under the IPCC SRES A2 emissions scenario (Nakićenović *et al.*, 2000) (Table 22.3). Nine of the RCM experiments were performed by nesting within the atmosphere-only high-resolution Global Climate Model (GCM) HadAM3H of the UK Hadley Centre. One RCM, HadRM3P, was nested within HadAM3P, a more recent version of the same atmosphere-only GCM; but HadRM3H and HadRM3P can be considered as essentially the same model for Europe (Moberg and Jones, 2004). The variable resolution global atmospheric model, Arpége, is nested directly within HadCM3. Additionally, two RCM integrations, HIRHAM and RCAO, are driven by lateral boundary conditions from two separate integrations of the ECHAM4/OPYC3 coupled ocean–atmosphere GCM.

Table 22.3. The thirteen RCM integrations used in this study. The first part of each acronym refers to the RCM, and the last letter to the GCM supplying the boundary conditions. All RCM integrations are from PRUDENCE.

Acronym	Institution/Model Origin and References	RCM	GCM
ARPH	French Meteorological Service; ARPEGE/IFS variable resolution global model. Model: Déqué *et al.* (1998).	Arpège	HadCM3
HADH	Hadley Centre, UK Met Office, Exeter; Regional model at the Hadley Centre. Model: Jones *et al.* (2004a)	HadRM3P	HadAM3P
HIRH	Danish Meteorological Institute, Copenhagen; Dynamical core from HIRLAM, Parameterisations from	HIRHAM	HadAM3H
HIRE	ECHAM4. Model: Christensen *et al.* (1996, 1998). Physiographic datasets: Hagemann *et al.* (2001), Christensen *et al.* (2001).	ECHAM4	
RCAOH	Swedish Meteorological and Hydrological Institute, Stockholm; Rossby Centre Atmosphere Ocean Model.	RCAO	HadAM3H
RCAOE	Model: Jones *et al.* (2004b), Meier *et al.* (2003), Döscher *et al.* (2002), Räisänen *et al.* (2004).		ECHAM4/ OPYC
CHRMH	Swiss Federal Institute of Technology (ETH), Zurich; Climate High-Resolution Model. Model: Lüthi *et al.* (1996), Vidale *et al.* (2003).	CHRM	HadAM3H
CLMH	GKSS, Institute for Coastal Research, Geesthatcht, Germany; Climate version of "Lokalmodell" of German Weather Service. Model: Steppeler *et al.* (2003).	CLM	HadAM3H
REMOH	Max Planck Institute for Meteorology, Hamburg, Germany; Dynamical core from "EuropaModell" of German Weather Service, Parameterisations from ECHAM4. Model: Jacob (2001), Roeckner *et al.* (1996).	REMO	HadAM3H
PROMH	Universidad Complutense de Madrid, Spain; Climate version of PROMES model. Model: Castro *et al.* (1993), Arribas *et al.* (2003).	PROMES	HadAM3H
REGH	The Abdus Salam International Centre for Theoretical Physics, Italy (ICTP); Dynamical core from MM5, Parameterisations from CCM3. Model: Giorgi *et al.* (1993a,b, 2000), Pal *et al.* (2000).	RegCM	HadAM3H

(*Continued*)

Table 22.3. (*Continued*)

Acronym	Institution/Model Origin and References	RCM	GCM
RACH	The Royal Netherlands Meteorological Institute (KNMI), Netherlands; Dynamical core from HIRLAM, Parameterisations from ECMWF physics. Model: Tiedtke (1989, 1993), Lenderink *et al.* (2003).	RACMO2	HadAM3H
METH	Norwegian Meteorological Institute; Version of HIRHAM. Model: Christensen *et al.* (2001), Hanssen-Bauer *et al.* (2003).	MetNo	HadAM3H

All the RCMs operate with grid spacing of ~0.5° longitude by ~0.5° latitude (~50 km spatial resolution) over a European domain and data were re-gridded to a regular 0.5° × 0.5° grid to allow direct comparison between models. Suffixes E and H denote RCMs driven by ECHAM4/OPYC3 and HadAM3H/P/HadCM3 GCMs respectively. More details of the experimental design of the PRUDENCE integrations can be found in Jacob *et al.* (2007).

A dataset of comparable spatial scale to the RCM outputs was produced by aggregating a daily observed precipitation 5 km by 5 km grid produced by the UK Meteorological Office (Perry and Hollis, 2005a,b) to the regular 0.5° × 0.5° grid. This was achieved by taking a daily average across the 5 km boxes contained within each 0.5° × 0.5° grid cell for each day of 1961–1990.

22.5.2. *Detection indices*

Seasonal Maximum (SM) series of 1, 5 and 10 day precipitation totals were extracted for each grid cell, for each RCM time-slice (the control period, 1961–1990, and future period, 2071–2100), and for observations. These SM series were standardised by their median (Rmed; following Fowler *et al.*, 2005) and the standardised SM were then pooled for each of the nine UK rainfall regions delineated by Wigley *et al.* (1984) (Figure 22.2). The homogeneity of these regions for extreme precipitation was confirmed previously by Fowler and Kilsby (2003a).

Next, a Generalised Extreme Value (GEV) distribution was fitted to each pooled SM sample using Maximum Likelihood Estimation (MLE) and return values of precipitation intensities with average recurrence of 10 and 50 years were estimated. The estimates were then rescaled using

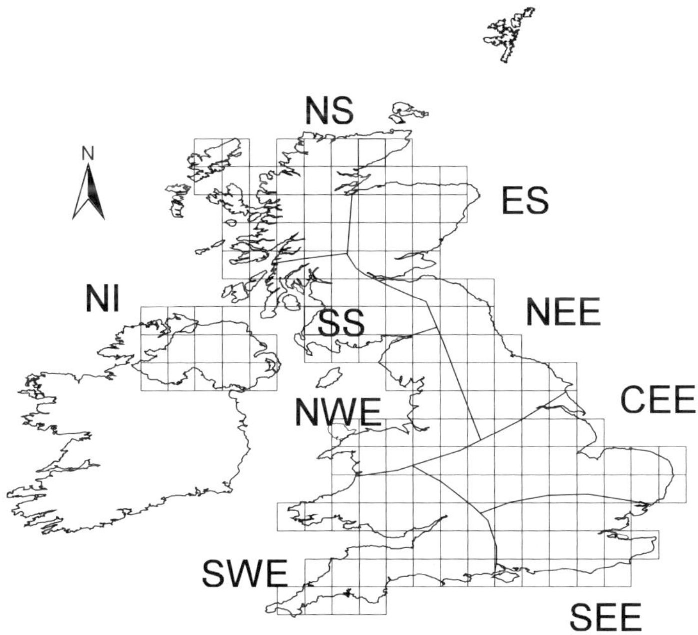

Figure 22.2. The RCM regular 0.5° by 0.5° grid and the nine coherent rainfall regions. The regions are: North Scotland (NS), East Scotland (ES), South Scotland (SS), Northern Ireland (NI), Northwest England (NWE), Northeast England (NEE), Central and Eastern England (CEE), Southeast England (SEE) and Southwest England (SWE) (from Fowler *et al.*, 2007).

the regional average Rmed from the original SM dataset. We also calculate the associated error by the 95% confidence intervals on the return value estimates via the delta method using three different estimates of variance to test the sensitivity: (i) RCM pooled sample variance; (ii) observed variance, 1961–1990; (iii) observed variance, 1958–2002 (full length of record). Note that the use of the full record of observed variance gives a more conservative estimate of confidence intervals compared to the RCM pooled sample variance.

A further step is needed to estimate transient changes in the return values of precipitation intensities for each RCM between the two time-slices, 1961–1990 and 2071–2100. We applied a conventional pattern scaling approach (Mitchell, 2003). The technique assumes that regional changes in extreme precipitation (or any climate variable) will occur in proportion to the projected change in global mean temperature, in this case, from the two GCMs providing lateral boundary conditions for the 13 PRUDENCE RCMs (i.e. HadCM3 and ECHAM4). Pattern scaling was based on the change in global mean temperature predicted for 30-year time slices centred on the years 1975, 2025, 2055 and 2085. Therefore, designating the global mean temperature as T_y where y is the central year (1975, 2025, 2055 and 2085 respectively), the scale factor for each GCM time slice may be written as:

$$SF_y = \frac{T_y - T_{Con}}{T_{Fut} - T_{Con}}, \tag{22.1}$$

where T_{Con} and T_{Fut} indicate the global mean temperature for the control (1961–1990) and future (2071–2100) time slices respectively. For intervening years the scale factors were linearly interpolated to provide transient scale factors. Therefore, for 1990 to 2100, *SF* varies from zero to one. The transient scale factors show that global mean temperature change will accelerate over the next 100 years. An alternative to pattern scaling on global mean temperature change would be to simply apply linear scaling between 1990 and 2100; this would imply earlier detection times. We use the more conservative non-linear pattern-scaling to estimate return values of precipitation intensities and the associated confidence intervals for each year between 1990 and 2100.

22.5.3. *Detection times*

We define a detectible increase in extreme precipitation, D_x, as the point (year) at which we would reject (at the $\alpha = 0.05$ or 95% significance level) the null hypothesis that the return level estimated for the 1961–1990 period, μ_c, and the return level estimated for a year x (where $x > 1990$), μ_x, are equal in favour of the alternative hypothesis that μ_x is greater (lower) than μ_c. This statistical test is based on the signal-to-noise ratio and provides a distribution that is approximately normally distributed with a mean of zero and a standard deviation of 1; *N(0,1)* (see Equation 22.2). We then use a two-tailed Student's t-test (assuming that the trend can go down as well as up) to estimate the point at which the return levels are shown to be from a significantly different population at the $\alpha = 0.05$ level, i.e. where $D_x \geq 1.96$:

$$D_x = \frac{|\mu_x - \mu_c|}{\sqrt{\sigma_f^2 + \sigma_c^2}} \geq 1.96 \tag{22.2}$$

where μ_x is the pattern-scaled return level for year x, μ_c is the estimated return level for the control period (1961–1990), σ_f^2 is the variance in the return level estimate for the RCM generated future period (2071–2100) and σ_c^2 is the variance in the return level estimate for RCM generated control period (1961–1990). Note that the observed variance, $2\sigma_o^2$, for 1961–1990 and 1958–2002 respectively, was substituted for the summed variance from the RCM control and future integrations, $\sigma_f^2 + \sigma_c^2$, in Equation (22.2) to establish the sensitivity of the test to assumed variance (where the delta confidence intervals on the return level estimates were also calculated using the same assumed variance estimate, either RCM generated or observed for 1961–1990 and 1958–2002 respectively, as previously stated).

This test defines the "detection year", D_x, as the first year at which $D_x \geq 1.96$ (the transient return level estimate for year x is significantly different to the return level estimate for the control period (1961–1990) at the $\alpha = 0.05$ level).

We apply the principle of equal weighting across different RCMs to produce probability distributions of estimated D_x for change to the return levels of seasonal extreme precipitation in the nine UK rainfall regions. This is reasonable because weighting RCMs in the PRUDENCE ensemble by their skill at reproducing observed extremes did not make any practical difference since most are driven by lateral boundary conditions from the same GCM, HadAM3H/CM3 (Fowler and Ekström, 2009; Manning *et al.*, 2009). However, we examine the sensitivity of D_x to the assumed variance in extremes and to the α level used in the detection test. The analysis is truncated at 2100 as we cannot assume that the pattern scaling relationship holds true after this time. Furthermore, although the analysis includes summer precipitation extremes, it should be noted that the present generation of RCMs do not adequately reproduce observed extremes during this season (Fowler and Ekström, 2009). Throughout the study we define the point beyond which the probability of detection is more likely than not as when more than 50% of the RCMs suggest a significant change in extreme precipitation.

22.5.4. *Results*

Cumulative Frequency Distributions (cdfs) of detection time by region and season are presented for 1 and 10 day precipitation totals at the 10-year return level. Figure 22.3 shows the detection times for significant change (at the $\alpha = 0.05$ level) to winter extreme precipitation. In these plots, the vertical black dotted line indicates the year 2050 and the horizontal red dashed line indicates the point at which the probability of detection is 0.5 (i.e. the chance of detection is 50%). Beyond this point, change to

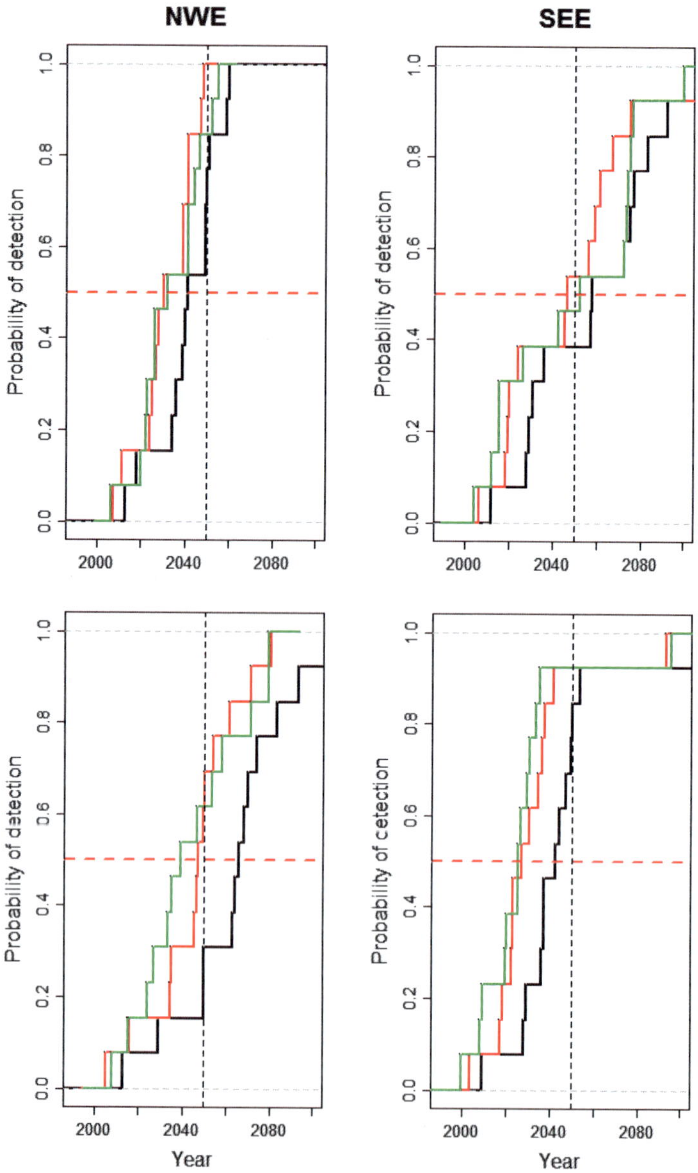

Figure 22.3. Detection year for significant change ($\alpha = 0.05$) in the return levels of maximum 1-day (upper panels) and 10-day (lower panels) winter precipitation totals with 10-year return period in NWE (left-hand column) and SEE (right-hand column). Data used to estimate natural variability were: observed 1958–2002 (red lines); RCM 1961–1990 (green lines); observed 1961–1990 (black lines).

these metrics is more likely than not to be detectible. Three individual cumulative frequency distributions are shown on each plot: the green cdf shows the detection time, D_x, using RCM internal variability (i.e. using Equation 22.2); the red cdf shows the same but for observed variability for 1961–1990 (i.e. substituting this into Equation 22.2); the black cdf shows the same but for the full record of observed variability from 1958–2002 (i.e. substituting this into Equation 22.2).

The detection time, D_x, is sensitive to (i) the assumed variance used to calculate confidence intervals on the return level estimates (95% confidence intervals are calculated in all cases) and in the detection test, and (ii) the α significance level used in testing "detectability" (note that the standard used is the $\alpha = 0.05$ level). We find that there are no consistent patterns (across regions) of earlier or later detection time for the different assumed variance estimates. For the two case study regions, use of the observed variability for 1961–1990 yielded longer detection times. Across all regions, for the winter 1 day 10-year return level, the mean detection time is 2045 for RCM-estimated variability, 2055 for observed variability 1961–1990, but 2042 for the longer observed variability 1958–2002. Note that even with regional pooling of data, confidence intervals for the return period estimate are wide due to the relatively large uncertainty (variability) in the estimate. In the case study regions, use of the observed variability available from the 5 km gridded dataset (for 1958–2002) provides the shortest detection times for 1-day totals. However, for the 10-day totals the RCM-estimated natural internal variability enables earlier detection. In these cases, low variance of seasonal maxima within the RCM integration (and/or outliers in the longer observed record) decreases (increases) the width of confidence intervals and thus produces shorter (longer) detection times.

Records prior to the early 19th century are thought to be less representative of the present climate regime (see for example, Marsh *et al.*, 2007) and so were not used to characterise the variance statistic. Even so, it is recognised that long-term rainfall and reconstructed runoff records can help to quantify multi-decadal variability *without* anthropogenic climate change (see Jones *et al.*, 2006). For example, a 200-year monthly precipitation series for the English Lake District varied by −17% to +11% of long-term average between the driest (1850s) and wettest (1920s) decades (Barker *et al.*, 2004).

For the two exemplar regions, future changes in some winter precipitation indices could be detectable relative to 1958–2002 conditions as early as the 2020s and 2030s (Table 22.4). The earliest detection year (2027) was found for 10-day winter precipitation totals in SEE, although even earlier

Table 22.4. Detection years for significant changes in 1- and 10-day seasonal precipitation totals with 10-year return period, using variance estimates based on observations for the period 1958–2002. Note that summer was excluded because of the low confidence in RCM projections of regional rainfall. See Fowler and Wilby (2010) for other regions.

	1-day		10-day	
Season	NWE	SEE	NWE	SEE
Dec–Feb	2030	2046	2047	2027
Mar–May	2047	2083	2052	2052
Sep–Oct	2039	2071	2060	2058

detection may be possible in south-west England (not shown). Detection years were earlier for 1- than 10-day precipitation totals in NWE; but the opposite applies in SEE. However, detection is earliest for 10-year return period winter precipitation indices, and earlier than for spring or autumn. At 50-year return periods it is unlikely that changes will be detectible by 2050, and for some regions changes are not even detectible by the end of the 21st century (not shown). Across all regions and seasons, the earliest detection times were generally found in winter, for longer duration (10-day) precipitation events, with shorter (10-year) return periods, based on estimates of variability taken from RCM output. The earliest mean detection times for variance estimates based on observations (1958–2002) were also found for winter 10-year return period events but were broadly similar for 1- and 10-day totals.

22.6. Concluding Remarks

Following extensive flooding in England during summer 2007, budgets for flood risk management were significantly increased. This raises the questions as to *when* and *where* to prioritise future investment in flood management measures. This chapter has summarised scientific evidence for the detection and attribution of extreme events to climate change. Although attribution is not yet possible at regional scales, techniques are emerging for detection of trends in flood indices at river catchment scales, and for estimating the time-horizons for formal detection. We set the scene by reviewing evidence of changing flood risk in the observational records of the UK and by identifying some of the main factors that confound detection. We then demonstrated a method for estimating detection times for changes in seasonal precipitation extremes projected by the EU PRUDENCE RCM ensemble. We showed that for selected regions and extreme precipitation indices, the climate change signal(s) in the PRUDENCE projections could be detectable as early as the 2020s. The next step is to broaden the analysis to a national scale, and to investigate a larger suite of extreme indicators. By identifying potential "hotspots" of emerging flood risk, a more targeted approach to monitoring and investment might then be feasible.

Acknowledgements

This research was supported by the Environment Agency and by a NERC fellowship award to Dr Hayley Fowler. The views expressed are those of the authors and are not necessarily indicative of the position held by the Environment Agency.

References

Alexander, L.V. and Jones, P.D. (2001). Updated precipitation series for the UK and discussions of recent extremes, *Atmos. Sci. Lett.*, **1**, 142–150.

Alexander, L.V., Zhang, X., Peterson, T.C. *et al.* (2006). Global observed changes in daily climate extremes of temperature and precipitation, *J. Geophys. Res.*, **111**, doi:10.1029/2005JD006290.

Archer, D.R. (2007). The use of flow variability analysis to assess the impact of land use change on the paired Plynlimon catchments, mid-Wales, *J. Hydrol.*, **347**, 487–496.

Arribas, A., Gallardo, C., Gaertner, M.A. *et al.* (2003). Sensitivity of Iberian Peninsula climate to land degradation, *Clim. Dynam.* **20**, 477–489.

Baines, P.G. and Folland, C.K. (2007). Evidence for a rapid global climate shift across the late 1960s, *J. Climate*, **20**, 2721–2744.

Bárdossy, A. and Filiz, F. (2005). Identification of flood producing atmospheric circulation patterns, *J. Hydrol.*, **313**, 48–57.

Barker, P.A., Wilby, R.L. and Borrows, J. (2004). A 200-year precipitation index for the central English Lake District, *Hydrol. Sci. J.*, **49**, 769–785.

Barnett, C., Hossell, J., Perry, M. *et al.* (2006). *A Handbook of Climate Trends Across Scotland*, SNIFFER project CC03, Scotland and Northern Ireland Forum for Environmental Research.

Barnett, T.P., Pierce, D.W. AchutaRao, K.M. *et al.* (2005). Penetration of human-induced warming into the world's oceans, *Science*, **309**, 284–287.

Barnett, T.P., Zwiers, F., Hegerl, G. *et al.* (2005). Detecting and attributing external influences on the climate system: a review of recent advances, *J. Climate*, **18**, 1291–1314.

Burke, E.J., Brown, S.J. and Christidis, N. (2006). Modelling the recent evolution of global drought and projections for the 21st century with the Hadley Centre climate model, *J. Hydrometeorol.*, **7**, 1113–1125.

Cai, W.J. and Cowan, T. (2006). SAM and regional rainfall in IPCC AR4 models: can anthropogenic forcing account for southwest Western Australian winter rainfall reduction, *Geophys. Res. Lett.*, **33**, L24708.

Castro, M., Fernández, C. and Gaertner, M.A. (1993). "Description of a Meso-Scale Atmospheric Numerical Model" in: Díaz, J.I. and. Lions, J.L (eds), *Mathematics, Climate and Environment*, Masson, London.

CEH & UKMO (2001). *To What Degree can the October/November 2000 Flood Events be attributed to Climate Change?*, Defra FD2304.

Christensen, J.H., Carter, T.R., Rummukainen, M. *et al.* (2007). Evaluating the performance and utility of regional climate models: the PRUDENCE project, *Climatic Change*, **81**, 1–6.

Christensen, J.H., Christensen, O.B., Lopez, P. (1996). *The HIRHAM4 Regional Atmospheric Climate Model*, DMI Technical Report 96-4. Available from DMI, Lyngbyvej 100, Copenhagen Ø.

Christensen, O.B., Christensen, J.H., Machenhauer, B. *et al.* (1998). Very high-resolution regional climate simulations over Scandinavia – present climate, *J. Climate*, **11**, 3204–3229.

Christensen, J.H., Christensen, O.B. and Schultz, J.P. (2001). *High Resolution Physiographic Data Set for HIRHAM4: An Application to a 50 km Horizontal Resolution Domain Covering Europe*, DMI Technical Report 01–15. Available from DMI, Lyngbyvej 100, Copenhagen Ø.

Christidis, N., Stott, P.A., Brown, S. *et al.* (2007). Human contribution to the lengthening of the growing season 1950–99, *J. Climate*, **20**, 5441–5454.

Dang, H., Gillett, N.P., Weaver, A.J. *et al.* (2007). Climate change detection over different land surface vegetation classes, *Int. J. Climatol.*, **27**, 211–220.

Davey, C.A. and Pielke, R.A. (2005). Microclimate exposures of surface-based weather stations: implications for the assessment of long-term temperature trends, *B. Am. Meteorol. Soc.*, **86**, 497–504.

Déqué, M., Marquet, P. and Jones, R.G. (1998). Simulation of climate change over Europe using a global variable resolution general circulation model, *Clim. Dynam.*, **14**, 173–189.

Dixon, H., Lawler, D.M. and Shamseldon, A.Y. (2006). Streamflow trends in western Britain, *Geophys. Res. Lett.*, **33**, L19406, doi:10.1029/2006GL027325.

Döscher, R., Willén, U., Jones, C. *et al.* (2002). The development of the coupled regional ocean–atmosphere model RCAO, *Boreal Environ. Res.*, **7**, 183–192.

Environment Agency (EA) (2007). *Review of Summer 2007 Floods*, Environment Agency, Bristol.

European Environment Agency (EEA) (2005). *Vulnerability and Adaptation to Climate Change: Scoping Report*, EEA Technical Report, Copenhagen.

Fowler, H.J. and Ekström, M. (2009). Multi-model ensemble estimates of climate change impacts on UK seasonal rainfall extremes, *Int. J. Climatol.*, **29**, 385–416.

Fowler, H.J., Ekström, M., Blenkinsop, S. *et al.* (2007). Estimating change in extreme European precipitation using a multi-model ensemble, *J. Geophys. Res. Atmos.*, **112**, D18104, doi:10.1029/2007JD008619.

Fowler, H.J., Ekström, M., Kilsby, C.G. *et al.* (2005). New estimates of future changes in extreme rainfall across the UK using regional climate model integrations. 1. Assessment of control climate, *J. Hydrol.*, **300**, 212–233.

Fowler, H.J. and Kilsby, C.G. (2003a). A regional frequency analysis of United Kingdom extreme rainfall from 1961 to 2000, *Int. J. Climatol.*, **23**, 1313–1334.

Fowler, H.J. and Kilsby, C.G. (2003b). Implications of changes in seasonal and annual extreme rainfall, *Geophys. Res. Lett.*, **30**, L1720, doi:10.1029/2003GL017327.

Fowler, H.J. and Wilby, R.L. (2010). Detecting changes in seasonal precipitation extremes using regional climate model projections: implications for managing fluvial flood risk, *Water Resour. Res.*, doi:10.1029/2008WR007636.

Frei, C. and Schar, C. (2001). Detection probability of trends in rare events: theory and application to heavy precipitation in the Alpine region, *J. Climate*, **14**, 1568–1584.

Frich, P., Alexander, L.V., Della-Marta, P. (2002). Observed coherent changes in climatic extremes during the second half of the twentieth century, *Clim. Res.*, **19**, 193–212.

Gillett, N.P., Weaver, A.J., Zwiers, F.W. *et al.* (2004a). Detecting the effect of climate change on Canadian forest fires, *Geophys. Res. Lett.*, **31**, L18211, doi:10.1029/2004GL020876.

Gillett, N.P., Weaver, A.J., Zwiers, F.W. *et al.* (2004b). Detection of volcanic influence on global precipitation, *Geophys. Res. Lett.*, **31**, L112217, doi:10.1029/2004GL020044.

Gillet, N.P., Zwiers, F.W., Weaver, A.J. *et al.* (2003). Detection of human influence on sea level pressure, *Nature*, **422**, 292–294.

Giorgi, F., Marinucci, M.R. and Bates, G.T. (1993a). Development of a second generation regional climate model (REGCM2). Part I: Boundary layer and radiative transfer processes, *Mon. Weather Rev.*, **121**, 2794–2813.

Giorgi, F., Marinucci, M.R., Bates, G.T. *et al.* (1993b). Development of a second generation regional climate model (REGCM2). Part II: Convective processes and assimilation of lateral boundary conditions, *Mon. Weather Rev.*, **21**, 2814–2832.

Gregory, J.M., Stott, P.A., Cresswell, D.J. *et al.* (2002). Recent and future changes in Arctic sea ice simulated by the HadCM3 AOGCM, *Geophys. Res. Lett.*, **29**, L2175, doi:10.1029/2001GL014575.

Groisman, P.Y., Knight, R.W., Easterling, D.R. *et al.* (2005). Trends in intense precipitation in the climate record, *J. Climate*, **18**, 1326–1350.

Hagemann, S., Botzet, M. and Machenhauer, B. (2001). The summer drying problem over south-eastern Europe: sensitivity of the limited area model HIRHAM4 to improvements in physical parametrization and resolution, *Phys. Chem. Earth B*, **26**, 391–396.

Hannaford, J. and Marsh, T. (2006). An assessment of trends in UK runoff and low flows using a network of undisturbed basins, *Int. J. Climatol.*, **26**, 1237–1253.

Hannaford, J. and Marsh, T.J. (2007). High-flow and flood trends in a network of undisturbed catchments in the UK, *Int. J. Climatol.*, doi: 10.1002/joc.1643.

Hanssen-Bauer, I., Førland, E. and Haugen, J.E. (2003). Temperature and precipitation scenarios for Norway: comparison of results from dynamical and empirical downscaling, *Clim. Res.*, **25**, 15–27.

Haylock, M. and Goodess, C. (2004). Interannual variability of European extreme winter rainfall and links with mean large-scale circulation, *Int. J. Climatol.*, **24**, 759–776.

Hegerl, G.C., Karl, T.R., Allen, M. *et al.* (2006). Climate change detection and attribution: beyond mean temperature signals, *J. Climate*, **19**, 5058–5077.

Hegerl., C.C., Zwiers, F.W., Braconnot, P. *et al.* (2007). "Understanding and Attributing Climate Change", in: Solomon, S., Qin, D., Manning, M. *et al.* (eds), *Climate Change 2007: The Physical Basis. Contribution of Working Group I to the Fourth Assessment of the Intergovernmental Panel on Climate Change*, Cambridge University Press, Cambridge.

Hegerl, G.C., Zwiers, F.W., Stott, P.A. and *et al.* (2004). Detectability of anthropogenic changes in annual temperature and precipitation extremes, *J. Climate*, **17**, 3683–3700.

Hosking, J.R.M. and Wallis, J.R. (1997). *Regional Frequency Analysis: An Approach Based on L-Moments*, Cambridge University Press, Cambridge.

Jacob, D. (2001). A note to the simulation of the annual and inter-annual variability of the water budget over the Baltic Sea drainage basin, *Meteorol. Atmos. Phys.*, **77**, 61–73.

Jacob, D., Bärring, L., Christensen, O.B. *et al.* (2007). An inter-comparison of regional climate models for Europe: model performance in present-day climate, *Climatic Change*, **81**, 31–52.

Jones, P.D., Lister, D.H., Wilby, R.L. *et al.* (2006). Extended river flow reconstructions for England and Wales, 1865–2002, *Int. J. Climatol.*, **26**, 219–231.

Jones, R.G., Noguer, M., Hassell, D.C. *et al.* (2004a). *Generating High Resolution Climate Change Scenarios Using PRECIS*, Tech report available from Met. Office, Hadley Centre, Exeter, UK.

Jones, C.G., Ullerstig, A., Willén, U. *et al.* (2004b). The Rossby Centre regional atmospheric climate model (RCA). Part I: Model climatology and performance characteristics for present climate over Europe, *Ambio*, **33**, 199–210.

Kalnay, E. and Cai, M. (2003). Impact of urbanization and land-use change on climate, *Nature*, **423**, 528–531.

Karoly, D.J. and Braganza, K. (2005). Attribution of recent temperature changes in the Australian region, *J. Climate*, **18**, 457–464.

Karoly, D.J., Braganza, K., Stott, P.A. *et al.* (2003). Detection of a human influence on North American climate, *Science*, **302**, 1200–1203.

Karoly, D.J. and Stott, P.A. (2006). Anthropogenic warming of central England temperature, *Atmos. Sci. Lett.*, DOI:10.1002/asl.136.

Katz, R.W. (1999). Extreme value theory for precipitation: sensitivity analysis for climate change, *Adv. Water Resour.*, **23**, 133–139.

Kendon, E.J., Rowell, D.P., Jones, R.G. *et al.* (2008). Robustness of future changes in local precipitation extremes, *J. Climate*, **4**, 4280–4297.

Kiktev, D., Sexton, D., Alexander, L. *et al.* (2003). Comparison of modelled and observed trends in indices of daily climate extremes, *J. Climate*, **16**, 3560–3571.

Kundzewicz, Z.W., Graczyk, D., Maurer, T. *et al.* (2005). Trend detection in river flow series: 1. Annual maximum flow, *Hydrol. Sci. J.*, **50**, 797–810.

Kundzewicz, Z.W. and Robson, A.J. (2004). Change detection in hydrological records – a review of the methodology, *Hydrol. Sci. J.*, **49**, 7–19.

Lambert, F.H., Stott, P.A., Allen, M.R. *et al.* (2004). Detection and attribution of changes in 20th century land precipitation, *Geophys. Res. Lett.*, **31**, L10203, doi:10.1029/2004GL019545.

Lane, S.N. (2008). Climate change and the summer 2007 floods in the UK, *Geography*, **93**, 91–97.

Lanzante, J.R. (1996). Resistant, robust and non-parametric techniques for the analysis of climate data: theory and examples, including applications to historical radiosonde data, *Int. J. Climatol.*, **16**, 1197–1226.

Legates, D.R., Lins, H.F. and McCabe, G.J. (2005). Comments on "Evidence for Global Runoff Increase Related to Climate Warming" by Labat *et al.*, *Adv. Water Res.*, **28**, 1310–1315.

Lenderink, G., van den Hurk, B., van Meijgaard, E. *et al.* (2003). *Simulation of Present-Day Climate in RACMO2: First Results and Model Developments*, KNMI Technical Report 252. Available from KNMI, Postbus 201, 3730 AE, De Bilt, the Netherlands.

Lüthi, D., Cress, A., Davies, H.C. *et al.* (1996). Interannual variability and regional climate simulations, *Theor. Appl. Climatol.*, **53**, 185–209.

Malby, A.R., Whyatt, J.D., Timmis, R.J. *et al.* (2007). Long-term variations in orographic rainfall: analysis and implications for upland catchments, *Hydrol. Sci. J.*, **52**, 276–291.

Manning, L., Hall, J.W., Fowler, H.J. *et al.* (2009). Using probabilistic climate change information from a multi-model ensemble for water resources assessment, *Water Resour. Res.*, **45**, doi:10.1029/2007WR006674.

Maraun, D., Osborn, T.J. and Gillett, N.P. (2008). United Kingdom daily precipitation intensity: improved early data, error estimates and an update from 2000 to 2006, *Int. J. Climatol.*, **28**, 833–842.

Marsh, T., Cole, G. and Wilby, R.L. (2007). Major droughts in England and Wales, 1800–2006, *Weather*, **62**, 87–93.

McCabe, G.J. and Wolock, D.M. (1991). Detectability of the effects of a hypothetical temperature increase on the Thornthwaite moisture index, *J. Hydrol.*, **125**, 25–35.

Meehl, G.A., Arblaster, J.M. and Tebaldi, C. (2007). Contributions of natural and anthropogenic forcing to changes in temperature extremes over the United States, *Geophys. Res. Lett.*, **19**, L19709, doi:10.1029/2007GL030948.

Meier, H.E.M., Döscher, R. and Faxén, T. (2003). A multiprocessor coupled ice-ocean model for the Baltic Sea. Application to the salt inflow, *J. Geophys. Res.*, **108**, 3273, doi:10.1029/2000JC000521.

Michaels, P.J., Knappenberger, P.C., Frauenfeld, O.W. *et al.* (2004). Trends in precipitation on the wettest days of the year across the contiguous USA, *Int. J. Climatol.*, **24**, 1873–1882.

Mitchell, T.D. (2003). Pattern scaling: an examination of the accuracy of the technique for describing future climates, *Climatic Change*, **60**, 217–242.

Moberg, A. and Jones, P.D. (2004). Regional climate models simulations of daily maximum and minimum near-surface temperatures across Europe compared with observed station data for 1961–1990, *Clim. Dynam.*, **23**, 695–715.

Nakićenović, N., Alcamo, J., Davis, G. *et al.* (2000). *Emissions Scenarios. A Special Report of Working Group III of the Intergovernmental Panel on Climate Change*, Cambridge University Press, Cambridge, UK.

Narisma, G.T., Foley, J.A., Licker, R. *et al.* (2007). Abrupt changes in rainfall during the twentieth century, *Geophys. Res. Lett.*, **34**, L06710, doi:10.1029/2006GL028628.

Osborn, T.J., Hulme, M., Jones, P.D. *et al.* (2000). Observed trends in the daily intensity of United Kingdom precipitation, *Int. J. Climatol.*, **20**, 347–364.

Pal, J.S., Small, E.E. and Eltahir, E.A.B. (2000). Simulation of regional-scale water and energy budgets: representation of subgrid cloud and precipitation processes within RegCM, *J. Geophys. Res.* **105**, 29579–29594.

Pall, P., Allen, M.R. and Stone, D.A. (2007). Testing the Clausius-Clapeyron constraint on changes in extreme precipitation under CO_2 warming, *Clim. Dynam.*, **28**, 351–363.

Perry, M. and Hollis, D. (2005a). The development of a new set of long-term climate averages for the UK, *Int. J. Climatol.*, **25**, 1023–1039.

Perry, M. and Hollis, D. (2005b). The generation of monthly gridded datasets for a range of climatic variables over the UK, *Int. J. Climatol.*, **25**, 1041–1054.

Pfizenmayer, A. and von Storch, H. (2001). Anthropogenic climate change shown by local wave conditions in the North Sea, *Clim. Res.*, **19**, 15–23.

Pitt Review (2007). *Learning Lessons from the 2007 Floods: An Independent Review by Sir Michael Pitt*, Cabinet Office, London.

Prudhomme, C., Jakob, D. and Svensson, C. (2003). Uncertainty and climate change impact on the flood regime of small UK catchments, *J. Hydrol.*, **277**, 1–23.

Pujol, N., Neppe, L. and Sabatier, R. (2007). Regional tests for trend detection in maximum precipitation series in the French Mediterranean region, *Hydrol. Sci. J.*, **52**, 956–973.

Räisänen, J., Hansson, U., Ullerstig, A. *et al.* (2004). European climate in the late 21st century: regional simulations with two driving global models and two forcing scenarios, *Clim. Dynam.* **22**, 13–31.

Robson, A. (2002). Evidence for trends in UK flooding, *Philos. T. R. Soc. Lond.*, **360**, 1327–1343.

Robson, A.J., Jones, T.K., Reed, D.W. *et al.* (1998). A study of national trend and variation in UK floods, *Int. J. Climatol.*, **18**, 165–182.

Roeckner, E., Arpe, K., Bengtsson, L. *et al.* (1996). *The Atmospheric General Circulation Model ECHAM4: Model Description and Simulation of Present-Day Climate*, Max Planck Institut für Meteorologie, Report No. 218, Hamburg, Germany, p. 90.

Santer, B.D., Mears, C., Wentz, F.J. *et al.* (2007). Identification of human-induced changes in atmospheric moisture content, *P. Natl. Acad. Sci. USA*, **104**, 15248–15253.

Santer, B.D., Wehner, M.F., Wigley, T.M.L. *et al.* (2003). Contributions of anthropogenic and natural forcing to recent tropopause height changes, *Science*, **301**, 479–483.

Seiler, R.A., Hayes, M. and Bressan, L. (2002). Using the standardized precipitation index for flood risk monitoring, *Int. J. Climatol.*, **22**, 1365–1376.

Sparks, T.H. and Tryjanowski, P. (2005). The detection of climate impacts: some methodological considerations, *Int. J. Climatol.*, **25**, 271–277.

Steppeler, J., Doms, G., Schättler, U. *et al.* (2003). Meso-gamma scale forecasts using the nonhydrostatic model LM, *Meteorol. Atmos. Phys.*, **82**, 75–96.

Stern, N. (2006). *The Economics of Climate Change: The Stern Review*, Cambridge University Press, Cambridge.

Stott, P.A. (2003). Attribution of regional-scale temperature changes to anthropogenic and natural causes, *Geophys. Res. Lett.*, **30**, 1724, doi:10.1029/2003GL017324.

Stott, P.A., Stone, D.A. and Allen, M.R. (2004). Human contribution to the European heatwave of 2003, *Nature*, **432**, 610–614.

Stott, P.A. and Tett, S.F.B. (1998). Scale-dependent detection of climate change, *J. Climate*, **11**, 3282–3294.

Stott, P.A., Tetts, S.F.B., Jones, G.S. *et al.* (2001). Attribution of twentieth century temperature change to natural and anthropogenic causes, *Clim. Dynam.*, **17**, 1–21.

Svensson, C. and Jones, D.A. (2002). Dependence between extreme sea surge, river flow and precipitation in eastern England, *Int. J. Climatol.*, **22**, 1149–1168.

Svensson, C., Kundzewicz, Z.W. and Maurer, T. (2005). Trend detection in river flow series: 2. Flood and low-flow index series, *Hydrol. Sci. J.*, **50**, 811–824.

Tebaldi, C., Hayhoe, K., Arblaster, J.M. *et al.* (2006). Going to extremes – an intercomparison of model-simulated historical and future changes in extreme events, *Climatic Change*, **79**, 185–211.

Tett, S.F.B., Betts, R., Crowley, T.J. *et al.* (2007). The impact of natural and anthropogenic forcings on climate and hydrology since 1550, *Clim. Dynam.*, **28**, 3–34.

Tett, S.F.B., Stott, P.A., Allen, M.R. *et al.* (1999). Causes of twentieth-century temperature change near the earth's surface, *Nature*, **339**, 569–572.

Tiedtke, M. (1989). A comprehensive mass flux scheme for cumulus parameterization in large-scale models, *Mon. Weather Rev.* **117**, 1779–1800.

Tiedtke, M. (1993). Representation of clouds in large-scale models, *Mon. Weather Rev.*, **121**, 3040–3061.

Timbal, B., Arblaster, J.M. and Power, S. (2005). Attribution of late 20th century rainfall decline in South-West Australia, *J. Climate*, **19**, 2046–2062.

Trenberth, K.E., Dai, A., Rasmussen, R.M. *et al.* (2003). The changing character of precipitation, *B. Am. Meteorol. Soc.*, **84**, 1205–1217.

Vidale, P.L., Lüthi, D., Frei, C. *et al.* (2003). Predictability and uncertainty in a regional climate model, *J. Geophys. Res.*, **108**, 4586, doi: 10.1029/2002JD002810.

Wade, S., Vidal, J-P., Dabrowski, C. *et al.* (2005). *Effect of Climate Change on River Flows and Groundwater Recharge. A Practical Methodology. Task 7. Trends in UK River Flows: 1970–2002.* UKWIR Report CL\04\C\Task7, London.

Werritty, A. (2002). Living with uncertainty: climate change, river flows and water resource management in Scotland, *Sci. Total Environ.*, **294**, 29–40.

Wigley, T.M.L., Lough, J.M., Jones, P.D. (1984). Spatial patterns of precipitation in England and Wales and a revised, homogeneous England and Wales precipitation series, *J. Climatol.*, **4**, 1–25.

Wilby, R.L. (2006). When and where might climate change be detectable in UK river flows? *Geophys. Res. Lett.*, **33**, L19407, doi: 10.1029/2006GL027552.

Wilby, R.L., Beven, K.J. and Reynard, N.S. (2008). Climate change and fluvial flood risk in the UK: more of the same?, *Hydrol. Process.*, **22**, 2511–2523.

Wild, R., O'Hare, G. and Wilby, R.L. (2000). An analysis of heavy snowfalls /blizzards /snowstorms greater than 13 cm across Great Britain between 1861 and 1996, *J. Meteorol.*, **25**, 41–49.

Willett, K.M., Gillett, N.P., Jones, P.D. *et al.* (2007). Attribution of observed surface humidity changes to human influence, *Nature*, **449**, 710–713.

Wolock, D.M. and Hornberger, G.M. (1991). Hydrological effects of changes in levels of atmospheric carbon dioxide, *J. Forecasting*, **10**, 105–116.

Wu, P., Wood, R. and Stott, P.A. (2005). Human influence on increasing Arctic river discharges, *Geophys. Res. Lett.*, **32**, L02703, doi:10.1029/2004GL 021570.

Zhang, X., Hegerl, G., Zwiers, F.W. *et al.* (2005). Avoiding inhomogeneity in percentile-based indices of temperature extremes, *J. Climate*, **18**, 1641–1651.

Zhang, X., Zwiers, F.W., Hegerl, G.C. *et al.* (2007). Detection of human influence on twentieth-century precipitation trends, *Nature*, **448**, 461–465.

Zhang, X.B., Zwiers, F.W. and Stott, P.A. (2006). Multimodel multisignal climate change detection at regional scale, *J. Climate*, **19**, 4294–4307.

Ziegler, A.D., Maurer, E.P., Sheffield, J. *et al.* (2005). Detection time for plausible changes in annual precipitation, evapotranspiration, and streamflow in three Mississippi River sub-basins, *Climatic Change*, **72**, 17–36.

Zwiers, F.W. and von Storch, H. (2004). On the role of statistics in climate research, *Internat. J. Climatol.*, **24**, 665–680.

CHAPTER 23

Flood Risk in Eastern Australia — Climate Variability and Change

Stewart W. Franks
School of Engineering, University of Tasmania, Australia

23.1. Introduction

The use of empirical flood frequency analysis is widespread throughout hydrological practice — however, flood risk estimation is typically achieved through relatively simple statistical analyses of relatively short data series without any regard to the climatological conditions that produce floods. The basic assumption underlying empirical flood frequency analysis is that annual maximum flood peaks are Independently and Identically Distributed (iid). The implied assumption is that the climate is statistically "static" at all timescales — the risk of a flood of a given magnitude is taken as being the same from one year to the next, irrespective of the underlying climate mechanisms. Whilst the iid assumption may hold in many locations, if violated this may lead to substantially biased estimates of both short-term and long-term risk.

Additionally, in recent decades there has been concern over the idea of anthropogenically-induced climate change — specifically that human inputs of carbon dioxide and other radiatively active (or "greenhouse") gases will, or even are already, changing the radiative energy balance of the planet in a detrimental way. Such a change in the planetary radiative balance is thought by many to lead irrevocably to "disastrous climate change". Over recent years, it has been speculatively suggested that regional climates will be marked by an "enhanced hydrological cycle" resulting in substantial changes to flood frequency and severity.

Such concerns over possible anthropogenic climate change have resulted in numerous studies that have sought to employ statistical analyses of trends in hydrologic time series in the hope of identifying a consistent "climate change signal" (see Kundzewicz and Robson, 2004). Typically these studies utilise simple statistical tests for trend and/or step changes in observed flood sequences. Key to climate change detection approaches is that if a trend is identified and cannot be attributed to other changes, for instance land use changes, then anthropogenic climate change is the most likely suspect.

A possible problem with these approaches is again an implicit assumption of hydrologic stationarity. In common with typical engineering approaches to flood risk estimation, hydrologic time series are assessed statistically without any understanding of the underlying climate processes that produce hydrologic variability for entirely natural reasons.

In this chapter, the empirical evidence for natural long-term changes in flood frequency is examined for eastern Australia. In particular, the causal climate mechanisms for changes in flood regimes are investigated using known documented climatological phenomena such as the El Niño-Southern Oscillation (ENSO). Subsequently, the assumption of stationarity of flood risk is explored with regard to the traditional engineering estimation of flood risk. The role of non-stationary flood risk in confounding climate change detection methodologies is also explored.

23.2. Notation

AEP	Annual Exceedance Probability
DDR	Drought Dominated Regime
ENSO	El Niño-Southern Oscillation
FDR	Flood Dominated Regime
iid	independent and identically distributed
IPO	Interdecadal Pacific Oscillation
ITCZ	Inter Tropical Convergence Zone
NINO3	area of equatorial Pacific Ocean used for monitoring ENSO events
PDO	Pacific Decadal Oscillation
SPCZ	South Pacific Convergence Zone
SST	Sea Surface Temperatures

23.3. Empirical Studies of Variability and Change in Eastern Australia

It is well known that Australia experiences one of the most markedly variable climates — numerous studies have previously documented regional climate shifts across Australia. Importantly, there is an abundance of evidence that climate variables affecting Australia shifted significantly during the 1940s. In particular, Cornish (1977) noted a marked and abrupt change in annual average rainfall across eastern Australia occurring around 1945. Related to these observations of rainfall change, Allan *et al.* (1995) showed that Indian Ocean Sea Surface Temperatures (SST) were cooler at mid-latitudes and warmer in the subtropical latitudes in the periods 1900–1941 when compared with the period 1942–1983. In addition they found similar anomalies in surface winds, concluding that the, "semi-permanent anticyclone in the mean flow field of the atmosphere over the southern Indian Ocean in the austral summer was weaker in the first 42 years of the 1900s."

In parallel with these studies noting change in standard meteorological rainfall, Erskine and Warner (1988) investigated floods, sediments and geomorphological changes in eastern Australia and identified what they termed Flood- and Drought- Dominated Regimes (FDR/DDR). The essence of the concept is that eastern Australia regularly experiences shifts from one climatological state to another — whereby high rainfall periods lead to marked variability in sediment delivery and accumulation.

In a recent study of instrumental annual maximum flood series, Franks and Kuczera (2001) demonstrated there is an apparent shift in flood frequency across eastern Australia. Utilising 41 gauges located across New South Wales, 37 of these (or 90%) could be shown to have experienced an increase in flood risk after 1945. Franks (2001) demonstrated that the change in the instrumental flood frequency record could be objectively identified as being broadly in line with the previous observations of a shift in annual rainfall and circulation patterns as noted above by Allan *et al.* (1995).

To demonstrate the marked shifts in eastern Australian flood risk, Figure 23.1 shows a typical annual maximum flood series from a representative flow gauge. As can be seen, a marked increase in flood risk is apparent around 1945. Of particular interest is that the mean annual maximum flood post-1945 is approximately twice that occurring in the pre-1945 period. Also apparent in Figure 23.1 is a return to lower flood risk occurring around

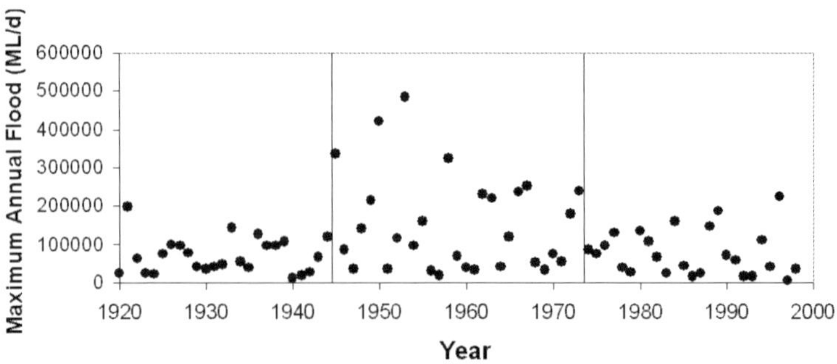

Figure 23.1. Typical time series of maximum annual flood for a catchment in eastern Australia.

the mid-1970s. Figure 23.1 clearly demonstrates that the assumption that individual annual maximum flood events are clustered and therefore do not satisfy the iid assumption. This record and the majority of others in eastern Australia indicate that some process or processes must be operating that dictate structure in terms of temporal patterns in the instrumental flood records.

Importantly, this time series clearly indicates that flood risk estimation will be biased if based on a sub-set of the data arising from one or other of the flood or drought dominated regimes. This therefore demonstrates the need to understand the underlying climatological causal factors that give rise to actual flood behaviour.

23.4. Climatological Phenomena and Flood Risk

In addition to empirical hydrological observations of changing climate risk, climatological insights into the mechanisms of climate variability point to the invalidity of a purely empirical approach to risk estimation. Indeed, numerous previous studies have shown that a strong relationship exists between streamflow and the ENSO phenomenon.

In terms of New South Wales climate, the warm El Niño events are associated with marked reductions in rainfall and increased air temperatures and evaporative demand, whereas the cool La Niña events typically deliver enhanced rainfall totals and cooler air temperatures. It is therefore clear that as individual wet and dry years are usually associated with ENSO extreme events, year-to-year flood and drought risk might also vary according to ENSO state.

Recent climatological studies have also revealed multi-decadal variability in the modulation of the magnitude of ENSO impacts. In particular, Power *et al.* (1999) have investigated marked temporal changes in ENSO correlations to Australian rainfall records. The temporal stratification of the rainfall sequences was achieved according to what has been termed the Inter-decadal Pacific Oscillation (IPO). The IPO was defined by anomalous warming (1920–1945 and 1975–2001) and cooling (1945–1975) in the Pacific Ocean. Power *et al.* (1999) showed how Australian ENSO correlations changed with the observed changes in persistent large-scale Pacific Ocean SST anomalies. Importantly, Power *et al.* (1999) demonstrated that individual ENSO events (i.e. El Niño, La Niña) had a stronger impact across Australia during the negative phase of the IPO, implying that there exists a multi-decadal modulation of the magnitude of ENSO events.

Figure 23.2 shows the derived IPO index. Also plotted is the Pacific Decadal Oscillation (or PDO) independently derived by Mantua *et al.* (1999) and similarly utilised in assessing ENSO impacts in Pacific salmon fisheries. Importantly, the IPO and PDO (hereafter referred to as IPO-PDO) state changes occur in the mid-1940s and the mid 1970s, broadly in line with the observations of change in eastern Australian rainfall and flood records.

To assess historic regional flood risk and its possible association with ENSO extremes, Kiem *et al.* (2003) employed a state-wide index and subsequently stratified according to ENSO classifications based on the

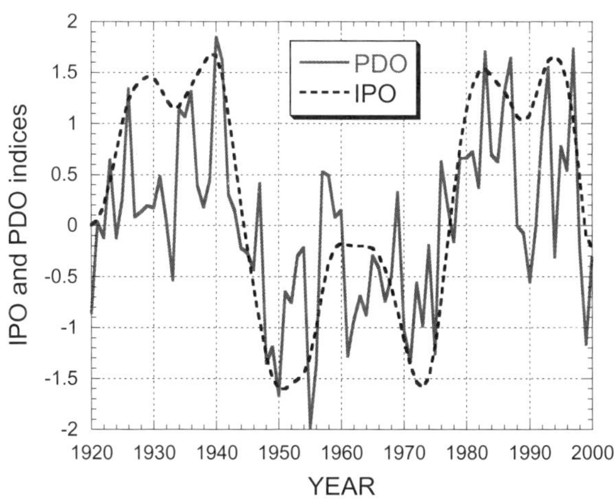

Figure 23.2. Time series of the IPO (dashed line) and PDO (solid line) climate indices.

NIÑO3 index — an index of SST anomalies in the eastern equatorial Pacific Ocean. Each year from 1924 to 1999 was given an ENSO classification of El Niño, La Niña or neutral, based on the six month October to March average NIÑO3 value. The index was then further stratified according to the multi-decadal IPO-PDO classifications. The stratified flood frequency data were analysed using Bayesian flood frequency analysis (Kuczera, 1999) to quantify uncertainty on quantiles and thus elucidate the key controls on New South Wales flood risk.

If the flood gauges were correlated perfectly, treating them as entirely uncorrelated would imply 41 independent records with any inferred change having substantial statistical support. In fact, if the gauges were correlated perfectly the true support would be equivalent to that of a single record. Developing a regional index of flooding, in effect collapses the flood data into a single time series, with any correlation between gauges being accounted for implicitly. Adopting this alternative extreme, if the gauges are not perfectly correlated (as would seem most likely), the statistical support of the 40 gauges, will be underestimated. This simplified approach is adopted here.

A regional index is derived through a purely statistical approach – the individual records are scaled by the mean of the log annual flood maximum discharge. The log is used in this study to weight each gauge equally under the assumption of a log-normal flood frequency model. For example:

$$x_t^j = \frac{\ln Q_t^j}{\sum_{i=1}^{N} \ln Q_i^j / N},$$

where x_t^j is the normalised index for gauge, j, at time, t, and where N is the total length of the flood record. For each year, the resultant 40 scaled time series are then averaged to provide a gauge-mean log maximum;

$$RI_t = \sum_{j=1}^{M} x_t^j / M \quad \text{for } t = 1, \ldots, N,$$

where RI_t is the normalised regional index, and M is the total number of gauges. The advantage of this approach is that each gauge is weighted equally in the derivation of the regional index as a function of the typical magnitude of the annual flood.

To assess the role of ENSO extremes, Figure 23.3 presents the flood frequency under El Niño and La Niña conditions along with the associated 90% confidence limits. From this plot it can be readily seen that a much

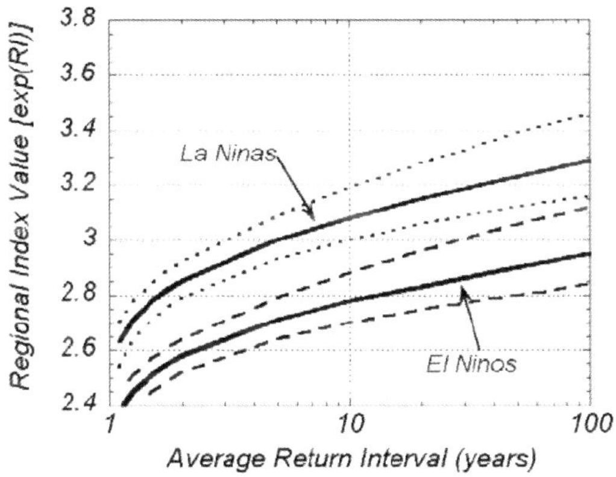

Figure 23.3. Flood frequency curves associated with El Niño and La Niña classified years.

higher flood risk must be associated with La Niña events than with El Niño events. It is also immediately apparent that the degree of separation of the confidence limits indicating a highly statistically significant difference between the two ENSO extremes.

Given the clear role of La Niña events in flood risk identified in Figure 23.3, to test the hypothesis that the IPO-PDO apparently modulates the magnitude of La Niña events, as suggested by Power *et al.* (1999), a stratification on La Niña under different IPO-PDO phases is required. To achieve this test, the regional index is stratified according to La Niña events occurring under negative IPO-PDO phase and then according to La Niña events occurring under neutral and positive IPO-PDO phases. Figure 23.4 shows the resultant flood frequency curves. As can be seen, the frequency curve associated with La Niña events under negative IPO-PDO is markedly higher than the flood frequency associated with all other La Niña events.

Finally, given the multi-decadal persistence of IPO-PDO phases, it is desirable to assess the variability of flood risk under the different IPO phases irrespective of inter-annual ENSO events. Figure 23.5 shows the flood frequency curves for IPO-PDO negative (IPO<-0.5) against non-negative IPO-PDO phases. Again, it can be seen that IPO-PDO negative phase corresponds to a much increased flood risk when compared to the non-negative phases of IPO-PDO. Of interest, the solid line plotted in Figure 23.5 indicates the 100 year (1% AEP) flood that would be derived through a solely statistical analysis of the entire time series.

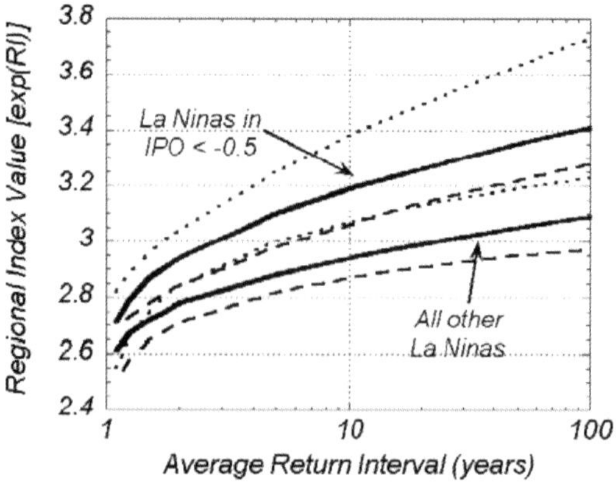

Figure 23.4. Flood frequency curves for La Niña events under IPO negative conditions and all other La Niña years.

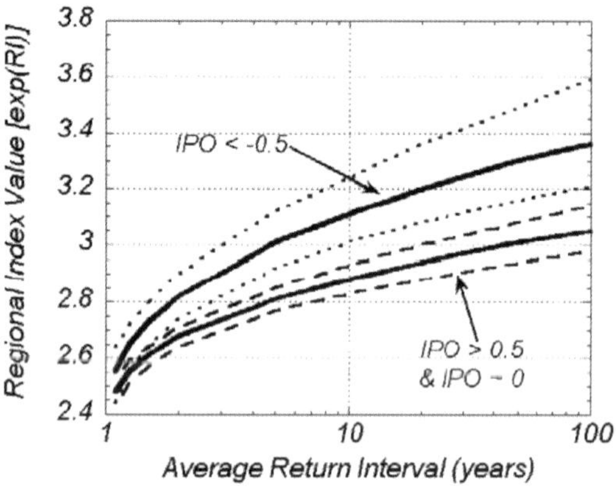

Figure 23.5. Flood frequency curves for IPO negative (<0.5) and IPO non-negative conditions.

As can be seen, the 100 year flood would occur on average every 16 years of IPO-PDO negative conditions. It is therefore clear that monitoring of the multi-decadal IPO-PDO phase may provide valuable insight into flood risk on multi-decadal scales, whilst the joint occurrence of inter-annual La Niña events within the IPO-PDO negative phase represents further elevated flood risk.

In assessing the apparent modulation of ENSO by the multi-decadal IPO-PDO mode, Kiem *et al.* (2003) also demonstrated that the frequency of El Niño and La Niña events also varied between the different phases of the IPO-PDO. Again utilising the NINO3 index of ENSO extremes, it was demonstrated that the negative (or cool) IPO-PDO phase was associated with a statistically significant increase in the number of La Niña events compared to the positive (or warm) phase of the IPO-PDO mode. Kiem *et al.* (2003) concluded that there was an apparent dual modulation of ENSO both in terms of magnitude of the impacts of individual events but also in terms of the frequency with which they occur. The reader is directed to Kiem *et al.* (2003) for more details.

23.5. Causes of Variability

Given the high degree of hydro-climatological variability explained by the combined ENSO and IPO-PDO indices, it is advantageous to develop a qualitative conceptual understanding of the physical mechanisms of ENSO and IPO-PDO processes. In particular, it is important to ask how IPO-PDO processes interact with ENSO to ultimately deliver the marked observations of variability.

23.5.1. *IPO modulation of ENSO event magnitude*

Whilst ENSO processes were initially identified using atmospheric pressure differences between Tahiti and Darwin, the most obvious indication of ENSO events is given by eastern equatorial Sea Surface Temperature (SST). The standard model of ENSO processes is given by the "delayed action oscillator". Individual ENSO events are seen as preferred states arising from the internal interaction of oceanic and atmospheric processes in the equatorial Pacific.

In essence, an anomalous perturbation in this coupled system, if sufficiently large, is magnified through the interaction of processes due to positive feedback reinforcing the anomalies in each. A longitudinal shift in equatorial circulation (i.e. Walker cell) is developed which subsequently interferes with the Inter Tropical Convergence Zone (ITCZ) and the South Pacific Convergence Zone (SPCZ). It has previously been demonstrated that relatively small shifts in the location of the SPCZ result in very large rainfall anomalies either side of the SPCZ (Salinger *et al.*, 1995).

The ITCZ-SPCZ is most significant as it delivers rain-bearing cloud bands across eastern Australia. The SPCZ is most active during November through to April (Folland *et al.*, 2002) which in general terms coincides with the season of highest and most variable rainfall over eastern Australia, and also corresponds to the period of greatest ENSO event impact (Kiem and Franks, 2001). Warm El Niño events disrupt the SPCZ, preventing its propagation to its usual southern latitude. Cold La Niña events represent an enhancement of the neutral ENSO state, with the effect that the SPCZ propagates further south than normal delivering more frequent rain-bearing cloud bands across south-east Australia (Salinger *et al.*, 1995).

In contrast to the equatorial nature of ENSO processes, the IPO processes are revealed in mid-latitude SST anomalies across the Pacific and Indian Oceans. Indeed the IPO index itself is derived from principal component analysis of the modes of SST variability revealing the predominance of the low frequency component in the mid-latitudes (Power *et al.*, 1999). A recent study by Folland *et al.* (2002) assessed the location of the SPCZ as a function of both ENSO and IPO. They demonstrated that the IPO SST anomalies affect the location of the SPCZ during the austral summer (November–April) in a manner similar to that induced by ENSO but on a multi-decadal timescale. Importantly, the results of Folland *et al.* (2002) showed that the SPCZ was at its southern-most during La Niña events under IPO negative conditions. This provides strong corroborative evidence of the enhancement of La Niña events under IPO negative conditions as originally suggested by Power *et al.* (1999), and as inferred from flood and water supply drought analyses (see also Kiem and Franks, 2004).

23.5.2. *IPO modulation of ENSO frequency*

Whilst recent climatological research points to the explanation of IPO modulation of the magnitude of individual ENSO events, the issue of the IPO modulation of ENSO frequency is less clear. A number of previous studies have observed and evaluated changes in the relative frequencies of El Niño and La Niña events using historical or long-term proxy data, although these studies did not have the advantage of IPO processes in their considerations. When IPO stratification of an ENSO time series is considered, it is immediately apparent that the IPO negative phase appears to be associated with a higher frequency of La Niña events (Kiem *et al.*, 2003; Kiem and Franks, 2004). Indeed between 1945 and 1976 almost 50% of all years in IPO negative states are classified as La Niña according to the NINO3 classification (Kaplan *et al.*, 1998).

To elucidate how IPO processes may affect La Niña frequency, consideration of the nature of the IPO signal and long-term changes in SST may be of some use. Recent climatological research has attempted to explain the observations of long-term change in equatorial SST anomalies. Kleeman and Power (2000) note that there are two key mechanisms by which equatorial SST anomalies may arise; (a) stochastic forcing by atmospheric transients or chaotic climate dynamics within the equatorial zone, and (b) extra-tropical/mid-latitude forcing due to gyres or thermohaline circulations, external to the equatorial zone. Of these two possibilities, Kleeman and Power (2000) suggest that observations of subtropic subduction and consequent upwelling in the equatorial Pacific point to an influence of the mid-latitude SST anomalies (and hence IPO) interacting with ENSO.

In terms of IPO-PDO, the negative state is associated with cooling mid-latitude SST anomalies. This water is subducted and transported to the equatorial Pacific where upwelling permits the anomalously cool water to return to the surface and influence the development of La Niña conditions. This circulation, known as a Sub Tropic Cell (STC) or gyre, provides a mechanism by which long-term anomalies represented by the IPO may prejudice equatorial SST and hence predispose ENSO processes towards a particular ENSO state. This oceanographic mechanism indicates the possibility of IPO SST anomalies influencing ENSO SST anomalies. As noted earlier, the instrumental record indicates a clear increased frequency of La Niña events under IPO negative conditions, however no statistically significant difference was observed for El Niño events. Nonetheless, the possibility of IPO modulation of ENSO through subtropical cells may have substantial basis.

To assess the possibility of frequency changes in El Niño and La Niña events occurring on multi-decadal timescales, a proxy record of derived ENSO extremes is analysed (D'Arrigo et al., 2005). A 30-year moving window is applied to the classified ENSO climate states and the number of individual El Niño and La Niña events was recorded. Figure 23.6 shows the variability of the frequency of El Niño and La Niña, as well as the difference between El Niño and La Niña counts. As can be seen, there is marked variability in ENSO event frequency occurring on multi-decadal timescales. It is therefore clear that ENSO event frequency is highly variable over long periods of time and will have likely led to marked variability in pre-instrumental flooding across eastern Australia.

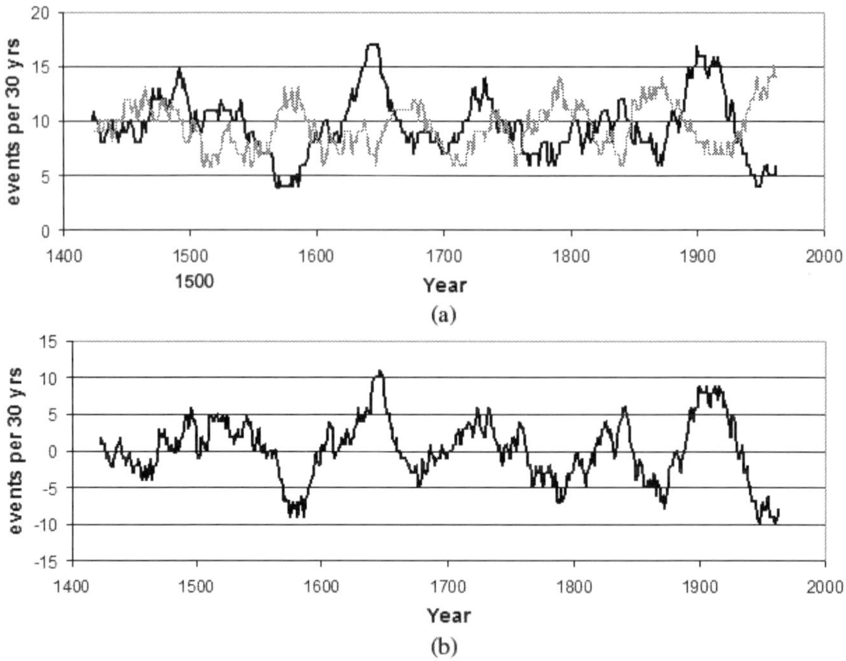

Figure 23.6. (a) Frequency of El Niño (dashed line) and La Niña (grey line) occurring within a 30 year moving window when applied to D'Arrigo *et al.* (2005) ENSO paleoreconstruction data. (b) The number of El Niño events minus the number of La Niña events, clearly indicating variable periods of alternating dominance.

23.6. Implication of Multi-Decadal Variability

It is clear from the instrument records both of rainfall and floods, as well as from sedimentary records, that eastern Australia has experienced marked changes in flood risk over decadal timescales. It should also be clear that known climatological processes such as ENSO and IPO-PDO appear to have played a strong role in determining such variability. There are a number of potential implications arising from these observations:

- Traditional flood frequency based on simple statistics and limited data may be significantly in error if no account of the prevailing climate state is included in flood risk estimation.
- Studies seeking to attribute changes in flood frequency as evidence of anthropogenic climate change may reveal highly significant trends that are in fact natural in origin.

The following sections seek to quantify the possible errors arising both in flood risk estimation and in climate change attribution.

23.6.1. *Assessing uncertainty in long-term flood risk estimation arising from limited samples of multi-decadal variability*

To assess the uncertainty associated with the iid assumption employed in traditional flood risk estimation, a simple Monte Carlo-based methodology is developed. The representative series of annual maximum flood data shown earlier in Figure 23.1 was employed. These data were stratified according to the positive and negative phases of the IPO-PDO. Log-normal flood frequency distributions associated with each of these phases were calculated. Figure 23.7 shows marked differences in the resultant flood frequency curves with the IPO negative phase. This figure demonstrates a statistically significant increase of a factor of 2.5 in the 1:100 year flood risk, an increase typical across eastern Australia (Micevski *et al.*, 2006).

To generate feasible Monte Carlo realisations of long-term flood risk, a pre-instrumental PDO reconstruction is employed. This series of IPO-PDO phase transitions was derived from assessing multiple proxy reconstructions for Pacific climate variability, whereby statistically significant step changes

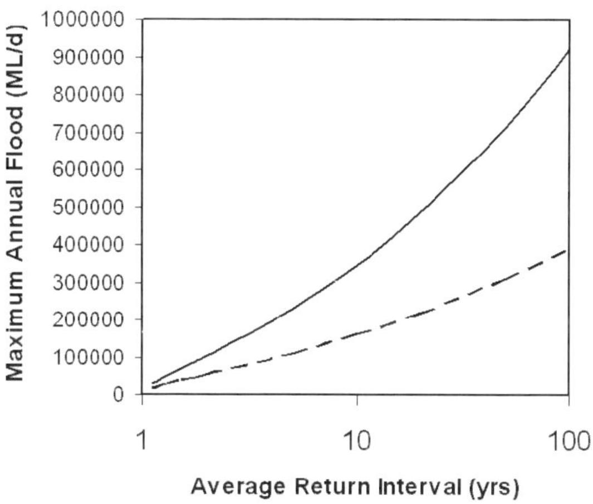

Figure 23.7. Derived flood frequency curves for IPO negative (solid) and IPO positive (dashed) periods.

in the derived time series were evaluated for coherence. A composite IPO-PDO index of phase transitions was then derived (see Verdon and Franks, 2006, for more details). Ten thousand individual sub-samples of the composite IPO-PDO time series were then randomly generated of length n years. Monte Carlo generated annual maximum flood series were then sampled from the flood frequency distributions associated with the corresponding IPO-PDO phase. The resultant distributions of the mean annual flood and the 1:100 year flood were then calculated. This process was repeated for values of n from 30–300 years.

To demonstrate the uncertainty of the mean annual flood risk as a function of available data length, n, and unknown IPO-PDO climate state, Figure 23.8 shows the simulated median and 90% uncertainty envelope of the Monte Carlo derived distributions. As can be seen, the uncertainty envelope is particularly wide for data lengths of 30 years, indicating a 10% chance of the true value being either under 60,000 Ml/d or over 180,000 Ml/d. Whilst Figure 23.8 does indicate a reduction in the upper and lower uncertainty limits as the available data length, n, increases it is worthwhile to note that even with 150 years of data the residual uncertainty remains very large.

Figure 23.9 demonstrates the corresponding uncertainty envelope for the estimated 1:100 year flood, a common criterion in practical engineering and planning processes. In this case, it can again be seen that the uncertainty envelopes are wide, even as the length of available data, n, increases.

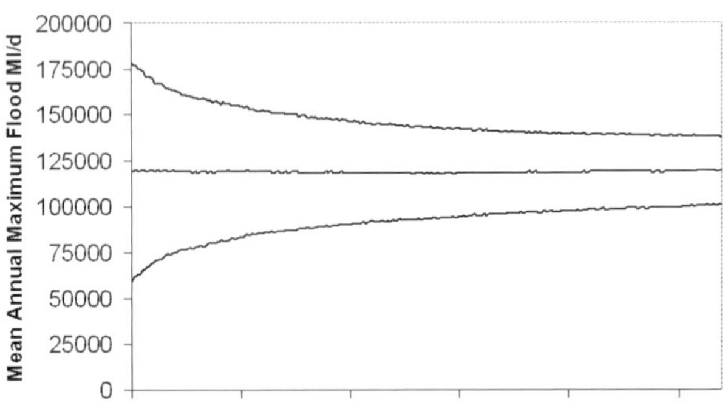

Figure 23.8. Uncertainty in mean annual maximum flood risk as a function of data length.

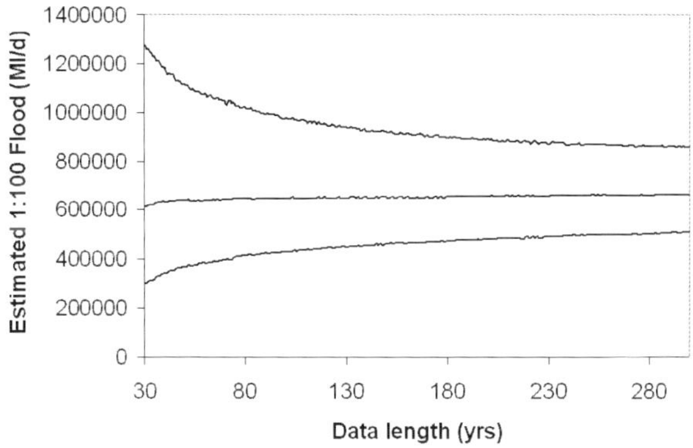

Figure 23.9. Uncertainty in 1:100 year flood risk as a function of data length.

23.6.2. *Implications for empirical climate change detection methodologies*

As noted earlier, the current interest in anthropogenically-induced climate change has led to a number of studies whereby trends and changes are assessed in hydrological records with the aim of climate change attribution. Perhaps the simplest form of trend detection is a simple linear trend applied to an available time series. To assess the potential for significant error in such methodologies (or misattribution) Figures 23.10a and b show the employed representative time series with (a) a linear trend applied to the entire data set, and (b) linear trends applied from 1920–1975 and 1945–1998.

Figure 23.10a clearly shows no discernable trend when all available data are available (gradient = −21 Ml/d/yr; p = 0.963). If we were back in 1975, as shown in Figure 23.10b, we would have identified a marked upward trend in flood risk (gradient = 1837 Ml/d/yr; p = 0.0528). However, if our records only began in 1945, then we would have identified a marked and significant downward trend in flood risk (gradient = −2471 Ml/d/yr; p = 0.0188). Clearly, in the presence of multi-decadal variability the length of sample available as well as the sampling of the prevailing multi-decadal climate states could lead to many false claims of statistically significant trends.

More typically, a Mann–Kendall non-parametric test for monotonic trends is employed in the identification of change in hydrological flood time series (see Kundzewicz and Robson, 2004, for an overview of alternative methodologies). Interestingly, when applied to the two sub-series, the identified trends become even more highly statistically significant

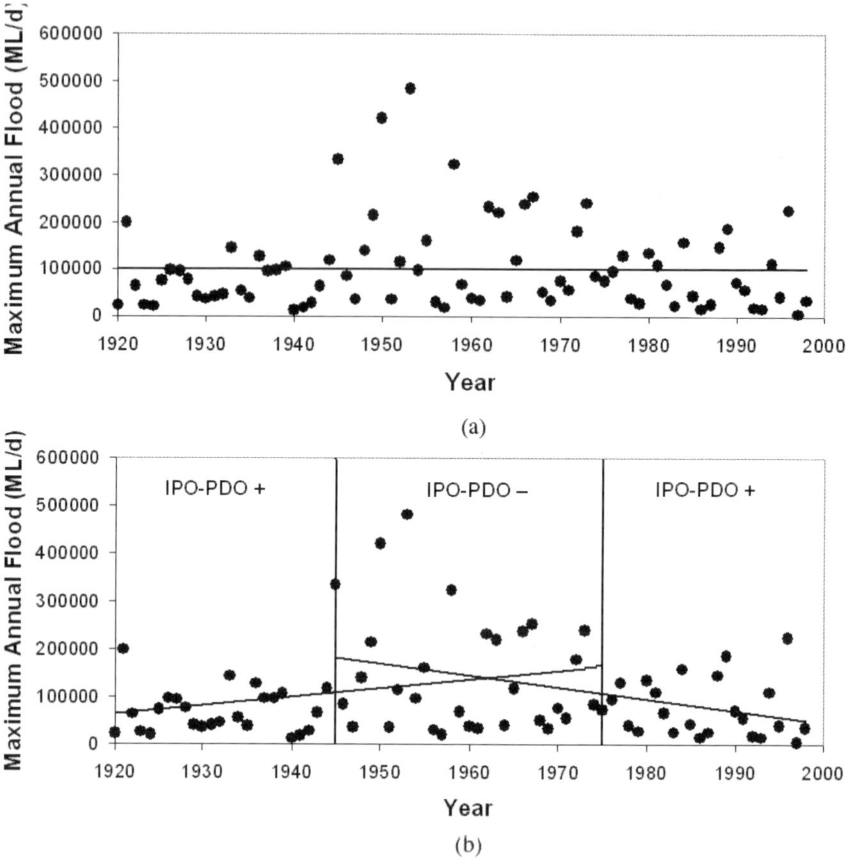

Figure 23.10. Trends in the annual maximum flood risk data, when (a) all data are employed, and (b) when data are subsampled.

(1920–1975 –p = 0.017; 1945–1998 –p = 0.008). Clearly, such high levels of statistical confidence are incorrect in the presence of multi-decadal variability. Put simply, the assumptions about the data employed in the tests are invalid.

To assess the possible errors associated with identifying statistically significant trends associated with the presence of multi-decadal variability in flood series, the Monte Carlo-derived sequences derived above were each individually tested for monotonic trends using the Mann–Kendall test. Each occurrence of significance was recorded at the 10%, 5% and 1% significance level. The fraction of Monte Carlo replicates corresponding to significant trends was then plotted as a function of sample length, n.

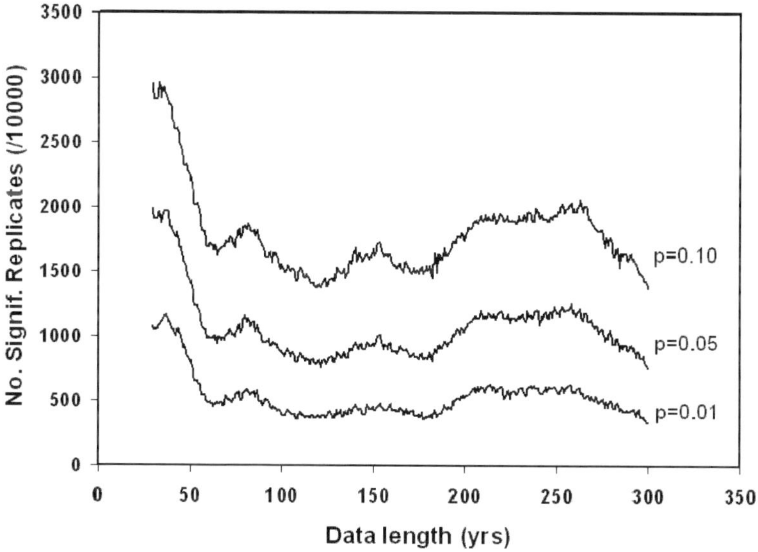

Figure 23.11. The number of replicates displaying significant trends according to the Mann–Kendall test. As can be seen, the statistical significance of the test is exaggerated under natural variability at multi-decadal timescales.

Figure 23.11 shows the number of replicates out of 10,000 that returned significant trends as a function of multi-decadal flood variability and sample data length. Figure 23.11 clearly shows that utilising 30 years of data, almost 30% of replicates are found to contain significant trends at the 10% level, 20% at the 5% level and 15% at the 1% level. These represent marked errors in the application of the test in the presence of multi-decadal variability. It should also be noted that as the length of data in a sample increases, the fraction of errors decrease but do not disappear even after 300 years.

As noted by Kundzewicz *et al.* (2004), hydrological time series that display marked variability may be effectively filtered through the analysis of serial correlation in the records. However, an open question remains as to whether one should arbitrarily select a threshold significance level of serial correlation on which to accept or reject individual time series. Moreover, it is entirely feasible that individual regional climate may display only aperiodic epochs of elevated/reduced flood risk as a complex function of the interaction of climate modes, unlike the apparently periodic time series employed here. This would further confound attempts to filter datasets prior to trend analysis.

Ultimately, the application of simple statistical techniques cannot provide a definitive analysis of the causes of changes in hydrological time series without robust understanding of the causal mechanisms that lead to floods. Furthermore, without such insights there may be a misattribution of change.

23.7. Discussion and Conclusions

Flood risk estimation is a cornerstone of practical engineering and planning processes. The consequences of significant error in flood risk estimation techniques are substantial. In this study, it has been demonstrated that traditional flood risk estimation techniques based on simple statistical analysis of relatively short flood histories may result in large biases when applied to catchments that experience marked multi-decadal climate variability.

It should be noted that under the traditional iid assumption, 30 years of data are typically deemed an adequate minimum for robust flood risk estimation. The results presented here demonstrate that uncertainty in both the mean annual maximum and the 1:100 year flood risk may be in error by as much as a factor of three when only limited data are available. Whilst this uncertainty reduces as a function of available data length, the residual uncertainty still remains high when compared to the invalid uncertainty estimates under the iid assumption. These results indicate the importance of understanding processes in providing appropriate flood risk estimates given both multi-decadal variability and limited instrumental records.

The approach adopted contains a number of specific assumptions. Above all, the approach adopted assumes that whilst flood risk varies on multi-decadal timescales according to the IPO-PDO, it also assumes that this association is itself stable. In other words, that the association between IPO-PDO and variable flood risk observed within the instrumental record (typically the 20th century), holds for the centuries prior. It is entirely possible that at longer timescales than the instrumental record other processes may mean a more or less variable association between the IPO-PDO and eastern Australian flood risk. From the viewpoint of the requirement to provide robust flood risk estimates and their uncertainty that may need to be equally applicable now as in 100 years time, the assumption seems suitably conservative.

This assumption can be tested through the development of appropriate and robust proxy measures of pre-instrumental climatic risk in eastern

Australia. This is the subject of current research and will be reported in future.

An additional assumption is that anthropogenic effects on climate have been negligible over the period of the instrumental record. This is a necessary and seemingly robust assumption. As noted by Kundzewicz and Robson (2004), analyses of long-term flood risk from over 195 quality gauged stations has not provided any general proof as to how anthropogenic climate change may or may not affect flood risk. It remains possible that anthropogenic emissions of radiatively active gases may affect flood extremes. However, it is clear that historic variability on multi-decadal timescales has been a destructive feature of natural Australian climate variability prior to recent concerns over climate change. On the basis of the available evidence it appears that it should remain the primary concern for the immediate future.

Whilst this chapter has focused on eastern Australian flood data, it should also be appreciated that ENSO is a quasi-global phenomena, affecting regional climates across the world. Significant impacts have been detected across Asia, the Americas and more recently there have been a number of studies indicating a complex role for ENSO in affecting European climates albeit interacting with the Arctic and North Atlantic Oscillation (Zanchettin *et al.*, 2008). It should therefore be considered that changes in flood frequency in regional climates, whether historically or into the future, can occur as a function of entirely natural climate variability. It is clear that a more sophisticated approach to climate change detection is required whereby causal physics are employed to evaluate changes in climatic regime. Until then, speculative claims of enhanced hydrological cycles arising through consideration of only simplified physics or through blind adoption of climate model output should be viewed with obvious caution.

Acknowledgements

This chapter represents an overview of a number of studies and has benefited from interactions with a number of colleagues and students. Thanks are due to Anthony Kiem, Danielle Verdon-Kidd, George Kuczera and Tom Micevski for their ongoing interest in natural modes of climate variability and their impact on Australian hydrology. These studies have been funded by the Australian Research Council, Hunter Water Corporation and Sydney Catchment Authority.

References

Allan, R.J., Lindesay, J.A. and Reason, C.J.C. (2005). Multi-decadal variability in the climate system over the Indian Ocean region during the austral summer, *J. Clim.*, **8**, 1853–1873.

Cornish, P.M. (1977). Changes in seasonal and annual rainfall in New South Wales, *Search*, **8**, 38–40.

D'Arrigo, R., Cook, E., Wilson R. *et al.* (2005). On the variability of ENSO over the past six centuries, *Geophys. Res. Lett.*, **32**, L03711, doi:10.1029/2004GL022055.

Erskine, W.D. and Warner, R.F. (1988). "Geomorphic Effects of Alternating Flood and Drought Dominated Regimes on a NSW Coastal River" in: Warner, R.F. (ed.), *Fluvial Geomorphology of Australia*, Academic Press, Sydney, Australia, pp. 223–244.

Folland, C.K., Renwick, J.A., Salinger, M.J. *et al.* (2002). Relative influences of the Interdecadal Pacific Oscillation and ENSO on the South Pacific Convergence Zone, *Geophys. Res. Lett.*, **29**, 1643.

Franks, S.W. (2002). Identification of a change in climate state using regional flood data, *Hydrol. Earth Syst. Sci.*, **6**, 11–16.

Franks, S.W. and Kuczera, G. (2002). Flood frequency analysis: evidence and implications of secular climate variability, *Water Resour. Res.*, **38**, doi:10.1029/2001WR000232.

Kaplan A., Cane M.A., Kushnir Y. *et al.* (1998). Analyses of global sea surface temperature 1856–1991, *J. Geophys. Res.-Oceans*, **103**, 18567–18589.

Kiem, A.S. and Franks, S.W. (2004). Multi-decadal variability of drought risk – eastern Australia, *Hydrol. Proc.*, **18**, doi:10.1002/hyp.1460.

Kiem, A.S., Franks, S.W. and Kuczera, G. (2003). Multi-decadal variability of flood risk, *Geophys. Res. Lett.*, **30**, 1035, doi:10.1029/2002GL015992.

Kuczera, G. (1999). Comprehensive at-site flood frequency analysis using Monte Carlo Bayesian inference, *Water Resour. Res.*, **35**, 1551–1558.

Kundzewicz, Z.W. and Robson, A.J. (2004). Change detection in river flow records — review of methodology, *Hydrol. Sci. J.*, **49**, 7–19.

Mantua, N.J., Hare, S.R., Zhang, Y. *et al.* (1997). A Pacific decadal climate with impacts on salmon, *Bull. Am. Meterol. Soc.*, **78**, 1069–1079.

Micevski T., Franks S.W., Kuczera G. (2006). Multidecadal variability in coastal eastern Australian flood data, *J. Hydrol.*, **327**, 219–225.

Power., S., Casey, T., Folland, C. *et al.* (1999). Inter-decadal modulation of the impact of ENSO on Australia, *Clim. Dyn.*, **15**, 319–324.

Salinger, M.J., Basher, R.E. and Fitzharris, B.B. (1995). Climate trends in the South West Pacific, *Int. J. Climatol.*, **15**, 285–302.

Verdon D.C., Franks S.W. (2006). Long-term behaviour of ENSO: interactions with the PDO over the past 400 years inferred from paleoclimate records, *Geophys. Res. Lett.*, **33**, doi:10.1029/2005GL025052.

Zanchettin D., Franks S.W., Traverso P. *et al.* (2008). On ENSO impacts on European wintertime rainfalls and their modulation by the NAO and the Pacific multi-decadal variability described through the PDO index, *Int. J. Climatol.*, **28** 995–1006.

SECTION VI

COMMUNICATING UNCERTAINTIES

CHAPTER 24

Translating Uncertainty in Flood Risk Science

Hazel Faulkner and Meghan Alexander
Flood Hazard Research Centre, University of Middlesex, UK

David Leedal
Lancaster Environment Centre, Lancaster University, UK

24.1. Introduction

It has long been recognised that effective communication is a central component of Flood Risk Management (FRM). The many professionals[a] charged with the delivery of flood risk reduction within communities frequently request that communications are clear, short and simple. But as the previous chapters in this book serve to testify, the modelling of future flood risk is increasingly associated with uncertainties of many types, not only those associated with the scientific tools selected.

Considerable international progress is being made to quantify the uncertainties surrounding the prediction and forecasting of floods and most other hydrometeorological risks. However, to those whose professional function requires them to deliver FRM on the ground, scientific formulations of future risk are articulated in a complex language which is relatively inaccessible and in some cases also impenetrable. The challenge of communicating these uncertainties in inter-professional communication discourse in everyday language remains a major international challenge (Frewer *et al.* 2003; Hall, 2006; Klir, 2006; Lavis *et al.* 2003). Kinzig *et al.*

[a]The term "professional" we take to include all those for whom FRM and communication is a part, even if only a small part, of their professional role and paid employment.

(2003) have called for "a new science-policy forum" between scientists and policy makers, but the choice of the appropriate language to use within this forum and the inclusion of uncertainty in it have proved problematic. This is not least because the "ownership" of the uncertainty remains contested at this interface.

The principal goal of this chapter is to explore some of the *translatory* challenges that are introduced into the professional agendas when flood risk managers are presented with increasingly complex scientific models that come with an embedded "health warning". First, we explore conceptualisations of "effective communication" in a range of fields in FRM. Second, we identify and consider the many interfaces at which communications can sit in FRM and the various constraints upon effective communication. Thirdly, the concept of uncertainty and its "ownership" within communications is discussed, referring to two examples of flood risk communication tools: the flood warning; and the flood risk map. Finally, examples of methods to communicate the inherent uncertainty in the message are discussed, using an example of an effective flood risk map with visualisations of the embedded uncertainty, currently under development (Leedal *et al.* 2010).

24.1.1. *What is effective communication?*

In order to comprehend the makings of effective communication there is a need to firstly grasp the characteristics of communication as a concept. Faulkner *et al.* (2007) outline five features of communication. The process of communicating is inherently purposive; that is to say, communication is driven by an underlying current of intentions. In its simplest form communication may be understood as a means of transferring information from one party to another; however, in reality this shallow form of communication rarely occurs free from other connotations. Instead communication might be viewed as a process of negotiation whereby we test, confirm and re-evaluate our ideas with one another. A deeper and perhaps hidden layer of communication is the influence upon the relationship between the communicating parties, in maintaining, building or eroding ties. In this light, communication is a social act. Thirdly, communication functions in a field of various expectations, with regards to the outcome of the communication and how this can be achieved, for example. Such expectations are influenced by the communicator's underlying assumptions about the other party (e.g. their agenda, capability). Content, tone, mode of message delivery can further influence expectations in the communication process. Furthermore, communication represents a learning process; facilitating an improved understanding beyond the subject matter at hand. Some exchange of mutual trust seems to be implicit in this, i.e. trust between the relevant parties involved in the exchange is required. The final characteristic of communication is the requirement for a system of transmission. Language is the most common mode of transfer, though advancement in information technologies has witnessed the construction of new communication tools, particularly employing visualisation techniques and thus broadening the scope of communication channels or the "tool kit" for communication.

In conceptualising the challenges of communicating across the science-to-professional interface Faulkner *et al.* (2007) argue that the fields of semantics and semiotics may have something to offer. Semantics acknowledges the agency of the individual in creating meaning from the world (Sarukkai, 2007). Semiotics is concerned with "signs", which may be linguistic, visual or auditory; semantics involves the study of meaning attached to these signs (Chandler, 2002). Meaning is to be understood and therefore identified within the social context of the speaker and listener, and so it is intrinsically bound to social and cultural norms (Stamper *et al.* 2000). One might consider signs then as "sign posts" which orientate communication, interpretation and assimilation of information; moreover, signs are considered as norm-triggers and therefore related to human behaviours (e.g. Stamper *et al.*, 2000). The debate itself is interesting theoretically but in taking a pragmatic approach to semiotics here, we understand *effective communication* to be simply defined as a meaningful "fit" between the mode of information transfer (communication "tool kit") to the range of communicative purposes. Seeking a "good fit" is not a case of simply finding the missing jigsaw piece; the social context of communication means that these pieces continue to shape-shift and therefore what may have worked successfully in the past is not necessarily meaningful in the "now". Effective communication should be viewed as a continuing process of seeking this good fit of language and signs between communicating parties, thereby ensuring that the message is understood in the manner it was intended (Faulkner *et al.*, 2007). Articulating risk and uncertainty therefore requires a sensitised domain for translation to avoid the pitfall of information being "lost in translation."

24.1.2. *Communicating in scientific language at a range of professional interfaces*

This chapter specifically engages with communication operating at the scientific community–professional stakeholder interface. Effective communication at this interface seeks the middle ground in satisfying the scientist's detail of knowledge, with the requirements of the end user of this information. In the context of FRM, information is passed through a complex web of communication channels, differing knowledge domains and contexts of the multiple stakeholders involved in FRM (Figure 24.1; Faulkner *et al.*, 2007; Morss *et al.*, 2005). This observation underlines our argument that translating complex and uncertain science into a set of different conceptualisations of risk, for purposes of communicating with stakeholders, is a considerable additional challenge for FRM professionals.

The overarching communication challenge is thus a process of sensitising communication to these apparently conflicting demands. The category of "professional stakeholders" accommodates multiple roles, professional obligations, organisational agendas and varying resource bases

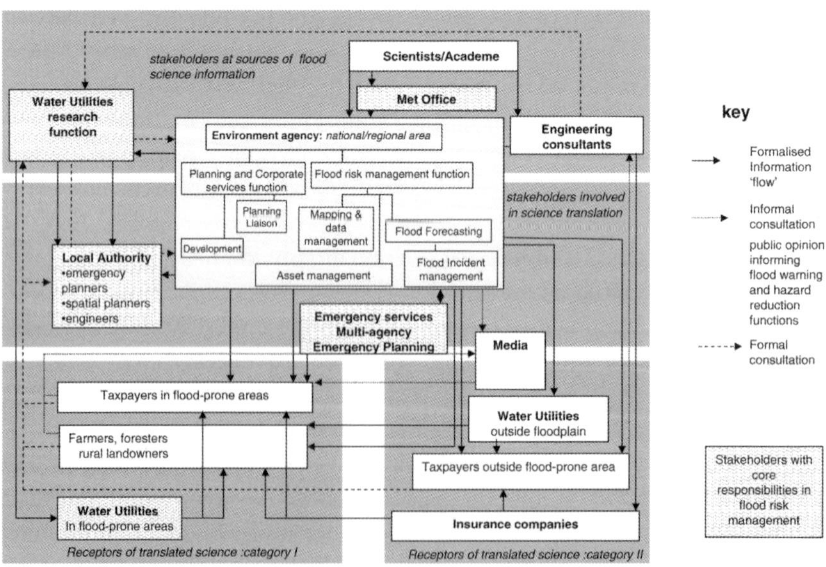

Figure 24.1. Pattern of information flow in FRM in the UK. The various professionals stakeholders involved are suggested to have an uncomfortable intermediary translatory function between the scientists at the source of flood risk science and the lay audience to whom they are formally or informally obliged to communicate about flood risk. (From Faulkner *et al.*, 2007.)

which collectively serve to complicate the already challenging task of communicating from the sciences: a "one size fits all" is clearly inappropriate (see Figure 24.4). It is crucial to consider the position of professional roles within the FRM cycle, functioning between flood events (e.g. emergency planning, asset management) or in real-time (e.g. forecasting and warning, emergency response), in order to optimise *appropriate* communication instruments. There is an apparent need for dialogue and some form of translational discourse to communicate uncertainty and encourage acceptance, ownership and adoption into professional roles.

24.1.3. *Uncertainty*

The sources of uncertainty in flood forecasting and runoff prediction models are largely associated with their assumptions, structure and boundary conditions. Articulation of the uncertainties in a single model is fraught with difficulty, but to complicate matters further the FRM problem involves a cascade of models. For example tide, wave and weather forecasts can be used as input to models of flood size, timing frequency, routing and inundation pattern. Those then provide the input to models of flood defence failure or become part of damage and economic and social assessments of flood impact. Faulkner *et al.* (2007) depict how uncertainties become cascaded in ways that can be only partially constrained. Additionally, longer-term flood forecasting in a non-stationary environment will also be necessarily uncertain.

It is only recently that methods have been considered for communicating these uncertainties to the end-user of flood science (e.g. Butts *et al.*, 2006; Leedal *et al.*, 2010a,b). Uncertainty communication represents a unique challenge, for while the understanding and responsibility for risk estimation and communication has been adopted and enveloped within the remits of professional roles, uncertainty remains in a battlefield of contested ownership and therefore largely neglected in everyday risk communication (see Table 24.1). This chapter hopes to shed some light on the context in which uncertainty must be deployed, alongside some of the barriers or constraints that might be faced in transferring uncertainty from science to practice.

24.2. Professional "Ownership" of Uncertain Science — a Step too Far?

The post-modern take on risk and risk communication alludes to transparency and building trust between science and practitioners. The corollary

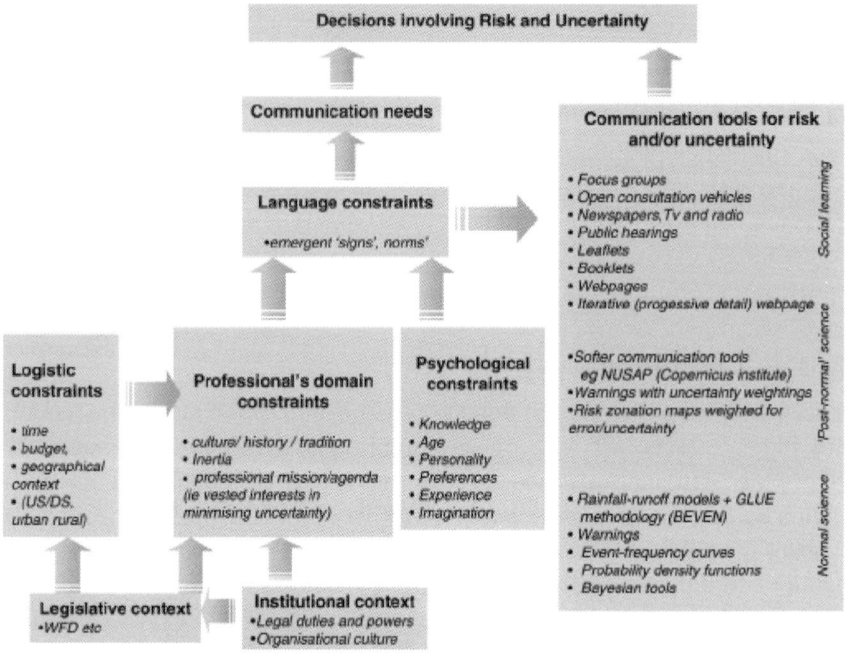

Figure 24.2. This highlights how a range of logistic, professional and psychological factors can constrain the effectiveness of the optimisation of language and "extant communication challenge" (Faulkner *et al.* 2007). In the UK setting, a range of tools are now available to choose from when approaching risk assessments and communicating findings to stakeholders.

to this has to be, however, that we need to enhance the communication of *uncertainty* as well as the message about risk. Part of the problem of the reluctance to embrace uncertainty is that professional contexts, and other constraining factors, diminish the effectiveness of the exchanges.

There is resonance in the concept of "the effective plan of application" with the views expressed by some consulting engineer practitioners at the Flood Risk Management Research Consortium's Risk and Uncertainty Practitioners' Workshop (held at Lancaster University in January 2006). At this meeting, the professionals' suggestion was that they would be happy to make some estimates of uncertainty in flood hazard and flood risk if they were provided with some recipes to use (and preferably recipes that would not be much more expensive to apply than existing procedures). However, unless more professional leadership was evident, they doubted that they would be able to provide estimates of uncertainty to clients,

Table 24.1. The uncertainty content of some current communication tools.

High-level methods	Intermediate methods	Detailed methods
Brainstorming	Analysis of Interconnected Decision Areas (AIDA)	Event tree analysis
Consuitation exercises		Figure of merit
Risk register	Decision trees	*Joint probability methods*
Screening	Expert judgment	Extreme value methods
	Pairwise comparisons	FMECA (Failure Mode, Element, and Criticality Analysis)
	Risk ranking matrix	*Fragility curves*
	S-P-R-C models	Cost-benefit analysis
	Uncertainty radial charts	Cost-effectiveness analysis
	NUSAP	Cross-impact analysis
		Bayesian analysis
		GLUE
		Monte Carlo analysis
		Probabilistic reliability
		Scenario modelling
		Sensitivity analysis
		Utility theory

unless clients demanded them. This suggested an initial unwillingness to embrace ownership of uncertainty, or at least a reluctance to be criticised for going out on a limb and deciding which methods to use.

A more careful reading of professionals' reluctance to embrace uncertainty could be that decisions in management situations are binary, and shades of grey are unwelcome. Hall (2002), Todini (2006), and Hall and Solomatine (2008) have argued that it is helpful when exploring the communication of uncertainty at the science–professional interface to distinguish between the decision uncertainty that preoccupies flood risk managers, and the scientific uncertainty of a flood risk assessment or within a warning. Whereas to the scientist, scientific uncertainty is a challenging part of the professional domain, for agency professionals and other flood managers decisions have to be made; decisions often with considerable implications for cost, well-being, and (not least) liability. Some of these decisions will be binary (on–off, yes–no), such as issuing a flood warning, raising demountable flood barriers or control gates, and deciding whether a development site is in or out of a flood risk zone. Decision uncertainty therefore, includes a far wider range of imponderables than the certainty of the scientific forecast alone. Because the articulation of these imponderables may be even less certain than the science, it is not surprising that scientific uncertainty is an unwelcome part of decision uncertainty from the perspective of a manager (Faulkner *et al.*, 2007, and see Maxim and van de Sluijs, 2011). Beven (2001) comments on the current mismatch in uncertainty articulation from the modeller to practitioner. The modeller's approach for quantifying model uncertainty is invariably construed in mathematical terms, whereas the practitioner's responsibility for communicating this onto lay stakeholders, or indeed the public, requires a qualitative language. Professional stakeholders have consistently reported the need for simplicity in scientific communication and accurate information (see McCarthy *et al.*, 2007). It appears that the need for clarity and conviction in decision making is at odds with the inherent uncertainty in flood inundation and forecast models; nevertheless, it is an ethical imperative to pass on these uncertainties to the decision-makers who will use their predictions.

24.3. The Need for a Conversation

While the scientific community and professional stakeholders operate in distinct structural and intellectual domains, effective communication necessitates interaction at the interface of these two groups. Interactive models of communication have been promoted at the professional-to-public interface and represent an equally appropriate approach here (e.g. Defra/Environmental Agency, 2004; Frewer *et al.*, 2003; Penning-Rowsell *et al.*, 2000). Interactive communication naturally implies a shared ownership of uncertainty. There are a series of questions that this poses, particularly surrounding issues of responsibility: who is responsible for promoting and facilitating this notion of shared ownership? Is the scientist required to tailor research outputs to the end-user context? Will uncertainties increase the burden on decision makers? Herein lays the niche and necessity for a formalised translational discourse.

The implication of contextual factors (i.e. varying institutional, social and knowledge domains across professional stakeholders) upon effective communication means that any mode of communication must be tailored to the "audience". Caution must be taken at this point in using these terms of "audience" or "end-user" which imply that scientific knowledge or tools are simply delivered as products to a passively receiving group. In reality, it is crucial to acknowledge that the end-users actively engage with these "products" which are assimilated within their existing frameworks for understanding (Morss *et al.*, 2005). The end-user must then be viewed as an active participant within the research process in order to facilitate

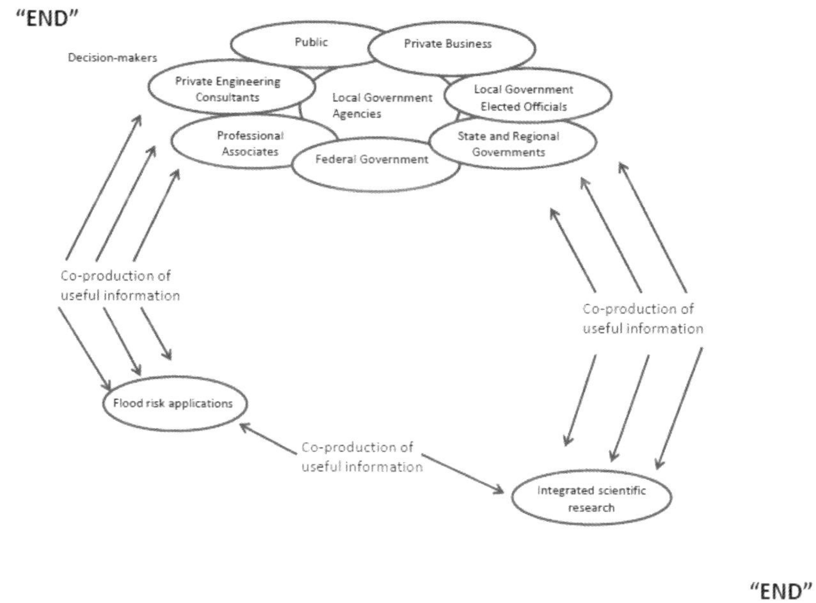

Figure 24.3. Integrated end-to-end research in FRM. A non-linear process whereby multiple communication pathways exist and must be sustained in order to facilitate co-knowledge production and the development and uptake of new tools (after Morss *et al.*, 2005).

knowledge exchange and encourage the uptake of new ideas and tools. Morss *et al.* (2005) present the evolution of end-to-end research perspectives and propose a desired, integrated and iterative model which not only connects the two ends (research and decision maker) but promotes a feedback cycle for the interchange of information (Figure 24.3).

In order to bridge semiotic divides it is crucial that the scientific and end-user communities build upon "conversations" (Colin Green, cited in Faulkner and Ball, 2007). FRMRC members reported that effective strategies for "building bridges" and enhancing risk and uncertainty communication may include for instance one day meetings, software libraries, manuals and guidance documents regarding the use of methods via a website and online "tool kit" (see Figure 24.4, and Pappenberger and Faulkner, 2004). Nobert *et al.* (2010) highlight the importance of early-stage engagement with end-users in steering initial project designs and outputs.

One of the concerns about "not talking" across the scientific–professional divide is the build-up of assumptions about the "other". For instance, Nobert *et al.* (2010) report on interviews with forecasters on their

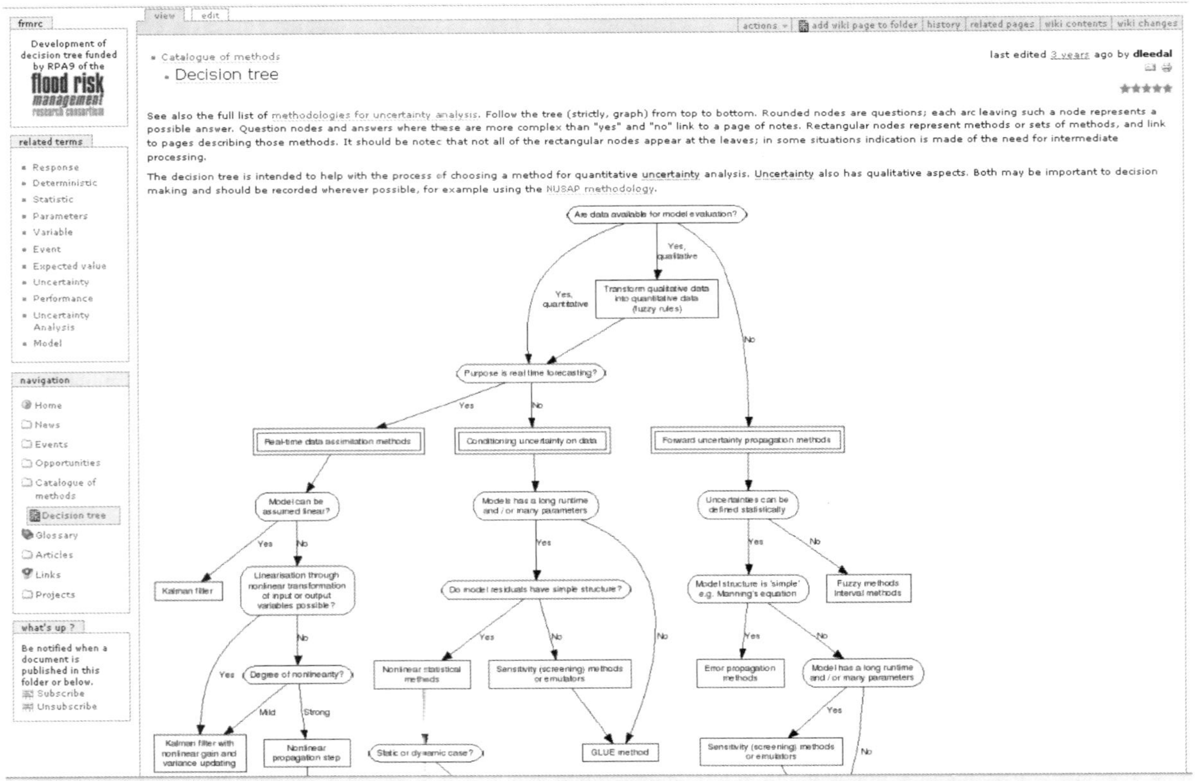

Figure 24.4. FRMRC Wiki pages. screen shot of decision tree to guide users in selecting an appropriate methodology for quantifying uncertainty (see http://www.floodrisknet.org.uk/methods/DecisionTree).

reservations about the use of EPS in civil protection agencies. Some of the following responses were recorded:

"*They want it black and white*" (Flood forecaster, Germany);

"*But these people simply don't understand... I don't care what the probability is. Give me the exact figure...*" (Flood forecaster, Serbia).

A post-modernist reflection on science means we should not consider translation as a process of simplification. We need to move away from assumptions of "the simple other" and instead acknowledge that the end-user is merely operating within differing institutional and social contexts to that of the scientific community. An understanding of these differences is required in order to obtain our goal for effective communication.

As the opening of this chapter explained, communication is a social act and a process of learning; communicating parties learn not just *from* each other, but also *about* the other. Suggestions made in Pappenburger and Faulkner (2004) may provide a useful step forward in targeting both aspects of this learning process. Initiatives such as the FRMRC Wiki pages on uncertainty enable a joint transfer of knowledge; from the scientific community advising on uncertainty quantification methods and from the end-user's ability to contribute and add to the pages (see Figure 24.4). The success of collaborative knowledge products has been assessed in the global environmental change research community. Weichselgartner and Kasperson (2009) build on Nowotny's term of "socially-robust knowledge" and extend this to emphasise the need for "context-sensitive knowledge" construction in order to facilitate the transmission of scientific knowledge into the policy arena (cited in Weichselgartner and Kasperson, 2009; Nowotny, 1999; Pennesi, 2007). This notion of "context-sensitive" knowledge construction is reiterated here as an important step forward in enhancing the communication from the sciences to practitioners in the FRM context, and in doing so facilitating the uptake of uncertainty assessment, integration and ownership.

24.4. In Practice — Flood Warning and Flood Risk Mapping

This section seeks to ground the discussion thus far in two current and contrasting areas of FRM in England and Wales; namely within the context of flood warning and floodplain planning. These two domains operate along differing temporal continuums and exhibit contrasting communication distances from the scientific interface.

24.4.1. *Flood warning in England and Wales*

In the UK, the debate about communicating under conditions of uncertainty has been increasing, as FRM professionals[b] seek to address their new remit of "joined-up" working. This requirement of connectivity between and within stakeholder groups has been reinforced most recently by the 2007 floods and the subsequent Pitt Review (e.g. recommendation 17: 2007). Here, predictions regarding the likelihood of flood occurrence are based on communications from meteorological and hydrological scientists at the Met Office to scientists and practitioners within the Environment Agency (EA). The response of river systems is presented using the EA's National Flood Forecasting System (NFFS) and is used to steer flood warning managers appropriating flood warnings; within the agency, to professional partners — for instance, other Category One and Two responders under the Civil Contingencies Act, 2004 — and members of the public. Cross-agency communication means that warning messages are translated into different institutional cultures and organisational structures, charged with various responsibilities and priority-settings.

Flood warning communications occupy a continuum of timescales (Faulkner *et al.*, 2007). In the long-term, working relationships may form between scientists and practitioners and discourses exchanged regarding aspects of risk, warning and effective response. Routine multi-agency meetings are conducted through Local Resilience Forums (LRF)[c] for example, and provide a forum for developing a common understanding of flood risk, warning codes and the relevant responses of the agencies involved

[b]In the UK, the Civil Contingencies Act (2004) provides a statutory framework for civil protection. "Category One" organisations as outlined in the Act include Local Authorities; emergency services (blue light services such as police, fire and rescue, ambulance); health agencies (Primary Care Trust, Health Protection Agency); and government agencies (Environment Agency, Scottish Environment Agency). These organisations function at the core of emergency response and are subject to the full set of civil protection duties. "Category Two" responders are referred to in the Act as "co-operating bodies" and are required to cooperate and share relevant information to aid emergency response; this category includes the Strategic Health Authority, Health and Safety Executive, transport and utility companies.

[c]Local Resilience Forums (LRF) aim to facilitate the requirement for a joined-up, multiagency approach in emergency management emphasised in the UK's Civil Contingencies Act (2004). Membership consists of representatives from Category One and Two Responder groups. Regional Resilience Forums (RRF) also exist in the UK and similarly aim to enhance co-ordination for regional-wide emergency preparedness (i.e. between central government and the region and from the region to local responders).

in flood incident management. Rehearsals and exercises are a regular part of this process.

At the short-term end of this continuum lies operational response to an occurring flood event. Depending on the nature of the flood (lead and lag time) timings for communications can be heavily constrained, and the effectiveness of planning and rehearsals are put to the test. Here radar models are conflated into weather prediction ensembles, yet these are again further translated by professional meteorologists and somehow quickly converted into short messages. At this scientific–professional interface, scientific information may be considerably conflated and any nuances in the message disappear. Whilst uncertainty estimations may be there in the radar-to-meteorological exchange, they are simplified or dropped as the pressure of real-time operational activity builds up. In this sense, communication and operationalisation of *uncertainty* in flood forecasting into flood warning, is currently neglected. Indeed, responding practitioners receive relatively the same level of detail as is made publicly available. EA warnings are tempered by confidence estimates, which lack clarity and are inadequately designed for professional application (Faulkner *et al.*, 2007). There is a need for practitioners to be educated about the concept of uncertainty and model assumptions from which uncertainty estimates may be based. McCarthy *et al.* (2007) have shown that there is an enthusiasm for this amongst practitioners when given the opportunity to "play" with new visualisations and to clarify meaning with scientists. In an operational setting it may well be useful to continue the dialogue from the scientist to practitioner and collectively discuss uncertainty outputs (e.g. ensembles) and implications for decision making. This could be achieved via a shared telephone or web video conference call. It is crucial to consider uncertainty communication in a practical setting such as this; uncertainty needs to be perceived as salient, useable and as a mechanism for robust decision making.

24.4.2. *Flood risk mapping and floodplain planning in England and Wales*

Floodplain planning clearly operates on a longer timescale, as such communication channels include face-to-face meetings, email exchange and presentations and are not subject to the time constraints of operational warning and flood response. Constraints from planning policy (PPS25) and recent EU Floods Directive Guidelines mean that planners are now required to develop flood risk maps, based on different exceedance probabilities.

While there is no stipulation for uncertainty boundaries, it is foreseeable that this will become a future recommendation. Research has shown, however, that this is an un-welcomed ownership issue (Faulkner *et al.*, 2007). The rationale accounting for this is the presumption that uncertainty boundaries will enable people to disregard or circumvent the imperatives of legislation. This represents an apparent constraint to the ownership of uncertainty and it is vital to remain mindful of these challenges facing the planning domain (Tunstall *et al.*, 2009). The inclusion of uncertainty in flood risk maps needs to be clear, understandable and relevant to this context in order to appease its integration into a planning setting.

24.5. "Knowing by Sight" — Visualisation Strategies for Communicating Uncertainty

This section moves away from "the talk" to describe how visualisation techniques may also facilitate the transfer of knowledge and uptake of new ideas. Any visualisation needs to be designed according to the task at hand: asking who is the user?; what are the users' goals?; and how does the visualisation tool seek to enhance goal attainment?

24.5.1. *"In the mind's eye": what is visualisation?*

Visualisation techniques are a means of evoking visualisation, but it is crucial to stress that visualisation itself is a cognitive process. Visualisation techniques may therefore be considered as a means of combining computer power with human vision, steering decision making and learning as well as scientific insight (MacEachren and Ganter 1990; MacEachren *et al.*, 2005). There is a vast literature base detailing "efficient vision" and the role played by existing, mental schemas and how these may in turn be utilised to design effective interfaces for visualisation tools (see MacEachren and Ganter, 1990; MacEachren and Kraak, 2001; Neisser, 1976). Subsequent work has expanded on the semiotics of visualisations (see MacEachren, 2001). Visualisation techniques have been employed within various disciplines and have thus manifested in a number of branches such as Geovisualisation (GVis), information visualisation and scientific visualisation. Geovisualisation stems from cartography and engages specifically with the exploration, analysis, synthesis and presentation of georeferenced information, with particular emphasis on the need for interactivity and active end-user engagement (MacEachren, *ibid*); and is therefore perhaps the closest counterpart from which flood science may draw. Effective visualisation requires a balancing act, between the need to simplify, abstract and represent detailed scientific knowledge so as to be comprehensible to the end-user; whilst simultaneously minimising the loss of detail and nuances in the data (Ishikawa *et al.*, 2005). The following sections will consider suggestions for visualisation strategies and presentation options for conveying uncertainty and, furthermore, will critique the purpose of visualisation and what it can hope to achieve.

24.5.2. *Mapping*

Mapping is a widely acknowledged, powerful tool for visualisation and has become the keystone for flood risk communication, evolving from mere outlines of flood extents to more sophisticated detailing of depth–velocity functions of flood waters. The EU Floods Directive (2007) stipulates that all EU countries must produce flood hazard and risk maps by 2013 towards the establishment of FRM plans by 2015 (van Alphen *et al.*, 2009; Moel *et al.*, 2009). Flood inundation modelling is the cornerstone of flood map production and as this book has testified is by no means a "perfect science". While considerable effort has been made by the scientific community to develop numerical methods for quantifying this uncertainty, there is no formal, legislative obligation to incorporate boundaries of uncertainty inherent within the flood risk maps to which FRM strategies will evolve.

The presentation of uncertain information or data accuracy has been explored widely in cartography (see MacEachren *et al.*, 2005 for a review). *Colour*, e.g. saturation, hue and texture, and the concept of "focussing" (Davis and Keller, 1997; MacEachren, 1992) and *object boundaries*, e.g. fuzziness, transparency (Pham and Brown, 2005) in particular have been highlighted as effective strategies for portraying uncertainty and reliability in data. In relation to flood science it has been widely voiced that the presentation of floodplain boundaries, traditionally depicted as single lines, should be alternatively presented as a series of probabilities (Bales and Wagner, 2009: Smemoe *et al.*, 2007; see Figure 24.5).

The use of video clips and real-time animations of flood inundation models, predicted hydrographs and ensembles of predicted hydrographs have also been implicated as a powerful means of communicating from the scientific domain to practitioners (Faulkner *et al.*, 2007). One concern for uncertainty presentation is that it should not over-complicate and confuse the picture (Beard and Mackaness, 1993). There are numerous approaches for avoiding a "messy" picture. MacEachren's work has paid

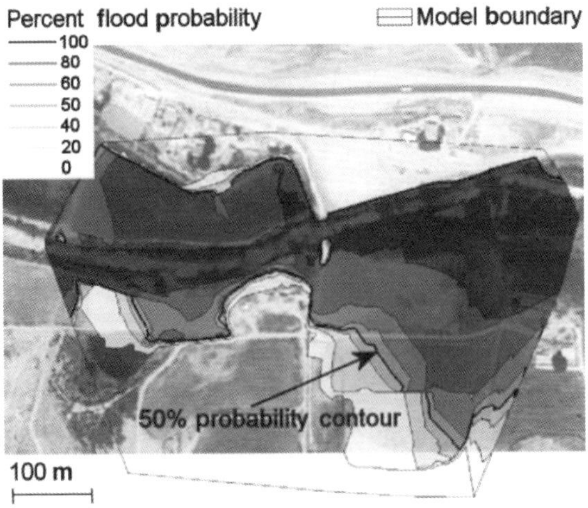

Figure 24.5. Flood probability map for the Virgin River in S.Utah (after Smemoe *et al.*, 2007).

attention to the user interface for interactive mapping and distinguishes two presentation styles; whereas bivariate representation pairs data and uncertainty (e.g. through bivariate colour in map presentation, or the option to view each in isolation or alongside each other; e.g. Cliburn *et al.*, 2002); data certainty/uncertainty can be alternatively represented through threshold mapping i.e. the user selects a threshold of certainty, past which data is masked (e.g. Leedal *et al.*, 2010b). These considerations are crucial for communicating uncertainty in a visual form.

24.5.3. *Visually explicit ensembles*

The evolution of modelling practices has obviously sparked new forms of visual presentation; this is particularly the case with the development of ensemble modelling for forecasting (otherwise termed Ensemble Prediction Systems (EPS); Nobert *et al.*, 2010). Ensemble forecasting is internationally well established and deployed to aid decision making in forecasting agencies (McCarthy, 2007); however, the roll-out of this tool into operational warning and responding agencies has not been witnessed within the UK. McCarthy (2007) discusses a number of reservations voiced by EA staff, which included concerns for a change in the way decisions are currently made, the need for further training and overburdening warning staff. On the other hand, Nobert *et al.* (2010) present a "success story" for operational EPS in Sweden, between the Swedish Meteorological and Hydrological Institute

(SMHI) and Civil Protection Authorities (CPAs). A number of reasons for this success are described and include the importance of tailoring both the "product" (from the initial design, "upstreaming") and training to end-user needs and professional cultures. One of the aims of this chapter sought to propose ideas to encourage the acceptance and adoption of uncertainty within decision making; these findings by Nobert *et al.* (2010) and the successful application of EPS in Sweden offer profound insights for the UK challenge. McCarthy *et al.* (2007) demonstrated how current professional stakeholders rely (and are happy to do so) on the "expert advice" when it comes to judging uncertainty. While ensembles were viewed as useful, some professionals were concerned about competency and responsibility in interpreting them. The four day workshop conducted by McCarthy *et al.* (2007; 2008) simulated a real-time event and showed the important value of interaction between scientists, communication and visualisation tools, and end-users. Similar fora for enabling professional stakeholders to engage with the science (the scientist and new tools) are one way of raising awareness of new communication and assessment tools and their efficacy, as well as the self-efficacy of end-users to employ such tools.

24.5.4. *Maps and visualisations as decision support tools*

The outputs from flood modelling and forecasting have been integrated into decision support instruments, targeting different end-users and operating at various stages within the FRM cycle, with more recent work focussed on real-time support. Decision support systems are understood as tools for supporting user decision making, through centralising resources, mining resources and weighing up and evaluating alternative decision choices (Rinner, 2003). Sophisticated DSS for flood management are being increasingly developed; for supporting forecasting and flood incident response (e.g. Abebe and Price, 2005; Butts *et al.*, 2006); as well as spatial planning (e.g. Milograna *et al.*, 2009; Pasche *et al.*, 2007). The option to centralise and query vast quantities of data is a central asset to DSS and enabling the user to integrate with the data in a manner suiting their decision needs (e.g. Zerger and Wealands, 2004). Multi-Criteria Decision Support Systems (MCDSS) have further sought to capture the complex frameworks in which flood management decisions must be made; with further exploration into group MCDSS for facilitating communication within and between professional stakeholder groups (see Levy *et al.*, 2007).

The decision support literature highlights the growth of risk assessment and communication, based on interactive interface designs and user-controls (e.g. Todini, 1999). Saying this however, there are limited examples where uncertainty is explicitly addressed; this may be partly due to a lack of clarity as to how uncertainty can improve decision making. Butts *et al.* (2006) acknowledge the need to address uncertainty presentation and encourage the development of understandable and consistent methods designed with the end user in mind. It is imperative that uncertainty is integrated intuitively within such systems to complement decision making, rather than complicate it. There are numerous design suggestions for the visualisation of uncertainty; assuming it does not "complicate the picture" but is presented in a readable form, users can begin to fully comprehend its meaning (within professional and decision-making realms) and this may go some length towards adoption and a shared ownership of uncertainty.

24.6. An Example: the Probabilistic Flood Inundation Visualisation Tool (PFIVT)

A Probabilistic Flood Inundation Visualisation Tool (PFIVT) has been developed by Leedal *et al.* (2010a). This tool, like all forms of visual communication, is an object composed of a series of signs (or semiotic systems). The viewer must decode these signs to form meaning. According to structuralist semiotic principles, the key question becomes: "what is the code required for two or more people to 'correctly' interpret the semiotic system?" This question can be recast from the point of view of the producer of the object, for example, "given the information I intend to communicate, what are the tools, knowledge and understanding required by the audience such that the intended message can be consistently recovered?" From this perspective, the success of the object is judged on its ability to provide a code that will allow the viewer to reconstitute the message intended by the producer. To achieve success, the code must be rich enough to span the information content of the message while being accessible enough to maintain the viewer's interest. If the viewer is confused by the code, he or she may reject it or develop a subjective interpretation of the message.

The sections below provide:

- a brief description of the modelling exercise used to produce the data for the PFIVT;
- an introduction to the PFIVT and the design process used in its production; and

- a very brief discussion of issues relating to the PFIVT from a social semiotic perspective.

24.6.1. *The Mexborough probabilistic flood inundation model*

A flood risk mapping study was commissioned to produce a model-based estimate of the magnitude of the 1% annual exceedance probability (1% AEP) event in Mexborough, South Yorkshire, UK. This region is vulnerable to flood damage with the last significant flood event occurring in June 2007. The study was carried out within a probabilistic framework using an ensemble of results generated by the TUFLOW 2D hydrodynamic flow routing model. The model incorporated a digital elevation model of the region and operated at a 10 m grid resolution. The experiment design took into account three key modelling uncertainties:

1. *The magnitude of the design event.* The 1% AEP design event was estimated from past data tables using standard Flood Estimation Handbook (FEH) flood frequency analysis techniques within the WinFAP software package. This analysis resulted in an estimate of the 95% uncertainty range for the 1% design event flow of between 76.7 and 92.9 m^3s^{-1}.
2. *The value of a global Manning's roughness coefficient (this is an empirical coefficient determined by channel roughness and other characteristics such as sinuosity).* This parameter is well known in hydrological modelling fields to exert a significant influence on simulated flow dynamics while also being difficult to define from observations of the system under study; as a result, this parameter is often used to tune a flow routing model in order to fit the observed data (Morvan *et al.*, 2008). For this study, a uniformly distributed random variable with 95th percentiles of 00.05 and 0.2 was used for Manning's roughness coefficient.
3. *The channel capacity.* For the type of hydrodynamic simulation used in this study, it is assumed that flows exceeding the channel capacity will spill onto the flood plane for storage and/or conveyance downslope. The channel capacity is often difficult to calculate and is therefore commonly estimated as being equal to the median annual maximum flood (QMED). However, this itself is difficult to estimate if data are scarce and, as here, often approximated using catchment descriptors and empirical equations. For this study, a uniformly distributed random

variable with 95th percentiles of 12 and $54\,m^3s^{-1}$ was used for channel capacity.

Five hundred model runs were made drawing the three parameters described above at random from their respective probability distribution functions. After completing the model runs, the probability that inundation would be observed in any one of the 10 m cells within the model domain was calculated. For example, if a cell became inundated in all 500 model runs, then that $10\,m \times 10\,m$ region of the floodplain was ascribed a 100% probability of being flooded during a flood of magnitude equal to the design event; if a cell was inundated in half of the ensemble members then the corresponding region of the floodplain was ascribed a probability of flooding of 50% and so on.[d] The raw experiment results took the form of large amounts of numerical data. We needed to apply some form of intermediate processing in order to make the data accessible to human users. The PFIVT was designed as a tool to provide this intermediate layer. In the discussions that follow, the PFIVT is the "object", the designers of PFIVT and the probabilistic study are the "producers" and the end user is the "viewer."

24.6.2. *The PFIVT*

We can assess what codes are available to the producers and how effective these are at communicating the desired message to the viewer by applying a simple structuralist semiotic analysis (van Leeuwen and Jewitt, 2001, pp. 140–152). A first step is to define the content of the message intended by the producers, and to make the simplifying assumption that this is an objective statement of the entire scope of the message.[e] A four-point description of the message carried by the PFIVT could take the following form:

- It is possible to use a computer together with some scientific principles to estimate the extent of a flood event of a given magnitude. It is possible to estimate the frequency of a flood of a given magnitude;

---

[d]Details of the experiment design were supplied by N. Hunter and J. Neal, personal communication, January 10, 2010.

[e]This task is made somewhat easier here by the fact that this chapter and the PFIVT object share an author. Otherwise an analysis would have to consider the additional layer of interpretation involved in the translation of message from model study to web tool author to chapter author!

- We, the producers of the object, estimate a flood of the magnitude of this study to have a 1 in a 100 chance of happening in any given year; the inundation modelling exercise has been performed within a probabilistic paradigm;
- It is important for us to communicate the estimate of flood extent for the design event together with the uncertainty associated with this estimate;
- We can address the above by overlaying, on a scalable map of the region, an outline that shows the extent of the flood inundation at a probability level chosen by the viewer.

These four points, together with the data collected from the probabilistic flood risk mapping exercise, form the producer's message. A successful PFIVT tool must allow viewers to consistently reconstruct the above message. *The task is therefore to define a semiotic system contained by the PFIVT to facilitate the unerring "take up" of this message by the viewer.*

24.6.3. *Model "the viewer"*

The first design stage for the PFIVT was to form a model of the viewer. This model incorporated assumptions about the high level objectives of the viewer i.e., the things the viewer would want to achieve using the PFIVT. The high-level objectives were then broken down into sub-tasks. This process of abstraction was repeated until the tasks were very simple and would require no problem solving on the part of the user. For example, a high-level task that the producers anticipated would be carried out by the viewer was: "to view the extent of uncertainty in flood inundation modelling". This objective was decomposed through sub-tasks such as "show inundation overlays on a map at chosen levels of probability". After several iterations of this process we arrived at, or close to, the *unit task* level of operation. An example of a unit task could be: move a slider to the 50th percentile value. This process can be very time consuming; also, the definition of the viewer's objectives and the granularity at which we assume the unit task has been achieved is inevitably rather subjective.

Having identified a large number of unit tasks, a Goals, Operators, Methods and Selection (GOMS) study was performed. This process is defined in detail by Dix *et al.* (2004, p. 422); but briefly, the GOMS study defines the following:

- Goals: the objective(s) that the user wishes to achieve. The goals need to have been reduced to a level close to the unit task defined above.

- Operators: these are the lowest level actions that can be carried out by the user or the user acting on the system, for example "move a slider" or "read the contents of a dialogue box" in order to achieve the goal. The purpose of defining the operators is to arrive at easily described operations that can be converted into design or programming elements.
- Methods: there is often more than one way to achieve the goal (different button/mouse click combinations, etc.). The methods section of the GOMS study contains a list of these.
- Selection: this is the assumed choice from the available methods taken by a user. In the case where more than one method is available it is usual to explicitly define which selection the viewer will take based on assumptions or known information about the viewer.

As an example, we can consider the goal of "choose an inundation probability within the PFIVT". A GOMS analysis of this task takes the form:

```
GOAL: CHOOSE-PROBABILITY
.       [select GOAL: USE-SLIDER-METHOD
.       .                  .       MOVE-MOUSE-TO-SLIDER
.       .                  .       DRAG-AND-RELEASE-SLIDER
.       .                 GOAL:    USE-TEXT-BOX-METHOD
.       .                          MOVE-FOCUS-TO-TEXT-BOX
.       .                          INPUT-PROBABILITY-VALUE]
```

Analysing the GOMS units (for example the depth of nesting of goals), provides indicators of the performance of the modelled viewer. The combination of many GOMS units was used to predict the requirements and actions of the viewer and translate these into the design elements of the PFIVT.

24.6.4. *Visual analysis*

In parallel with the GOMS approach, a more aesthetic design process was followed using principles taken from the field of visual analysis and informed by the sort of concerns discussed above in relation to cartographic message content. This type of analysis aided the design process in terms of supporting decisions for element placement, text layout etc. As well as this, the method also encouraged a reflective and critical analysis of the production process. This forced us to develop an awareness of the culturally situated motivations behind design decisions, such as the choice of text and the map viewpoint. The visual analysis considers the PFIVT to be an object made-up of a set of semiotic resources where these resources not only provide the tools to allow the viewer to decode the message, but also to consciously, or otherwise, reproduce the social relations between the producer and viewer. The implication of the social semiotics is not pursued in great detail in this chapter as the PFIVT is a small-scale and speculative research project; however, it is important to reflect that these issues will be significant if probabilistic flood risk visualisation tools become widely-used by large organisations.

The PFIVT object needed to integrate representational mechanisms from entirely different families, most significantly, text fields together with images of both naturalistic and scientific modality. Text was used to provide instruction and guidance to the viewer, naturalistic modality (photo-realistic representation) was used to communicate the location of the study site,[f] scientific modality was used to make the results of computer simulation visible.

The salience of different components of the object were considered and manipulated to provide an effective balance of focus for the viewer. For the PFIVT, the background map image, with its bright colour content, becomes the immediate focus of the object. This was considered correct as it provides a familiar and interesting introduction for the viewer. Attention is then drawn away to the user interaction controls situated top-left. This mechanism takes advantage of the top-to-bottom left-to-right flow of text on the printed page. According to Kress and van Leeuwen (1996) the left–right placement of items within a Roman script environment not only provides a familiar guide for the viewers' focus, but also provides a sense of "given" (on the left-hand side) and "new" (on the right-hand side). Figure 24.6a shows the PFIVT web interface together with added indicators of the grouping of related elements within the page and the anticipated flow of the viewers' focus.

This flow is used within the PFIVT design to draw the attention of the viewer from the map image to the primary dynamic interaction tool which is a slider for selecting probability thresholds. The operation of the slider tool is illustrated in Figure 24.6b. The slider is used to select from the model study data the maximum inundation extent forecast to occur at the chosen probability level. From the slider tool, the left-to-right flow draws

[f] An abstraction of this is also an option by using a map-based representation.

the viewer past a text-based description of the chosen scenario and on to a popup menu inviting the viewer to "choose a definition" from a list of words that may be unfamiliar but important for developing an understanding of concepts within probabilistic FRM.

Similarly, top-to-bottom placement of elements can be used to generate a sense of movement between the "ideal" or general essence of information, to the "real" or more "down to earth" formulation of the information. Here, the ideal took the form of both the numerical text field giving the chosen inundation probability and the text field describing the chosen scenario. The real includes the Google map and overlay showing where the flood outline is situated and a text field providing a definition in plain English of the terms and concepts used by the PFIVT object.

The control elements, or widgets, (shown in grouping 2 within Figure 24.6a) were framed together to signify a common purpose and make use of the principle of consistency, i.e. the re-use of familiar design elements. In the field of Human–Computer Interaction (HCI), consistency is cited as being an important mechanism in aiding the viewers' understanding and learning. It takes advantage of any familiarity the viewer may have with common design elements (Reiman, 1994). The viewer does not need to travel too far with eye or mouse within the group of control elements. This should allow the viewer to operate the page efficiently. Also, no elements are introduced that are not commonly found on standard webpages.

Google Maps™ were chosen for the mapping interface; the map pane is labelled as grouping 1 in Figure 24.6a. This choice takes advantage of the consistency principle by reusing a map navigation interface that is already familiar to a large number of potential viewers.

Both distance and point of view are important semiotic resources and it is wise to be sensitive to their use. When the PFIVT webpage loads, the default distance between the viewer and the study site was set such that the inundation region roughly fills the available map pane. The point of view is always from directly above the site. This inevitably defines the viewer as an actor with a god-like overview of our simulated disaster. We feel this is a concern as the chosen distance and point of view have the secondary effect of imparting the producers' message with an innate sense of authority which, however unintentionally, may be misplaced. However, from the default distance the viewer is free to use familiar interaction methods to zoom and pan within the study site taking a close-up tour of regions of interest. It is worth bearing in mind that this process could be emotive for the viewer as he or she may be viewing their own home.

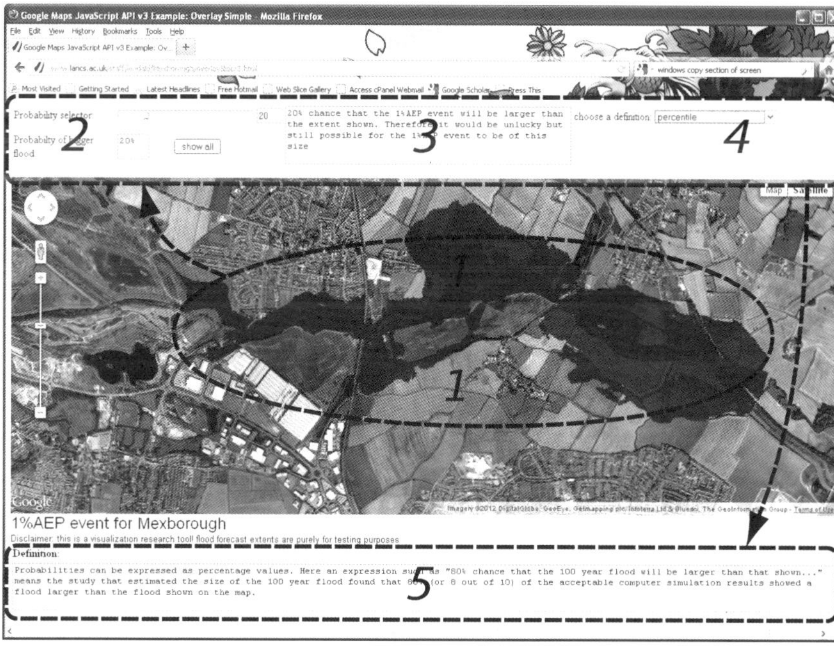

Figure 24.6a. The PFIVT web interface. The anticipated focus of attention and order with which the user assimilates the information on the page is indicated by the dashed shapes, arrows and sequence numbers that have been added to the figure.

The juxtaposition of images of natural modality (the satellite option of Google Maps^{TM}) with an image of scientific modality (the inundation overlay) is also problematic. The inundation overlay has been produced using best practice within the remit of the modelling study and the image has been accurately georeferenced to the underlying map; however, by linking these two types of image we are taking advantage of the sense of *realness* afforded by photorealistic representation in order to impart the same realness to the results of the modelling experiment. This process provides an illusionary link between the virtual world of computer-based simulations and the real world captured by aerial photography. In effect, we are placing equal information value on the content of the images. As with considerations of point of view and distance, the overlay approach may provide a good practical solution to the visualisation problem; however, the implications of such design choices should always be carefully considered.

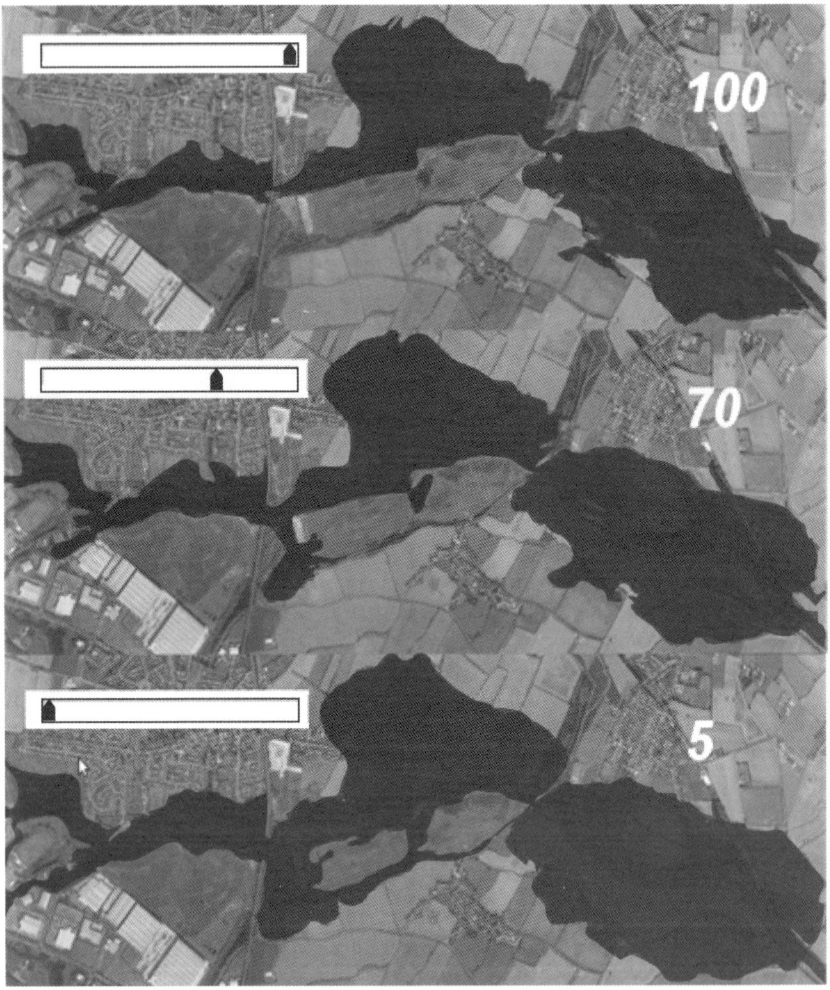

Figure 24.6b. This mosaic of elements from the PFIVT (actual layout is shown in Figure 24.6a) illustrates the connection between the interactive slider tool and the inundation overlay. The viewer moves the slider control to select the inundation overlay image. The connection between slider, text and overlay is accomplished with DHTML.[g] The figure shows results for a 100, 70 and 5% chance of an actual 1% AEP event inundation exceeding the extent shown.

[g]A combination of HTML, JavaScript and, in this case, access to elements of the Google Maps™ API.

24.6.5. *Words and pictures*

The interactive element of the PFIVT, i.e. the dynamic linking between the slider tool, the text fields and the inundation overlay, together with the ability to zoom and navigate within the map pane, provide a simple and intuitive mechanism for communicating spatial probabilistic flood inundation information. The PFIVT makes use of two prominent text-based elements to augment the message presented by the map figure. The first of these elements was positioned directly to the right of the slider tool. The text within the field is selected automatically by the page code as the viewer moves the slider. The text adds a descriptive element to the chosen scenario. The choice of wording is designed to add to the information content by using qualitative expressions such as “very unlucky”, “a little unlucky” etc. The probability values are expressed as percentages as research has shown that this is the most readily understood format (Biehl and Halpern-Felsher, 2001). The actual text is shown in Table 24.2. Example definition text is shown in Table 24.3.

24.6.6. *Trials and testing*

At the time of writing, the PFIVT is in the early stages of user-testing. The tool has been demonstrated at a number of workshops where users have been able to try the tool out for themselves. The aim of these trials is to assess how far it is possible to optimise the tool for the end users’ needs. A sub-goal is to explore the manner in which the uncertainty inherent in a modelling exercise of flood inundation can be *owned* jointly by the producer and viewer using methods such as the PFIVT. An active part of the new research associated with the PFIVT is to modify and enhance the design inline with the results of trials and testing.

24.6.7. *Summary on the Mexborough tool*

The PFIVT object provides a set of semiotic resources to communicate a message from the producer to the viewer. These resources include familiar mechanisms such as webpage widgets (sliders, text boxes and popup menus, etc.) but also less obvious mechanisms such as point of view, information value and left-to-right top-to-bottom placement. Given that both the producer and viewer are culturally situated, the power relations present in the wider cultural settings of these actors will be reproduced to a greater or lesser extent by the semiotic resources used within the object. With this

Table 24.2. Dynamically selected text fields applied to the string: "<field 1> chance that the 1% AEP event will be larger than the extent shown. Therefore <field 2> for the 1% AEP event to be <field 3>" (where the fields are replaced by the appropriate text in the table depending on the position of the slider).

Text <field 1>	Text <field 2>	Text <field 3>
5%	it would be very unlucky but still possible	as large as this
10%	it would be very unlucky but still possible	as large as this
20%	it would be unlucky but still possible	of this size
30%	it would be unlucky but still possible	of this size
40%	it would be a little unlucky but still possible	of this size
50%	there is an even chance	smaller or larger than this size
60%	it would be a little unlucky but still possible	of this size
70%	it would be lucky but still possible	of this size
80%	it would be lucky but still possible	of this size
90%	it would be very lucky but still possible	as small as this
95%	it would be very lucky but still possible	as small as this

Table 24.3. Example definitions available to the PFIVT viewer via a popup menu linked to a text area.

Concept	Definition
Return period	The return period is the average amount of time in years that you would expect a flood of a particular size to occur once. For example, a flood with a return period of 100 years would be expected to occur 10 times in a century. It is very important to realise that this does not mean that if a flood with that return period has just happened that there will definitely not be another one for 100 years! In addition to this, the accuracy with which the return period can be calculated is not perfect, so there is always some degree of uncertainty in this value. It is generally advised to use the annual exceedance probability (which is 1 divided by the return period) when expressing flood risk.
Percentile	Probabilities can be expressed as percentage values. Here an expression such as "80% chance that the 1% AEP event will be larger than that shown..." means the study that estimated the size of the 1% AEP flood found that 80% (or 8 out of 10) of the acceptable computer simulation results showed a flood larger than the flood shown on the map.
Probability of exceedance	This webpage shows that flood extent forecasting can never be exact. This is because flood forecasting is based on computer estimates of what might happen during a real flood. One way to communicate the range of possibilities for what might happen is to specify the chance that a flood will be bigger than the one shown on the map. For example, a probability of exceedance of 20% means that the computer simulation estimates that the 1% AEP event has a 20% (or 1 in 5) chance of being bigger than the one shown on the map.

in mind, the designer of the resource can take care not to communicate in a way that alienates, confuses or misleads the viewer.

24.7. Towards a Translational Discourse

Visualisation should not be considered as the end product but as a prompt for inciting learning, decision making and new modes of thinking. For instance, the forecast example by McCarthy *et al.* (2007) shows that user-interaction with new visualisation tools and with tools designers can alter initial assumptions and facilitate the want for new techniques, and most importantly, the self-efficacy of the end-user to utilise new techniques. The expressed desire for "simplicity" voiced by end-users arguably stems from an uneasiness with new and seemingly complicated tools. For new tools to be fully integrated and not viewed as "the shiny new toy" (only to be neglected soon after) it is crucial that any new visualisation tool presented is user-friendly and self-explanatory; if not its use needs to be explained by an interpreter. The function of the translation should be to effectively translate the science in a way that meets the end-user's declared needs.

While risk communication has been well-established from science to practitioner, boundaries of uncertainty remain obscured in issues of responsibility and ownership. This chapter has highlighted the need for translation across the scientific and practitioner domains. These two domains sit in differing social, institutional and knowledge contexts so as to make the goal of effective communication, a communication challenge. Effective communication requires satisfaction of both parties by optimising the "fit" of communication tools (language, computational tools and visualisations) transferring the message (with embedded uncertainties) to a declared professional purpose; whilst simultaneously satisfying the needs of the scientist to minimise the loss of information associated with the translation of the signs and norms of science. Faulkner *et al.* (2007) argue that information loss on translation may be reduced by more effective discourse and exchanges between those scientists developing risk prediction models and those professional agencies charged with managing risk-prone settings. Improving the effectiveness in risk and uncertainty communication in FRM could be facilitated by constructing a "translational discourse" which straddles these divided domains of science and practitioner (Figure 24.7).

In order to build, adopt and maintain a translational discourse we must strive to meet a series of objectives. Firstly there is a need to promote a

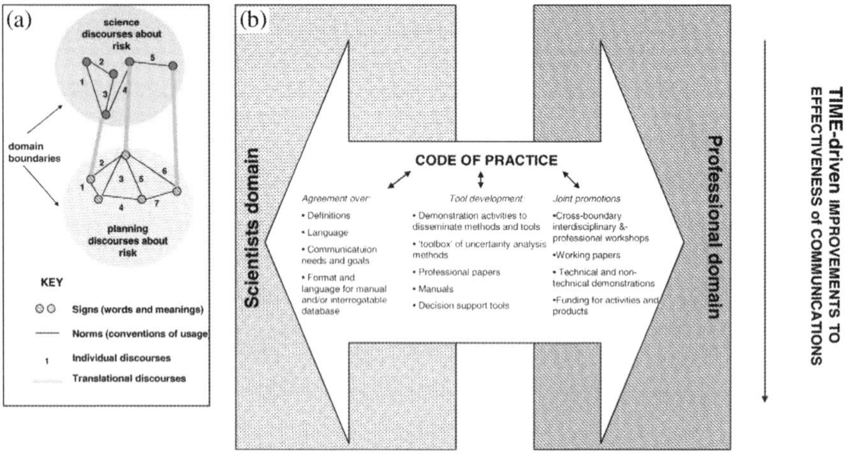

Figure 24.7. Towards a translational discourse between scientific and professional domains. (a) The positions of translational discourse straddling domains. (b) The functional content of translational discourse, suggesting the probable components of the debate and the tools to be developed and promotions involved (Faulkner *et al.*, 2007).

definitional exchange, acknowledging that meanings assigned to key terms are variable between and within the professional and scientific domain. These semantic differences need to be understood in the context they were constructed. The search for "neutral" definitions is perhaps futile and would only strip terms of their meaning and context in which they must be applied; therefore, the semiotic differences in these terms should be articulated and appreciated in conversation. This feeds into the next objective to enhance respective understanding of differing domains associated with different professional groups (and indeed within science itself). Understanding relates not only to the context in which these groups operate (e.g. institutional frameworks) but also suggests a need for comprehending the assumptions and expectations constructed between and within professional groups and with/from the scientific community. It is vital that all communicate their desired needs and goals.

These objectives support the call for a collaborative discourse between these seemingly opposing domains. There is a need to strip away the sentiment of "the other" in favour of a united "we" (Pellizzon, 2003). There are numerous strategies in which this may be achieved, such as the creation of virtual institutions (Mansilla *et al.*, 2006) and expanding the number of joint-conferences, seminars and workshops. The need for a more open, transparent and collaborate research process has been voiced as an important step towards more salient knowledge production (Weichselgartner and Kasperson, 2009). It is a matter of true end-to-end research for sustaining working relationships and knowledge exchange (Morss *et al.*, 2005); erasing technocratic traditionalism and instead considering the stakeholders of science as active participants in the research process (Voinov and Bousquet, 2010).

Ultimately, such efforts require funding and leadership to promote and motivate a collaborative community effort; Faulkner *et al.* (2007) suggest the Environment Agency and Defra are the natural partners for the role, though some academic leadership should also be sought. In terms of driving uncertainty communication forward, incentives and disincentives should be considered to ensure a professional obligation and standard is met. Ethical standards for uncertainty analysis and communication must be appraised (Faulkner *et al.*, 2007). These suggestions nestle within the code of practice for risk and uncertainty communication proposed by Pappenberger and Beven (2006). In the short-term a guidance manual to outline the objectives for cultivating a translational discourse is required to steer this forward in the near future.

24.8. Conclusion

This book recognises and seeks to describe and measure the uncertainty in the science of flood risk. The present chapter has discussed how on top of the modelling, parameter and validation uncertainties discussed elsewhere in this volume, decision makers face a wider range of operational (often binary) choices of which these scientific uncertainties are merely one subset. This chapter has explored the manner in which effective communications about risk and uncertainty get translated. We argue the tools used should and can be tailored for particular exchanges and needs; an argument we develop using material from the field of semiotics. The material we debate here recognises that the communication needs are very different in differing FRM settings and we use the fields of warnings and flood risk mapping as two examples of communication and translation of science.

In particular, we have discussed how the ownership of the uncertainty in the communication is often contested or ignored. Getting the use of new tools to describe into professional practice may need an improved conversation at the science–professional interface. To address the considerable challenge of embracing and owning uncertainty for those undertaking these exchanges remains considerable. There is a need for a new professional function — the science–professional interpreter, working to develop the translational discourses that are going to be increasing needed as non-stationarity continues to confound our abilities to validate models. A negotiated ownership and articulation of uncertainty is a good starting point.

Flood science has thus far focussed on the enhancement of communication, i.e. to transfer "what we know", but this chapter has suggested that new technologies, especially interactive animation and visual depiction of how confident we are about what we know, might prompt new thinking in users. Whether this new generation of visualisations and decision support instruments will change the data requirements of end-users in the long-term remains to be seen, as does whether uncertainty will influence the act of decision making itself and decision outcomes. We conclude that visualisation techniques and forums for conducting "conversations" between scientists and stakeholders go hand-in-hand, and are key drivers in seeking a translational discourse.

References

Abebe, A.J. and Price, R.K. (2005). Decision support system for urban flood management, *J. Hydroinform.*, **7**, 3–15.

Bales, J.D. and Wagner, C.R. (2009). Sources of uncertainty in flood inundation maps, *J. Flood Risk Manage.*, **2**, 139–147.

Beard, K. and Mackaness, W. (1993). Visual access to data quality in geographic information systems, *Cartographica*, **30**, 37–45.

Biehl, M. and Halpern-Felsher, B.L. (2001). Adolescents' and adults' understanding of probability expressions, *J. Adolescent Health*, **28**, 30–33.

Butts, M., Klinting, A., Ivan, M. *et al.* (2006). *A Flood Forecasting System: Integrating Web, GIS and Modelling Technology*, ESRI International User Conference Proceedings, August 7–11, 2006.

Cabinet Office (2004). Chapter 2: Co-operation in Emergency Preparedness. Guidance on Part 1 of the Civil Contingencies Act 2004, its associated regulations and non-statutory arrangements. Available at http://www.cabinetoffice.gov.uk/ukresilience/preparedness/cooperation.aspx (Accessed 01/11/2010).

Chandler, D. (2002). *Semiotics: The Basics*, Routledge, London.

Civil Contingencies Secretariat, Cabinet Office (2004). *Civil Contingencies Act 2004: A Short Guide (Revised)*, Cabinet Office, London.

Cliburn, D.C., Fedderna, J.J., Miller, J.R. *et al.* (2002). Design and evaluation of a decision support system in a water balance application, *Comput. Graph.-UK*, **26**, 931–949.

Davis, T.J. and Keller, C.P. (1997). Modelling and visualising multiple spatial uncertainties, *Comput. Geosci.*, **23**, 397–408.

Defra/Environment Agency (2004). *Risk Performance and Uncertainty in Flood and Coastal Defence, a Review*, Defra, London, R&D Technical Report FD23022/TR.

Defra/Environment Agency (2004). *Making Space for Water: Developing a New Government Strategy for Flood and Coastal Erosion Risk Management in England*, Defra consultation document, London, p. 154.

Dix, A., Finlay, J., Abowd, G.D. *et al.* (2004). *Human–Computer Interaction 3rd Edition*, Prentice Hall, London.

Faulkner, H. and Ball, D. (2007). Environmental hazards and risk communication, *Environ. Haz.*, **7**, 71–78.

Faulkner, H., Parker, D., Green, C. *et al.* (2007). Developing a translational discourse to communicate uncertainty in flood risk between science and the practitioner, *Ambio*, **36**, 692–703.

Frewer, J., Hunt, S., Brennan, M. *et al.* (2003). The views of scientific experts on how the public conceptualise uncertainty, *J. Risk Res.*, **6**, 75–85.

Gluck, M., Yu, L., Ju, B. *et al.* (1999). *Augmented Seriation: Usability of a Visual and Auditory Tool for Geographic Pattern Discovery with Risk Perception Data*, GeoComputation 99, Mary Washington College, Fredericksburg, Virginia, 25–28 July, 1999.

Hall, J.W. (2002). A contingency approach to choice, *Civ. Eng. Environ. Syst.*, **19**, 87–118.

Hall, J.W. (2006). *Principles of Uncertainty Estimation and Decision Making under Uncertainty*, paper presented at the Annual Floodsite Partner workshop in Braunschweig, Germany, 13–15 February, 2006.

Hall, J.W. and Solomatine, D. (2008). A framework for uncertainty analysis in flood risk management decisions, *Int. J. River Basin Manage.*, **6**, 85–98.

Ishikawa, T., Barnston, A.G., Kastens, K.A. *et al.* (2005). Climate forecast maps as a communication and decision-support tool: an empirical test with prospective policy makers, *Cartogr. Geogr. Inform. Sci.*, **32**, 3–16.

Kinzig, A.P., Starrett, D., Arrow, K. *et al.* (2003). Coping with uncertainty: a call for a new science-policy forum, *Ambio*, **32**, 330–335.

Klir, G.J. (2006). *Uncertainty and Information: Foundations of Generalized Information Theory*, John Wiley, Hoboken, NJ.

Kress, G. and van Leeuwen, T. (1996). *Reading Images — The Grammar of Visual Design*, London, Routledge.

Lavis, J.N., Robertson, D., Woodside, J.M. *et al.* (2003). How can research organizations more effectively transfer research knowledge to decision makers?, *Milbank Q.*, **81**, 221–48.

Leedal, D., Hunter, N., Neal, J. *et al.* (2010a). *A Case Study of Tools for Manipulating and Visualizing Large Flood Risk Management Data Sets*, Proceeding of BHS 2010: Role of Hydrology in Managing Consequences of a Changing Global Environment, Newcastle University, 19–23 July, 2010.

Leedal, D., Neal, J., Bevan, K. *et al.* (2010b). Visualisation approaches for communicating real-time flood forecasting level and inundation information, *J. Flood Risk Manage.*, **3**, 140–150.

Leiss, W. (2004). Effective risk communication practice, *Toxicol. Lett.*, **149**, 399–404.

Levy, J.K., Hartmann, J., Li., K.W. *et al.* (2007). Multi-criteria decision support systems for flood hazard mitigation and emergency response in urban watersheds, *J. Am. Water Resour. As.*, **43**, 346–358.

MacEachren, A.M. (1992). Visualising uncertain information, *Cartogr. Perspect.*, **13**, 10–19.

MacEachren, A.M. (2001). An evolving cognitive-semiotic approach to geographic visualisation and knowledge construction, *Inform. Design J.*, **10**, 26–36.

MacEachren, A.M. and Ganter, J.H. (1990). A pattern identification approach to cartographic visualisation, *Cartographica*, **27**, 64–81.

MacEachren, A.M. and Kraak, M.-J. (2001). Research challenges in geovisualisation, *Cartogr. Geogr. Inform. Sci.*, **28**, 23–12.

MacEachren, A.M., Robinson, A., Hopper, S. *et al.* (2005). Visualising geospatial information uncertainty: what we know and what we need to know, *Cartogr. Geogr. Inform. Sci.*, **32**, 139–161.

Mansilla, V.B., Dillon, D. and Middlebrooks, K. (2006). *Building Bridges across Disciplines: Organisational and Individual Qualities of Exemplary Interdisciplinary Work*, "Goodwork Project", Harvard Graduate School of Education, Cambridge, MA.

Maxim, L. and van der Sluijs, J.P. (2011). Quality in environmental science for policy: assessing uncertainty as component of policy analysis, *Environ. Sci. Policy*, **14**, 482–492.

McCarthy, S. (2007). Contextual influences on national level flood risk communication, *Environ. Haz.*, **7**, 128–140.

McCarthy, S., Tunstall, S. and Faulkner, H. (2008). *Risk Communication: Inter-Professional Flood Risk Management*, RPA Stakeholder and Policy, final report WP 7.3 to FRMRC1.

McCarthy, S., Tunstall, S., Parker, D. *et al.* (2007). Risk communication in emergency response to extreme floods, *Environ. Haz.*, **7**, 179–192.

Milograna, J., Campana, N.A. and Baptista, M.B. (2009). *Choice of Flood Control Measures in Urban Areas – A Decision Aid Tool*, "Road Map towards a Flood Resilient Urban Environment", Final Conference COST Action, C22, Paris, 26–27 November, 2009.

Moel, H.D., Alphen, J.V. and Aerts, J.C.J.H. (2009). Flood maps in Europe — methods, availability and use, *Nat. Haz. Earth Syst.*, **9**, 289–301.

Morss, R.E., Wilhelmi, O.A., Downton, M.W. *et al.* (2005). Flood risk, uncertainty and scientific information for decision making: lessons from an interdisciplinary project, *B. Am. Meteorol. Soc.*, **86**, 1593–1601.

Morvan, H.P., Knight, D.K., Wright, N.G. *et al.* (2008). The concept of roughness in fluvial hydraulics and its formulation in 1d, 2d and 3d numerical simulation models, *J. Hydraul. Res.*, **46**, 191–208.

Neisser, U. (1976). *Cognition and Reality*, Freeman, New York.

Norbert, S., Dermeritt, D. and Cloke, H. (2010). Informing operational flood management with ensemble predictions: lessons from Sweden, *J. Flood Risk Manage.*, **3**, 72–79.

Pappenburger, F. and Beven, K. (2006). Ignorance is bliss: seven reasons not to use uncertainty analysis, *Water Resour. Res.*, **42**, W05032.

Pappenburger and Faulkner, H. (2004). *Responses to Flood Risk Questionnaire*, report to FRMRC1.

Pasche, E., Küpferle, C. and Manojlovic, N. (2007). *Capacity Building of Spatial Planners for Flood Risk Management in Urban Environment through Decision Support Systems and Interactive Learning*, International Symposium on New Directions in Urban Water Management, UNESCO, Paris, 12–14 September, 2007.

Pellizzoni, L. (2003). Uncertainty and Participatory democracy, *Environ. Value.*, **12**, 195–224.

Pennesi, K. (2007). Improving forecast communication — linguistic and cultural considerations, *B. Am. Meteorol. Soc.*, **88**, 1033–1044.

Penning-Rowsell, E.C., Tunstall, S.M. and Tapsell, S.M. (2000). The benefits of flood warnings: real but elusive and politically significant, *J. Chart. Inst. Water E.*, **14**, 7–14.

Pham, B. and Brown, R. (2005). Visualisation of fuzzy systems: requirements, techniques and framework, *Future Gener. Comp. Sy.*, **21**, 1199–1212.

Pitt, M. (2007). *The Pitt review, Interim Report*, HM Government Cabinet Office, London.

Rieman, J., Lewis, C., Young, R.M. *et al.* (1994). "Why is a Raven Like a Writing Desk?: Lessons in Interface Consistency and Analogical Reasoning from Two Cognitive Architectures" in: Adelson, B., Dumais, S., and Olson, J. (eds), *Proceedings of the SIGCHI Conference on Human Factors in Computing Systems: Celebrating Interdependence*, Boston, Massachusetts, United States, 24–28 April, 1994, ACM: New York, pp. 438–444.

Rinner, C. (2003). Web-based spatial decision support: status and research directions, *J. Geogr. Inform. Decis. Anal.*, **7**, 14–31.

Sarukkai, S. (2007). Mathematics, nature and cryptography: insights from philosophy of information, *Curr. Sci.*, **92**, 1690–1696.

Smemoe, C.M., Nelson, J.E., Zundel, A.K. *et al.* (2007). Demonstrating floodplain uncertainty using flood probability maps, *J. Am. Water Resour. As.*, **43**, 359–371.

Stamper, R., Liu, K., Hafkamp, M. *et al.* (2000). Understanding the roles of signs and norms in organisations — a semiotic approach to information systems design, *J. Behave. Inform. Technol.*, **19**, 15–27.

Todini, E. (1999). An operational decision support system for flood risk mapping, forecasting and management, *Urban Water J.*, **1**, 131–143.

Todini, E. (2006). *The Nature and Role of Uncertainty in Operational Flood Management*, paper presented at the Annual Floodsite Partner Workshop in Braunschweig, Germany, 13–15 February, 2006.

van Alphen, J., Martini, F., Loat, R. *et al.* (2009). Flood risk mapping in Europe, experiences and best practices, *J. Flood Risk Manage.*, **2**, 285–292.

van Leeuwen, T., and Jewitt, C. (eds) (2001). *Handbook of Visual Analysis*, Sage, London.

Voinov, A. and Bousquet, F. (2010). Modelling with stakeholders, *Environ. Modell. Softw.*, **25**, 1268–1281.

Weichselgartner, J. and Kasperson, R. (2010). Barriers in the science-policy-practice interface: toward a knowledge-action system in global environmental change research, *Global Environ. Change*, **20**, 266–277.

Zerger, A. and Wealands, S. (2004). Beyond modelling: linking models with GIS for flood risk management, *Nat. Haz.*, **33**, 191–208.

Index